# NANOCONJUGATE NANOCARRIERS FOR DRUG DELIVERY

# NANOCONJUGATE NANOCARRIERS FOR DRUG DELIVERY

*Edited by*

**Raj K. Keservani, MPharm**
**Anil K. Sharma, MPharm, PhD**

AAP | APPLE ACADEMIC PRESS

Apple Academic Press Inc.
3333 Mistwell Crescent
Oakville, ON L6L 0A2
Canada

Apple Academic Press Inc.
9 Spinnaker Way
Waretown, NJ 08758
USA



First issued in paperback 2021

*Exclusive worldwide distribution by CRC Press, a member of Taylor & Francis Group*
No claim to original U.S. Government works

ISBN 13: 978-1-77463-163-8 (pbk)
ISBN 13: 978-1-77188-677-2 (hbk)



CIP data on file with Canada Library and Archieves

**Library and Archives Cataloguing in Publication**

Names: Keservani, Raj K., 1981- editor. | Sharma, Anil K., 1980- editor.

Title: Nanoconjugate nanocarriers for drug delivery / editors, Raj K. Keservani, Anil K. Sharma.

Description: Toronto ; New Jersey : Apple Academic Press, [2018] | Includes bibliographical references and index.

Identifiers: LCCN 2018027318 (print) | LCCN 2018028137 (ebook) | ISBN 9781351171045 (ebook) | ISBN 9781771886772 (hardcover : alk. paper)

Subjects: | MESH: Nanoconjugates--chemistry | Nanoconjugates--therapeutic use | Drug Delivery Systems

Classification: LCC RS192 (ebook) | LCC RS192 (print) | NLM QV 785 | DDC 615.1/9--dc23

LC record available at https://lccn.loc.gov/2018027318

Apple Academic Press also publishes its books in a variety of electronic formats. Some content that appears in print may not be available in electronic format. For information about Apple Academic Press products, visit our website at **www.appleacademicpress.com** and the CRC Press website at **www.crcpress.com**

*The Present Book Is Dedicated to our Beloved*

*Aashna*

*Anika*

*Atharva*

*and*

*Vihan*

# ABOUT THE EDITORS

**Raj K. Keservani**

Raj K. Keservani, MPharm, is in the Faculty of B. Pharmacy, CSM Group of Institutions, Allahabad, India. He has more than ten years of academic (teaching) experience from various institutes of India in pharmaceutical education. He has published 35 peer-reviewed papers in the field of pharmaceutical sciences in national and international journals, 16 book chapters, 2 co-authored books, and 10 edited books. He is also active as a reviewer for several international scientific journals. Mr. Keservani graduated with a pharmacy degree from the Department of Pharmacy, Kumaun University, Nainital, Uttarakhand, India. He received his Master of Pharmacy (MPharm) (specialization in pharmaceutics) from the School of Pharmaceutical Sciences, Rajiv Gandhi Proudyogiki Vishwavidyalaya, Bhopal, India. His research interests include nutraceutical and functional foods, novel drug delivery systems (NDDS), transdermal drug delivery/ drug delivery, health science, cancer biology, and neurobiology.

**Anil K. Sharma, MPharm, PhD**

Anil K. Sharma, MPharm, PhD, is currently working as lecturer at Delhi Institute of Pharmaceutical Sciences & Research, University of Delhi, New Delhi, India. He has more than 9 years of academic experience in the pharmaceutical sciences. He has published 28 peer-reviewed papers in the field of pharmaceutical sciences in national and international reputed journals and has authored 15 book chapters and edited 10 books. His research interest encompass nutraceutical and functional foods, novel drug delivery systems (NDDS), drug delivery, nanotechnology, health science/ life science, biology/cancer, and biology/neurobiology. He graduated (B. Pharmacy) from the University of Rajasthan, Jaipur, India, and then qualified GATE in the same year conducted by IIT Mumbai. He received his postgraduate degree (M. Pharmacy) from the School of Pharmaceutical Sciences, Rajiv Gandhi Proudyogiki Vishwavidyalaya, Bhopal, Madhya Pradesh, India, with a specialization in pharmaceutics. He has earned his PhD from the University of Delhi.

# CONTENTS

# LIST OF CONTRIBUTORS

**B. A. Aderibigbe**
Department of Chemistry, University of Fort Hare, Alice Campus, Eastern Cape, South Africa

**Ayuob Aghanejad**
Research Center for Pharmaceutical Nanotechnology, Tabriz University of Medical Science, Tabriz, Iran

**Onur Alpturk**
Department of Chemistry, Istanbul Technical University, 34469, Maslak, Istanbul, Turkey

**Subham Banerjee**
Centre for Bio-design & Diagnostics (CBD), Translational Health Science & Technology Institute (THSTI), Faridabad, Haryana, India
Department of Pharmaceutics, National Institute of Pharmaceutical Education and Research (NIPER)-Guwahati, Assam, India

**Jaleh Barar**
Research Center for Pharmaceutical Nanotechnology, Tabriz University of Medical Science, Tabriz, Iran
Department of Pharmaceutics, Faculty of Pharmacy, Tabriz University of Medical Science, Tabriz, Iran

**Dariane Jornada Clerici**
Nanotechnology Laboratory, Centro Universitário Franciscano, Santa Maria, Rio Grande do Sul, Brazil

**Bhaskar Das**
Department of Biosciences and Bioengineering, Indian Institute of Technology Guwahati, North Guwahati 781039, Assam, India
Centre for the Environment, Indian Institute of Technology, Guwahati, North Guwahati 781039, Assam, India

**Márcia Ebling de Souza**
Nanotechnology Laboratory, Centro Universitário Franciscano, Santa Maria, Rio Grande do Sul, Brazil

**Arju Dhawan**
Department of Pharmaceutics, Chitkara College of Pharmacy, Chitkara University, Rajpura 140401, Patiala, Punjab, India

**Gazali**
Department of Pharmaceutics, Chitkara College of Pharmacy, Chitkara University, Rajpura 140401, Patiala, Punjab, India

**Manashjit Gogoi**
Department of Biomedical Engineering, North-Eastern Hill University, Shillong 793022, Meghalaya, India

**Nayan A. Gujarathi**
Sandip Foundation, College of Pharmacy, Nashik, India

**Ayushi Gupta**
Indian Institute of Information Technology, Devghat, Jhalwa, Allahabad 211012, India

**Satakshi Hazra**
Department of Biosciences and Bioengineering, Indian Institute of Technology Guwahati, North Guwahati 781039, Assam, India

**Snežana Ilić-Stojanović**
University of Niš, Faculty of Technology, Bulevar oslobodjenja 124, 16000 Leskovac, Republic of Serbia

**Ashish S. Jain**
Shri D. D. Vispute, College of Pharmacy & Research Center, Mumbai University, New Panvel, India

**Jun-ichi Kadokawa**
Department of Chemistry, Biotechnology, and Chemical Engineering, Graduate School of Science and Engineering, Kagoshima University, 1-21-40 Korimoto, Kagoshima 890-0065, Japan

**Sandeep Kaur**
Department of Pharmaceutics, Chitkara College of Pharmacy, Chitkara University, Rajpura 140401, Patiala, Punjab, India

**Raj K. Keservani**
Faculty of B. Pharmacy, CSM Group of Institutions, Allahabad, India

**Preeti Khulbe**
School of Pharmaceutical Sciences, Jaipur National University, Jaipur 302017, India

**Sabyasachi Maiti**
Department of Pharmaceutics, Gupta College of Technological Sciences, Ashram More, G. T. Road, Asansol 713301, West Bengal, India

**Nidhi Mishra**
Indian Institute of Information Technology, Devghat, Jhalwa, Allahabad 211012, India

**Montserrat Mitjans**
Department Bioquimica i Fisiologia, Facultat de Farmàcia i Ciències de l'Alimentació, Universitat de Barcelona, Av. Joan XIII, 27-31, Barcelona 08028, Spain

**Mangal S. Nagarsenker**
Department of Pharmaceutics, Bombay College of Pharmacy, Mumbai 400098, India

**Preethi Naik**
Department of Pharmaceutics, Bombay College of Pharmacy, Mumbai 400098, India

**Ljubiša Nikolić**
University of Niš, Faculty of Technology, Bulevar oslobodjenja 124, 16000 Leskovac, Republic of Serbia

**Vesna Nikolić**
University of Niš, Faculty of Technology, Bulevar oslobodjenja 124, 16000 Leskovac, Republic of Serbia

**Daniele R. Nogueira-Librelotto**
Department Farmácia Industrial, Universidade Federal de Santa Maria,
Av. Roraima 1000, 97105-900 Santa Maria, Rio Grande do Sul, Brazil
Programa de Pós-Graduação em Ciências Farmacêuticas, Universidade Federal de Santa Maria,
Av. Roraima 1000, 97105-900 Santa Maria, Rio Grande do Sul, Brazil

**Yadollah Omidi**
Research Center for Pharmaceutical Nanotechnology, Tabriz University of Medical Science, Tabriz, Iran
Department of Pharmaceutics, Faculty of Pharmacy, Tabriz University of Medical Science, Tabriz, Iran

**Sanjukta Patra**
Department of Biosciences and Bioengineering, Indian Institute of Technology Guwahati, North Guwahati 781039, Assam, India

**Sanja Petrović**
University of Niš, Faculty of Technology, Bulevar oslobodjenja 124, 16000 Leskovac, Republic of Serbia

**Jonathan Pillai**
Centre for Bio-design & Diagnostics (CBD), Translational Health Science & Technology Institute (THSTI), Faridabad, Haryana, India

**Bhushan Rajendra Rane**
Shri D. D. Vispute, College of Pharmacy & Research Center, Mumbai University, New Panvel, India

**Clarice M. B. Rolim**
Department Farmácia Industrial, Universidade Federal de Santa Maria, Av. Roraima 1000, 97105-900 Santa Maria, Rio Grande do Sul, Brazil
Programa de Pós-Graduação em Ciências Farmacêuticas, Universidade Federal de Santa Maria, Av. Roraima 1000, 97105-900 Santa Maria, Rio Grande do Sul, Brazil

**Roberto Christ Vianna Santos**
Laboratory of Oral Microbiology Research, Universidade Federal de Santa Maria, Santa Maria, Rio Grande do Sul, Brazil

**Ceyda Tuba Sengel-Turk**
Department of Pharmaceutical Technology, Ankara University Faculty of Pharmacy, 06100, Ankara, Turkey

**Shalmoli Seth**
Department of Pharmaceutics, Gupta College of Technological Sciences, Ashram More, G.T. Road, Asansol 713301, West Bengal, India

**Komal Sharma**
Indian Institute of Information Technology, Devghat, Jhalwa, Allahabad 211012, India

**Inderbir Singh**
Department of Pharmaceutics, Chitkara College of Pharmacy, Chitkara University, Rajpura 140401, Patiala, Punjab, India

**Ana Tačić**
University of Niš, Faculty of Technology, Bulevar oslobodjenja 124, 16000 Leskovac, Republic of Serbia

**Arushi Verma**
Indian Institute of Information Technology, Devghat, Jhalwa, Allahabad 211012, India

**M. Pilar Vinardell**
Department Bioquimica i Fisiologia, Facultat de Farmàcia i Ciències de l'Alimentació, Universitat de Barcelona, Av. Joan XIII, 27-31, Barcelona 08028, Spain

**V. Wiwanitkit**
Hainan Medical University, Hainan, China

**S. Yasri**
KMT Primary Care Center, Bangkok, Thailand

**Parham Sahandi Zangabad**
Research Center for Pharmaceutical Nanotechnology, Tabriz University of Medical Science, Tabriz, Iran
Department of Materials Science and Engineering, Sharif University of Technology, Tehran, Iran

# LIST OF ABBREVIATIONS

| | |
|---|---|
| Abs | antibodies |
| ACT | artemisinin-based combination therapy |
| AD | Alzheimer's disease |
| ADME | adsorption, distribution, metabolism, and excretion |
| AFM | atomic force microscopy |
| AlPcS | aluminum phthalocyanines |
| AmB | amphotericin B |
| AmB-PLGA-NS | amphotericin B-loaded PLGA-based nanosphere |
| apo | apolipoprotein |
| Apo b | apolipoprotein B |
| APs | aptamers |
| ARPES | angle-resolved photoelectron spectroscopy |
| ART | artemisinin |
| ASA | amino salicylic acid |
| AsODN | antisense oligonucleotides |
| ATD | anti-tubercular drugs |
| AuCM | gold-cysteamine |
| BBB | blood–brain barrier |
| BCG | bacillus Calmette–Guerin |
| BOB | blood–ocular barrier |
| BPEI | branched poly-(ethylenimine) |
| BPGDD | biotin-PEG-GNR-DNA/DOX |
| BphEA | 2-(butylamino)-1-phenyl-1-ethanethiosulfuric acid |
| BSA-NP | bovine serum albumin nanoparticle |
| CBD | Centre for Bio-design & Diagnostics |
| C-dots | carbon quantum dots |
| $CD-C_{10}$ | cyclodextrin derivatives grafted with decanoic alkyl chain |
| Ce6 | chlorin e6 |
| CM | cerebral malaria |
| CMC | critical micelle concentration |
| CNT | carbon nanotube |
| CP | NIR dye (Cy5.5)-labeled-MMP-14 substrate peptide |

| | |
|---|---|
| CPT | camptothecin |
| CRISPR | clustered regularly interspaced short palindromic repeats |
| CS | chitosan |
| CSC | cancer stem cell |
| CTLM | circular transmission line model |
| CVD | chemical vapor deposition |
| DDFR | dihydrofolate reductase |
| DDS | drug delivery system |
| DHA | dihydroartemisinin |
| DHA-SLN | DHA-loaded solid lipid nanoparticle |
| DHPS | dihydropteroate synthase |
| DI-TSL | DOX/ICG-loaded temperature sensitive liposomes |
| DIVEMA | divinyl ether-maleic anhydride |
| DLS | dynamic light scattering |
| DMF | N-N dimethyl formamide |
| DOPC | dioleyl phosphatidylcholine |
| DOPG | dioleoyl phosphatidylgycerol |
| DOX | doxorubicin |
| DOX@PSS-GNRs | DOX molecules onto the GNRs through PSS |
| DPPA | dipalmitoyl phosphatidic acid |
| DPPC | 1,2-dipalmitoyl-sn-glycero-3-phosphocholine |
| DPPC | dipalmitoyl phosphatidylcholine |
| DPPE | dipalmitoyl phosphatidylethanolamine |
| DPPS | dipalmitoyl phosphoserine |
| DSPE-PEG2000 | 1,2-distearoyl-sn-glycero-3-glycero-3-phosphoethanolamine-N-[(polyethylene glycol)-2000] |
| DSPE-PEG2000 | 1,2-distearoyl-sn-glycero-3-phosphoethanolamine-N-[methoxy(polyethylene glycol)-2000] |
| dsRNA | double-stranded RNA |
| DT-DTPA | dithiolated diethylenetriamine pentaacetic acid |
| DTPA | diethylentriaminepentaacetic dianhydride |
| DV | dengue virus |
| EGF | epidermal growth factor |
| EGFR | epidermal growth factor receptor |
| EMA | European Medicines Agency |
| EOEOVE-ODVE | poly[2-(2-ethoxy) ethoxyethyl vinyl ether-block-octadecyl vinyl ether] |
| EPR | enhanced permeability and retention |

| | |
|---|---|
| f-CNT | functional CNT |
| FA | folic acid |
| FA-ICG//RAPA-TSL | RAPA-loaded and folate-conjugated ICG TSL |
| FCC | face-centered cubic |
| FRET | fluorescence resonance energy transfer |
| FTIR | Fourier-transform infrared spectroscopy |
| G− | Gram-negative species |
| G+ | Gram-positive bacteria |
| GCSCs | gastric cancer stem cells |
| Gd | gadolinium |
| GD | gramicidin |
| Gd-nanoLE | Gd-containing lipidenanoemulsion |
| GDS | gene delivery system |
| GFN | graphene nanoparticle |
| GFP | green fluorescent protein |
| GI | gastrointestinal |
| GL-COO-β-CD/CA | graphene-β-cyclodextrin/chlorhexidine acetate |
| GNCgs | gold nanocages |
| GNP | gold nanoparticle |
| GNPs/AuNPs | gold nanoparticles |
| GNR | gold nanorods |
| GO | graphene oxide |
| GQD | graphene quantum dot |
| GrO | reduced graphene oxide |
| GSH | glutathione |
| HA | hyaluronic acid |
| HbF | fetal hemoglobin |
| HBsAg | hepatitis B surface antigen |
| HGNS | hollow gold nanosphere |
| His-co-Phe | poly(histidine)-co-phenylalanine-b-PEG |
| hMSC | human mesenchymal stem cell |
| HNE | human neutrophil elastase |
| HPMA | N-(2-hydroxypropyl) methacrylamide |
| HSPC | hydrogenated L-α-phosphatidylcholine |
| HUVEC | human umbilical vein endothelial cells |
| ICG | indocyanine green |
| IFNγ | interferon γ |
| INH | isoniazid |

| | |
|---|---|
| iRGD | internalized RGD |
| iRGD-ICG-LPs | iRGD-modified ICG liposomes |
| IT | itraconazole |
| ITO | indium tin oxide |
| IUV | unilamellar vesicles |
| IV | intravenous |
| KET | ketoconazole |
| LCST | lower critical solution temperature |
| LDC | lipid drug conjugate |
| LDH | lactate dehydrogenase |
| LET | linear energy transfer |
| LP | lipid peroxidation |
| LSPR | localized surface plasmon resonance |
| LUV | large unilamellar vesicles |
| MCT | medium-chain triglyceride |
| MDA | malondialdehyde |
| MDR | multidrug resistance |
| MDR-TB | multidrug-resistant TB |
| miRNA | micro-RNA |
| MLV | multilamellar vesicles |
| MMP | matrix-metalloproteinases |
| MMP-2 | matrix metalloproteinase-2 |
| MMP-14 | matrix metalloproteinase-14 |
| MPEG-SS-PLA | redox-responsive PEG-b-poly (lactic acid) |
| MPS | mononuclear phagocyte system |
| MRC | Medical Research Council |
| MRI | magnetic resonance imaging |
| MSN | mesoporous silica |
| MSOT | multispectral optoacoustic tomography |
| MSP-119 | merozoite surface protein-119 |
| MVV | multivesicular vesicles |
| MWCN | multiple-walled carbon nanotube |
| NCs | nanoconjugates |
| NECT | eflornithine/nifurtimox combination therapy |
| NGO | nanoscale GO |
| NIR | near-infrared |
| NLC | nanostructured lipid carrier |
| NPs | nanoparticles |

| | |
|---|---|
| NRF | National Research Foundation |
| NSCLC | non-small-cell lung carcinoma |
| NSs | nanosystems |
| NTD | neglected tropical disease |
| OASIS | Organic and Sustainable Industry Standards |
| ODN | oligonucleotides |
| p(OEGMA-co-MEMA) | poly(oligo(ethylene oxide) methacrylate-co-2-(2-methoxyethoxy)ethyl methacrylate) |
| P-gp | P-glycoprotein |
| PAA | poly(acrylic acid) |
| PAA | polyamidoamines |
| PbAE | poly(β-amino esters) |
| PBS | phosphate-buffered saline |
| PC | cyclophosphamide |
| PC | phosphatidylcholine |
| PC-12 | pheochromocytoma-derived |
| PC-Chol-SSG | SSG-loaded PC-cholesterol liposomes |
| PC-SA-SSG | SSG-loaded phosphatidylcholine stearylamine liposomes |
| PCL | PEG-b-polycaprolactone |
| PCL-SSPEEP | copolymer of PCL and poly (ethylethylene phosphate) |
| PCMX | parachlorometaxylenol |
| PD | Parkinson's disease |
| pDNA | plasmid DNA |
| PDT | photodynamic therapy |
| PEG | polyethylene glycol |
| PEG-b-PBD | PEG-b-poly (butadiene) |
| PEG-b-PLA | poly(ethylene glycol) with poly(lactic acid) |
| PEG-b-PPO-b-PEG | polyethylene glycol-b-polypropylene oxide-b-polyethylene glycol |
| PEI | polyethylenimine |
| PEO-b-PMABC | poly(ethyleneoxide)-b-poly {N-methacryloyl-N-(t-butyloxycarbonyl)cystamine} |
| PEO-PbAE | PEO-modified PbAE |
| PERL | polyunsaturated ER-targeting liposome |
| PF-DNA | proton-fueled DNA |
| PGA | poly(glutamic acid) |

| | |
|---|---|
| PG-b-PCL | poly(glycidol-block-ε-caprolactone) |
| PIT | phase inversion temperature |
| PLA | poly(dl-lactide) |
| PLA | polylactides |
| PLG | poly(lactideco-glycolide) |
| PLGA | poly(d,l-lactide-co-glycolide) |
| PLL | poly-L-lysine |
| PM-MTH | mithramycin encapsulated in polymeric micelles |
| PMA | poly(methacrylic acid) |
| PMMA | PZQ in poly(methyl methacrylate) |
| pNIPAAm | poly(N-isopropylacrylamide) |
| PNP | polymeric nanoparticle |
| PNP | RSV-encapsulated lipid nanoparticle |
| PPa | pyropheophorbide-a |
| PPEI-EI | poly(propionylethyleneimine-coethyleneimine) |
| PPI | polypropylenimine |
| PQP | piperaquine |
| pri-miRNA | primary miRNA |
| PSs | photosensitizers |
| PSS | poly(sodium 4-styrene sulfonate) |
| pTAT | polypeptide derived from the transactivator of transcription |
| PTT | photothermal therapy |
| PTX | paclitaxel |
| PVA | poly(vinyl alcohol) |
| PVA | polyvinyl acrylonitrile |
| PVA-R | polyvinyl alcohol conjugated hydrophobic anchors |
| PVK | poly-N-vinyl carbazole |
| PVP | poly(vinylpyrrolidone) |
| PZA | pyrazinamide |
| PZQ | praziquantel |
| QD | quantum dots |
| QN | quinine |
| RAPA | anti-angiogenesis agent rapamycin |
| rb | Bohr's radius |
| RBCs | red blood cells |
| RCPN | Research Center for Pharmaceutical Nanotechnology |
| RGD | arginine-glycine-aspartate |

| | |
|---|---|
| RGD-GNR | RGD peptides-conjugated GNRs |
| rGO | reduced graphene oxide |
| rGO-GNRVe | rGO-loaded ultrasmall plasmonic GNR vesicles |
| RIF | rifampicin |
| RISC | RNA-induced silencing complex |
| RNA | ribonucleic acid |
| RNAi | RNA interference |
| ROS | reactive oxygen species |
| RSV | resveratrol |
| SA | stearylamine |
| SC | stratum corneum |
| SDBS | sodium dodecyl benzene sulfonate |
| SDS | sodium dodecyl sulfate |
| SDS-PAGE | sodium dodecyl sulphate polyacryl amide gel electrophoresis |
| SEM | scanning electron microscope/microscopy |
| shRNA | short hairpin RNA |
| siRNA | small interfering RNA |
| SLN | solid lipid nanoparticle |
| SPIONs | superparamagnetic iron oxide nanoparticles |
| SPN | solid lipid nanoparticle |
| SPR | surface plasmon resonance |
| SRD | sustained release system |
| SRN | stimuli-responsive nanocarrier |
| ssDNA | single-stranded DNA |
| SSG | sodium stibogluconate |
| ssLips | submicron-sized liposomes |
| STEM | scanning electron transmission microscopy |
| STM | scanning tunneling microscope |
| STS | scanning tunneling spectroscopy |
| SUV | small unilamellar vesicles |
| SWCN | single-walled carbon nanotube |
| SWCNT | single-wall carbon nanotubes |
| TAT | transactivator of transcription |
| TCPS | tissue culture polystyrene substrates |
| TEM | transmission electron microscope/microscopy |
| TGA | thermogravimetric analysis |
| THF | tetrahydrofuran |

| | |
|---|---|
| THSTI | Translational Health Science & Technology Institute |
| TME | tumor microenvironment |
| TNBC | triple-negative breast cancer |
| TPGS | polyethylene glycol-succinate |
| Trp | l-tryptophan |
| TSL | thermal sensitive liposome |
| TZB | trastuzumab |
| TZB-GNP-$^{111}$In | $^{111}$In-labeled TZB-modified GNP |
| US FDA | United States Food and Drug Administration |
| UV | ultraviolet |
| VL | visceral leishmaniasis |
| WHO | World Health Organization |
| XRD | X-ray diffraction |

# PREFACE

The association of medicine and mankind is age old. There has been a constant evolution in treatment strategies aimed at curing the variety of ailments. Academicians as well researchers from industry have yielded many viable formulations that are being used as drug carriers. However, the conventional dosage forms suffer from a few pitfalls, such as non-specific drug delivery, dose dumping, poor patient compliance, toxicities linked with higher doses, etc. The past several decades have witnessed an emergence of nanotechnology-based products. Nanotechnology has been observed to uplift the level of sophistication through a variety of ways. Its uses embrace material science, engineering, medical, dentistry, drug delivery, etc.

In light of the drug delivery context, the present book is an attempt to provide the requisite information to its readers. The content of this book has been written by highly skilled, experienced, and renowned scientists and researchers from all over the world. They provide updated knowledge and drug delivery information to readers, researchers, academicians, scientists, and industrialists around the globe.

*Nanoconjugate Nanocarriers for Drug Delivery* comprises 16 chapters divided into 4 sections that present an introduction of nanocarriers, physicochemical features, and generalized and specific applications dealing with drug delivery in particular. The materials used as well as formulation and characterization have been discussed in detail. The nanocarriers covered are nanoparticles, vesicular carriers, carriers having carbon as core constituent, dispersed systems, etc.

## SECTION I: INTRODUCTION TO NANOCARRIERS

Chapter 1, written by *B. A. Aderibigbe,* provides an overview of different nanobiomaterials used for delivery of drug molecules. The nanobiomaterials are further classified as polymeric, metallic, and derived from carbon. The author has summarized the nanobiomaterials with their striking features.

Chapter 2, written by *M. Pilar Vinardell* and *colleagues,* describes to what extent the surfactants are essential for preparation of numerous drug delivery systems. The chapter begins with a brief preamble of nanotechnolgy followed by classification of surfactants. The discussion of different polymeric and lipidic nanoparticles is provided in ensuing sections. In addition, the chapter covers delivery systems meant via various routes of application containing surfactants. Interestingly, there is mention of utility of surfactant-based nanocarriers in cancer chemotherapy.

The details of general principles and methods of preparation of smart nanoconjugates have been presented in Chapter 3, written by *Subham Banerjee* and *Jonathan Pillai.* The authors have reviewed various therapeutic applications of such stimuli-responsive systems in brief. Finally, the authors conclude with mentioning future perspectives on the therapeutic uses of smart polymeric nanocarriers.

Chapter 4, written by *Ayuob Aghanejad* and *colleagues,* gives an exhaustive description of gold nanoparticles, beginning with introduction going across various synthesis methods and applications in different fields. Further, smart gold nanopartculates are mentioned with relevance to cancer and gene delivery.

## SECTION II: VESICLE-BASED DRUG CARRIERS

Different aspects of drug delivery via vesicular nanocarriers are described in Chapter 5, written by *Sanja Petrović* and *colleagues.* The authors have given an overview of the delivery systems that represent multifunction carriers of active substances for controlled, delayed, and targeted drug substance delivery, with a special review of the vesicular drug carriers. The liposomes have been the key formulations with an elaborated discussion.

Chapter 6, written by *Preethi Naik* and *Mangal S. Nagarsenker,* deals with issues related to biological systems in particular. The authors have strived to provide information about different attempts for improving intracellular delivery of siRNA, with special focus on excipient-driven liposomal nanocarrier systems, clinical relevance of siRNA delivery vehicles, and the toxicity concerns associated with such carrier systems. In addition, the toxicity and clinical perspectives of liposomal delivery for siRNA have been discussed.

The description of liposomes applications in therapy as well as in diagnosis is provided in Chapter 7, written by *Sabyasachi Maiti* and *Shalmoli Seth.* The authors have highlighted recent progress in developing modified liposomes for either diagnostic or therapeutic as well as theranostic applications. Initially, an introduction of liposomes is given followed by their applications. Further, theranostic uses of particular liposomes have been described in detail.

Chapter 8, written by *Bhushan Rajendra Rane* and *colleagues,* gives a focused view of ceramic nanocarriers explored for drug delivery. The authors have availed the current status of these nanotechnology-based drug formulations. The pros and cons, mechanism, and their rationale in drug delivery have been mentioned.

## SECTION III: NANOCARRIERS DERIVED FROM CARBON

The uses of quantum dots as drug carriers and in other fields of medicine have been discussed in Chapter 9, written by *Komal Sharma* and *associates*. The authors have discussed properties, synthesis methods, and potential toxicity of quantum dots along with their applications in the biomedical field in light of their promising potential as a drug delivery carrier. To conclude, the discussion of future clinical application potential of these nanocarriers is mentioned.

Chapter 10, written by *Gazali* and *colleagues,* has exhaustively discussed numerous applications of graphene and its derivatives in gene and drug delivery, including their uses in biomaging and biosensing. Further, methods of synthesis of all these approaches, including their characterization techniques along with toxicity concern, have also been discussed. To conclude, future perspectives of graphene and graphene-generated forms have been mentioned.

The customized applications of graphenes in control of microorganisms are given in Chapter 11, written by *Dariane Jornada Clerici* and *associates*. The authors have endeavored to provide a general introduction of infections and graphene followed by its potential for drug delivery and specialized uses as antimicrobials in detail.

Chapter 12, written by *Ceyda Tuba Sengel-Turk* and *Onur Alpturk,* discussed extensively many aspects of these materials, including functionalization, drug-loading capacity, and mechanisms. Subsequently, the literature on recent patents is surveyed to shed light on the current status of the field. Lastly, their potential applications as different therapeutic modalities in drug delivery, their toxicology profiles, and the developed strategies to overcome their cytotoxicities are also discussed with a detailed perspective.

## SECTION IV: MISCELLANEOUS

An overview of nanoemulsions and their different applications has been presented in Chapter 13, written by *Preeti Khulbe*. The author has attempted to cover different applications, advantages, disadvantages, and methods of preparation of nanoemulsions. The future scopes of nanoemulsions are also discussed in this chapter.

Chapter 14 is written by *S. Yasri* and *V. Wiwanitkit*. The authors have summarized and discussed a number of nanoconjugate nanocarriers for drug delivery in management of tropical diseases. To conclude, the authors have raised concern of cost-effectiveness and safety of the techniques in curing such ailments.

The details of nanocarriers employed for treatment of neglected tropical diseases have been provided in Chapter 15, written by *Bhaskar Das* and *colleagues.* The chapter has endorsed the current status of application of nano-based carriers in combating the challenges associated with control of some selected neglected tropical diseases. The first part has discussed various nanocarriers and their general applications. This is followed by application of these particles in selected neglected tropical diseases, such as tuberculosis, leishmaniasis, malaria, etc.

Chapter 16 is written by *Jun-ichi Kadokawa*. The author has described the hierarchically organized self-assembly of sucrose and trehalose alkyl ether amphiphiles to construct nanoparticles and nanorods under aqueous conditions. The chapter has dealt in detail about preparation and evaluation of nanocomposites.

# PART I
# Introduction to Nanocarriers

**CHAPTER 1**

# NANOBIOMATERIALS FOR DRUG DELIVERY

B. A. ADERIBIGBE*

*Department of Chemistry, University of Fort Hare, Alice Campus, Eastern Cape, South Africa*

**E-mail: blessingaderibigbe@gmail.com*

## CONTENTS

## ABSTRACT

Conventional administration of drugs is limited by poor water solubility and bioavailability, serious side effects, high toxicity, nonspecific biodistribution, and low therapeutic indices. These limitations have resulted in several researchers investigating different nanobiomaterials that can enhance the therapeutic effects of the incorporated bioactive agents. The application of nanobiomaterials in the design of drug delivery systems offers several benefits such as increased surface area resulting in enhanced binding of the drug molecules, exhibit sizes smaller than eukaryotic or prokaryotic cells, ability to penetrate inaccessible areas, enhanced permeability and retention effect, and suitability for administration of genes and proteins. They can target the reticuloendothelial cells by passive targeting of drug to the macrophages of liver and spleen, thus useful for treatment intracellular infections. This chapter is focused on the recent trend of different nanobiomaterials used for drug delivery systems and the biological evaluation results (in vitro and in vivo).

## 1.1 INTRODUCTION

Biomaterials interact well with the biological environment. Nanobiomaterials are nanosized and are used in drug delivery systems because they offer several advantages when compared to the conventional administration of drug molecules. The limitations associated with conventional administration of drugs are high toxicity, poor bioavailability, serious side effects, lack of cell/tissue specificity, poor water solubility, low cure rate and high drug dosage is required (Chidambaram et al., 2011; Rani and Paliwal, 2014). These limitations have resulted in several researchers investigating nanobiomaterials which can overcome the aforementioned limitations. Nanobiomaterials used in the design of drug delivery systems have the following advantages: increased surface area, resulting in enhanced binding of the drug molecules, exhibit sizes smaller than eukaryotic or prokaryotic cells, ability to penetrate inaccessible areas, enhanced permeability and retention effect, and suitability for administration of genes and proteins. They target the reticuloendothelial cells by passive targeting of drug to the macrophages of liver and spleen, thus useful for treatment intracellular infections (Hubbell and Chilkoti, 2012; Jain et al., 2015; Yadav et al., 2011). In the selection of nanomaterials for the preparation of drug delivery systems, selected properties must be taken in to consideration such as: they must be soluble, nontoxic, biocompatible, and readily available (Mukherjee et al., 2014; Webster et al., 2013). They must have the capacity to protect the incorporated drug from degradation, that is, enzymatic or hydrolytic in the gastrointestinal tract, thereby enhancing the drug circulation for longer time and must enhance the half-lives of the drugs (Mukherjee et al., 2014). Several nanobiomaterials have been reported to be useful in the design of drug delivery systems (Fig. 1.1). This chapter will be focused on recent polymer, metal, ceramic, and carbon-based nanobiomaterials used in drug delivery systems.

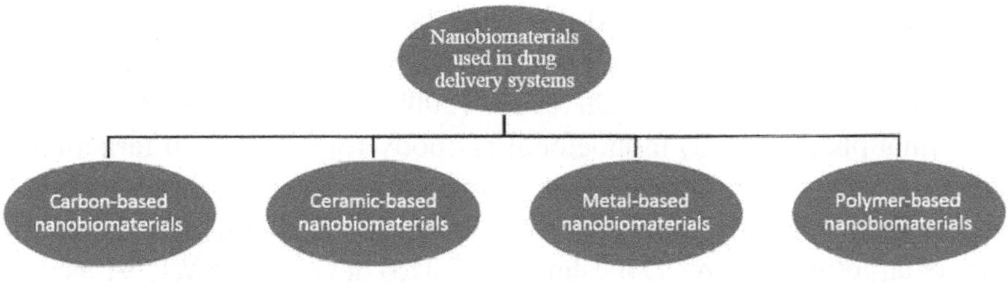


**FIGURE 1.1** Classification of nanobiomaterials used in drug delivery systems.

## 1.2 CARBON-BASED NANOBIOMATERIALS

Carbon-based nanobiomaterials are attractive materials for drug delivery and they include carbon nanotubes (CNTs), carbon nanodots, carbon nanohorns, graphene oxide, nanodiamonds, and fullerenes (Fig. 1.2). They exhibit unique properties such as good mechanical properties, excellent electrical conductivity, high energy, unparalleled thermal conductivity, high elasticity and surface area, good power densities, and are easy to functionalize (Huang et al., 2012; Tang et al., 2009; Wang et al., 2009). They can adsorb varieties of aromatic biomolecules which make them suitable materials for biomedical applications such as drug delivery, etc. (Aderibigbe et al., 2015; Huang et al., 2012).

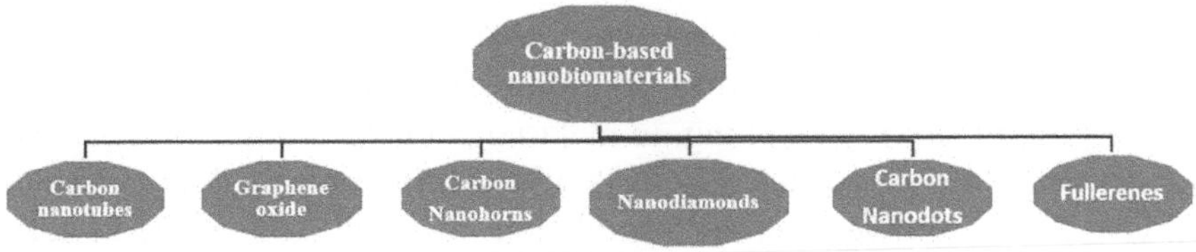


**FIGURE 1.2** Classification of carbon-based nanobiomaterials used in drug delivery systems.

### 1.2.1 CARBON NANOTUBE

CNTs are useful for the delivery of bioactive agents. They exist as single-walled, double-walled, and multiwalled CNT. Panchapakesan et al. (2005) prepared nanobombs from single-wall carbon nanotubes (SWCNT) for killing breast cancer cells. In vitro studies indicated that the nanobombs were active against BT474 breast cancer cells. The nanobombs acted by exploding into fragments without signs of toxicity. In a report by Lay et al. (2010), paclitaxel was loaded onto polyethylene glycol (PEG)-grafted SWCNTs. The release profile of paclitaxel from the formulation was sustained over a period of 40 days and in vitro cytotoxicity evaluation on HeLa cells and MCF-7 cells showed that these systems exhibited low cytotoxicity in both cells. Ou et al. (2009) developed SWCNT-based integrin alpha(v)beta(3) monoclonal antibody for cancer cell targeting in vitro. In vitro study showed that incorporating integrin alpha(v)beta(3) monoclonal antibody onto SWCNT enhanced its therapeutic effects in cancer targeting therapy. Daneshmehra (2015) developed SWCNT as drug carriers for high selective delivery of quercetin, a flavonoid. Configuration between SWCNT and quercetin was studied at gas phase and polarizable continuum model. The gas phase or polarizable continuum model did not have any significant effect on the systems. Ren et al. (2012) prepared PEGylated oxidized multiwalled CNTs loaded with doxorubicin and angiopep-2 for the treatment of brain glioma. The system was able to accumulate in tumors. Intracellular tracking and fluorescence imaging were used to evaluate the dual-targeting ability of the system to brain glioma. In vivo evaluation on glioma bearing mice revealed that the system exhibited good biocompatibility, low toxicity, and is a promising dual-targeting carrier for the treatment of brain tumor. Rezaei et al. (2011) incorporated artemisinin onto CNTs. In vitro analysis on K562 cancer cell lines proved that multiwalled CNTs enhanced the cytotoxicity of artemisinin. Singh and Konwar (2013) reported the conjugation of single-walled CNT with artemisinin and its derivatives against translationally controlled tumor protein of *Plasmodium falciparum*. CNT was found to be an effective system for the delivery of antimalarial drug and inhibiting translationally controlled tumor protein enzyme resulting, in malarial parasite death, *P. falciparum*. Chen et al. (2015) modified SWCNTs with folate and loaded doxorubicin. In vivo studies showed that the formulation was shut in the acidic vesicles after entering the lung cancer cells. Redistribution of doxorubicin from the formulation enhanced antitumor efficacy in tumor-bearing mice and no significant side effects were observed. Spinato et al. (2016) synthesized CNT-based carrier containing radioactive agent. SWCNTs were filled with radioactive agents and sealed at high temperature to afford closed-end CNTs. External functionalization of the filled CNTs was performed, followed by incorporation of cetuximad, a monoclonal antibody. The drug-loaded CNTs internalize significantly in EGFR-positive cancer cells. Nowacki et al. (2015) prepared cisplatin doped with SWCNT. In vivo evaluation on BALB/c nude mice induced with renal cancer showed that the drug-doped SWCNT inhibited reoccurrence of cancer and is suitable as hemostatic dressings for chemoprevention. Beydokhti et al. (2014) incorporated cisplatin onto oxidized and nonoxidized SWCNT. Cytotoxicity evaluation of the systems was performed on mice-bearing C26 colon carcinoma and the therapeutic activity of the systems was better than the free drug, suggesting that these systems are potential delivery systems for cancer treatment. Mehra et al. (2014) synthesized a multiwalled CNT-based nanocarrier containing vitamin E and doxorubicin. In vitro evaluation in human breast cancer cell lines showed that doxorubicin-loaded multiwalled CNTs significantly extended the survival time than the free doxorubicin solution. In vivo analysis showed that drug-loaded CNTs cancer targeting efficacy in tumor-bearing mice was higher than the free drug. Ghoshal and Mishra (2014) loaded 6-mercaptopurine onto CNTs by solvent method. The drug release was found to follow a zero-order release pattern. Risi et al. (2014) loaded mitoxantrone, an anticancer drug to MWCNT. In vitro cytotoxicity was evaluated on NIH3T3, a nonneoplastic fibroblast cell line and MDA 231, a breast cancer cell line. CNT loaded with mitoxantrone cytotoxic effects was dependent on time and dosage. Yu et al. (2015a) loaded doxorubicin onto nanofibrous scaffold prepared from CNTs. The release profile of doxorubicin from the composite nanofibers was prolonged and sustained with enhanced antitumor efficacy. Iannazzo et al. (2015) explored CNTs as HIV inhibitors through structure-based design by incorporating the drugs onto MWCNTs. The nanomaterial containing the drugs was highly hydrophilic with good water dispersibility and exhibited enhanced antiviral activity. Banerjee et al. (2012) conjugated porphyrin onto multiwalled CNTs resulting in effective antiviral agents. The antiviral agent reduced influenza A virus ability to infect mammalian cells, significantly suggesting that they can be used for influenza virus causing diseases without developing resistance. Subbiah et al. (2011) fabricated CNT–silver nanoparticle/polymer-based nanocomposites with antimicrobial properties. Various types of polymers were used and the nanocomposites exhibited good antimicrobial activity. There are other reports on the application of CNT for delivery of bioactive agents for the treatment of infectious diseases (Kang et al., 2007, 2008; Yandar et al., 2008). Mirazi et al. (2015) prepared metformin-conjugated nanotubes and studied their effects on the blood glucose level in the streptozotocin-induced male diabetic rats. Drug-conjugated nanotubes reduced the blood glucose levels in the diabetic rats significantly for a longer period of time more than the free metformin.

CNT has also been investigated as vaccines. Parra et al. (2013) prepared azoxystrobin-based conjugates by coupling azoxystrobin derivative containing a carboxylated functionality covalently to bovine serum albumin. The resultant conjugate was covalently incorporated onto functionalized CNTs. New Zealand rabbits and BALB/c mice were immunized with the CNT conjugates; enhanced titers and excellent IgG responses were significant, suggesting that CNTs have self-adjuvanting capability. Other researchers have reported the potential application of CNT in vaccines (Meng et al., 2008; Pantarotto et al., 2003a,b).

### *1.2.2 FULLERENES*

Fullerenes have unique physical, chemical, and biological properties, making them potential therapeutic agent. The fullerene family, especially C60, has been exploited in various medical fields (Fig. 1.3). Raza et al. (2015) investigated the potential of C60-fullerene conjugated with docetaxel for treatment of cancerous cells. The system was compatible with erythrocytes and the release profile of docetaxel was controlled. The cytotoxic effects of the system on MCF-7 and MDA-MB231 cell lines were significant. Panchuk et al. (2015) used pristine C60 fullerene for the delivery of doxorubicin. Doxorubicin conjugation onto C60 fullerene killed tumor cells by apoptosis induction. In vivo work showed that the tumor volume decreased 2.5 times, suggesting the fullerenes are potential delivery systems for cancer chemotherapy. Shi et al. (2014) developed doxorubicin-loaded poly(ethyleneimine) (PEI) derivatized fullerene for combination chemotherapy. Doxorubicin was conjugated onto fullerene C60-PEI via a pH-sensitive hydrazone linkage. Doxorubicin release profiles from the system were dependent on the pH and the system exhibited a high tumor targeting efficiency with good antitumor efficacy without any significant toxic effects. It also showed enhanced antitumor efficacy. Shi et al. (2013) also reported a PEI fullerene modified with folic acid via an amide linker followed by encapsulation of docetaxel within the nanoparticles. C60-based drug delivery system exhibited a good antitumor efficacy without any significant toxic effects and prolonged blood circulation. Fan et al. (2013) designed a multifunctional prodrug system from fullerene (C60) aggregates loaded with doxorubicin and folic acid. The prodrug entered folate receptor-positive cancer cells killing the cells via intracellular drug release. The prodrug had a combined therapeutic effect of photodynamic and chemotherapy. Hu et al. (2012) investigated the tumor-targeting ability of water-soluble C60 derivatives on HeLa cells. C60 was modified with folic acid, L-arginine, and L-phenylalanine. The uptake of C60 derivative into HeLa cells was more enhanced than in normal cells, suggesting good selectivity to tumor cells. Caspase-3 activity and induced apoptotic death were enhanced. Zhang et al. (2015a) prepared multifunctional drug delivery system by grafting hyaluronic acid onto fullerene followed by incorporation of transferrin for tumor-targeting and photodynamic therapy capacity. Artesunate was adsorbed to the system and its antitumor effects were significant. Wilson et al. (2004) invented a fullerene–antibiotic conjugate containing a minimum of one antibiotic molecule per fullerene moiety. Du et al. (2012) synthesized fullerene-liposome system incorporated with bioactive agent with good antiviral activity. In vivo studies on influenza virus infected mouse model indicated decreased lung index, reduced mean pulmonary virus yields, prolonged mean time to death, and lowered death rate with low toxicity. Dostalova et al. (2016) incorporated peptide to fullerene C60 nanocrystals. The antiviral activity of these peptides bound to fullerene C60 nanocrystals was more than the free peptides. The amount of peptide bound on the surface of fullerene C60 nanocrystals influenced the antiviral activity of the system. Nakamura and Isobe (2010) reported efficient delivery of *enhanced green fluorescent protein* and insulin-2 gene using amino fullerene to mice in vivo. Maeda-Mamiya et al. (2010) reported in vivo gene delivery using tetra(piperazino)fullerene epoxide on pregnant female ICR mice. The prodrug organ selectivity was distinct with no acute toxicity. Increased plasma insulin levels with reduced blood glucose concentrations were observed in the animal model used, revealing the potential of tetra(piperazino)fullerene epoxide based for gene delivery.

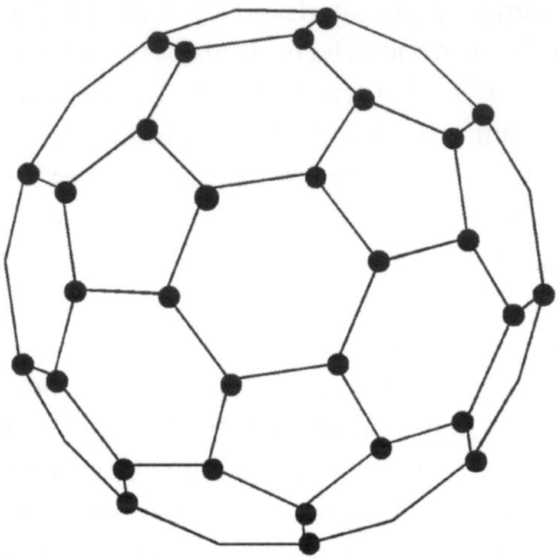

**FIGURE 1.3** Fullerenes.

### *1.2.3 GRAPHENE OXIDE*

Graphene oxide, a carbon-based material has unique properties, thus emerging as competitive drug delivery system for systemic and targeting drug delivery (Fig. 1.4). Zheng et al. (2016) prepared reduced graphene oxide-based nanocarriers functionalized and conjugated with anti-HER2 antibody and poly-L-lysine for the delivery of doxorubicin to the nucleus of HER2 overexpressed cancer cells. Cellular uptake of the drug-loaded nanocarriers into MCF7/HER2 cells was high. In vitro cytotoxicity effects of the drug-loaded nanocarriers revealed that graphene oxide is a potential system for efficient delivery of chemotherapeutic agents to HER2 overexpressed tumors. Yang et al. (2011) developed multifunctionalized graphene oxide for dual-targeting drug delivery. The system was pH sensitive; folic acid and $Fe_3O_4$ were conjugated onto it. Doxorubicin was loaded via $\pi$–$\pi$ stacking. The multifunctionalized delivery system exhibited targeted drug delivery mechanism. Zhang et al. (2010a) developed folic acid-conjugated graphene oxide system for targeted delivery of anticancer drugs. Doxorubicin and camptothecin were incorporated onto the folic acid-conjugated graphene oxide. The drug-loaded system displayed specific targeting to MCF-7 cells with high cytotoxicity effects. Zhi et al. (2013) prepared multifunctional nanocomplex as a carrier for Adriamycin. The carrier was prepared from polyethylenimine/poly(sodium

4-styrenesulfonates)/graphene oxide. The drug-loaded carrier enhanced therapeutic efficacy such as overcoming multidrug resistance mechanism. Wang et al. (2014) developed graphene oxide-based glioma-targeted drug delivery system loaded with doxorubicin and chlorotoxin. Doxorubicin release from the system was dependent on the pH and sustained. Cytotoxicity analysis showed that the rate of death of the glioma cells was high when compared with the free doxorubicin or graphene oxide loaded with doxorubicin only. Chlorotoxin enhanced doxorubicin accumulation within glioma cells. Tian et al. (2014) employed a PEG-functionalized graphene oxide for the delivery of cisplatin. The delivery system exhibited a sustained release profile of cisplatin over a period of 72 h with high cytotoxic effects on oral adenosquamous carcinoma and human breast cancer cells. Xu et al. (2015) conjugated paclitaxel to graphene oxide via covalent bonds. The drug conjugated system exhibited extended blood circulation time, excellent water solubility, high tumor-targeting ability, good biocompatibility, and suppressing effects than the free drug. Park et al. (2015) prepared functionalized reduced graphene oxide loaded with doxorubicin and CdSe quantum dot. The system was selective toward MDA-MB 231 cells with high cytotoxicity. Yin et al. (2016) functionalized graphene oxide with PEG and polyethylenimine for the delivery of plasmid-based Stat3 siRNA. Liu et al. (2015b) prepared graphene oxide dual-based drug delivery system loaded with dihydroartemisinin, transferrin, and ferric ion carrier. In vitro and in vivo experiment showed that the system exhibited enhanced tumor delivery specificity, cytotoxicity, and minimal side effects with complete tumor cure in the mice.

GO have been reported to be designed and use as vaccine. Xu et al. (2016a) designed graphene oxide as a vaccine adjuvant for immunotherapy using urease B as a model antigen. PEG and various types of polyethylenimine were used as coating polymers. It was found to be a highly effective vaccine nanoadjuvant. Yue et al. (2015) developed an intelligent vaccine system from graphene oxide. The vaccine exhibited high adsorption of antigen and acted as a cytokine self-producer and antigen reservoir promoting, the activation of antigen presenting cells and subsequent antigen cross-presentation. The result suggested that GO is a potential vaccine for prevention of cancer. Cao et al. (2014) designed ultrasmall graphene oxide–gold nanoparticles loaded with ovalbumin, a model antigen as an immune adjuvant to improve immune responses. In vivo studies demonstrate that the GO-based vaccine improved proliferation of CD8+ T cells, secretion of different cytokines, and robust ovalbumin-specific antibody response. The results suggested that the vaccine can stimulate potent humoral and cellular immune responses against antigens with potential applications for cancer and viral vaccines. Ni et al. (2012) demonstrated GO ability to adsorb anti-IL10 receptor antibodies. The release of anti-IL10R antibodies from GO was slow and dependent on the pH. The adsorption of anti-IL10R antibodies was more efficient and enhanced lipopolysaccharide-stimulated CD8 T-cell responses better than the free antibodies revealing GO potential as an adjuvant for vaccination. Li et al. (2016a) adsorb proteins to GO nanosheets resulting in an efficient uptake of the adsorb proteins by the dendritic cells and improved cross-presentation of antigen to CD8 T cells. Hong et al. (2011) designed GO loaded with model protein antigen, ovalbumin. The adsorbed proteins were released sequentially. In vitro toxicity of the materials on proliferating hematopoietic stem cells indicated limited cytotoxic effects.

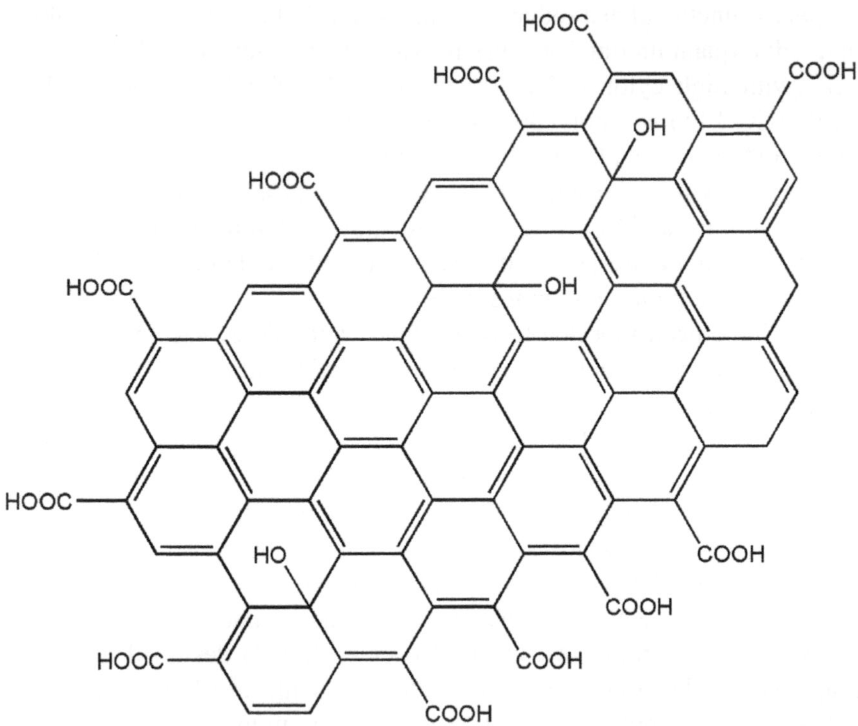


**FIGURE 1.4** Graphene oxide.

There are also reports on the application of GO for delivery of antimicrobial agents. Kavitha et al. (2015) functionalized graphene oxide with poly(4-vinyl pyridine) via atom transfer radical polymerization. The system exhibited high anticancer activity in vitro with antimicrobial properties against *Staphylococcus aureus* and *Escherichia coli*. Huang et al. (2015) reported GO-based drug delivery system cross-linked with PEI and loaded with ciprofloxacin. The ciprofloxacin-loaded GO had antibacterial effects, improved by ciprofloxacin release profile. Ghosh et al. (2012) loaded tetracycline to graphene oxide-*para*-amino benzoic acid nanosheet. Antimicrobial activity of tetracycline loaded onto the nanosheet confirmed that GO nanosheet has unique features useful for treatment of drug resistant bacteria infections.

## 1.3 CERAMIC-BASED NANOBIOMATERIALS

Nanobased ceramic materials have been widely used in biomedical applications. Presently, it is being explored in drug delivery systems. They have unique size, structural advantages, and active surface areas with good physical and chemical properties with ease of modification which suggest that they can be good materials for controlled and prolonged drug release systems. They can be classified based on their architectural design as ceramic-based nanoparticles and nanoscaffolds (Yang et al., 2010). It is noteworthy to mention that they also exist as hybrid compounds such as ceramic–polymer drug delivery systems that combine the benefits of the other types of materials.

### *1.3.1 CERAMIC NANOPARTICLES*

Ceramic nanoparticles are prepared from inorganic compounds with porous characteristics. Examples of inorganic materials used are silica, alumina, and titania (Bamrungsap et al., 2012; Rawat et al., 2009). Silica NPs are biocompatible, easy to synthesize and their surfaces can be modified such as mesoporous silica NPs (Bamrungsap et al., 2012; Slowing et al., 2007). Mesoporous silica NPs contain hundreds of empty channels known as mesopores which are arranged in a 2D network of a honeycomb-like porous structure (Bamrungsap et al., 2012). Amorphous silica materials exhibit low biocompatibility unlike mesoporous silica NPs that exhibit superior biocompatibility suitable for pharmacological applications (Bamrungsap et al., 2012; Trewyn et al., 2007). Several researchers have reported the recent application of mesoporous silica NPs for drug delivery. Singh et al. (2011) reported polymer-coated mesoporous silica NPs loaded with doxorubicin in both the core and shell domains. The NPs were stimulus responsive and the release of doxorubicin was dependent on the proteases at a tumor site resulting in cellular apoptosis. Popat et al. (2012) reported functionalized mesoporous silica nanoparticles containing sulfasalazine and sulfapyridine. They were enzyme-responsive nanocarriers and in vitro studies showed that the drug release was influenced by the presence of colon-specific enzyme, azo-reductase. Zhang et al. (2015b) prepared mesoporous silica nanoparticles modified with poly(2-(diethylamino)ethyl methacrylate) by surface-initiated atom transfer radical polymerization. Doxorubicin was loaded onto the nanoparticles with a high loading capacity and the release of the drug was pH dependent. The drug-loaded nanoparticles were internalized by HeLa cancer cells with good cytotoxic effects against the cells. Yang et al. (2016) reported mesoporous silica nanorods with transcription factor responsive for delivery of doxorubicin. The nanorods were stable with high drug loading capacity and cell uptake capability. In vitro evaluation on HeLa cells showed that the nanorods exhibited a transcription factor-triggered drug release with cytotoxic effects. Martínez-Carmona et al. (2016) incorporated topotecan onto mesoporous silica nanoparticles. In vitro tests showed that the nanosystems were efficient at killing tumor cells. Li et al. (2016b) prepared mesoporous silica nanoparticles loaded with doxorubicin. The nanospheres exhibited cyto-compatibility with unique sustained release kinetics. In vitro study showed that the drug-loaded nanoparticles were active against MCF-7 human breast cancer cells. Maggini et al. (2016) reported redox responsive mesoporous organo-silica nanoparticles with disulphide bridges loaded with anticancer drug. In vitro analysis on glioma C6 cells showed that the NPs are potential systems for target drug delivery. Liu et al. (2015d) designed mesoporous silica nanoparticles, that is, metalloproteinases responsive and biocompatible. Doxorubicin hydrochloride was encapsulated to the NPs with lactobionic acid acting as a targeting moiety. In vivo experiments proved that the system effectively inhibited the growth of tumor with no significant side effects. Quan et al. (2015) developed lactosaminated mesoporous silica nanoparticles for targeted delivery of docetaxel. In vitro cytotoxicity studies on HepG2 and

SMMC7721 cancer cell lines showed that the NPs inhibited the growth of cancer cell lines which was dependent on time and concentration. Chen et al. (2014) functionalized hollow mesoporous silica nanoparticle for drug delivery to the tumor. Sunitinib and doxorubicin were loaded inside the NPs. In vitro studies on HUVEC and MCF-7 human breast cancer cells showed that the NPs had excellent stability and target specificity. Liu et al. (2015c) reported hollow mesoporous silica nanoparticles–tLyp-1 peptide loaded with doxorubicin for dual-targeting drug delivery to tumor cells and angiogenic blood vessel cells. The drug-loaded nanoparticles inhibited growth of MDA-MB-231 and HUVECs cell lines. Wang et al. (2015) developed mesoporous silica nanospheres and nanorods loaded with doxorubicin. The mesoporous silica nanospheres exhibited better controlled release profiles than the nanorods. Koneru et al. (2015) reported mesoporous silica nanoparticles loaded with tetracycline for controlled release mechanism. The drug-loaded nanoparticles were more effective than the free tetracycline against *E. coli* over a 24-h period. Pasqua et al. (2013) reported mesoporous silica nanoparticles loaded with stable isotope, $^{165}$Ho for treatment of ovarian cancer metastasis. Intraperitoneal administration of $^{166}$Ho-mesoporous silica nanoparticles to SKOV-3 ovarian tumor-bearing mice showed that the radioactive-loaded nanoparticles accumulated predominantly in tumors and were useful as a radiotherapeutic agent for ovarian cancer metastasis. Halamová and Zeleňák (2012) prepared mesoporous silica material loaded with naproxen. The release profile of the drug from the system was sustained. Zhai (2013) prepared silica-based nanoparticles loaded with propranolol hydrochloride. The release of the drug from the prepared nanoparticle in simulated body fluid indicated a drug release over a period of 32 h. Siavashani et al. (2014) loaded insulin onto mesoporous silica nanoparticles by hydrothermal process. Mesoporous silica particles showed a high insulin loading and release capacity. Mahkam et al. (2012) modified silica nanoparticles for oral delivery of insulin with sustained release profile. He et al. (2011) developed a pH-responsive mesoporous silica nanoparticles multidrug delivery system with excellent monodispersity loaded with doxorubicin. The nanoparticle exhibited a pH-dependent drug release with a high anticancer activity against drug-resistant MCF-7/ADR cells. Meng et al. (2015) loaded gemcitabine and paclitaxel to mesoporous silica nanoparticle. The synergistic effects of the nanoparticles on pancreatic cells and tumors were evaluated in vivo by intravenous administration on mice carrying subcutaneous PANC-1 xenografts. Effective tumor shrinkage was significant on mice administered with nanoparticles loaded with both drugs than the free drugs with no sign of toxicity.

Few researchers have reported recent findings on the application of titania nanoparticles in drug delivery. Wu et al. (2011) prepared mesoporous titania nanoparticles loaded with doxorubicin for excellent drug delivery and intracellular bioimaging in human breast cancer cells, BT-20. Yin et al. (2014) reported mesoporous $TiO_2$ nanoparticles shell for synergistic targeted cancer therapy. The system exhibited controlled drug release profile and high rate of death of breast carcinoma cells. Ganesh et al. (2012) reported mesoporous $TiO_2$ prepared solvothermally and loaded with duloxetine by wet method. The drug release studies occurred in two stages: burst release over a period of 12 h and a sustained release. Liu et al. (2015a) prepared titanium dioxide nanoparticles modified with hyaluronic acid containing cisplatin for the treatment of ovarian cancer. The release of cisplatin from the nanoparticles was pH dependent. Drug-loaded nanoparticles accumulation in A2780 ovarian cancer cells was increased with improved anticancer activity.

The application of calcium phosphate nanoparticles has also been reported by some researchers. Chen et al. (2010) loaded methazolamide to calcium phosphate nanoparticles. In vitro drug release from the nanoparticles was controlled and proved that the nanoparticles intraocular pressure lowering effects lasted over 18 h, suggesting that the nanoparticles can be effective for treatment of glaucoma. Li et al. (2013) reported amphiphilic gelatin–iron oxide core/calcium phosphate shell nanoparticle for efficient delivery of anticancer drug. Doxorubicin release from the nanoparticles was pH-dependent. Cheng and Kuhn (2007) prepared calcium phosphate nanoparticles conjugated with cisplatin. In vitro drug release studies showed a sustained release of cisplatin from the nanoparticles. The cytotoxicity of the nanoparticles on cisplatin drug resistant A2780c human ovarian cancer cell line had an increased $IC_{50}$ value when compared to the free drug. Paul and Sharma (2012) developed nanosized zinc-modified calcium phosphates for delivery of bioactive agents such as: insulin, antibiotics, etc. The insulin-loaded nanoparticles were coated with pH-sensitive alginate thereby enhancing the stability of these nanoparticles. In vivo studies proved that the blood glucose level of diabetic rats was normal on administration of the formulation.

Calcium phosphate has also been evaluated as vaccine adjuvant. Olmedo et al. (2014) studied the adjuvant activity of calcium phosphate. The antibody response in mice toward the venom of the snake, *Bothrops asper*, showed that the adjuvants release of venom was slow. Venom adsorption on calcium phosphate enhanced a higher antibody response toward all tested HPLC fractions of the venom. Volkova et al. (2014) studied the adjuvant activity of calcium phosphate particles. In vivo studies were performed on commercial chickens with inactivated Newcastle disease virus vaccine in combination with calcium phosphate particles. There was increase in the antibody titers in blood and mucosal samples of the chickens. Sahdev et al. (2013) investigated the potential of calcium phosphate nanoparticles for transcutaneous vaccination. The nanoparticles were prepared by nano-precipitation method with sequential adsorption of sugars and ovalbumin. In vitro release of ovalbumin from nanoparticles was 60% within 24 h. In vivo administration of the nanoparticle by intradermal injection showed significantly higher antibody titers when compared to ovalbumin alone (Loomba and Sekhon, 2015).

## 1.4 METAL-BASED NANOBIOMATERIALS

This section will be focused on metal-based nanoparticles. Metal-based nanoparticles exhibit several outstanding features such as high stability, enhanced oral bioavailability, biocompatibility, high surface area, reduced toxicity, and ability to deliver to specific sites resulting in the overall drug therapeutic effects. Metal-based nanoparticles have been used for delivery of bioactive agents such as anticancer, antimalarial, and delivery to the brain, etc.

### *1.4.1 GOLD-BASED NANOPARTICLES*

Gold nanoparticles have good physical, optical, chemical, and electronic properties making it a good platform for delivery of therapeutic agents. Their large size makes them to accumulate at tumor sites and in inflamed tissues. They are used to deliver drugs with low solubility and poor pharmacokinetics. Ganeshkumar et al. (2012) synthesized modified gold nanoparticles using folic acid coupled with 6-mercaptopurine for targeted drug delivery to treat laryngeal cancer. In vitro cytotoxicity studies of the nanoparticles against HEp-2 cells showed that the nanoparticles are promising drug delivery system. Murawala et al. (2012) reported methotrexate loaded bovine serum albumin capped gold nanoparticles. In vitro analysis against MCF-7 showed that the nanoparticles exhibited higher cytotoxicity on MCF-7 cells than the free drug. Rastogi et al. (2012) prepared protein capped gold nanoparticles functionalized with various amino-glycosidic antibiotics as drug delivery vehicles. Antibacterial activity of the nanoparticles against Gram-negative and Gram-positive bacterial strains by well diffusion assay was studied; the results showed that they exhibited enhanced antibacterial activity than the free drug. Adeyemi and Sulaiman (2015) conjugated diminazene aceturate, a drug used for the treatment of trypanosomiasis to gold nanoparticles. Evaluation for cytotoxic actions in vitro demonstrated no significance difference between free drug and the nanoparticles. Ahangari et al. (2013) loaded gentamicin onto gold-based drug system for delivery to *Staphylococcal*-infected foci. The nanoparticles exhibited enhanced antibacterial effect in mouse model. Hahn et al. (2013) invented gold nanoparticles for liver targeted drug delivery. The system was biocompatible and biodegradable with liver tissue-specific delivery property. Malathi et al. (2013) synthesized gold nanoparticles loaded with rifampicin. In vitro studies showed that the nanoparticles were effective against bacillus subtils and *Pseudomonas aeruginosa* bacteria. Dixon et al. (2014) prepared composite of PEG-functionalized gold nanoparticles dispersed within poly(lactic-*co*-glycolic) films for the treatment of tumors such as gliomas. The composites exhibited zero-order drug release profile. Chakravarthy et al. (2010) developed gold nanorods to deliver an innate immune activator. The nanorod was effective against type A influenza virus. Zhang et al. (2010b) developed gold nanoparticles modified with glutathione with conjugated folic acid. In vitro analysis against HeLa cells signified that the nanoparticles are effective drug delivery system. Bhattacharya et al. (2007) prepared gold nanoparticles containing folic acid with different PEG backbones. They were useful as targeted delivery of anticancer drugs. You et al. (2010) reported a dual-functional hollow gold nanosphere loaded with doxorubicin. The release of doxorubicin was influenced by the pH and a high anticancer effect studied in vitro against MDA-MB-231. Venkatpurwar et al. (2011) prepared porphyran capped gold NPs loaded with doxorubicin with enhanced cytotoxicity against human glioma cell line, LN-229. The nanoparticles demonstrated higher cytotoxicity on LN-229 cell line than the free drug. Choi et al. (2010) prepared PEGylated-gold nanoparticles with varied amounts of human transferrin. The effects of transferrin content on nanoparticles were investigated on mice-bearing tumors. The targeted nanoparticles provided intracellular delivery of therapeutic agents to the cancer cells within solid tumors. Chanda et al. (2010) developed drug-loaded gum arabic glycoprotein-functionalized gold nanoparticles with antitumor effect. In vivo therapeutic investigations of the nanoparticles on mice bearing human prostate tumor xenografts revealed tumor regression and controlled growth of prostate tumors over a period of 30 days.

### *1.4.2 SILVER-BASED NANOPARTICLES*

Silver nanoparticles have also been reported as useful delivery systems for bioactive agents. Benyettou et al. (2015) reported silver nanoparticle for simultaneous intracellular delivery of doxorubicin and alendronate. The nanoparticles exhibited significantly greater anticancer activity in vitro than either or the free drugs. Fayaz et al. (2010) prepared silver nanoparticles loaded with antibiotics: ampicillin, kanamycin, erythromycin, and chloramphenicol. In vitro analysis showed that the incorporation of antibiotics onto the silver nanoparticles resulted in better antimicrobial effects. Martinez-Gutierrez et al. (2010) evaluated the antimicrobial activity of silver and titanium nanoparticles. Brown et al. (2012) loaded ampicillin onto silver nanoparticles. The drug-loaded NPs were effective at subverting antibiotic resistance mechanisms of multiple drug resistant bacteria.

### *1.4.3 IRON-BASED NANOPARTICLES*

Iron-based nanoparticles have unique properties and are employed as drug delivery systems. Recent reports have shown their great potentials in drug delivery. Aires et al. (2016) loaded anti-CD44 antibody and gemcitabine derivatives to multifunctionalized iron oxide magnetic nanoparticles. The system was biocompatible and selective for CD44-positive cancer therapy. Kossatz et al. (2015) developed iron oxide nanoparticles functionalized with either nucant multivalent pseudopeptide, doxorubicin, or both. The nanoparticles cytotoxic effects toward breast cancer cells was significant than the free drugs. Majeed et al. (2013) reported water-soluble magnetic iron oxide nanoparticles conjugated with the anticancer drug, doxorubicin. The efficacy of the nanoparticles was found to be significantly higher than that of the free drug. Sundaresan et al. (2014) reported dual-responsive iron-based nanoparticles for targeted drug delivery and imaging applications. It was loaded with doxorubicin and it exhibited dose-dependent cellular uptake of the nanoparticles in the presence of 1.3-T magnetic field. The nanoparticles demonstrated cytocompatibility over a period of 24 h with human dermal fibroblasts and normal prostate epithelial cells. Hsiao et al. (2015) developed iron oxide nanoparticles loaded with paclitaxel for targeted delivery to human glioblastoma cells. The nanoparticles exhibited sustained drug release profile. Quan et al. (2011) developed a human serum albumin-coated iron oxide nanoparticle loaded with doxorubicin. The drug was released in a sustained fashion. Su et al. (2015) reported magneto-responsive drug carrier that enhanced deep tumor penetration with a porous nanocomposite incorporating a tumor-targeted lactoferrin on mesoporous iron oxide nanoparticles loaded with paclitaxel. The system showed high efficiency for targeting tumors and suppressed subcutaneous tumors in 16 days after administration. Agostoni et al. (2013) prepared highly porous iron(III) trimesate MIL-100 nanoparticles loaded with azidothymidine triphosphate, an antiviral drug. Madrid et al. (2015) reported the synthesis of $Fe_3O_4$@*m*$SiO_2$ nanostructures loaded with ibuprofen. The nanoparticle exhibited good biocompatibility and drug-holding capacity ideal for targeted drug delivery. Kayal et al. (2010) prepared magnetic drug-targeting nanoparticle loaded with doxorubicin. They were found to be promising for magnetically targeted drug delivery.

### *1.4.4 ZINC-BASED NANOPARTICLES*

Zinc-oxide-nanostructured materials are unique because of their low toxicity and cost. They are not only insoluble in physiological environment but can dissolve as nontoxic ions in acidic environment, such as the late endosome and lysosome of the tumor cells (Abdelmonem et al., 2015). Palanikumar et al. (2013) prepared zinc oxide nanoparticles loaded with amoxicillin. The nanoparticles showed good antibacterial activity against infectious Gram-positive and Gram-negative bacteria. Wahab et al. (2013) synthesized ZnO nanoparticles that induced cell death in selected cancer cell lines namely: malignant human T98G gliomas, KB epithermoids, and HEK normal nonmalignant kidney cells. Wahab et al. (2014) in another research report prepared zinc oxide nanoparticles, biocompatible and active against HepG2 and MCF-7 cells. The nanoparticles also exhibited antibacterial activities. Yun et al. (2015) developed iron oxide–zinc oxide core–shell nanoparticles for antigen delivery onto dendritic cells. In vivo analysis on C57BL/6 mice showed that the nanoparticles are potential antigen carriers for DC-based immunotherapy. Baskar et al. (2015) developed nanobiocomposite of zinc oxide nanoparticles conjugated with L-asparaginase. The nanobiocomposite was active against MCF-7 cell line. Hackenberg et al. (2012) loaded paclitaxel and cisplatin onto zinc oxide nanoparticles. The nanoparticles loaded with both drugs at low concentrations resulted in an increase in cytotoxicity effects revealing a synergistic effect (Vinardell and Mitjans, 2015). Liu et al. (2016) loaded doxorubicin onto zinc oxide nanoparticles. The drug-loaded NPs enhanced the cell uptake of the drug intracellularly and decreased cell efflux in MDR cancer cells. Shokri and Javar (2015) prepared zinc oxide nanoparticles and calcium phosphate nanoparticles for dermal delivery of albumin. Maximum amount of permeated albumin occurred in drug-loaded zinc oxide NPs than calcium phosphates NPs over a period of 1 h and 30 min.

### *1.4.5 COPPER OXIDE NANOPARTICLES*

Copper oxide nanoparticles have been reported to be effective against cancer cell lines. Sun et al. (2012) evaluated the toxicity of metal oxide nanoparticles such as copper oxide, silica, titanium oxide, and ferric oxide on A549, H1650, and CNE-2Z cancer cell lines. Copper oxides nanoparticles induced autophagy in A549 cells. Wongrakpanich et al. (2016) reported that the size of the nanoparticles can influence the amount of its intracellular dissolution and impact on its cytotoxicity. Laha et al. (2015) prepared copper oxide nanoparticles modified with folic acid for targeted delivery. Targeting efficacy evaluated on human breast cancer cells MCF7 indicated cell death occurred through apoptosis. In vivo study by peritoneal injection on tumor bearing mice showed effective killing of the tumor cells. Wang et al. (2013) explored the antitumor properties of copper oxide nanoparticles in vivo on mice with subcutaneous melanoma and metastatic lung tumors by intratumoral and systemic injections. The drug-loaded nanoparticles reduced the growth of melanoma, inhibited the metastasis of B16–F10 cells, and increased the survival rate of tumor-bearing mice. Maity et al. (2014) synthesized spheroidal cuprous oxide nanoparticles conjugated with L-tryptophan. They were toxic to different cancer cells with controlled release profiles. Trickler et al. (2012) evaluated copper nanoparticles ability to induce the release of proinflammatory mediators which can restrict characteristics of the blood–brain barrier. In vivo studies on confluent rat brain microvessel endothelial cells treated with the nanoparticles revealed an increase in the cellular proliferation of rat brain microvessel endothelial cells and enhanced release of prostaglandin $E_2$.

## 1.5 POLYMER-BASED NANOBIOMATERIALS

Polymer-based nanobiomaterials used for delivery of therapeutic agents can be classified as natural, synthetic, and pseudosynthetic polymers (Duncan, 2003). The commonly used synthetic polymers are divinylethermaleic anhydride, poly(vinylpyrrolidone) (PVP), (polyethylene glycol, *N*-(2-hydroxypropyl) methacrylamide (HPMA) copolymers, PEI, and polyamidoamines, etc. Examples of natural polymers are chitosan, dextran, and hyaluronic acid while examples of pseudosynthetic polymers are poly(glutamic acid), poly(amino acids) poly(L-lysine), poly(malic acid), and poly(aspartamides) (Duncan, 2003). These polymers have been used to prepare nanosized drug delivery systems such as polymer–drug conjugates, dendrimers, nanomicelles, nanovesicles, etc.

### *1.5.1 POLYMER–DRUG CONJUGATES*

Polymer–drug conjugates are nanocarriers which were first proposed by Ringsdorf in 1975 for delivery of therapeutic agents. They are composed of five major components (Fig. 1.5): (1) polymeric backbone, (2) incorporated drug, (3) drug spacer, (4) targeting moiety, and (5) solubilizing agent. The type of polymers used is very important and the polymers used must be water soluble, nontoxic, nonimmunogenic, biodegradable and should have functional groups for conjugation of the drug or spacer. The conjugation of the drug can be directly or via suitable linkers onto the polymer backbone to enhance the release of the drug from the conjugate at the target site (Markovsky et al., 2012). The type of polymers selected for polymeric backbone influences the pharmacokinetics and pharmacodynamics of the conjugated drug. Different types of polymers have been used for the preparation of polymer–drug conjugates. Polymer–drug conjugates offer several advantages such as (Aderibigbe and Mukaya, 2016) (1) it enhances the water solubility and bioavailability of the conjugated drugs;

(2) it protects the drug temporary from serum protein binding, enzymatic attack, and other scavenging processes resulting in minimized renal clearance, extended serum life time and drug bioavailability; (3) it enhances the pharmacokinetics of the drugs; and (4) it protects the patient's immunity. In this section, the focus will be only on polymer–drug conjugates.

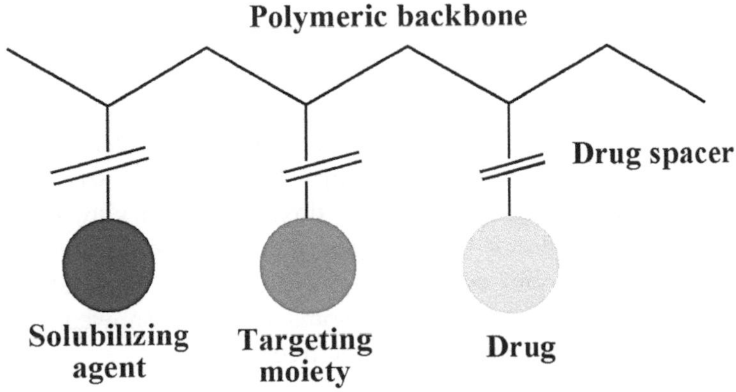


**FIGURE 1.5** **(See color insert.)** A schematic diagram of polymer–drug conjugates.

#### *1.5.1.1 POLYASPARTAMIDES*

There are few recent reports on the application of polyaspartamides for polymer–drug conjugates. Končič et al. (2011) prepared polyaspartamide-based conjugates containing lipophilic estrogen hormone. They conjugates were water soluble with improved bioavailability and are potential estrogen prodrugs. Paolino et al. (2015) conjugated aminobisphosphonate and neridronate to polyaspartamide with or without a spacer. In vivo studies suggested that the selectivity of conjugates toward the bone tissues was time dependent (Fresta et al., 2012). Ngoy et al. (2011) prepared polyaspartamide-based conjugates containing folic acid. The bioreversible binding of conjugates was studied. Aderibigbe and Neuse (2012) reported polyaspartamide conjugates containing 4- and 8-aminoquinolines and anticancer drugs (Aderibigbe et al., 2011). Aderibigbe and Ray (2016) reported the polyaspartamides containing aminoquinoline and ferrocene derivatives. The release mechanism of 4-aminoquinoline from the carrier was a case II. Zero-order release mechanism dominated the first 24 h suggesting that they can overcome drug resistance that is common with the presently used antimalarials. Nkazi et al. (2012) prepared polyaspartamide conjugates containing curcumin and an analogue of ferrocene via hydrazone linker. In a report by Mufula et al. (2012), coconjugation of ferrocene with methotrexate or folic acid unto polyaspartamides was performed. In another research by Aderibigbe et al. polyaspartamide conjugates containing different analogues of ferrocene were reported. The release profile of the analogues was an anomalous (non-Fickian) diffusion (Nkazi et al., 2013). Aderibigbe et al. (2014) reported the in vitro antiplasmodial activity of polyaspartamide-based conjugates containing 4- and 8-aminoquinoline. Some of the polymeric conjugates were found to be most active against the chloroquine-sensitive strain of *P. falciparum* which was dependent on the solubilizing group. Di Meo et al. (2015) prepared polyaspartamide conjugates containing doxorubicin. In vitro evaluation on MCF-7 and T47D cell lines suggested that the conjugates retarded cytotoxic effect on tumor cells. In vivo evaluation on NOD-SCID mice bearing a MCF-7 human breast carcinoma xenograft showed improved antitumor activity when compared to the free drug.

### *1.5.1.2 POLYETHYLENE GLYCOL*

The incorporation of drugs onto PEG increases the solubility and the bioavailability of the incorporated drug. However, the limitation with the use of PEG is its poor drug-loading efficiency due to the presence of limited number of reactive groups. To overcome this limitation, it is modified with functionalities to enhance the drug loading efficiency. Presently, there are several research reports on PEG-based polymer–drug conjugates.

Veronese et al. (2005) prepared a series of PEG–doxorubicin conjugates with different peptidyl linkers. Some of the conjugates exhibited sustained release profile and were more than 10-fold less toxic than the free drug. The highest molecular weight exhibited the longest plasma residence time with enhanced tumor targeting ability. Anbharasi et al. (2010) incorporated doxorubicin to D-alpha-tocopheryl PEG succinate and folic acid for targeted chemotherapy to enhance the therapeutic effects and reduce the side effects of the drug. The in vitro drug release was found to be pH dependent and the cytotoxic effects on MCF-7 cancer cells were more enhanced than the free drug. Zhu et al. (2010, 2011) demonstrated controlled release of free doxorubicin in weak acidic lysosomes using PEG-based conjugates. The conjugates exhibited improved therapeutic effects against murine B16 melanoma. Sapra et al. (2011) designed a multi-arm PEG-10-hydroxy-7-ethyl-camptothecin conjugate for longer lasting exposure of tumors. The conjugates exhibited enhanced antitumor activity. Calvo et al. (2010) prepared pegylated form of docetaxel with sustained release profile and superior antitumor activity. Li et al. (2012) prepared a methoxy-poly(ethylene glycol)-succinyl-5′-*O*-zidovudine conjugates that varied in molecular weight. The conjugates showed good stability with good anti-HIV activity and low cytotoxicity in MT-4 cells. Some of the conjugates showed good inhibition to wild-type viruses (strains IIIB and ROD) and in vivo oral pharmacokinetic study in rats suggested that the half-life time of all conjugates was prolonged compared to the free drug. Sedlák et al. (2001) prepared methoxypoly(ethylene glycol)-4-nitrophenyl carbonate conjugates containing amphotericin B. In vitro antifungal activity suggested that the conjugates exhibited similar effect as the conventional amphotericin B formulated with sodium desoxycholate. Sui et al. (2014) prepared doxorubicin-conjugated Y-shaped copolymers with sustained release behaviors under physiological conditions over a period of 72 h. The antitumor efficacy of the copolymers was enhanced against HeLa cells. Yang et al. (2012) conjugated curcumin molecules to poly(lactic acid) via tris(hydroxymethyl)aminomethane linker using methoxy-poly(ethylene glycol) as the hydrophilic block. The conjugates exhibited enhanced cytotoxicity effects against human hepatocellular carcinoma (HepG2) cells. Jin et al. (2011) synthesized photo-triggered biocompatible drug delivery system from coumarin-functionalized block copolymers by atom transfer radical polymerization. 5-Fluorouracil was conjugated to the coumarin copolymers. The conjugates exhibited controlled drug release profile with low side effects on normal cells. Zhang et al. (2014a) prepared PEG-based conjugates containing paclitaxel and 9-fluorenylmethoxycarbonyl as a functional building block to interact with drug molecules. The conjugates exhibited sustained release kinetics, and in vivo evaluation of the conjugates showed that they exhibited improved antitumor activity than the free drug. Zhang et al. (2015c) reported PEG-based nanocarriers for codelivery of doxorubicin and dasatinib. The carriers exhibited excellent antitumor activity with a 95% tumor growth inhibition rate at a dose of 5 mg/kg for both drugs in a murine breast cancer model. Mikhail and Allen (2010) incorporated docetaxel onto PEG-based nanocarriers and the drug release mechanism was sustained over a period of 1 week. Li et al. (2014) reported the design of a single polymer–drug conjugate containing acetyl-11-keto-β-boswellic acid and methotrexate. Yu et al. (2015b) prepared PEG-based conjugates containing doxorubicin. Significant tumor accumulation was observed in conjugates with molecular weight between ~40 and ~80 kDa. Gao et al. (2014) prepared PEG macromonomer containing doxorubicin. The macromonomer exhibited simultaneous degradation and release of the incorporated drug. Clementi et al. (2011) conjugated paclitaxel and alendronate onto PEG-based carriers. The conjugate demonstrated great binding affinity to the bone mineral with an $IC_{50}$ similar to that of the free drugs combination in human adenocarcinoma prostate cell lines, PC3. The findings suggested that the conjugates are potential drugs for treatment of bone cancer metastasis.

#### *1.5.1.3 N-(2-HYDROXYPROPYL) METHACRYLAMIDE COPOLYMERS*

Etrych et al. (2010) conjugated paclitaxel and docetaxel to HPMA copolymer carrier. The conjugates formed exhibited prolonged blood circulation and targeted drug delivery with good antitumor efficacy in 4T1 model of mammary carcinoma than the free drugs. Lomkova et al. (2016) reported micellar polymer–betulinic acid conjugates based on HPMA copolymer. The drug release was pH dependent. A high cytostatic activity of the conjugates against DLD-1, HT-29, and HeLa carcinoma cell lines enhanced tumor accumulation in HT-29 xenograft in mice. Subr et al. (2014) conjugated doxorubicin to HPMA copolymers with good cytostatic activity. Journo-Gershfeld et al. (2012) reported hyaluronan–HMPA copolymers loaded with paclitaxel. The conjugates exhibited excellent cytotoxic effect toward CD44-overexpressing cells. Williams et al. (2014) prepared a series of HMPA polymer–drug conjugates containing 7-ethyl-10-hydroxycamptothecin that retained the cytotoxicity of the incorporated drug. Zhang et al. (2014b) prepared HPMA copolymer conjugates for sequential combination therapy against A2780 human ovarian carcinoma xenografts. Conjugates prepared were HPMA copolymer–paclitaxel and HPMA copolymer–gemcitabine. The conjugates exhibited improved tumor accumulation, enhanced internalization, and effective intracellular release than the free drug. Pan et al. (2013) incorporated doxorubicin onto HPMA copolymer. The conjugates exhibited antitumor activity toward human ovarian A2780/AD carcinoma in nude mice. Chytil et al. (2015) reported HMPA conjugates containing doxorubicin via hydrazone bond.

Incorporation of the drugs to the polymers prolonged its blood circulation and tumor accumulation than the free drug. Russell-Jones et al. (2012) used vitamin B12 as a targeting moiety for the delivery of HPMA-conjugated daunomycin in murine tumor models. The conjugates increased the survival time of tumor-bearing mice.

### *1.5.1.4 POLY(VINYLPYRROLIDONE)*

There is very little recent report on the polymer–drug conjugates of PVP. In a study by Manju and Sreenivasan (2011), polyvinylpyrrolidone–curcumin conjugates were prepared. The conjugates exhibited cytotoxic effects on L929 fibroblast cells.

### *1.5.1.5 POLY(ETHYLENEIMINE)*

PEI has been employed in drug delivery systems. Chen et al. (2016) incorporated doxorubicin onto PEI-based conjugate for targeted delivery of different anticancer drugs. Hao et al. (2015) prepared multifunctional polymer–drug conjugate from low molecular weight PEI and candesartan for codelivery of drug and siRNA for treatment of lung cancer. Kim et al. (2013) developed PEI–methyl acrylate–PEI with high gene transfection efficiency and low toxicity. Other researchers also reported the application of PEI for gene delivery (Hashemi et al., 2011; Kim et al., 2013; Rekha and Sharma, 2011; Sawant et al., 2012).

### *1.5.1.6 POLYAMIDOAMINES*

Urbán et al. (2014) developed PAA for the delivery of antimalarial drugs to pRBCs. These polymers exhibited no specific toxicity, high biodegradability, and selective internalization into pRBCs, but not in healthy erythrocytes for human and rodent malarias. Xu et al. (2016b) incorporated angiopep-2 and doxorubicin for drug accumulation in glioma cells. The drug accumulation in the tumor site was increased due to the targeting effects of angiopep-2 via LDL receptor-related protein-mediated endocytosis resulting in a reduction in the volume of the vascular C6 glioma spheroids in vitro. Mekuria et al. (2015) incorporated doxorubicin onto polyamidoamine dendrimer. The conjugates exhibited higher drug loading with fast rate of drug release and excellent cytotoxic effects against cervical cancer cells. Zhong et al. (2016) conjugated doxorubicin to carboxyl-terminated poly(amidoamine) dendrimers. The in vivo activity of the dendrimers against melanoma (B16-F10) lung metastasis mouse model showed decreased tumor burden and increased rates of survival in the animals. Zhong and Rocha (2016) prepared pseudosolution formulations in propellant-based metered-dose inhalers from PEGylated poly(amidoamine) dendrimer loaded with doxorubicin. The cellular entry of the conjugates was improved when compared to the free drug. Choksi et al. (2013) conjugated dexamethasone to poly(amidoamine) dendrimer. Kisil et al. (2013) prepared alpha-fetoprotein-based conjugates containing doxorubicin-loaded poly(amidoamine) dendrimer with enhanced cytotoxic activity.

## 1.6 CONCLUSION

Nanobiomaterials have been employed in drug delivery systems for targeted, dual, controlled, and sustained release of therapeutic agents. In vitro and in vivo evaluations of these systems have proved that they are potential systems with enhanced therapeutic effects. Incorporation of drugs onto delivery systems prepared from nanobiomaterials have resulted in enhanced cellular uptake, reduced toxicity, enhanced cell/tissue specificity of the drugs, improved drug stability, and enhanced solubility of the drug, increased blood circulation, and bioavailability of the drug. The release of the incorporated drug was influenced by several factors such as pH, time, selected enzymes, magnetic field, and temperature depending on the nature of materials used. In most in vivo evaluations, no significant side effect was reported and the few cases of side effects reported were mild. Systems developed for cancer chemotherapy were in most cases suitable for dual applications such as drug delivery and bioimaging. Nanobiomaterial-based systems loaded with drugs were effective in vivo and in vitro for the treatment of diseases such as cancer, malaria, diabetes, bacterial infections, viral infections, other protozoan infections and as vaccine more than the free drugs. The findings reported so far have confirmed without doubt that these biomaterials have the potentials to solve many of today's challenging medical limitations. However, there is a pressing and urgent need for further research on these materials before they can be evaluated in clinical trials. There is also a need to perform a thorough evaluation of the toxicity of the nanosized biomaterials in short- and long-term applications.

## ACKNOWLEDGMENTS

The financial assistance of the National Research Foundation (NRF) and Medical Research Council (MRC) South Africa are hereby acknowledged.

## KEYWORDS


- **drugs**
- **carbon nanotubes**
- **bioavailability**
- **nanobiomaterials**
- **nanocarriers**
- **diseases**


## REFERENCES


Abdelmonem, A. M.; Pelaz, B.; Kantner, K.; Bigall, N. C.; del Pino, P.; Parak, W. J. Charge and Agglomeration Dependent In Vitro Uptake and Cytotoxicity of Zinc Oxide Nanoparticles. *J. Inorg. Biochem.* **2015,** *153*, 334–338.

Aderibigbe, B. A.; Jacques, K. D.; Neuse, E. W. Polymeric Conjugates of Selected Aminoquinoline Derivatives as Potential Drug Adjuvants in Cancer Chemotherapy. *J. Inorg. Organomet. Polym. Mater.* **2011,** *21*, 336–345.

Aderibigbe, B. A.; Mukaya, H. E. Nanobiomaterials Architectured for Improved Delivery of Antimalaria Drugs. In *Nanoarchitectonics for Smart Delivery and Drug Targeting*; Holban, A. M., Grumezescu, A. M., Ed.; Elsevier: Amsterdam, 2016; pp 169–193.

Aderibigbe, B. A.; Neuse, E. W. Macromolecular Conjugates of 4- and 8-Aminoquinoline Compounds. *J. Inorg. Organomet. Polym. Mater.* **2012,** *22*, 429–438.

Aderibigbe, B. A.; Neuse, E. W.; Sadiku, E. R.; Shina Ray, S.; Smith, P. J. Synthesis, Characterization, and Antiplasmodial Activity of Polymer-Incorporated Aminoquinolines. *J. Biomed. Mater. Res., A* **2014,** *102*, 1941–1949.

Aderibigbe, B. A.; Owonubi, S. J.; Jayaramudu, J.; Sadiku, E. R.; Ray, S. S. Targeted Drug Delivery Potential of Hydrogel Biocomposites Containing Partially and Thermally

Reduced Graphene Oxide and Natural Polymers Prepared Via Green Process. *Colloid Polym Sci.* **2015,** *293*, 409–420.
Aderibigbe, B. A.; Ray, S. S. Preparation, Characterization and In Vitro Release Kinetics of Polyaspartamide-Based Conjugates Containing Antimalarial and Anticancer Agents for Combination Therapy. *J. Drug Deliv. Sci. Technol.* **2016,** *36*, 34–45.
Adeyemi, O. S.; Sulaiman, F. A. Evaluation of Metal Nanoparticles for Drug Delivery Systems. *J. Biomed. Res.* **2015,** *29*, 145.
Agostoni, V.; Horcajada, P.; Rodriguez-Ruiz, V.; Willaime, H.; Couvreur, P.; Serre, C.; Gref, R.; Green Fluorine-Free Mesoporous Iron(III) Trimesate Nanoparticles for Drug Delivery. *Green Mater*. **2013,** *1*, 209–217.
Ahangari, A.; Salouti, M.; Heidari, Z.; Kazemizadeh, A. R.; Safari, A. A. Development of Gentamicin–Gold Nanospheres for Antimicrobial Drug Delivery to Staphylococcal Infected Foci. *Drug Deliv.* **2013,** *20*, 34–39.
Aires, A.; Ocampo, S. M.; Simões, B. M.; Josefa Rodríguez, M.; Cadenas, J. F.; Couleaud, P.; Spence, K.; Latorre, A.; Miranda, R.; Somoza, Á.; Clarke, R. B.; Carrascosa, J. L.; Cortajarena, A. L. Multifunctionalized Iron Oxide Nanoparticles for Selective Drug Delivery to CD44-Positive Cancer Cells. *Nanotechnology* **2016,** *27*, 065103.
Anbharasi, V.; Cao, N.; Feng, S. S. Doxorubicin Conjugated to D-α-Tocopheryl Polyethylene Glycol Succinate and Folic Acid as a Prodrug for Targeted Chemotherapy. *J. Biomed. Mater. Res., A* **2010,** *94*, 730–743.
Bamrungsap, S.; Zhao, Z.; Chen; T.; Wang; L.; Li; C.; Fu; T.; Tan, W. Nanotechnology in Therapeutics. *Nanomedicine* **2012,** *7*, 1253–1271.
Banerjee, I.; Douaisi, M. P.; Mondal, D.; Kan, R. S. Light-Activated Nanotube–Porphyrin Conjugates as Effective Antiviral Agents. *Nanotechnology* **2012,** *23*, 7 pages.
Baskar, G.; Chandhuru, J.; Sheraz Fahad, K.; Praveen, A. S.; Chamundeeswari, M.; Muthukumar, T. Anticancer Activity of Fungal L-Asparaginase Conjugated with Zinc Oxide Nanoparticles. *J. Mater. Sci. Mater. Med.* **2015,** *26*. DOI:10.1007/s10856-015-5380-z.
Benyettou, F.; Rezgui, R.; Ravaux, F.; Jaber, T.; Blumer, K.; Jouiad, M.; Motte, L.; Olsen, J. C.; Platas-Iglesias, C.; Magzoub, M.; Trabolsi, A. Synthesis of Silver Nanoparticles for the Dual Delivery of Doxorubicin and Alendronate to Cancer Cells. *J. Mater. Chem. B* **2015,** *3*, 7237–7245.
Beydokhti, A. K.; Heris, S. Z.; Jaafari, M. R.; Nikoofal-Sahlabadi, S.; Tafaghodi, M.; Hatamipoor, M. Microwave Functionalized Single-Walled Carbon Nanotube as Nanocarrier for the Delivery of Anticancer Drug Cisplatin: In Vitro and In Vivo Evaluation. *J. Drug Deliv Sci. Technol.* **2014,** *24*, 572–578.
Bhattacharya, R.; Patra, C. R.; Earl, A.; Wang, S.; Katarya, A.; Lu, L.; Kizhakkedathu, J. N.; Yaszemski, M. J.; Greipp, P. R.; Mukhopadhyay, D.; Mukherjee, P. Attaching Folic Acid on Gold Nanoparticles Using Noncovalent Interaction via Different Polyethylene Glycol Backbones and Targeting of Cancer Cells. *Nanomed. Nanotechnol. Biol. Med.* **2007,** *3*, 224–238.
Brown, A. N.; Smith, K.; Samuels, T. A.; Lu, J.; Obare, S. O.; Scott, M. E. Nanoparticles Functionalized with Ampicillin Destroy Multiple-Antibiotic-Resistant Isolates of *Pseudomonas aeruginosa* and *Enterobacter aerogenes* and Methicillin-Resistant *Staphylococcus aureus*. *Appl. Environ. Microbiol.* **2012,** *78*, 2768–2774.
Calvo, E.; Hoch, U.; Maslyar, D. J.; Tolcher, A. W. Dose-Escalation Phase I Study of NKTR-105, a Novel Pegylated Form of Docetaxel. *ASCO Annu. Meet. Proc.* **2010,** *28*, TPS160.

Cao, Y.; Ma, Y.; Zhang, M.; Wang, H.; Tu, X.; Shen, H.; Dai, J.; Guo, H.; Zhang, Z. Ultrasmall Graphene Oxide Supported Gold Nanoparticles as Adjuvants Improve Humoral and Cellular Immunity in Mice. *Adv. Funct. Mater.* **2014,** *24*, 6963–6971.

Chakravarthy, K. V.; Bonoiu, A. C.; Davis, W. G.; Ranjan, P.; Ding, H.; Hu, R.; Bowzard, J. B.; Bergey, E. J.; Katz, J. M.; Knight, P. R.; Sambhara, S.; Prasad, P. N.; Turro, N. J. *Proc. Natl. Acad. Sci. U.S.A.* **2010,** *107*, 10172–10177.

Chanda, N.; Kan, P.; Watkinson, L. D.; Shukla, R.; Zambre, A.; Carmack, T. L.; Engelbrecht, H.; Lever, J. R.; Katti, K.; Fent, G. M.; Casteel, S. W. Radioactive Gold Nanoparticles in Cancer Therapy: Therapeutic Efficacy Studies of GA-198 AuNP Nanoconstruct in Prostate Tumor-Bearing Mice. *Nanomed. Nanotechnol. Biol. Med.* **2010,** *6*, 201–209.

Chen, A.; Xu, C.; Li, M.; Zhang, H.; Wang, D.; Xia, M.; Meng, G.; Kang, B.; Chen, H.; Wei, J. Photoacoustic Nanobombs Fight against Undesirable Vesicular Compartmentalization of Anticancer Drugs. *Sci. Rep.* **2015,** *5*, 15527.

Chen, C.; Zhou, B.; Zhu, X.; Shen, M.; Shi, X. Branched Polyethyleneimine Modified with Hyaluronic Acid via a PEG Spacer for Targeted Anticancer Drug Delivery. *RSC Adv.* **2016,** *6*, 9232–9239.

Chen, F.; Hong, H.; Shi, S.; Valdovinos, H.; Barnhart, T.; Cai, W. Tumor Targeted Drug Delivery with Hollow Mesoporous Silica Nanoparticles. *J. Nucl. Med.* **2014,** *55*, 547–547.

Chen, R.; Qian, Y.; Li, R.; Zhang, Q.; Liu, D.; Wang, M.; Xu, Q. Methazolamide Calcium Phosphate Nanoparticles in an Ocular Delivery System. *Yakugaku Zasshi* **2010,** *130*, 419–424.

Cheng, X.; Kuhn, L. Chemotherapy Drug Delivery from Calcium Phosphate Nanoparticles. *Int. J. Nanomed.* **2007,** *2*, 667–674.

Chidambaram, M.; Manavalan, R.; Kathiresan, K. Nanotherapeutics to Overcome Conventional Cancer Chemotherapy Limitations. *J. Pharm. Pharm. Sci.* **2011,** *14*, 67–77.

Choi, C. H. J.; Alabi, C. A.; Webster, P.; Davis, M. E. Mechanism of Active Targeting in Solid Tumors with Transferrin-Containing Gold Nanoparticles. *Proc. Natl. Acad. Sci.* **2010,** *107*, 1235–1240.

Choksi, A.; Sarojini, K. V. L.; Vadnal, P.; Dias, C.; Suresh, P. K.; Khandare, J. Comparative Anti-inflammatory Activity of Poly(Amidoamine) (PAMAM) Dendrimer–Dexamethasone Conjugates with Dexamethasone–Liposomes. *Int. J. Pharm.* **2013,** *449*, 28–36.

Chytil, P.; Sírová, M.; Koziolová, E.; Ulbrich, K.; Říhová, B.; Etrych, T. The Comparison of In Vivo Properties of Water-Soluble HPMA-Based Polymer Conjugates with Doxorubicin Prepared by Controlled RAFT or Free Radical Polymerization. *Physiol. Res.* **2015,** *64*, S41–S49.

Clementi, C.; Miller, K.; Mero, A.; Satchi-Fainaro, R.; Pasut, G. Dendritic Poly(Ethylene Glycol) Bearing Paclitaxel and Alendronate for Targeting Bone Neoplasms. *Mol. Pharm.* **2011,** *8*, 1063–1072.

Daneshmehra, S. Carbon Nanotubes for Delivery of Quercetin as Anticancer Drug: Theoretical Study. *Proc. Mater. Sci.* **2015,** *11*, 131–136.

Di Meo, C.; Cilurzo, F.; Licciardi, M.; Scialabba, C.; Sabia, R.; Paolino, D.; Capitani, D.; Fresta, M.; Giammona, G.; Villani, C.; Matricardi, P. Polyaspartamide–Doxorubicin Conjugate as Potential Prodrug for Anticancer Therapy. *Pharm. Res.* **2015,** *32*, 1557–1569.

Dixon, D.; Meenan, B. J.; Manson, J. PEG Functionalized Gold Nanoparticle loaded PLGA Films for Drug Delivery. *J. Nano Res.* **2014,** *27*, 83–94.

Dostalova, S.; Moulick, A.; Milosavljevic, V.; Guran, R.; Kominkova, M.; Cihalova, K.; Heger, Z.; Blazkova, L.; Kopel, P.; Hynek, D.; Vaculovicova, M. Antiviral Activity of Fullerene C60 Nanocrystals Modified with Derivatives of Anionic Antimicrobial Peptide Maximin H5. *Monatsh. Chem.* **2016,** *147*, 905–918.

Du, C. X.; Xiong, H. R.; Ji, H.; Liu, Q.; Xiao, H.; Yang, Z. Q. The Antiviral Effect of Fullerene–Liposome Complex against Influenza Virus (H1N1) In Vivo. *Sci. Res. Essays* **2012,** *7*, 705–711.

Duncan, R. The Dawning Era of Polymer Therapeutics. *Nat. Rev. Drug Discov.* **2003,** *2*, 347–360.

Etrych, T.; Šírová, M.; Starovoytova, L.; Říhová, B.; Ulbrich, K. HPMA Copolymer Conjugates of Paclitaxel and Docetaxel with pH-Controlled Drug Release. *Mol. Pharm.* **2010,** *7*, 1015–1026.

Fan, J.; Fang, G.; Zeng, F.; Wang, X.; Wu, S. Water-Dispersible Fullerene Aggregates as a Targeted Anticancer Prodrug with both Chemo- and Photodynamic Therapeutic Actions. *Small* **2013,** *9*, 613–621.

Fayaz, A. M.; Balaji, K.; Girilal, M.; Yadav, R.; Kalaichelvan, P. T.; Venketesan, R. Biogenic Synthesis of Silver Nanoparticles and their Synergistic Effect with Antibiotics: A Study against Gram-Positive and Gram-Negative Bacteria. *Nanomed. Nanotech. Biol. Med.* **2010,** *6*, 103–109.

Fresta, M.; Giammona, G.; Cavallaro, G.; Licciardi, M.; Paolino, D. Bisphosphonate–Polyaspartamide Conjugates as Polymeric Carriers for Drug Targeting to Bones. Patent WO 2012098222 A1, 26 July 2012.

Ganesh, M.; Hemalatha, P.; Peng, M. M.; Cha, W. S.; Palanichamy, M.; Jang, H. T. Drug Release Evaluation of Mesoporous $TiO_2$: A Nanocarrier for Duloxetine. *Computer Applications for Modelling, Simulation, and Automobile*; Springer: Berlin-Heidelberg, 2012; pp 237–243.

Ganeshkumar, M.; Sastry, T. P.; Kumar, M. S.; Dinesh, M. G.; Kannappan, S.; Suguna, L. Sun Light Mediated Synthesis of Gold Nanoparticles as Carrier for 6-Mercaptopurine: Preparation, Characterization and Toxicity Studies in Zebrafish Embryo Model. *Mater. Res. Bull.* **2012,** *47*, 2113–2119.

Gao, A. X.; Liao, L.; Johnson, J. A. Synthesis of Acid-Labile PEG and PEG-Doxorubicin–Conjugate Nanoparticles via Brush-First ROMP. *ACS Macro Lett.* **2014,** *3*, 854–857.

Ghosh, D.; Chandra, S.; Chakraborty, A.; Ghosh, S. K.; Pramanik, P. A Novel Graphene Oxide-Para Amino Benzoic Acid Nanosheet as Effective Drug Delivery System to Treat Drug Resistant Bacteria. *Int. J. Pharm. Sci. Drug Res.* **2012,** *2*, 127–133.

Ghoshal, S.; Mishra, M. K. Release Kinetic Profiles of 6-Mercaptopurine Loaded Covalently Functionalized Multiwalled Carbon Nanotubes. *Am. J. Adv. Drug Deliv.* **2014,** *2*, 110–119.

Hackenberg, S.; Scherzed, A.; Harnisch, W.; Froelich, K.; Ginzkey, C.; Koehler, C.; Hagen, R.; Kleinsasser, N. Antitumor Activity of Photo-Stimulated Zinc Oxide Nanoparticles Combined with Paclitaxel or Cisplatin in HNSCC Cell Lines. *J. Photochem. Photobiol. B* **2012,** *114*, 87–93.

Hahn, S. K.; Lee, M. Y. Yang, J.; Jung, H. S. Liver Targeted Drug Delivery Systems Using Metal Nanoparticles and Preparing Method Thereof. WO 2013176468 A1. 28 November 2013.

Halamová, D.; Zeleňák, V. NSAID Naproxen in Mesoporous Matrix MCM-41: Drug Uptake and Release Properties. *J. Incl. Phenom. Macrocycl. Chem.* **2012,** *72*, 15–23.

Hao, S.; Yan, Y.; Ren, X.; Xu, Y.; Chen, L.; Zhang, H. Candesartan-Graft-Polyethyleneimine Cationic Micelles for Effective Co-Delivery of drug and gene in Anti-Angiogenic Lung Cancer Therapy. *Biotechnol. Bioprocess Eng.* **2015,** *20*, 550–560.

Hashemi, M.; Parhiz, B. H.; Hatefi, A.; Ramezani, M. Modified Polyethyleneimine with Histidine–Lysine Short Peptides as Gene Carrier. *Cancer Gene Ther.* **2011,** *18*, 12–19.

He, Q.; Gao, Y.; Zhang, L.; Zhang, Z.; Gao, F.; Ji, X.; Li, Y.; Shi, J. A pH-Responsive Mesoporous Silica Nanoparticles-Based Multi-Drug Delivery System for Overcoming Multi-Drug Resistance. *Biomaterials* **2011,** *32*, 7711–7720.

Hong, J.; Shah, N. J.; Drake, A. C.; DeMuth, P. C.; Lee, J. B.; Chen, J.; Hammond, P. T. Graphene Multilayers as Gates for Multi-Week Sequential Release of Proteins from Surfaces. *ACS Nano* **2011,** *6*, 81–88.

Hsiao, M. H.; Mu, Q.; Stephen, Z. R.; Fang, C.; Zhang, M. Hexanoyl-Chitosan-PEG Copolymer Coated Iron Oxide Nanoparticles for Hydrophobic Drug Delivery. *ACS Macro Lett.* **2015,** *4*, 403–407.

Hu, Z.; Zhang, C.; Huang, Y.; Sun, S.; Guan, W.; Yao, Y. Photodynamic Anticancer Activities of Water-Soluble C(60) Derivatives and Their Biological Consequences in a HeLa Cell Line. *Chem. Biol. Interact.* **2012,** *195*, 86–94.

Huang, T.; Zhang, L.; Chena, H.; Gao, C. A Cross-Linking Graphene Oxide Polyethyleneimine Hybrid Film Containing Ciprofloxacin: One-Step Preparation, Controlled Drug Release and Antibacterial Performance. *J. Mater. Chem. B.* **2015,** *3*, 1605–1611.

Huang, X.; Qi, X.; Boey, F.; Zhang, H. Critical Review: Graphene-Based Composites. *Chem. Soc. Rev.* **2012,** *41*, 666–686.

Hubbell, J. A.; Chilkoti, A. Nanomaterials for Drug Delivery. *Science* **2012,** *337*, 303–305.

Iannazzo, D.; Pistone, A.; Galvagno, S.; Ferro, S.; Luca, L. D.; Monforte, A. M.; Ros, T. D.; Hadad, C.; Prato, M.; Pannecouque, C. Synthesis and Anti-HIV Activity of Carboxylated and Drug-Conjugated Multi-Walled Carbon Nanotubes. *Carbon* **2015,** *82*, 548–561.

Jain, V.; Jain, S.; Mahajan, S. C. Nanomedicines Based Drug Delivery Systems for Anti-Cancer Targeting and Treatment. *Curr. Drug Deliv.* **2015,** *12*, 177–192.

Jin, Q.; Mitschang, F.; Agarwal, S. Biocompatible Drug Delivery System for Photo-Triggered Controlled Release of 5-Fluorouracil. *Biomacromolecules* **2011,** *12*, 3684–3691.

Journo-Gershfeld, G.; Kapp, D.; Shamay, Y.; Kopeček, J.; David, A. Hyaluronan Oligomers–HPMA Copolymer Conjugates for Targeting Paclitaxel to CD44-Overexpressing Ovarian Carcinoma. *Pharm. Res.* **2012,** *29*, 1121–1133.

Kang, S. Herzberg, S.; Rodrigues, D. F.; Elimelech, M. Antibacterial Effects of Carbon Nanotubes: Size Does Matter! *Langmuir* **2008,** *24*, 6409–6413.

Kang, S. Pinault, M.; Pfefferle, L. D.; Elimelech, M. Single-Walled Carbon Nanotubes Exhibit Strong Antimicrobial Activity. *Langmuir* **2007,** *23*, 8670–8673.

Kavitha, T.; Kang, I. K.; Park, S. Y. Poly(4-vinyl pyridine) -Grafted Graphene Oxide for Drug Delivery and Antimicrobial Applications. *Polym. Int.* **2015,** *64*, 1660–1666.

Kayal, S.; Ramanujan, R. V. Anti-Cancer Drug Loaded Iron–Gold Core–Shell Nanoparticles (Fe@Au) for Magnetic Drug Targeting. *J. Nanosci. Nanotechnol.* **2010,** *10*, 5527–5539.

Kim, T. H.; Choi, H.; Yu, G. S.; Lee, J.; Choi, J. S. Novel Hyperbranched Polyethyleneimine Conjugate as an Efficient Non-Viral Gene Delivery Vector. *Macromol. Res.* **2013,** *21*, 1097–1104.

Kisil, S. I.; Chernikov, V. A.; Danilevskiy, M. I.; Savvateeva, L. V.; Gorokhovets, N. V. Conjugation of the Recombinant Third Domain of Human Alpha-Fetoprotein with

Doxorubicin Using Poly(Amidoamine) Dendrimers and Study of Its Cytotoxic Activity. *Engineering* **2013,** *4*, 80–83.

Končič, M.; Zorc, B.; Novak, P. Macromolecular Prodrugs. XIII. Hydrosoluble Conjugates of 17β-Estradiol and Estradiol-17β-Valerate with Polyaspartamide Polymer. *Acta Pharm.* **2011,** *61*, 465–472.

Koneru, B.; Shi, Y.; Wang, Y. C.; Chavala, S. H.; Miller, M. L.; Holbert, B.; Conson, M.; Ni, A.; Di Pasqua, A. J. Tetracycline-Containing MCM-41 Mesoporous Silica Nanoparticles for the Treatment of *Escherichia coli*. *Molecules* **2015,** *20*, 19690–19698.

Kossatz, S.; Grandke, J.; Couleaud, P.; Latorre, A.; Aires, A.; Crosbie-Staunton, K.; Ludwig, R.; Dähring, H.; Ettelt, V.; Lazaro-Carrillo, A.; Calero, M. Efficient Treatment of Breast Cancer Xenografts with Multifunctionalized Iron Oxide Nanoparticles Combining Magnetic Hyperthermia and Anti-cancer Drug Delivery. *Breast Cancer Res*. **2015,** *17*, 1.

Laha, D.; Pramanik, A.; Chattopadhyay, S.; Kumar Dash, S.; Roy, S.; Pramanik, P.; Karmakar, P. Folic Acid Modified Copper Oxide Nanoparticles for Targeted Delivery in In Vitro and In Vivo Systems. *RSC Adv.* **2015,** *5*, 68169–68178.

Lay, C. L.; Liu, H. Q.; Tan, H. R.; Liu, Y. Delivery of Paclitaxel by Physically Loading onto Poly(Ethylene Glycol) (PEG)-Graft-Carbon Nanotubes for Potent Cancer Therapeutics. *Nanotechnology* **2010,** *21*, 065101.

Li, H.; Fierens, K.; Zhang, Z.; Vanparijs, N.; Schuijs, M. J.; Van Steendam, K.; Feiner Gracia, N.; De Rycke, R.; De Beer, T.; De Beuckelaer, A.; De Koker, S. Spontaneous Protein Adsorption on Graphene Oxide Nanosheets Allowing Efficient Intracellular Vaccine Protein Delivery. *ACS Appl. Mater. Interfaces* **2016,** *8*, 1147–1155.

Li, W.; Wu, J.; Zhan, P.; Chang, Y.; Pannecouque, C.; De Clercq, E.; Liu, X. Synthesis, Drug Release and Anti-HIV Activity of a Series of PEGylated Zidovudine Conjugates. *Int. J. Biol. Macromol.* **2012,** *50*, 974–980.

Li, W. M; Chen, S. Y.; Liu, D. M. In Situ Doxorubicin–CaP Shell Formation on Amphiphilic Gelatin–Iron Oxide Core as a Multifunctional Drug Delivery System with Improved Cytocompatibility, pH-Responsive Drug Release and MR Imaging. *Acta Biomater.* **2013,** *9*, 5360–5368.

Li, Y.; Dong, H.; Li, X.; Shi, D.; Li, Y. Single Polymer–Drug Conjugate Carrying Two Drugs for Fixed-Dose Co-delivery. *Med. Chem.* **2014,** *4*, 672–683.

Li, Y.; Zhou, Y.; Li, X.; Sun, J.; Ren, Z.; Wen, W.; Yang, X.; Han, G. A Facile Approach to Upconversion Crystalline $CaF_2$:$Yb^{3+}$, $Tm^{3+}$@*m*$SiO_2$ Nanospheres for Tumor Therapy. *RSC Adv.* **2016,** *6*, 38365–38370.

Liu, E.; Zhou, Y.; Liu, Z.; Li, J.; Zhang, D.; Chen, J.; Cai, Z. Cisplatin Loaded Hyaluronic Acid Modified $TiO_2$ Nanoparticles for Neoadjuvant Chemotherapy of Ovarian Cancer. *J. Nanomater.* **2015a,** *16*, 275.

Liu, J.; Ma, X.; Jin, S.; Xue, X.; Zhang, C.; Wei, T.; Guo, W.; Liang, X. J. Zinc Oxide Nanoparticles as Adjuvant to Facilitate Doxorubicin Intracellular Accumulation and Visualize pH-Responsive Release for Overcoming Drug Resistance. *Mol. Pharm.* **2016,** *13*, 1723–1730.

Liu, L.; Wei, Y.; Zhai, S.; Chen, Q.; Xing, D. Dihydroartemisinin and Transferrin Dual-Dressed Nano-Graphene Oxide for a pH-Triggered Chemotherapy. *Biomaterials* **2015b,** *62*, 35–46.

Liu, Y.; Chen, Q.; Xu, M.; Guan, G.; Hu, W.; Liang, Y.; Zhao, X.; Qiao, M.; Chen, D; Liu, H. Single Peptide Ligand-Functionalized Uniform Hollow Mesoporous Silica

Nanoparticles Achieving Dual-Targeting Drug Delivery to Tumor Cells and Angiogenic Blood Vessel Cells. *Int. J. Nanomed.* **2015c,** *10*, 1855–1867.

Liu, Y.; Ding, X.; Li, J.; Luo, Z.; Hu, Y.; Liu, J.; Dai, L.; Zhou, J.; Hou, C.; Cai, K. Enzyme Responsive Drug Delivery System Based on Mesoporous Silica Nanoparticles for Tumor Therapy In Vivo. *Nanotechnology* **2015d,** *26*, 145102.

Lomkova, E. A.; Chytil, P.; Janoušková, O.; Mueller, T.; Lucas, H.; Filippov, S. K.; Trhlíková, O.; Aleshunin, P. A.; Skorik, Y. A.; Ulbrich, K.; Etrych, T. Biodegradable Micellar HPMA-Based Polymer–Drug Conjugates with Betulinic Acid for Passive Tumor Targeting. *Biomacromolecules* **2016**. DOI:10.1021/acs.biomac.6b00947.

Loomba, L.; Sekhon, B. S. Calcium Phosphate Nanoparticles and their Biomedical Potential. *Nanomater. Molecul. Nanotechnol.* **2015**. DOI:10.4172/2324-8777.1000154.

Madrid, S. I. U.; Pal, U.; Kang, Y. S.; Kim, J.; Kwon, H.; Kim, J. Fabrication of $Fe_3O_4$@ *m*$SiO_2$ Core–Shell Composite Nanoparticles for Drug Delivery Applications. *Nanoscale Res. Lett.* **2015,** *10*, 1–8.

Maeda-Mamiya, R.; Noiri, E.; Isobe, H.; Nakanishi, W.; Okamoto, K.; Doi, K.; Sugaya, T.; Izumi, T.; Homma, T.; Nakamura, E. *In vivo* Gene Delivery by Cationic Tetraamino Fullerene. *Proc. Natl. Acad. Sci. USA* **2010,** *107*, 5339–5344.

Maggini, L.; Cabrera, I.; Ruiz-Carretero, A.; Prasetyanto, E. A.; Robinet, E.; De Cola, L. Breakable Mesoporous Silica Nanoparticles for Targeted Drug Delivery. *Nanoscale* **2016,** *8*, 7240–7247.

Mahkam, M.; Hosseinzadeh, F.; Galehassadi, M. Preparation of Ionic Liquid Functionalized Silica Nanoparticles for Oral Drug Delivery. *J. Biomater. Nanobiotechnol.* **2012,** *3*, 391–395.

Maity, M.; Pramanik, S. K.; Pal, U.; Banerji, B.; Maiti, N. C. Copper(I) Oxide Nanoparticle and Tryptophan as Its Biological Conjugate: A Modulation of Cytotoxic Effects. *J. Nanopart. Res.* **2014,** *16*, 1–13.

Majeed, M. I.; Lu, Q.; Yan, W.; Li, Z.; Hussain, I.; Tahir, M. N.; Tremel, W.; Tan, B. Highly Water-Soluble Magnetic Iron Oxide ($Fe_3O_4$) Nanoparticles for Drug Delivery: Enhanced In Vitro Therapeutic Efficacy of Doxorubicin and MION Conjugates. *J. Mater. Chem. B* **2013,** *1*, 2874–2884.

Malathi, S.; Balakumaran, M. D.; Kalaichelvan, P. T.; Balasubramanian, S. Green Synthesis of Gold Nanoparticles for Controlled Delivery. *Adv. Mater Lett.* **2013,** *4*, 933–940.

Manju, S.; Sreenivasan, K. Synthesis and Characterization of a Cytotoxic Cationic Polyvinylpyrrolidone–Curcumin Conjugate. *J. Pharm. Sci.* **2011,** *100*, 504–511.

Markovsky, E.; Baabur-Cohen, H.; Eldar-Boock, A.; Omer, L.; Tiram, G.; Ferber, S.; Ofek, P.; Polyak, D.; Scomparin, A.; Satchi-Fainaro, R. Administration, Distribution, Metabolism and Elimination of Polymer Therapeutics. *J. Control. Release* **2012,** *161*, 446–460.

Martínez-Carmona, M.; Lozano, D.; Colilla, M.; Vallet-Regí, M. Selective Topotecan Delivery to Cancer Cells by Targeted pH-Sensitive Mesoporous Silica Nanoparticles. *RSC Adv.* **2016,** *6*, 50923–50932.

Martinez-Gutierrez, F.; Olive, P. L.; Banuelos, A.; Orrantia, E.; Nino, N.; Sanchez, E. M.; Ruiz, F.; Bach, H.; Av-Gay, Y. Synthesis, Characterization, and Evaluation of Antimicrobial and Cytotoxic Effect of Silver and Titanium Nanoparticles. *Nanomed. Nanotechnol. Biol. Med.* **2010,** *6*, 681–688.

Mehra, N. K.; Verma, A. K.; Mishra, P. R.; Jain, N. K. The Cancer Targeting Potential of D-α-Tocopheryl Polyethylene Glycol 1000 Succinate Tethered Multi-Walled Carbon Nanotubes. *Biomaterials* **2014,** *35*, 4573–4588.

Mekuria, S. L.; Debele, T. A.; Chou, H. Y.; Tsai, H. C. IL-6 Antibody and RGD Peptide Conjugated Poly(Amidoamine) Dendrimer for Targeted Drug Delivery of HeLa Cells. *J Phys. Chem. B* **2015,** *120*, 123–130.

Meng, H.; Wang, M.; Liu, H.; Liu, X.; Situ, A.; Wu, B.; Ji, Z.; Chang, C. H.; Nel, A. E. Use of a Lipid-Coated Mesoporous Silica Nanoparticle Platform for Synergistic Gemcitabine and Paclitaxel Delivery to Human Pancreatic Cancer in Mice. *ACS Nano* **2015,** *9*, 3540–3557.

Meng, J.; Meng, J.; Duan, J.; Kong, H.; Li, L.; Wang, C.; Xie, S.; Chen, S.; Gu, N.; Xu, H.; Yang, X. D. Carbon Nanotubes Conjugated to Tumor Lysate Protein Enhance the Efficacy of an Antitumor Immunotherapy. *Small* **2008,** *4*, 1364–1370.

Mikhail, A. S.; Allen, C. Poly(Ethylene Glycol)-b-Poly(ε-Caprolactone) Micelles Containing Chemically Conjugated and Physically Entrapped Docetaxel: Synthesis, Characterization, and the Influence of the Drug on Micelle Morphology. *Biomacromolecules* **2010,** *11*, 1273–1280.

Mirazi, N.; Shoaei, J.; Khazaei, A.; Hosseini, A. A Comparative Study on Effect of Metformin and Metformin-Conjugated Nanotubes on Blood Glucose Homeostasis in Diabetic Rats. *Eur. J. Drug Metab. Pharmacokinet.* **2015,** *40*, 343–348.

Mufula, A. I.; Aderibigbe, B. A.; Neuse, E. W.; Mukaya, H. E. Macromolecular Co-conjugates of Methotrexate and Ferrocene in the Chemotherapy of Cancer. *J. Inorg. Organomet. Polym. Mater.* **2012,** *22*, 423–428.

Mukherjee, B.; Dey, N. S.; Maji, R.; Bhowmik, P.; Das, P. J.; Paul, P. Current Status and Future Scope for Nanomaterials in Drug Delivery. In *Application of Nanotechnology in Drug Delivery*; Sezer, A. D., Ed.; InTech Publishers: Croatia, 2014; pp 525–544.

Murawala, P.; Tirmale, A.; Shiras, A.; Prasad, B. L. V. In Situ Synthesized BSA Capped Gold Nanoparticles: Effective Carrier of Anticancer Drug Methotrexate to MCF-7 Breast Cancer Cells. *Mater. Sci. Eng. C* **2014,** *34*, 158–167.

Nakamura, E.; Isobe, H. In Vitro and In Vivo Gene Delivery with Tailor-Designed Aminofullerenes. *Chem. Rec.* **2010,** *10*, 260–270.

Ngoy, J. M.; Iyuke, S. E.; Yah, C. S.; Neuse, W. E. Kinetic Optimization of Folic Acid Polymer Conjugates for Drug Targeting. *Am. J. Appl. Sci.* **2011,** *8*, 508–519.

Ni, G.; Wang, Y.; Wu, X.; Wang, X.; Cheng, S.; Liu, X. Graphene Oxide Absorbed Anti-IL10R Antibodies Enhance LPS Induced Immune Responses In Vitro and In Vivo. *Immunol. Lett.* **2012,** *148*, 126–132.

Nkazi, B. D.; Neuse, E. W.; Aderibigbe, B. A. Polymeric Co-conjugates of Curcumin. *J. Inorg. Organomet. Polym. Mater.* **2012,** *22*, 886–891.

Nkazi, B. D.; Neuse, E. W.; Sadik, E. R.; Aderibigbe, B. A. Synthesis, Characterization, Kinetic Release Study and Evaluation of Hydrazone Linker in Ferrocene Conjugates at Different pH Values. *J. Drug Deliv. Sci. Technol.* **2013,** *23*, 537–545.

Nowacki, M.; Wiśniewski, M.; Werengowska-Ciećwierz, K.; Terzyk, A. P.; Kloskowski, T.; Marszałek, A.; Bodnar, M.; Pokrywczyn′ska, M.; Nazarewski, L.; Pietkun, K.; Jundziłł, A.; Drewa, T. New Application of Carbon Nanotubes in Haemostatic Dressing Filled with Anticancer Substance. *Biomed. Pharmacother.* **2015,** *69*, 349–354.

Olmedo, H.; Herrera, M.; Rojas, L.; Villalta, M.; Vargas, M.; Leiguez, E.; Teixeira, C.; Estrada, R.; Gutiérrez, J. M.; León, G.; Montero, M. L. Comparison of the Adjuvant Activity of Aluminum Hydroxide and Calcium Phosphate on the Antibody Response towards *Bothrops asper* Snake Venom. *J. Immunotoxicol.* **2014,** *11*, 44–49.

Ou, Z.; Wu, B.; Xing, D.; Zhou, F.; Wang, H.; Tang, Y. Functional Single-Walled Carbon Nanotubes Based on an Integrin Alpha v Beta 3 Monoclonal Antibody for Highly Efficient Cancer Cell Targeting. *Nanotechnology* **2009,** *11*, 105102.

Palanikumar, L.; Ramasamy, S.; Hariharan, G.; Balachandran, C. Influence of Particle Size of Nano Zinc Oxide on the Controlled Delivery of Amoxicillin. *Appl. Nanosci.* **2013,** *3*, 441–451.

Pan, H.; Sima, M.; Yang, J.; Kopeček, J. Synthesis of Long-Circulating, Backbone Degradable HPMA Copolymer–Doxorubicin Conjugates and Evaluation of Molecular-Weight-Dependent Antitumor Efficacy. *Macromol. Biosci.* **2013,** *13*, 155–160.

Panchapakesan, B.; Lu, S.; Sivakumar, K.; Teker, K.; Cesarone, G.; Wickstrom, E. Single-Wall Carbon Nanotube Nanobomb Agents for Killing Breast Cancer Cells. *Nanobiotechnology* **2005,** *1*, 133–139.

Panchuk, R. R.; Prylutska, S. V.; Chumak, V. V.; Skorokhyd, N. R.; Lehka, L. V.; Evstigneev, M. P.; Prylutskyy, Y. I.; Berger, W.; Heffeter, P.; Scharff, P.; Ritter, U. Application of C60 Fullerene–Doxorubicin Complex for Tumor Cell Treatment In Vitro and In Vivo. *J. Biomed. Nanotechnol.* **2015,** *11*, 1139–1152.

Pantarotto, D.; Partidos, C. D.; Graff, R.; Hoebeke, J.; Briand, J. P.; Prato, M.; Bianco, A. Synthesis, Structural Characterization, and Immunological Properties of Carbon Nanotubes Functionalized with Peptides. *J. Am. Chem. Soc.* **2003a,** *125*, 6160–6164.

Pantarotto, D.; Partidos, C. D.; Hoebeke, J.; Brown, F.; Kramer, E.; Briand, J. P.; Muller, S.; Prato, M.; Bianco, A. Immunization with Peptide-Functionalized Carbon Nanotubes Enhances Virus-Specific Neutralizing Antibody Responses. *Chem. Biol.* **2003b,** *10*, 961–966.

Paolino, D.; Licciardi, M.; Celia, C.; Giammona, G.; Fresta, M.; Cavallaro, G. Bisphosphonate–Polyaspartamide Conjugates as Bone Targeted Drug Delivery Systems. *J. Mater Chem. B* **2015,** *3*, 250–259.

Park, Y. H.; Park, S. Y.; In, I. Direct Noncovalent Conjugation of Folic Acid on Reduced Graphene Oxide as Anticancer Drug Carrier. *J. Ind. Eng. Chem.* **2015,** *30*, 190–196.

Parra, J.; Abad-Somovilla, A.; Mercader, J. V.; Taton, T. A.; Abad-Fuentes, A. Carbon Nanotube–Protein Carriers Enhance Size-Dependent Self-Adjuvant Antibody Response to Haptens. *J. Control. Release* **2013,** *170*, 242–251.

Pasqua, A. J. D.; Yuan, H.; Chung, Y.; Kim, J. K.; Huckle, J. E.; Li, C.; Sadgrove, M.; Tran, T. H.; Jay, M.; Lu, X. Neutron-Activatable Holmium-Containing Mesoporous Silica Nanoparticles as a Potential Radionuclide Therapeutic Agent for Ovarian Cancer. *J. Nucl. Med.* **2013,** *54*, 111–116.

Paul, W.; Sharma, C. P. Synthesis and Characterization of Alginate Coated Zinc Calcium Phosphate Nanoparticles for Intestinal Delivery of Insulin. *Proc. Biochem.* **2012,** *47*, 882–886.

Popat, A.; Ross, B. P.; Liu, J.; Jambhrunkar, S.; Kleitz, F.; Qiao, S. Z. Enzyme-Responsive Controlled Release of Covalently Bound Prodrug from Functional Mesoporous Silica Nanospheres. *Angew. Chem. Int. Ed.* **2012,** *51*, 12486–12489.

Quan, G.; Pan, X.; Wang, Z.; Wu, Q.; Li, G.; Dian, L.; Chen, B.; Wu, C. Lactosaminated Mesoporous Silica Nanoparticles for Asialoglycoprotein Receptor Targeted Anticancer Drug Delivery. *J. Nanobiotechnol.* **2015,** *13*, 1.

Quan, Q.; Xie, J.; Gao, H.; Yang, M.; Zhang, F.; Liu, G.; Lin, X.; Wang, A.; Eden, H. S.; Lee, S.; Zhang, G. HSA Coated Iron Oxide Nanoparticles as Drug Delivery Vehicles for Cancer Therapy. *Mol. Pharm.* **2011,** *8*, 1669–1676.

Rani, K.; Paliwal, S. A Review on Targeted Drug Delivery: Its Entire Focus on Advanced Therapeutics and Diagnostics. *Sch. J. Appl. Med. Sci.* **2014,** *2*, 328–331.

Rastogi, L.; Kora, A. J.; Arunachalam, J. Highly Stable, Protein Capped Gold Nanoparticles as Effective Drug Delivery Vehicles for Amino-Glycosidic Antibiotics. *Mater. Sci. Eng. C* **2012,** *32*, 1571–1577.

Rawat, M.; Singh, D.; Saraf, S.; Saraf, S. Nanocarriers: Promising Vehicle for Bioactive Drugs. *Biol. Pharm. Bull.* **2006,** *29*, 1790–1798.

Raza, K.; Thotakura, N.; Kumar, P.; Joshi, M.; Bhushan, S.; Bhatia, A.; Kumar, V.; Malik, R.; Sharma, G.; Guru, S. K.; Katare, O. P. C 60-Fullerenes for Delivery of Docetaxel to Breast Cancer Cells: A Promising Approach for Enhanced Efficacy and Better Pharmacokinetic Profile. *Int. J. Pharm.* **2015,** *495*, 551–559.

Rekha, M. R.; Sharma, C. P. Hemocompatible Pullulan–Polyethyleneimine Conjugates for Liver Cell Gene Delivery: In Vitro Evaluation of Cellular Uptake, Intracellular Trafficking and Transfection Efficiency. *Acta Biomater*. **2011,** *7*, 370–379.

Ren, J.; Shen, S.; Wang, D.; Xi, Z.; Guo, L.; Pang, Z.; Qian, Y.; Sun, X.; Jiang, X. The Targeted Delivery of Anticancer Drugs to Brain Glioma by PEGylated Oxidized Multi-walled Carbon Nanotubes Modified with Angiopep-2. *Biomaterials* **2012,** *33*, 3324–3333.

Rezaei, B.; Majidi, N.; Noori, S.; Hassan, Z. H. Multiwalled Carbon Nanotubes Effect on the Bioavailability of Artemisinin and Its Cytotoxicity to Cancerous Cells. *J. Nanopart. Res.* **2011,** *13*, 6339–6346.

Risi, G.; Bloise, N.; Merli, D.; Icaro-Cornaglia, A.; Profumo, A.; Fagnoni, M.; Quartarone, E.; Imbriani, M.; Visai, L. *In vitro* Study of Multiwall Carbon Nanotubes (MWCNTs) with Adsorbed Mitoxantrone (MTO) as a Drug Delivery System to Treat Breast Cancer. *RSC Adv*. **2014,** *4*, 18683–18693.

Russell-Jones, G.; McTavish, K.; McEwan, J.; Thurmond, B. Increasing the Tumoricidal Activity of Daunomycin–pHPMA Conjugates Using Vitamin B12 as a Targeting Agent. *J. Cancer Res. Update* **2012,** *1*, 203–211.

Sahdev, P.; Podaralla, S.; Kaushik, R. S.; Perumal, O. Calcium Phosphate Nanoparticles for Transcutaneous Vaccine Delivery. *J. Biomed. Nanotechnol*. **2013,** *9*, 132–141.

Sapra, P.; Kraft, P.; Pastorino, F.; Ribatti, D.; Dumble, M.; Mehlig, M.; Wang, M.; Ponzoni, M.; Greenberger, L. M.; Horak, I. D. Potent and Sustained Inhibition of HIF-1α and Downstream Genes by a Polyethyleneglycol–SN38 Conjugate, EZN-2208, Results in Anti-angiogenic Effects. *Angiogenesis* **2011,** *14*, 245–253.

Sawant, R. R.; Sriraman, S. K.; Navarro, G.; Biswas, S.; Dalvi, R. A.; Torchilin, V. P. Polyethyleneimine–Lipid Conjugate-Based pH-Sensitive Micellar Carrier for Gene Delivery. *Biomaterials* **2012,** *33*, 3942–3951.

Sedlák, M.; Buchta, V.; Kubicová, L.; Šimůnek, P.; Holčapek, M.; Kašparová, P. Synthesis and Characterisation of a New Amphotericin B–Methoxypoly (Ethylene Glycol) Conjugate. *Bioorg. Med. Chem. Lett.* **2001,** *11*, 2833–2835.

Shi, J.; Liu, Y.; Wang, L.; Gao, J.; Zhang, J.; Yu, X.; Ma, R.; Liu, R.; Zhang, Z. A Tumoral Acidic pH-Responsive Drug Delivery System Based on a Novel Photosensitizer (Fullerene) for In Vitro and In Vivo Chemo-Photodynamic Therapy. *Acta Biomater.* **2014,** *10*, 1280–1291.

Shi, J.; Zhang, H.; Wang, L.; Li, L.; Wang, H.; Wang, Z.; Li, Z.; Chen, C.; Hou, L.; Zhang, C.; Zhang, Z. PEI-Derivatized Fullerene Drug Delivery Using Folate as a Homing Device Targeting to Tumor. *Biomaterials* **2013,** *34*, 251–261.

Shokri, N.; Javar, H. A. Comparison of Calcium Phosphate and Zinc Oxide Nanoparticles as Dermal Penetration Enhancers for Albumin. *Ind. J. Pharm. Sci.* **2015,** *77*, 694–704.

Siavashani, A. Z.; Nazarpak, M. H.; Bakhsh, F. F.; Toliyat, T.; Solati-Hashjin, M. Preparation of Mesoporous Silica Nanoparticles for Insulin Drug Delivery. *Adv. Mater. Res.* **2014,** *829*, 251–257.

Singh, N.; Karambelkar, A.; Gu, L.; Lin, K.; Miller, J. S.; Chen, C. S.; Sailor, M. J.; Bhatia, S. N. Bioresponsive Mesoporous Silica Nanoparticles for Triggered Drug Release. *J. Am. Chem. Soc.* **2011,** *133*, 19582–19585.

Singh, S. P.; Konwar, B. K. Carbon Nanotube Assisted Drug Delivery of the Anti-Malarial Drug Artemesinin and Its Derivatives—A Theoretical Nanotechnology Approach. *J. Bionanosci.* **2013,** *7*, 630–636.

Slowing, I. I.; Trewyn, B. G.; Lin, V. S. Mesoporous Silica Nanoparticles for Intracellular Delivery of Membrane-Impermeable Proteins. *J. Am. Chem. Soc.* **2007,** *129*, 8845–8849.

Spinato, C.; Ruiz de Garibay, A. P.; Kierkowicz, M.; Pach, E.; Martincic, M.; Klippstein, R.; Bourgognon, M.; Wang, J. T-W.; Ménard-Moyon, C.; Al-Jamal, K. T.; Ballesteros, B.; Tobias, G.; Bianco; A. Design of Antibody-Functionalized Carbon Nanotubes Filled with Radioactivable Metals towards a Targeted Anticancer Therapy. *Nanoscale* **2016.** DOI:10.1039/C5NR07923C.

Su, Y-L.; Fang, J-H.; Liao, C-H.; Lin, C-T.; Li, Y-T.; Hu, S-H. Targeted Mesoporous Iron Oxide Nanoparticles-Encapsulated Perfluorohexane and a Hydrophobic Drug for Deep Tumor Penetration and Therapy. *Theranostics* **2015,** *5*, 1233–1248.

Subbiah, R.; Veerapandian, M.; Sadhasivam, S.; Yun, K. Triad CNT-NPs/Polymer Nanocomposites: Fabrication, Characterization, and Preliminary Antimicrobial Study. *Synth. React Inorg. Met-Org. Nano-Met. Chem.* **2011,** *41*, 345–355.

Subr, V.; Sivák, L.; Koziolová, E.; Braunova, A.; Pechar, M.; Strohalm, J.; Kabesova, M.; Ríhova, B.; Ulbrich, K.; Kovar, M. Synthesis of Poly[*N*-(2-hydroxypropyl) methacrylamide] Conjugates of Inhibitors of the ABC Transporter that Overcome Multidrug Resistance in Doxorubicin-Resistant P388 Cells In Vitro. *Biomacromolecules* **2014,** *15*, 3030–3043.

Sui, B.; Xu, H.; Jin, J.; Gou, J.; Liu, J.; Tang, X.; Zhang, Y.; Xu, J.; Zhang, H.; Jin, X. Self-Assembled Micelles Composed of Doxorubicin Conjugated Y-Shaped PEG-poly(Glutamic Acid) 2 Copolymers via Hydrazone Linkers. *Molecules* **2014,** *19*, 11915–11932.

Sun, T.; Yan, Y.; Zhao, Y.; Guo, F.; Jiang, C. Copper Oxide Nanoparticles Induce Autophagic Cell Death in A549 Cells. *PLoS ONE* **2012,** *7*, e43442.

Sundaresan, V.; Menon, J. U.; Rahimi, M.; Nguyen, K. T.; Wadajkar, A. S. Dual-Responsive Polymer-Coated Iron Oxide Nanoparticles for Drug Delivery and Imaging Applications. *Int. J. Pharm.* **2014,** *466*, 1–7.

Tang, L.; Wang, Y.; Li, Y.; Feng, H.; Lu, J.; Li, J. Preparation, Structure, and Electrochemical Properties of Reduced Graphene Sheet Films. *Adv. Funct. Mater.* **2009,** *19*, 2782–2789.

Tian, L.; Pei, X.; Zeng, Y.; He, R.; Li, Z.; Wang, J.; Wan, Q.; Li, X. Functionalized Nanoscale Graphene Oxide for High Efficient Drug Delivery of Cisplatin. *J. Nanopart. Res.* **2014,** *16*, 1–14.

Trewyn, B. G.; Slowing, I. I.; Giri, S.; Chen, H. T.; Lin, V. S. Y. Synthesis and Functionalization of a Mesoporous Silica Nanoparticle Based on the Sol–Gel Process and Applications in Controlled Release. *Acc. Chem. Res.* **2007,** *40*, 846–853.

Trickler, W. J.; Lantz, S. M.; Schrand, A. M.; Robinson, B. L.; Newport, G. D.; Schlager, J. J.; Paule, M. G.; Slikker, W.; Biris, A. S.; Hussain, S. M.; Ali, S. F. Effects of Copper Nanoparticles on Rat Cerebral Microvessel Endothelial Cells. *Nanomedicine* **2012,** *7*, 835–846.

Urbán, P.; Valle-Delgado, J. J.; Mauro, N.; Marques, J.; Manfredi, A.; Rottmann, M.; Ranucci, E.; Ferruti, P.; Fernàndez-Busquets, X. Use of Poly(Amidoamine) Drug Conjugates for the Delivery of Antimalarials to *Plasmodium*. *J. Control. Release* **2014,** *177*, 84–95.

Venkatpurwar, V.; Shiras, A.; Pokharkar, V. Porphyran Capped Gold Nanoparticles as a Novel Carrier for Delivery of Anticancer Drug: In Vitro Cytotoxicity Study. *Int. J. Pharm.* **2011,** *409*, 314–320.

Veronese, F. M.; Schiavon, O.; Pasut, G.; Mendichi, R.; Andersson, L.; Tsirk, A.; Ford, J.; Wu, G.; Kneller, S.; Davies, J.; Duncan, R. PEG–Doxorubicin Conjugates: Influence of Polymer Structure on Drug Release, In Vitro Cytotoxicity, Biodistribution, and Antitumor Activity. *Bioconj. Chem.* **2005,** *16*, 775–784.

Vinardell, M. P.; Mitjans, M. Antitumor Activities of Metal Oxide Nanoparticles. *Nanomaterials* **2015,** *5*, 1004–1021.

Volkova, M. A.; Irza, A. V.; Chvala, I. A.; Frolov, S. F.; Drygin, V. V.; Kapczynski, D. R. Adjuvant Effects of Chitosan and Calcium Phosphate Particles in an Inactivated Newcastle Disease Vaccine. *Avian Dis.* **2014,** *58*, 46–52.

Wahab, R.; Kaushik, N. K.; Kaushik, N.; Choi, E. H.; Umar, A.; Dwivedi, S.; Musarrat, J.; Al-Khedhairy, A. A. ZnO Nanoparticles Induces Cell Death in Malignant Human T98G Gliomas, KB and Non-malignant HEK Cells. *J. Biomed. Nanotechnol.* **2013,** *9*, 1181–1189.

Wahab, R.; Siddiqui, M. A.; Saquib, Q.; Dwivedi, S.; Ahmad, J.; Musarrat, J.; Al-Khedhairy, A. A.; Shin, H. S. ZnO Nanoparticles Induced Oxidative Stress and Apoptosis in HepG2 and MCF-7 Cancer Cells and their Antibacterial Activity. *Colloids Surf. B* **2014,** *117*, 267–276.

Wang, C.; Li, D.; Too, C. O.; Wallace, G. G. Electrochemical Properties of Graphene Paper Electrodes Used in Lithium Batteries. *Chem. Mater.* **2009,** *21*, 2604–2606.

Wang, H.; Gu, W.; Xiao, N.; Ye, L.; Xu, Q. Chlorotoxin-Conjugated Graphene Oxide for Targeted Delivery of an Anticancer Drug. *Int. J. Nanomed.* **2014,** *9*, 1433–1442.

Wang, J.; Wang, Z.; Chen, H.; Zong, S.; Cui, Y. The Synthesis and Application of Two Mesoporous Silica Nanoparticles as Drug Delivery System with Different Shape. In *Proc. SPIE 9543, Third International Symposium on Laser Interaction with Matter*, 954318, May 4, 2015.

Wang, Y.; Yang, F.; Zhang, H. X.; Zi, X. Y.; Pan, X. H.; Chen, F.; Luo, W. D.; Li, J. X.; Zhu, H. Y.; Hu, Y. P. Cuprous Oxide Nanoparticles Inhibit the Growth and Metastasis of Melanoma by Targeting Mitochondria. *Cell Death Dis.* **2013,** *4*, e783.

Webster, D. M.; Sundaram, P.; Byrne, M. E. Injectable Nanomaterials for Drug Delivery: Carriers, Targeting Moieties, and Therapeutics. *Eur. J. Pharm. Biopharm.* **2013,** *84*, 1–20.

Williams, C. C.; Thang, S. H.; Hantke, T.; Vogel, U.; Seeberger, P. H.; Tsanaktsidis, J.; Lepenies, B. RAFT-Derived Polymer–Drug Conjugates: Poly (Hydroxypropyl Methacrylamide)(HPMA)–7-Ethyl-10-Hydroxycamptothecin (SN-38) Conjugates. *ChemMedChem* **2012,** *7*, 281–291.

Wilson, L.; Mirakyan, A.; Cubbage, M. Fullerene (C60) Vancomycin Conjugates as Improved Antibiotics. US 20040241173 A1, December 2, 2004.

Wongrakpanich, A.; Mudunkotuwa, I. A.; Geary, S. M.; Morris, A. S.; Mapuskar, K. A.; Spitz, D. R.; Grassian, V. H.; Salem, A. K. Size-Dependent Cytotoxicity of Copper Oxide Nanoparticles in Lung Epithelial Cells. *Environ. Sci. Nano* **2016,** *3*, 365–374.

Wu, K. C. W.; Yamauchi, Y.; Hong, C. Y.; Yang, Y. H.; Liang, Y. H.; Funatsu, T.; Tsunoda, M. Biocompatible, Surface Functionalized Mesoporous Titania Nanoparticles for Intracellular Imaging and Anticancer Drug Delivery. *Chem. Comm.* **2011,** *47*, 5232–5234.

Xu, H.; Fan, M.; Elhissi, A. M.; Zhang, Z.; Wan, K. W.; Ahmed, W.; Phoenix, D. A.; Sun, X. PEGylated Graphene Oxide for Tumor-Targeted Delivery of Paclitaxel. *Nanomedicine* **2015,** *10*, 1247–1262.

Xu, L.; Xiang, J.; Liu, Y.; Xu, J.; Luo, Y.; Feng, L.; Liu, Z.; Peng, R. Functionalized Graphene Oxide Serves as a Novel Vaccine Nano-Adjuvant for Robust Stimulation of Cellular Immunity. *Nanoscale* **2016a,** *8*, 3785–3795.

Xu, Z.; Wang, Y.; Ma, Z.; Wang, Z.; Wei, Y.; Jia, X. A Poly(Amidoamine) Dendrimer-Based Nanocarrier Conjugated with Angiopep-2 for Dual-Targeting Function in Treating Glioma Cells. *Polym. Chem.* **2016b,** *7*, 715–721.

Yadav, A.; Ghune, M.; Jain, D. K. Nano-Medicine Based Drug Delivery System. *J. Adv. Pharm. Ed. Res.* **2011,** *1*, 201–213.

Yandar, N.; Pastorin, G.; Prato, M.; Bianco, A.; Patarroyo, M. E.; Lozano, J. M. Immunological Profile of a *Plasmodium vivax* AMA-1 Nterminus Peptide–Carbon Nanotube Conjugate in an Infected *Plasmodium berghei* Mouse Model. *Vaccine* **2008,** *26*, 5864–5873.

Yang, L.; Sheldon, B. W.; Webster, T. J. Nanophase Ceramics for Improved Drug Delivery: Current Opportunities and Challenges. *Am. Ceram. Soc. Bull.* **2010,** *89*, 24–32.

Yang, R.; Zhang, S.; Kong, D.; Gao, X.; Zhao, Y.; Wang, Z. Biodegradable Polymer–Curcumin Conjugate Micelles Enhance the Loading and Delivery of Low-Potency Curcumin. *Pharm. Res.* **2012,** *29*, 3512–3525.

Yang, X.; He, D.; He, X.; Wang, K.; Tang, J.; Zou, Z.; He, X.; Xiong, J.; Li, L.; Shangguan, J. Synthesis of Hollow Mesoporous Silica Nanorods with Controllable Aspect Ratios for Intracellular Triggered Drug Release in Cancer Cells. *ACS Appl. Mater. Interfaces* **2016,** *8*, 20558–20569.

Yang, X.; Wang, Y.; Huang, X.; Ma, Y.; Huang, Y.; Yang, R.; Duan, H.; Chen, Y. Multi-functionalized Graphene Oxide Based Anticancer Drug-Carrier with Dual-Targeting Function and pH Sensitivity. *J. Mater. Chem.* **2011,** *21*, 3448–3454.

Yin, D.; Li, Y.; Guo, B.; Liu, Z.; Xu, Y.; Wang, X.; Du, Y.; Xu, L.; Meng, Y.; Zhao, X.; Zhang, L. Plasmid-Based Stat3 siRNA Delivered by Functional Graphene Oxide Suppresses Mouse Malignant Melanoma Cell Growth. *Oncol. Res.* **2016,** *23*, 229–236.

Yin, M.; Ju, E.; Chen, Z.; Li, Z.; Ren, J.; Qu, X. Upconverting Nanoparticles with a Mesoporous $TiO_2$ Shell for Near-Infrared-Triggered Drug Delivery and Synergistic Targeted Cancer Therapy. *Chem. Eur. J.* **2014,** *20*, 14012–14017.

You, J.; Zhang, G.; Li, C. Exceptionally High Payload of Doxorubicin in Hollow Gold Nanospheres for Near-Infrared Light-Triggered Drug Release. *ACS Nano* **2010,** *4*, 1033–1041.

Yu, Q.; Wei, Z.; Shi, J.; Guan, S.; Du, N.; Shen, T.; Tang, H.; Jia, B.; Wang, F.; Gan, Z. Polymer–Doxorubicin Conjugate Micelles Based on Poly(Ethylene Glycol) and Poly(*n*-(2-Hydroxypropyl) Methacrylamide): Effect of Negative Charge and Molecular Weight on Biodistribution and Blood Clearance. *Biomacromolecules* **2015a,** *16*, 2645–2655.

Yu, Y.; Kong, L.; Li, L.; Li, N.; Yan, P. Antitumor Activity of Doxorubicin-Loaded Carbon Nanotubes Incorporated Poly(Lactic-*co*-Glycolic Acid) Electrospun Composite Nanofibers. *Nanoscale Res. Lett.* **2015b,** *10*, 1044–1047.

Yue, H.; Wei, W.; Gu, Z.; Ni, D.; Luo, N.; Yang, Z.; Zhao, L.; Garate, J. A.; Zhou, R.; Su, Z.; Ma, G. Exploration of Graphene Oxide as an Intelligent Platform for Cancer Vaccines. *Nanoscale* **2015,** *7*, 19949–19957.

Yun, J. W.; Yoon, J. H.; Kang, B. C.; Cho, N. H.; Seok, S. H.; Min, S. K.; Min, J. H.; Che, J. H.; Kim, Y. K. The Toxicity and Distribution of Iron Oxide–Zinc Oxide Core–Shell Nanoparticles in C57BL/6 Mice after Repeated Subcutaneous Administration. *J. Appl. Toxicol.* **2015,** *35*, 593–602.

Zhai, Q. Z. Preparation and Controlled Release of Mesoporous MCM-41/Propranolol Hydrochloride Composite Drug. *J. Microencapsul.* **2013,** *30*, 173–180.

Zhang, H.; Hou, L.; Jiao, X.; Ji, Y.; Zhu, X.; Zhang, Z. Transferrin-Mediated Fullerenes Nanoparticles as $Fe^{2+}$-Dependent Drug Vehicles for Synergistic Anti-Tumor Efficacy. *Biomaterials* **2015,** *37*, 353–366.

Zhang, L.; Xia, J.; Zhao, Q.; Liu, L.; Zhang, Z. Functional Graphene Oxide as a Nanocarrier for Controlled Loading and Targeted Delivery of Mixed Anticancer Drugs. *Small* **2010,** *6*, 537–544.

Zhang, P.; Huang, Y.; Liu, H.; Marquez, R. T.; Lu, J.; Zhao, W.; Zhang, X.; Gao, X.; Li, J.; Venkataramanan, R.; Xu, L. A PEG-Fmoc Conjugate as a Nanocarrier for Paclitaxel. *Biomaterials* **2014,** *35*, 7146–7156.

Zhang, P.; Li, J.; Ghazwani, M.; Zhao, W.; Huang, Y.; Zhang, X.; Venkataramanan, R.; Li, S. Effective Co-delivery of Doxorubicin and Dasatinib Using a PEG-Fmoc Nanocarrier for Combination Cancer Chemotherapy. *Biomaterials* **2015,** *67*, 104–114.

Zhang, R.; Yang, J.; Sima, M.; Zhou, Y.; Kopeček, J. Sequential Combination Therapy of Ovarian Cancer with Degradable *N*-(2-hydroxypropyl) Methacrylamide Copolymer Paclitaxel and Gemcitabine Conjugates. *Proc. Natl. Acad. Sci.* **2014,** *111*, 12181–12186.

Zhang, Y.; Ang, C. Y.; Li, M.; Tan, S. Y.; Qu, Q.; Luo, Z.; Zhao, Y. Polymer-Coated Hollow Mesoporous Silica Nanoparticles for Triple-Responsive Drug Delivery. *ACS Appl. Mater. Interfaces* **2015,** *7*, 18179–18187.

Zhang, Z.; Jia, J.; Lai, Y. Ma, Y. Weng, J.; Sun, L. Conjugating Folic Acid to Gold Nanoparticles Through Glutathione for Targeting and Detecting Cancer Cells. *Bioorg. Med. Chem.* **2010,** *18*, 5528–5534.

Zheng, X. T.; Ma, X. Q.; Li, C. M. Highly Efficient Nuclear Delivery of Anti-Cancer Drugs Using a Bio-Functionalized Reduced Graphene Oxide. *J. Colloid Interface Sci.* **2016,** *467*, 35–42.

Zhi, F.; Dong, H.; Jia, X.; Guo, W.; Lu, H.; Yang, Y.; Ju, H.; Zhang, X.; Hu, Y. Functionalized Graphene Oxide Mediated Adriamycin Delivery and miR-21 Gene Silencing to Overcome Tumor Multidrug Resistance In Vitro. *PLoS ONE* **2013,** *8*, e60034.

Zhong, Q.; Bielski, E. R.; Rodrigues, L. S.; Brown, M. R.; Reineke, J. J.; da Rocha, S. R. Conjugation to Poly(Amidoamine) Dendrimers and Pulmonary Delivery Reduce Cardiac Accumulation and Enhance Antitumor Activity of Doxorubicin in Lung Metastasis. *Mol. Pharm.* **2016,** *13*, 2363–2375.

Zhong, Q.; da Rocha, S. R. Poly(Amidoamine) Dendrimer–Doxorubicin Conjugates: In Vitro Characteristics and Pseudosolution Formulation in Pressurized Metered-Dose Inhalers. *Mol. Pharm.* **2016,** *13*, 1058–1072.

Zhu, S.; Hong, M.; Zhang, L.; Tang, G.; Jiang, Y.; Pei, Y. PEGylated PAMAM Dendrimer–Doxorubicin Conjugates: In Vitro Evaluation and In Vivo Tumor Accumulation. *Pharm. Res.* **2010,** *27*, 161–174.

Zhu, S.; Qian, L.; Hong, M.; Zhang, L.; Pei, Y.; Jiang, Y. RGD-Modified PEG–PAMAM–DOX Conjugate: In Vitro and In Vivo Targeting to both Tumor Neovascular Endothelial Cells and Tumor Cells. *Adv. Mater.* **2011,** *23*, H84–H89.

## CHAPTER 2

# ROLE OF SURFACTANTS IN NANOTECHNOLOGY-BASED DRUG DELIVERY

M. PILAR VINARDELL[1*], DANIELE R. NOGUEIRA-LIBRELOTTO[2,3], CLARICE M. B. ROLIM[2,3], and MONTSERRAT MITJANS[1]

[1]*Department Bioquimica i Fisiologia, Facultat de Farmàcia i Ciències de l'Alimentació, Universitat de Barcelona, Av. Joan XIII, 27-31, Barcelona 08028, Spain*

[2]*Department Farmácia Industrial, Universidade Federal de Santa Maria, Av. Roraima 1000, 97105-900 Santa Maria, Rio Grande do Sul, Brazil*

[3]*Programa de Pós-Graduação em Ciências Farmacêuticas, Universidade Federal de Santa Maria, Av. Roraima 1000, 97105-900 Santa Maria, Rio Grande do Sul, Brazil*

**Corresponding author. E-mail: mpvinardellmh@ub.edu*

## CONTENTS

## ABSTRACT

The delivery of drugs using nanocarriers has several advantages, such as protection from degradation, enhanced solubility and stability, increased surface area, and controlled and/or triggered drug release. To stabilize emulsions and reduce particle aggregation during particle synthesis, storage, and use, the surfaces of drug-loaded polymeric particles are usually coated with surfactants. Surfactants can influence particle size, morphology, encapsulation efficiency, and drug release kinetics. Different kinds of surfactants are used in these formulations, depending on their properties.

In this chapter, we focus on surfactant-based nanocarriers and describe the influence of surfactants on characteristics such as the size and structure of the nanocarriers and whether they facilitate drug delivery.


## 2.1 INTRODUCTION

Nanotechnology has emerged as a very important field in medicine in recent years, as illustrated by the increasing number of papers on this subject in the literature. It is very important in the area of drug delivery, as nanoparticles (NPs) can be generated that allow drugs to enter the body and control drug release, resulting in greater drug efficacy and safety. Due to their small size, nanostructures have unique properties such as a capacity to cross cell and tissue barriers. For this reason, they are suitable for biomedical applications (Wilczewska et al., 2012).

Different types of nanocarriers have been developed to securely deliver drugs and various other therapeutic agents specifically into target sites. Many of the conventional nanodrug delivery systems (e.g., liposomes, micelles, and polymer-based nanodevices) have reached the late stages of development, and some of them are already on the market (Marianecci et al., 2014). Other novel inorganic nanocarriers have been designed for the delivery of genetic materials (Loh et al., 2016), especially for use in the treatment of cancer and viral diseases (Varshosaz and Taymouri, 2015).

According to the National Nanotechnology Initiative's definition, NPs are structures with sizes ranging from 1 to 100 nm in at least one dimension. However, the prefix "nano" is commonly used for particles that are up to several hundred nanometers in size, as in NPs used in drug delivery through different methods. These NPs can dissolve, entrap, encapsulate, or attach therapeutic agents, while their nanomeric size means that they can be taken up by cells more easily than larger molecules.

The delivery of drugs using nanocarriers has several advantages, such as protection from degradation, enhanced solubility and stability, increased surface area, and controlled and/or triggered drug release.

To stabilize emulsions and reduce particle aggregation during particle synthesis, storage, and use, the surfaces of drug-loaded polymeric particles are usually coated with surfactants. Surfactants can influence particle size, morphology, encapsulation efficiency, and drug release kinetics (Mitra and Sin, 2003).

Surfactants such as wetting agents are widely used in pharmaceutical formulations to enhance the dissolution and absorption of poorly soluble drugs. Different kinds of surfactants are used in these formulations, depending on their properties. Low-molecular-weight ionic surfactants like sodium lauryl sulfate are very widely used. Nevertheless, in some studies they have been found to be more harmful to the biological membrane than nonionic surfactants (Rouse et al., 1994).

Furthermore, nonionic surfactants that are lipophilic in nature have a better capacity to dissolve poorly soluble moieties. Thus, nonionic surfactants are more effective than ionic surfactants in the dissolution of poorly soluble drugs. Nonionic surfactants are extremely proficient emulsifiers that can be used in self-emulsifying drug delivery systems. Nonionic surfactants may affect various characteristics of particles such as size, shape, and stability. The most common nonionic surfactants that can affect various characteristics of particles are Tween 80, Tween 20, and Brij 97. In earlier studies, it was concluded that the molecular configuration of surfactants can alter the self-assembly prototype of nanostructures (Suttipong et al., 2011).

Through continuing research to improve surfactant properties, novel structures have emerged recently that affect synergistic relations or improve surface and aggregation effects (Singh et al., 2014). These new surfactants are highly significant and consist of catanionics, bolaforms, gemini (or dimeric) surfactants, polymeric and polymerizable surfactants (Kaur et al., 2016).

Nonionic surfactants are the most common type of surface active agent used in preparing vesicles. They have greater benefits—including stability, compatibility, and lower toxicity—than their anionic, amphoteric, or cationic counterparts. They are generally less toxic, less hemolytic, and less irritating to eye and skin. In addition, they tend to maintain near physiological pH in solution. They have many functions, including acting as solubilizers, wetting agents, emulsifiers, and permeability enhancers (Kumar and Rajeshwarrao, 2011).

Surfactants molecules may be classified based on the nature of the hydrophobilic group within the molecule. The four main groups of surfactants with some examples are shown in Table 2.1.

**TABLE 2.1** Surfactant Classification.

| Type of surfactants | Examples |
|---|---|
| Nonionic | Polysorbates (Tween) |
| Anionic | Sodium dodecyl sulfate, sodium propionate |
| Cationic | Benzalkonium chloride, cethyl trimethyl ammonium bromide |
| Amphoteric | Lecithins |

Some nanocarriers are based on surfactants, whilst others are not. In this chapter, we focus on surfactant-based nanocarriers and describe the influence of surfactants on characteristics such as the size and structure of the nanocarriers and whether they facilitate drug delivery.

## 2.2 NPs AND NANOVESICLES BASED ON SURFACTANTS

### *2.2.1 SOLID LIPID NPs*

Solid lipid NPs (SLNs) are a generation of submicron-sized lipid emulsions. They comprised 0.1–30% (w/w) physiological solid lipid (i.e., solid at room and body temperatures) dispersed in an aqueous medium or surfactant solution. The latter creates a surfactant (tensed, emulsifier) shell over the NP surface that stabilizes the dispersion (Desmet et al., 2016). SLNs are formed using three different models: (1) solid solution model in which hydrophobic drug molecules are dissolved within the lipid matrix; (2) drug-enriched shell model in which drugs are dispersed around a central lipid core; and (3) drug-enriched core model, characterized by drug selectively located at the core of the SLN (Singhal et al., 2011).

Some of the disadvantages of the SLN system are the drug expulsion phenomenon during storage, limited particle concentration in the aqueous dispersion, load limitations due to the solubility of the drug in the solid lipid, and the highly ordered structure of the SLN. These drawbacks were overcome by the formation of a lipid particle with a controlled nanostructure, called nanostructured lipid carriers (NLCs). NLCs are tailored SLN composed of a mix of solid and fluid lipids. The fluid phase is embedded into the solid lipid matrix or localized on the surface of solid platelets and the surfactant layer (Schäfer-Korting et al., 2007).

### *2.2.2 NIOSOMES*

Niosomes are multilamellar vesicles prepared by hydration of nonionic single-chain surfactant, instead of phospholipids. The surfactants are components of liposomes that have cholesterol or another excipient incorporated as a stabilizer (Rajera et al., 2011). The behavior of niosomes in vivo is similar to that of liposomes: they increase the circulation time of entrapped drugs and alter their organ distribution and metabolic stability (Azmin et al., 1985). Niosomes are preferred over liposomes because they have high chemical stability, a lower cost, and intrinsic penetration-enhancing properties (Mahale et al., 2012; Nasr et al., 2008).

Niosomes mainly contain two types of components: nonionic surfactants and additives. Nonionic surfactants form the vesicular layer, and the additives used in niosome preparation are cholesterol and charged molecules. Cholesterol improves the rigidity of the bilayer, because it is an important component of the cell membrane and its presence affects bilayer fluidity and permeability. Niosome protects drugs from degradation and inactivation due to immunological and/or pharmacological effects. In recent years, niosomes have been extensively studied for their potential to serve as carriers for the delivery of different type of drugs, antigens, hormones, and even other bioactive agents (Manosroi et al., 2013; Rajera et al., 2011).

Cholesterol influences the physical properties and structure of niosomes, possibly due to its interaction with nonionic surfactants (Talsma et al., 1994).

Despite the advantages of niosomes, they suffer from physical instability comparable to that of liposomes. During suspension vesicles can fuse, aggregate, or leak, and the drug may hydrolyze, which limits the shelf life. To reduce or eliminate these disadvantages of niosomes, several derivatives have been developed including proniosomes, aspasomes, and aqueous and elastic niosomes (Hu and Rhodes, 2000).

Nonionic surfactants are the basic components of niosomes that form lamellar microscopic and nanoscopic vesicles upon hydration. Nonionic surfactants are preferred, because of their properties and ability to form stable formulations. They are useful in the formulation of noisome because of their stability, compatibility and low toxicity, their ability to maintain physiological pH, and their functions as solubilizers, wetting agents, and permeability enhancers.

One way of producing niosomes is to coat a water-soluble carrier such as sorbitol with a surfactant. The result of the coating process is a dry formulation named proniosome, in which each water-soluble particle is covered with a thin film of dry surfactant. Niosomes are obtained by the addition of aqueous phase at a water temperature higher than the mean phase transition temperatures.

Nonionic surfactants are the basic components of niosomes that form lamellar microscopic and nanoscopic vesicles upon hydration. Nonionic surfactants are preferred, because of their properties and ability to form stable formulations. They are useful in the formulation of noisome because of their stability, compatibility, and low toxicity, their ability to maintain physiological pH, and their functions as solubilizers, wetting agents, and permeability enhancers.

The temperature of the hydration medium plays a major role in the formation of vesicles and affects their shape and size. Temperature affects the assembly of surfactants into vesicles and also induces changes in vesicle shape (Fig. 2.1).

### *2.2.3 POLYESTER NANOPARTICLES*

Polyesters such as PLA and PLGA have been used extensively in drug delivery because of their biodegradable and mechanical properties, which can be adjusted. Most NP formulations based on PLA and PLGA have been focused on drug delivery to target tumors (Danhier et al., 2012).

Poly(D,L-lactide-*co*-glycolide) (PLGA) nano/microparticles are typically prepared by an emulsion solvent evaporation technique and nanoprecipitation method, in which surfactants stabilize the dispersed droplets, reduce their surface tension, and inhibit coalescence.

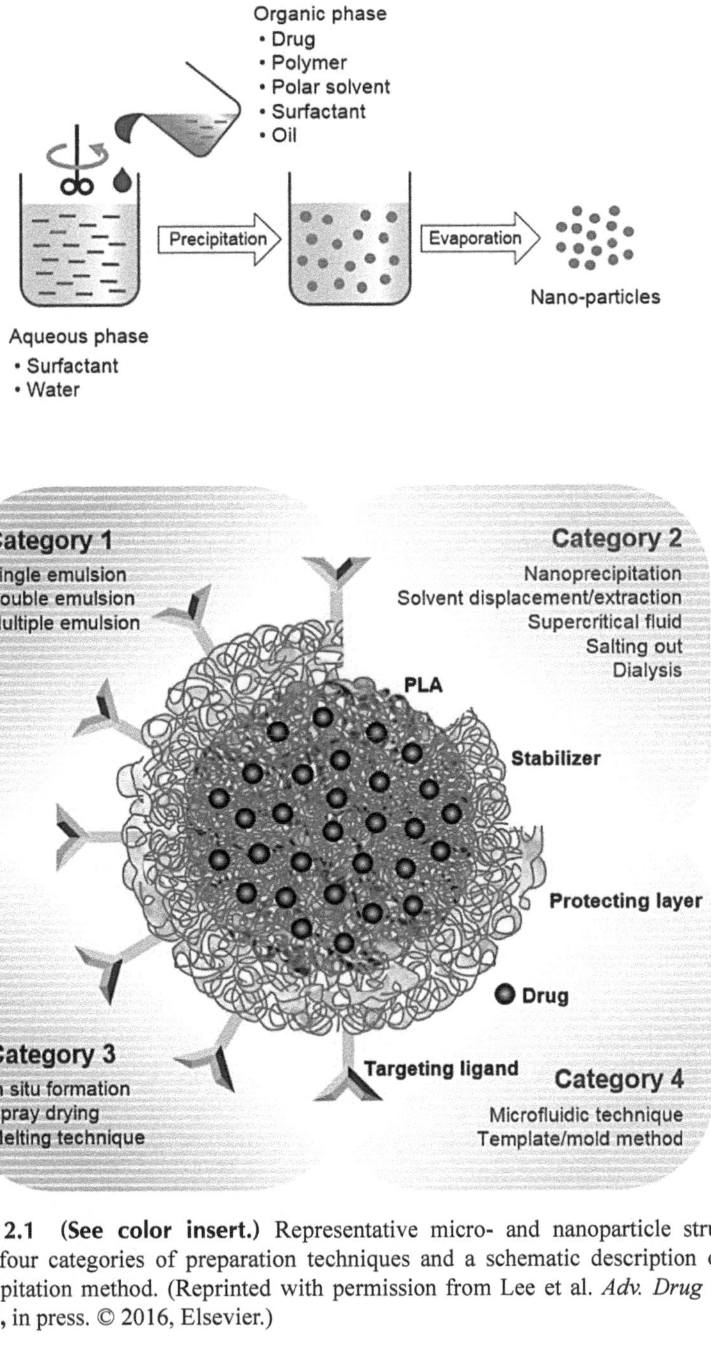


**FIGURE 2.1** **(See color insert.)** Representative micro- and nanoparticle structure with the four categories of preparation techniques and a schematic description of the nanoprecipitation method. (Reprinted with permission from Lee et al. *Adv. Drug Deliv. Rev.* **2016,** in press. © 2016, Elsevier.)

Several techniques can be used to prepare PLA-based microparticles and NPs. They can be classified into four categories. Category 1 includes traditional, emulsion-based methods such as single emulsion, double emulsion, and multiple emulsions. Category 2 contains precipitation-based methods, such as nanoprecipitation, rapid expansion of supercritical fluid into liquid, salting out, and dialysis. Category 3 covers direct methods, such as the melting technique, spray drying, supercritical fluid, and in situ forming microparticles. Category 4 includes new approaches, for example, the microfluidic technique and the template-/mold-based technique. Other criteria depend on the mode of drug encapsulation (Fig. 2.2). The drug is either entrapped inside the particles of "capsules" or dispersed in polymer matrices (Lee et al., 2016).

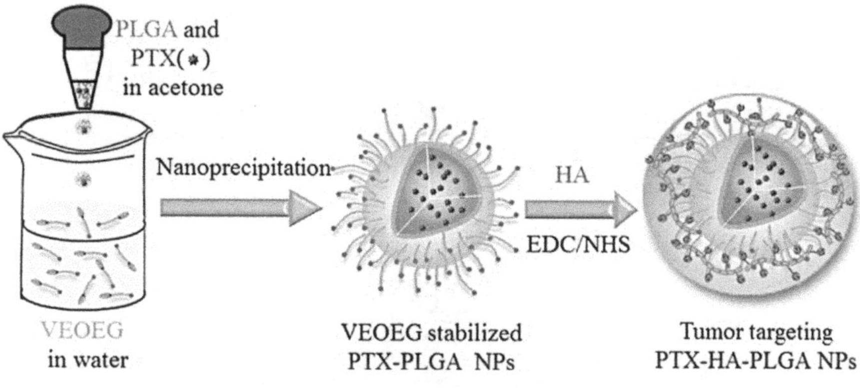


**FIGURE 2.2** Scheme of the preparation of PTX-loaded HA-PLGA NPs by nanoprecipitation method using VEOEG as a novel functional surfactant followed by HA coating for active tumor-targeting delivery of PTX. (Reprinted with permission from Wu et al. *Biomacromolecules* **2016,** *17*, 2367. © 2016, American Chemical Society.)

Vitamin E polyethylene glycol-succinate (TPGS) has recently appeared as a versatile and biocompatible surfactant (Zhang et al., 2012). TPGS has been shown to inhibit drug efflux, thus enhancing the cytotoxicity of anticancer drugs like doxorubicin, docetaxel, and paclitaxel (PTX) in multidrug resistant cancer cells.

A recent study has developed a novel class of biocompatible and functional surfactants based on vitamin E-oligo(methyl diglycol L-glutamate). They can be used for easy preparation of multifunctional PLGA NPs for active tumor-targeting PTX delivery in vivo, as presented in Figure 2.3.

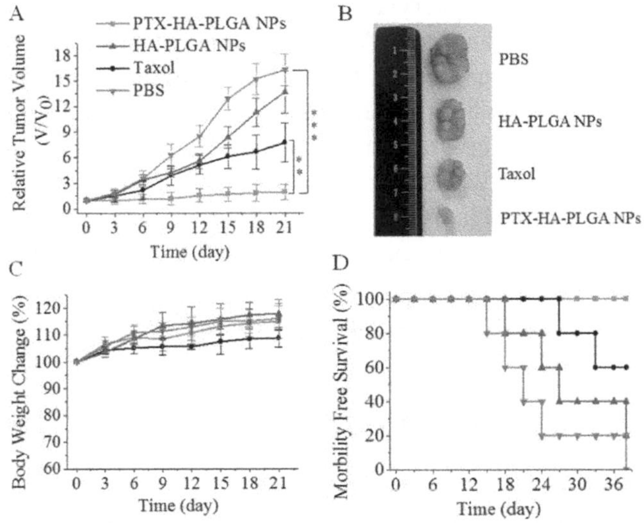


**FIGURE 2.3 (See color insert.)** In vivo antitumor activity of PTX-HA-PLGA NPs in MCF-7 tumor-bearing mice. Tumor volume changes of mice and photographs of typical tumor blocks isolated on day 21. (Reprinted with permission from Wu et al. *Biomacromolecules* **2016,** *17*, 2367. © 2016, American Chemical Society.)

In vitro antitumor activity has been demonstrated, as well as low cytotoxicity to normal cells. The in vivo antitumor efficacy of PTX–PLGA NPs was assessed using MCF-7 human breast cancer xenografts. The PTX-loaded HA-PLGA NPs inhibited tumor growth more effectively, led to a better survival rate and had fewer adverse effects than Taxol, a clinical PTX formulation (Wu et al., 2016).

There are variable key parameters for forming NPs, such as the mixing rate that affects particle size and drug encapsulation yield and the type of organic solvent. Other factors include the drug: polymer ratio, the surfactant, and the antisolvent phase volume ratio. The difficulty in this method is the choice of combination of the drug/polymer/solvent/nonsolvent system (in which NPs would be formed with high drug encapsulation efficiency).

## *2.2.4 POLYMERIC NANOPARTICLES*

Polymeric nanoparticles (PNPs) are structures with a diameter ranging from 10 to 100 nm. PNPs are obtained from synthetic polymers, such as poly-*e*-caprolactone, polyacrylamide, and polyacrylate or natural polymers, for example, albumin, DNA, chitosan, and gelatin. PNPs are usually coated with nonionic surfactants in order to reduce immunological interactions (e.g., opsonization or presentation PNPs to CD8 T-lymphocytes) as well as intermolecular interactions between the surface chemical groups of PNPs (e.g., van der Waals forces, hydrophobic interaction, or hydrogen bonding). Drugs can be immobilized on the surface of PNPs after a polymerization reaction or can be encapsulated on the PNP structure during a polymerization step. Moreover, drugs may be released by desorption, diffusion, or nanoparticle erosion in target tissue (Wilczewska et al., 2012).

## 2.3 ROLE OF SURFACTANTS IN THE DELIVERY OF DRUGS BY DIFFERENT ROUTES

### *2.3.1 ROLE OF SURFACTANTS IN DERMAL DRUG DELIVERY*

Dermal drug delivery is the topical application of drugs to the skin in the treatment of skin diseases and also for cosmetic purposes. The dermal drug delivery route has several advantages over conventional oral and parental routes such as avoidance of the risk and inconvenience of intravenous therapy, avoidance of first pass hepatic metabolism, which leads to an increase in drug bioavailability and efficacy, no gastrointestinal (GI) degradation, and an alternative to oral administration when such a route is unsuitable. However, transdermal drug delivery has the major disadvantage of a low penetration rate due to the skin's barrier properties. Currently, only a limited number of drugs can be formulated as transdermal delivery systems due to the functions of the stratum corneum (SC), which provides the main barrier to permeation (Marianecci et al., 2014). However, the SC barrier may be modified using suitable surfactants in nanoemulsions, which make the SC layer looser and more permeable (Hussain et al., 2013). Research in this area has increased in recent years to obtain better formulations with better dermal drug delivery.

One recent review looks at the role of nanocarriers containing surfactants in topical drug delivery (Desmet et al., 2016). It presents a list of lipid- and surfactant-based drugs that are currently on the market or in clinical trials for topical or transdermal treatment of skin disorders.

To increase the penetration capacity, the composition of liposomes is altered by adding edge activator(s). The edge activator, a single-chain surfactant (e.g., sorbitan monooleate or polysorbate), destabilizes the lipid bilayer, thereby increasing its fluidity, elasticity, and deformability. As a result, liposomes can easily squeeze through intercellular regions and diffuse through the skin (Sarmah et al., 2013).

Niosomes have been developed to increase drug transfer through the skin. Several mechanisms have been proposed to explain their ability to cross this barrier: (1) they diffuse as a whole; (2) smaller vesicles are formed in the skin; (3) they interact with the stratum cornea with a high concentration gradient at the vesicle-stratum cornea surface, which drives lipophilic drug penetration across it; (4) they modify the stratum cornea structure, altering its intercellular lipid barrier; and (5) the nonionic surfactant enhances drug permeation (Desmet et al., 2016). Recent studies have developed formulations of resveratrol-entrapped niosomes for dermatological preparations (Pando et al., 2015) and topical methotrexate-loaded niosomes for the management of psoriasis (Abdelbary and AbouGhaly, 2015).

Topical administration of anti-inflammatory drugs such as Ibuprofen is an alternative to oral administration in the treatment of diseases such as osteoarthritis, rheumatoid arthritis, ankylosing spondylitis, gout, extra-articular disorders, bursitis, tendonitis, and nonarticular rheumatic condition. Oral administration is associated with severe adverse effects in the GI tract, such as epigastric pain, heartburn, nausea, diarrhea, vomiting, peptic ulcer, and hepatic impairment (Pellicano, 2014). Different strategies have been used to increase the local bioavailability of topically administered Ibuprofen. Due to the penetration of phospholipid molecules or nonionic surfactants into the lipid layers of the SC and epidermis, vesicular formulations may act as penetration enhancers and facilitate dermal delivery, leading to higher localized drug concentrations. The effects of a wide range of surfactants also need to be studied (Ghanbarzadeh et al., 2015).

Specific pH-sensitive, nonionic surfactant vesicles have been proposed for the topical delivery of ibuprofen. However, only niosomes with Sp 60 and cholesteryl hemisuccinate led to a significant increase of in vitro skin permeation of the drug (Carafa et al., 2009).

Another case of an anti-inflammatory drug is diclofenac sodium salt loaded in nonionic niosomes and incorporated in pluronic gel. This could be a successful topical anti-inflammatory formulation, because its percutaneous permeation across the skin can be modified by changing the pluronic concentration (Antunes et al., 2011).

Topical immunization is a novel, needle-free strategy for vaccine delivery through topical application of antigen and adjuvants directly or via a suitable carrier system on intact skin. Topical immunization, that is, noninvasive vaccination enabling the use of a variety of antigens and adjuvant on the skin, provides a robust and novel approach to vaccination. Topical immunization using a niosome-based delivery system with application of the nontoxic cell-binding B subunit of cholera toxin as an adjuvant is simple, stable, and potentially safe for topical immunization. It would increase the rate of vaccine compliance and greatly facilitate successful implementation of worldwide mass vaccination campaigns against infectious diseases (Maheshwari et al., 2011).

Another application of nanomaterials in which the addition of surfactants is important is dermal delivery of amphotericin B, a potent polyene macrolide used as a broad-spectrum antibiotic to treat local fungal infections. This antibiotic has restricted therapeutic efficacy, owing to poor water solubility. Moreover, topical delivery of amphotericin B has several advantages over conventional delivery. It can achieve site-specific, targeted delivery to the infected site to maximize therapeutic efficacy and reduce undesirable effects. In a recent study, the addition of a surfactant such as labrasol with inherent antifungal activities increases the therapeutic activity of the antibiotic, improves drug solubility, and increases amphotericin B release (Hussain et al., 2016).

### *2.3.2 ROLE OF SURFACTANT ON DRUG MUCOSAL DELIVERY*

Mucus secretions serve as the body's first line of defense against pathogens, toxins, and environmental ultrafine particles at exposed surfaces not covered by skin. Examples are the eyes and the respiratory, GI, and cervicovaginal tracts, which are protected by a highly viscoelastic, adhesive mucus layer that traps most foreign particles, including conventional drug and gene nanocarriers (Cone, 2009).

Trapped particles are typically rapidly eliminated via natural mucus clearance mechanisms, thus limiting their effective exposure to mucosal tissues. Therefore, mucus represents a critical difficulty for drug and gene delivery to mucosal tissues (Lai et al., 2009).

Mucus penetration can be engineered by carefully tuning the surface properties of nanoparticles, to reduce their affinity to mucus constituents. A particular challenge in formulating these nanoparticles is that many commonly used surfactants either yield mucoadhesive particles or do not provide efficient drug encapsulation. One strategy is to replace chitosan and other mucoadhesive surfactants commonly used in particle formulation with surfactants that contain a low-molecular-weight PEG moiety (Yang et al., 2014).

Another study has presented the use of surfactants based on vitamin E–PEG conjugates to facilitate the formulation of drug-loaded biodegradable mucus penetrating particles (Collnot et al., 2006). This novel surfactant has several advantages, such as rapid penetration of undiluted human mucus, good nanoparticle dispersity, low porosity, and a smooth surface, high loading of a small molecule drug such as PTX and sustained release of the drug over several days. These advantages cannot easily be simultaneously achieved using conventional surfactants (Mert et al., 2012).

### *2.3.3 ROLE OF SURFACTANTS IN OCULAR DRUG DELIVERY*

Topical administration is the most common approach to drug delivery into the anterior segment of the eye, while the corneal route is the primary site for absorption of topically administered therapeutics. Conventional dosage forms, such as eye drops, represent 90% of marketed ophthalmic formulations. This may be attributed to ease of administration and patient compliance. However, ocular bioavailability is very low with topical drop administration, if we consider anatomical and physiological limitations such as tear turnover, nasolacrimal drainage, reflex blinking, and ocular static and dynamic barriers that make deeper ocular drug permeation difficult (Patel et al., 2013). The anatomical structure of the eye and the cornea in particular restricts the entry of any drug/molecule (Fig 2.4). The tightly bound corneal epithelium formed by about six layers of epithelial tissues acts as the primary barrier against the entrance of drugs (especially hydrophilic molecules) into the aqueous humor (DelMonte and Kim, 2011).

The penetration of drug molecules into the eye depends on the physicochemical properties of both the drug and vehicle. Vesicular systems provide prolonged action at the corneal surface by preventing ocular metabolism by enzymes in the lacrimal fluid (Allam et al., 2011). Nanocarriers such as polymer–surfactant nanoparticles are the preferred choice for ocular drug delivery. The ability of the nanoparticulate system to encapsulate both hydrophilic and hydrophobic drugs is a good strategy to delivery drugs across the ocular barriers and it reduces the toxicity of the encapsulated drug. In a recent study, a novel polymer–surfactant nanoparticulate formulation was developed, using the anionic surfactant aerosol OT and a polysaccharide polymer gellan gum for ophthalmic delivery of the water-soluble drug doxycycline hydrochloride, a broad-spectrum semisynthetic antibiotic for the treatment of ocular infections (Pokharkar et al., 2015). Nanosuspensions are colloidal dispersions of submicron drug particles stabilized by polymer(s) or surfactant(s). They are emerging as a promising strategy for the ocular delivery of hydrophobic drugs, because they have several advantages such as sterilization, ease of eye drop formulation, less irritation, increased precorneal residence, and enhancement of the ocular bioavailability of drugs that are insoluble in the tear fluid (Patravale et al., 2004).

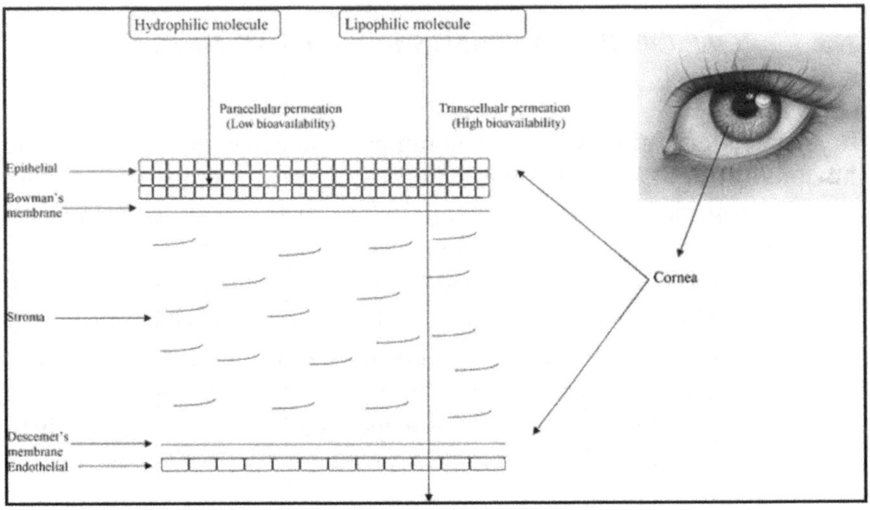


**FIGURE 2.4** Mode of transport of hydrophilic and lipophilic molecules through the multilayer cornea. (Reprinted with permission from Kumar and Rajeshwarrao, *Acta Pharmaceut Sin. B* **2011,** *1*, 208. © 2016, Springer.)

Glucocorticoids are recommended drugs for the treatment of inflammatory processes in the anterior segment of ocular tissues. Efforts have been made to improve the ocular bioavailability of glucocorticoids by formulating them as nanosuspensions. The results obtained in many research studies indicate that nanosuspensions could be an efficient ophthalmic drug delivery system for a range of poorly soluble drugs (Patel et al., 2013).

### 2.3.4 ROLE OF SURFACTANTS IN ORAL DRUG DELIVERY

One of the problems with oral administration is the low bioavailability of some drugs, especially poor water-soluble drugs. One of the strategies is to develop self-nanoemulsifying drug delivery systems. The potential advantages of these systems are their capacity to bypass the hepatic portal route and promote the lymphatic transport of lipophilic drugs, reducing metabolism by the cytochrome P450 family of enzymes present in the gut enterocytes (Date et al., 2010).

Self-nanoemulsifying drug delivery systems are defined as isotropic mixtures of natural or synthetic oils, solid or liquid surfactants, or, alternatively, one or more hydrophilic solvents and cosolvents/surfactants that have a unique ability to form fine oil-in-water (o/w) emulsions upon mild agitation, followed by dilution in aqueous media, such as GI fluids (Pouton, 2000; Singh et al., 2009).

Transresveratrol is a potent antioxidant with poor water solubility. It undergoes rapid first-pass metabolism by CYP3A4 in the liver and enterohepatic recirculation, leading eventually to a marked reduction in the orally bioavailable fraction of the drug (almost zero) in humans and animals (Singh et al., 2014). Recently, self-nanoemulsifying drug delivery was developed for resveratrol, and surfactants were studied to determine the best one for optimal delivery. The system increased the oral bioavailability of the drug, possibly by reducing metabolism by CYP3A4 in the liver and overcoming enterohepatic circulation (Singh and Pai, 2015).

Due to the mechanisms involved in intestinal transport, good delivery after oral administration of peptides and proteins is a challenge. A recent review of lipid-based nanocarriers for peptide delivery has shown interesting results, and the efficacy of different surfactants incorporated in the formulation has been studied. Lipid excipients with amphiphilic properties are commonly employed in the formation of emulsion-type systems changing the ratios of different lipids, surfactants, and drugs to achieve optimal solubilization capacity. The capacity of triglycerides to entrap proteins could also be enhanced by increasing the polarity of the oil phase through the incorporation of mono- and diglycerides and/or higher proportions of hydrophilic surfactants (Niu et al., 2016).

Surfactants are known to enhance transcellular transport by temporarily disrupting the lipid bilayer and transport through the intestinal epithelial membranes or altering tight junctions and increasing paracellular transport of peptides (Fig. 2.5).

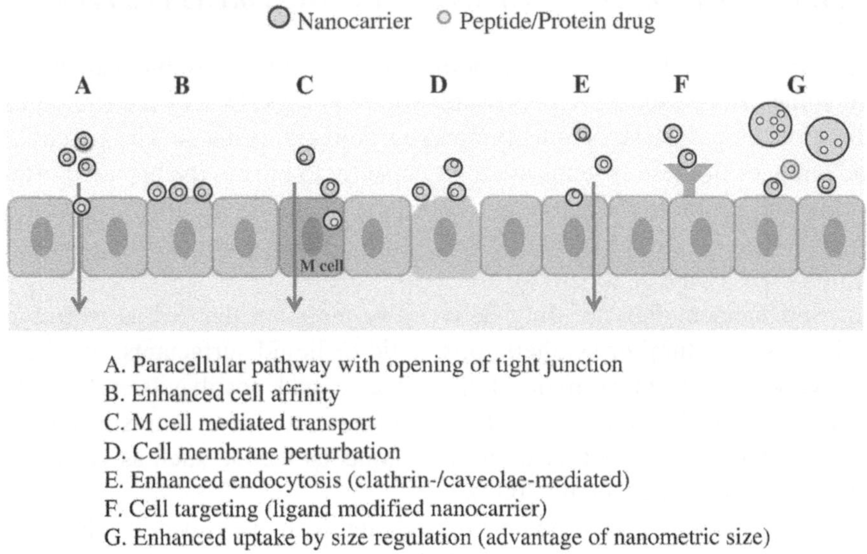


**FIGURE 2.5** Approaches how the lipid-based nanocarriers enhance peptide permeability through the intestinal barrier. (Reprinted with permission from Niu et al. *Adv. Drug Deliv. Rev.* © 2016, Elsevier).

An alternative strategy to increase peptide loading has been reported in which lipophilic surfactants are used to coat the peptide, which is then dispersed in the SMEDDS to produce solid-in-oil suspension/dispersions (Toorisaka et al., 2003).

Efficient peptide incorporation into the inner aqueous core of liposomes has been achieved, although the loading capacity of these nanocarriers clearly depends on the interaction of peptide molecules with lipids, surfactants, and potential polymers involved in the process.

Finally, another example is problems related to the oral administration of insulin. A combination of microemulsion systems with other nanocarriers has led to interesting in vivo results. For example, the incorporation of the insulin–chitosan nanocomplex into the oleic acid-based W/O emulsion, containing polyglyceryl-3 dioleate and PEG-8 C8/C10 glycerides as surfactants, resulted in 7% pharmacological bioavailability in diabetic rats following oral administration (Elsayed et al., 2009).

## 2.4 ROLE OF SURFACTANTS ON THE EFFICACY OF CANCER CHEMOTHERAPY

Most conventional chemotherapeutic agents used for cancer chemotherapy are affected by the multidrug resistance of tumor cells and have poor antitumor efficacy. Based on physiological differences between normal tissue and tumor tissue, one effective approach to improve the efficacy of cancer chemotherapy is to develop pH-sensitive polymeric micellar delivery systems based on surfactants.

It is known that the pH of primary and metastasized tumors is lower than the pH of normal tissue. The extracellular pH of normal tissues and blood are about 7.4, and by contrast, the measured tumor extracellular pH values of most solid tumors range from pH 6.5 to 7.2. Moreover, changes in pH are also observed in endosomes (pH 5.0–6.0) and lysosomes (pH 4.5–5.0), as shown in Figure 2.6.

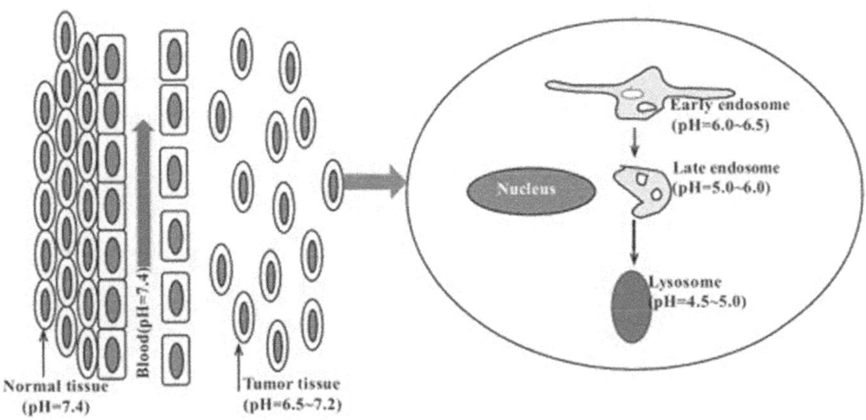


**FIGURE 2.6** Schematic illustrations of pH gradients on normal tissue, blood and tumor tissue and in the endosome–lysosome process. (Adapted with permission from Liu et al. *Asian J. Pharm. Sci.* **2013,** *8*, 159.)

The existing pH of tumor tissue has been considered an ideal trigger for the selective release of anticancer drugs in tumor tissues and/or within tumor cells. Different approaches have been taken to develop pH-sensitive delivery devices, including the incorporation of pH-responsive polymers, peptides, surfactants, and fusogenic lipids. The pH sensitivity strategy of polymeric micelles facilitates specific drug delivery with reduced systemic side effects and improved chemotherapeutical efficacy and is a novel, promising platform for tumor-targeting drug delivery (Liu et al., 2013).

Taking advantage of the altered pH gradients in tumor extracellular environments and in intracellular compartments, pH-sensitive polymeric micelles have made a significant impact on targeted drug delivery and overcome limitations of conventional nanodrug delivery systems. They increase target drug accumulation at the tumor site or in intracellular cytosolic compartments of tumor cells, with less drug distribution to normal tissues and organs.

Furthermore, pH-sensitive polymeric micelles can be triggered to release the drug in endosomes or lysosomes by pH-controlled micelle hydrolysis or dissociation after uptake by cells via the endocytic pathway (Fig 2.7).

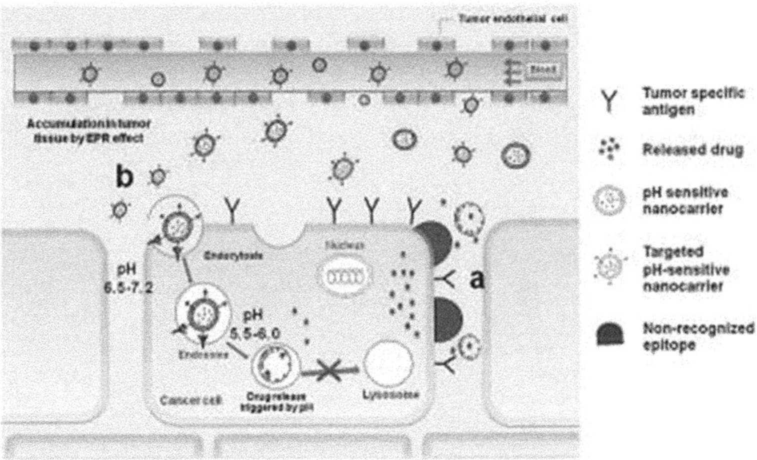


**FIGURE 2.7** **(See color insert.)** Schematic representation of the proposed mechanism of pH-responsive polymeric micelles. (Reprinted with permission from Liu et al. *Asian J. Pharm. Sci.* **2013,** *8*, 159.)

The pH-dependent release of drugs can be established by protonation of pH-sensitive polymers that form the hydrophobic core of polymeric micelles at physiological pH. Destabilization occurs when the protonatable groups become charged below the p*K*a, leading to repulsion between the polymer chains and micellar dissociation. Many examples of protonation of polymers that trigger micelle destabilization have been reported, including poly(histidine) (polyHis), poly(acrylic acid), and polysulfonamides. Among them, polyHis is the most commonly used pH-sensitive component in micelle-based pH-triggered release systems, since this polymer contains an imidazole ring endowing it with pH-dependent amphoteric properties (Oh et al., 2008).

Recently, we have explored whether chitosan nanoparticles with a biocompatible pH-sensitive surfactant derived from the amino acid lysine promote increased release of methotrexate in the acidic tumor environment. The capacity of the surfactant to disrupt the cell membrane was studied using a standard hemolysis assay as a model for endosomal membranes. The surfactant showed pH-sensitive hemolytic activity and improved kinetics at the endosomal pH range (Nogueira et al., 2011).

The chitosan nanoparticles released methotrexate under acidic conditions. This triggered release might be due to the pH-dependent activity of the incorporated surfactant. This specific activity also enhanced cytotoxicity in tumor cell lines, which may be attributable to the combined effect of increased protonation of surfactant and chitosan chemical groups on the nanoparticle surface at tumor pH (Nogueira et al., 2013).

In a recent study, we found that surfactants were needed to obtain enough nanoparticles to deliver doxorubicin as an antitumoral drug. In the presence of the surfactant, nanoparticle production was about 95%, but this figure dropped to less than 4% in the absence of the surfactant. In vitro studies have demonstrated the versatility of the nanoparticle system for facilitating drug internalization by lung carcinoma cells, which optimizes the cytotoxic effect. Furthermore, in vivo investigations have indicated that the nanoformulation leads to promising control over tumor-size progression and associated side effects. For these reasons, the doxorubicin-loaded nanoparticle system may help improve the current role of doxorubicin in the therapeutic management of lung cancer patients (Melguizo et al., 2015).

## 2.5 CONCLUSION

Surfactants are widely used in pharmaceutical formulations to enhance the dissolution and absorption of poorly soluble drugs and are incorporated into nanoparticles to promote the delivery of drugs with different therapeutic activities. Different kinds of surfactants are used in these formulations, depending on their properties. However, the most important factor is the low toxicity of the surfactant used. The role of different surfactants on the delivery of drugs by different routes has been demonstrated based on the results on the literature.

## KEYWORDS

- **nanotechnology**
- **nanoparticles**
- **surfactants**
- **drug delivery**
- **chemotherapy**
- **absorption**

## REFERENCES

Abdelbary, A. A.; AbouGhaly, M. H. H. Design and Optimization of Topical Methotrexate Loaded Niosomes for Enhanced Management of Psoriasis: Application of Box–Behnken Design, In-Vitro Evaluation and In Vivo Skin Deposition Study. *Int. J. Pharm.* **2015,** *485* (1–2), 235–243.

Allam, A. N.; Gamal, E. I.; Naggar, V. F. Formulation and Evaluation of Acyclovir Niosomes for Ophthalmicuse. *Asian J. Pharm. Biol. Res.* **2011,** *1*, 28–40.

Antunes, F. E.; Gentile, L.; Rossi, C. O.; Tavano, L.; Ranieri, G. A. Gels of Pluronic F127 and Nonionic Surfactants from Rheological Characterization to Controlled Drug Permeation. *Colloids Surf. B: Biointerfaces* **2011,** *87*, 42–48.

Azmin, M. N.; Florence, A. T.; Handjani-Vila, R. M.; Stuart, J. F.; Vanlerberghe, G.; Whittaker, J. S. The Effect of Nonionic Surfactant Vesicle (Niosome) Entrapment on the Absorption and Distribution of Methotrexate in Mice. *J. Pharm. Pharmacol.* **1985,** *37* (4), 237–242.

Carafa, M.; Marianecci, C.; Rinaldi, F.; Santucci, E.; Tampucci, S.; Monti, D. J. Span and Tween Neutral and pH-Sensitive Vesicles: Characterization and In Vitro Skin Permeation. *Liposome Res.* **2009,** *19*, 332–340.

Collnot, E. M.; Baldes, C.; Wempe, M. F.; Hyatt, J.; Navarro, L.; Edgar, K. J.; Schaefer, U. F.; Lehr, C. M. Influence of Vitamin E TPGS Poly(Ethylene Glycol) Chain Length on Apical Efflux Transporters in Caco-2 Cell Monolayers. *J. Control. Release* **2006,** *111*, 35–40.

Cone, R. Barrier Properties of Mucus. *Adv. Drug Deliv. Rev.* **2009,** *61*, 75–85.

Danhier, F.; Ansorena, E.; Silva, J. M.; Coco, R.; Le Breton, A.; Preat, V. PLGA-Based Nanoparticles: An Overview of Biomedical Applications. *J. Control. Release* **2012,** *161*, 505–522.

Date, A. A.; Desai, N.; Dixit, R.; Nagarsenker, M. Self-Nanoemulsifying Drug Delivery Systems: Formulation Insights, Applications and Advances. *Nanomedicine* **2010,** *5*, 1595–616.

DelMonte, D. W.; Kim, T. Anatomy and Physiology of the Cornea. *J. Cataract Refract. Surg.* **2011,** *37*, 588–598.

Desmet, E.; Van Gele, M.; Lambert, J. Topically Applied Lipid and Topically Applied Lipid- and Surfactant-Based Nanoparticles in the Treatment of Skin Disorders. *Expert. Opin. Drug.* **2016,** in press.

Elsayed, A.; Remawi, M. A.; Qinna, N.; Farouk, A.; Badwan, A. Formulation and Characterization of an Oily-Based System for Oral Delivery of Insulin. *Eur. J. Pharm. Biopharm.* **2009,** *73* (2), 269–279.

Ghanbarzadeh, S.; Khorrami, A.; Arami, S. Nonionic Surfactant-Based Vesicular System for Transdermal Drug Delivery. *Drug Deliv.* **2015,** *8*, 1071–1077.

Hu, C.; Rhodes, D. G. Proniosomes: A Novel Drug Carrier Preparation. *Int. J. Pharm.* **2000,** *206* (1–2), 110–122.

Hussain, A.; Samad, A.; Nazish, I.; Ahmed, F. J. Nanocarrier-Based Topical Drug Delivery for an Antifungal Drug. *Drug Dev. Ind. Pharm.* **2013,** *40*, 527–541.

Hussain, A.; Singh, V. K.; Singh, O. P.; Shafaat, K.; Kumar, S.; Ahmad, F. H. Formulation and Optimization of Nanoemulsion Using Antifungal Lipid and Surfactant for Accentuated Topical Delivery of Amphotericin B. *Drug Deliv.* **2016**. DOI:10.3109/10717544.2016.1153747.

Kaur, P.; Garg, T.; Rath, G.; Murthy, R. S. R.; Goyal, A. K. Surfactant-Based Drug Delivery Systems for Treating Drug-Resistant Lung Cancer. *Drug Deliv.* **2016,** *23* (3), 717–728.

Kumar, G. P.; Rajeshwarrao, P. Nonionic Surfactant Vesicular Systems for Effective Drug Delivery—An Overview. *Acta Pharm. Sin. B* **2011,** *1* (4), 208–219.

Lai, S. K.; Wang, Y. Y.; Hanes, J. Mucus-Penetrating Nanoparticles for Drug and Gene Delivery to Mucosal Tissues. *Adv. Drug Deliv. Rev.* **2009,** *61*, 158–171.

Lee, B. K.; Yun, Y.; Park, K. PLA Micro- and Nano-Particles. *Adv. Drug Deliv. Rev.* **2016,** in press.

Liu, Y.; Wang, W.; Yang, J.; Zhou, Ch.; Sun, J. pH-Sensitive Polymeric Micelles Triggered Drug Release for Extracellular and Intracellular Drug Targeting Delivery. *Asian J. Pharm. Sci.* **2013,** *8*, 159–167.

Loh, X. J.; Lee, T. C.; Dou, Q.; Deen, G. R. Utilising Inorganic Nanocarriers for Gene Delivery. *Biomater. Sci.* **2016,** *4* (1), 70–86.

Mahale, N. B.; Thakkar, P. D.; Mali, R. G.; Walunj, D. R.; Chaudhari, S. R. Niosomes: Novel Sustained Release Nonionic Stable Vesicular Systems—An Overview. *Adv. Colloid Interface Sci.* **2012,** *183–184*, 46–54.

Maheshwari, C.; Pandey, R. S.; Chaurasiya, A.; Kumar, A.; Selvam, D. T.; Prasad, G. B.; Dixit, V. K. Non-ionic Surfactant Vesicles Mediated Transcutaneous Immunization Against Hepatitis B. *Int. Immunopharmacol.* **2011,** *11*, 1516–1522.

Manosroi, A.; Chankhampan, C.; Ofoghi, H.; Manosroi, W. Low Cytotoxic Elastic Niosomes Loaded with Salmon Calcitonin on Human Skin Fibroblasts. *J. Hum. Exp. Toxicol.* **2013,** *32*, 31–44.

Marianecci, C.; Di Marzio, L.; Rinaldi, F.; Celia, C.; Paolino, D.; Alhaique, F.; Esposito, S.; Carafa M. Niosomes from 80s to Present: The State of the Art. *Adv. Colloid Interface Sci.* **2014,** *205*, 187–206.

Melguizo, C.; Cabeza, L.; Prados, J.; Ortiz, R.; Caba, O.; Rama, A. R.; Delgado, A. V.; Arias, J. L. Enhanced Antitumoral Activity of Doxorubicin Against Lung Cancer Cells Using Biodegradablepoly(Butylcyanoacrylate) Nanoparticles. *Drug Des. Dev. Ther.* **2015,** *9*, 6433–6444.

Mert, O.; Lai, S. K.; Ensign, L.; Yang, M.; Wang, Y. Y.; Wood, J.; Hanes, J. A Poly(Ethyleneglycol)-Based Surfactant for Formulation of Drug Loaded Mucus Penetrating Particles. *J. Control. Release* **2012,** *157* (3), 455–460.

Mitra, A.; Lin, S. Effect of Surfactant on Fabrication and Characterization of Paclitaxel-Loaded Polybutylcyanoacrylate Nanoparticulate Delivery Systems. *J. Pharm. Pharmacol.* **2003,** *55*, 895–902.

Nasr, M.; Mansour, S.; Mortada, N. D.; Elshamy A. A. Vesicular Aceclofenac Systems: A Comparative Study Between Liposomes and Niosomes. *J. Microencapsul.* **2008,** *25* (7), 499–512.

Niu, Z.; Conejos-Sánchez, I.; Griffin, B. T.; O'Driscoll, C. M.; Alonso, M. J. Lipid-Based Nanocarriers for Oral Peptide Delivery. *Adv. Drug Deliv. Rev.* **2016**. pii: S0169-409X(16)30100-4. DOI:10.1016/j.addr.2016.04.001.

Nogueira, D. R.; Mitjans, M.; Infante, M. R.; Vinardell, M. P. The Role of Counterions in the Membrane-Disruptive Properties of pH-Sensitive Lysine-Based Surfactants. *Acta Biomater.* **2011,** *7* (7), 2846–2856.

Nogueira, D. R.; Tavano, L.; Mitjans, M.; Pérez, L.; Infante, M. R.; Vinardell, M. P. In Vitro Antitumor Activity of Methotrexate via pH-Sensitive Chitosan Nanoparticles. *Biomaterials* **2013,** *34*, 2758–2772.

Oh, K. T.; Lee, E. S.; Kim, D.; Bae, H. Y. L-Histidine-Based pH-Sensitive Anticancer Drug Carrier Micelle: Reconstitution and Brief Evaluation of Its Systemic Toxicity. *Int. J. Pharm.* **2008,** *358*, 177–183.

Pando, D.; Matos, M.; Gutiérrez, G.; Pazos, C. Formulation of Resveratrol Entrapped Niosomes for Topical Use. *Colloids Surf. B: Biointerfaces* **2015,** *128*, 398–404.

Patel, A.; Cholkar, K.; Agrahari, V.; Mitra, A. K. Ocular Drug Delivery Systems: An Overview. *World J. Pharmacol.* **2013,** *2* (2), 47–64.

Patravale, V. B.; Date, A. A.; Kulkarni, R. M. Nanosuspensions: A Promising Drug Delivery Strategy. *J. Pharm. Pharmacol.* **2004,** *56*, 827–840.

Pellicano, R. Gastrointestinal Damage by Non-Steroidal Anti-Inflammatory Drugs: Updated Clinical Considerations. *Miner. Gastroenterol. Dietol.* **2014,** *60* (4), 255–256.

Pokharkar, V.; Patil, V.; Mandpe, L. Engineering of Polymer–Surfactant Nanoparticles of Doxycycline Hydrochloride for Ocular Drug Delivery. *Drug Deliv.* **2015,** *22* (7), 955–968.

Pouton, C. W. Lipid Formulations for Oral Administration of Drugs: Non-Emulsifying, Self-Emulsifying and 'Self-Microemulsifying' Drug Delivery Systems. *Eur. J. Pharm. Sci.* **2000,** *11*, S93–S98.

Rajera, R.; Nagpal, K.; Singh, S. K.; Mishra, D. N. Niosomes: A Controlled and Novel Drug Delivery System. *Biol. Pharm. Bull.* **2011,** *34* (7), 945–953.

Rouse, J. D.; Sabatini, D. A.; Suflita, J. M.; Harwell, J. H. Influence of Surfactants on Microbial Degradation of Organic Compound. *Crit. Rev. Environ. Sci. Technol.* **1994,** *24*, 325–370.

Sarmah, P. J.; Kalita, B.; Sharma, A. K. Transfersomes Based Transdermal Drug Delivery: An Overview. *IJAPR* **2013,** *4* (12), 2555–2563.
Schäfer-Korting, M.; Mehnert, W.; Korting, H-C. Lipid Nanoparticles for Improved Topical Application of Drugs for Skin Diseases. *Adv. Drug Deliv. Rev.* **2007,** *59* (6), 427–443.
Singh, B.; Bandopadhyay, S.; Kapil, R.; Singh, R.; Katare, O. Self-emulsifying Drug Delivery Systems (SEDDS): Formulation Development, Characterization, and Applications. *Crit. Rev. Ther. Drug Carrier Syst.* **2009,** *26*, 427–521.
Singh, G.; Pai, R. S. Trans-resveratrol Self-Nano-Emulsifying Drug Delivery System (SNEDDS) with Enhanced Bioavailability Potential: Optimization, Pharmacokinetics and In Situ Single Pass Intestinal Perfusion (SPIP) Studies. *Drug Deliv.* **2015,** *22* (4), 522–530.
Singh, G.; Pai, R. S.; Pandit, V. In Vivo Pharmacokinetic Applicability of a Simple and Validated HPLC Method for Orally Administered Transresveratrol Loaded Polymeric Nanoparticles to Rats. *J. Pharm. Invest.* **2014,** *44*, 69–78.
Singhal, G. B.; Patel, R. P.; Prajapati, B. G.; Patel, N. A. Solid Lipid Nanoparticles and Nano Lipid Carriers: As Novel Solid Lipid Based Drug Carrier. *Inter. Res. J. Pharm.* **2011,** *2* (2), 40–52.
Suttipong, M.; Tummala, N. R.; Kitiyanan, B.; Striolo, A. Role of Surfactant Molecular Structure on Self-Assembly: Aqueous SDBS on Carbon Nanotubes. *J. Phys. Chem. C* **2011,** *115*, 17286–17296.
Talsma, H.; Van Stenberg, M. J.; Brochert, J. H. C.; Crommelin, D. A Novel Technique for the One-Step Preparation of Liposomes and Nonionic Surfactant Vesicles Without the Use of Organic Solvents. Liposome Formation in a Continuous Gas Stream: The 'Bubble' Method. *J. Pharm. Sci.* **1994,** *83*, 276–280.
Toorisaka, E.; Ono, H.; Arimori, K.; Kamiya, N.; Goto, M. Hypoglycemic Effect of Surfactant-Coated Insulin Solubilized in a Novel Solid-in-Oil-in-Water (s/o/w) Emulsion. *Int. J. Pharm.* **2003,** *252* (1–2), 271–274.
Varshosaz, J.; Taymouri, S. Hollow Inorganic Nanoparticles as Efficient Carriers for siRNA Delivery: A Comprehensive Review. *Curr. Pharm.* **2015,** *21* (29), 4310–4328.
Wilczewska, A. Z.; Niemirowicz, K.; Markiewicz, K. H.; Car, H. Nanoparticles as Drug Delivery Systems. *Pharmacol. Rep.* **2012,** *64*, 1020–1037.
Wu, J.; Zhang, J.; Deng, Ch.; Meng, F.; Zhong, Z. Vitamin E-Oligo(Methyl Diglycol L-Glutamate) as a Biocompatible and Functional Surfactant for Facile preparation of Active Tumor-Targeting PLGA Nanoparticles. *Biomacromolecules* **2016,** *17*, 2367–2374.
Yang, M,; Lai, S. K.; Yu, T.; Wang, Y. Y.; Happe, C.; Zhong, W.; Zhang, M.; Anonuevo, A.; Fridley, C.; Hung, A.; Fu, J.; Hanes, J. Nanoparticle Penetration of Human Cervicovaginal Mucus: The Effect of Polyvinyl Alcohol. *J. Control Release*. **2014,** *192*, 202–208.
Zhang, Z.; Tan, S.; Feng, S.-S. Vitamin E TPGS as a Molecular Biomaterial for Drug Delivery. *Biomaterials* **2012,** *33* (19), 4889–4906.

# CHAPTER 3

# SMART POLYMERIC NANOCARRIERS FOR DRUG DELIVERY

SUBHAM BANERJEE[1,2*] and JONATHAN PILLAI[1]

[1]*Centre for Bio-design & Diagnostics (CBD), Translational Health Science & Technology Institute (THSTI), Faridabad, Haryana, India.*

[2]*Department of Pharmaceutics, National Institute of Pharmaceutical Education and Research (NIPER)-Guwahati, Assam, India.*

*Corresponding author. E-mail: banerjee.subham@yahoo.co.in*

## CONTENTS

## ABSTRACT

Smart polymeric nanocarriers are an important topic of interest and emerging trends in the arena of drug-delivery research. Their uniqueness stems from their payload in response to a particular stimulus applied either internally or externally. In this chapter, we discuss some of the most important and widely studied stimuli-responsive smart polymeric nanocarriers, with special emphasis on their in vitro and in vivo preclinical evaluation. The most common internally applied stimuli such as pH, enzymes, and oxidation potential are described in detail. Other evolving applications of smart nanocarriers responsive to externally applied stimulus like temperature, magnetic field, ultrasound, and light are discussed as well. Further, various therapeutic applications of such stimuli-responsive systems are briefly reviewed. Finally, we conclude this chapter with future perspectives on the therapeutic uses of this class of smart polymeric nanocarriers.


## 3.1 INTRODUCTION

Polymeric biomaterials offer several distinct advantages for biomedical drug-delivery applications such as excellent biocompatibility, biodegradability, and, in some cases, remarkable in vivo biomimetic potential. In the arena of drug-delivery applications, so-called smart biopolymeric materials have garnered significant interest because of their ability to facilitate targeted, site-specific drug delivery at the desired organ or tissue of interest, often with excellent therapeutic efficacy. The basic criteria for developing polymeric nanocarrier-based delivery systems are determined by several important functional parameters. These include the formulation of uniformly sized particles, particle surface characteristics, permeation enhancement, retention rate, solubility, and ability to control and prolong delivery of the drug cargo to achieve desired therapeutic concentrations at specific sites of interest. Lately, these criteria have been employed in numerous medical and biological applications of nanoscale structures to achieve site-specific, sustained delivery and improved bioavailability of water-insoluble drugs (Allemann et al., 1993; Kawashima, 2001; Panyam and Labhasetwar, 2003; Soppimath et al., 2001).

Recent advances in nanotechnology have led to dramatic improvements in the capabilities and performance of biomaterials. As a result, synthesis of nanoscale formulations has culminated in the integrated design of smart, stimuli-responsive nanocarriers (SRNs) that may be customized for specific therapeutic outcomes or capable of operating under specific physiological conditions. For instance, nanosizing of polymeric carriers has particularly proven to be a boon for targeting tumors with therapeutically active moieties, preliminary results from which appear to be encouraging (Brigger et al., 2002; Fleige et al., 2012). Such carriers are able alter their composition or surface properties and morphology or both, in response to various internal or external stimuli, further leading to a modulation of drug release.

There is a lack of a clear consensus in the community on a common definition of "smart" biomaterials and nanocarriers derived from them. It appears that any material or system that shows some activity or response as compared to an inert or passive material or system may be considered to be "smart." From the engineering and structural perspective, shape memory materials were perhaps the earliest to be acknowledged as "smart" systems (Langer and Tirelli, 2004). Biomimetic nanomaterials, such as those inspired by gecko footpads or shark skin, may also be considered smart, as these are able to partly modulate their morphology and surface characteristics to perform a desired function.

In contrast to passive materials, active materials and organisms possess the ability to respond to the local environment or to stimulus, stress, or injury by changing intrinsic characteristics such as morphology, confirmation, surface properties, or interaction with self and nonself entities within a defined time period, thereby demonstrating a degree of "smartness." This response may be repeatable, reproducible, and proportionate to the input provided by a particular environment or set of conditions. Hence, from a biological and drug-delivery perspective, it may be argued that stimuli-responsive biomaterials and their nanoformulations are the most relevant in terms of being classified as "smart" systems.

This chapter broadly discusses the current developments in this arena and reviews a selection of smart SRNs that have shown exceptional promise for drug delivery. In the physiological environment within which most drug-delivery systems operate, there are a number of stimuli and environmental conditions that may modulate the performance and effectiveness of smart systems. Stimuli that SRNs are designed to respond to may be broadly differentiated as internal or external stimuli, on the basis of their origin with respect to the local environment. The most common internal stimuli which SRNs are designed to respond to for drug delivery are the in vivo pH and numerous enzymes that are able to modulate a controlled and localized response. For example, the reduced levels of glutathione in the cytosol are frequently used for the design and synthesis of reduction-sensitive polymeric nanocarriers (Adamo et al., 2014). Examples of externally applied stimuli include heat, magnetic fields, ultrasound, and light, which have also been reviewed briefly (Fig. 3.1). Notable among these is the case of anatomical sites that are difficult to reach via systemic delivery, where thermoresponsive polymers have been employed in designing polymeric nanocarriers (Soga et al., 2004a,b). These are employed mostly to induce nanoparticle self-assembly by means of a temperature-triggered polymer polarity change, which can be disassembled by external hyper- or hypothermia. Finally, functionalization of nanocarriers is often able to impart a specific aspect of stimuli-responsiveness to otherwise passive delivery systems and is briefly tabulated to provide an overview of available strategies and their therapeutic applicability. However, programmable smart materials such as microfabricated devices or shape memory alloys are beyond the scope of this chapter, even though they have been successfully deployed for implementing smart drug-delivery systems.

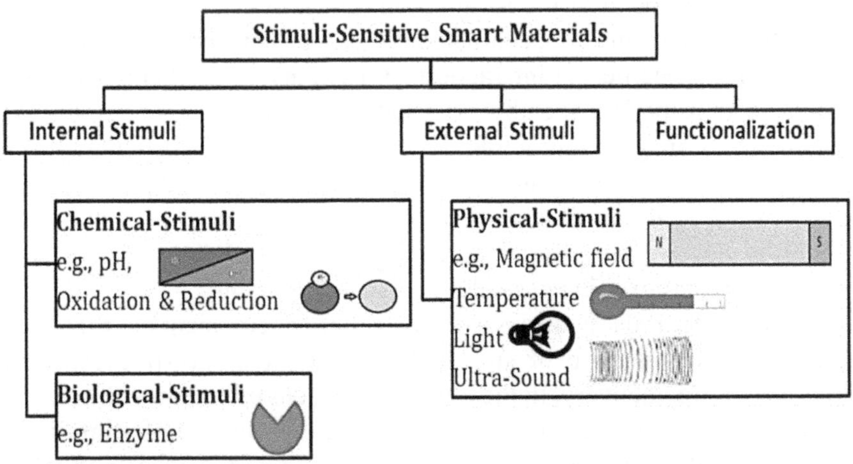


**FIGURE 3.1** Various classification of stimuli-sensitive smart materials.

## 3.2 SMART NANOCARRIERS RESPONSIVE TO INTERNAL STIMULI

Traditional or "passive" drug-delivery systems try and mitigate or minimize the deleterious effects of internal stimuli such as gastric pH, enzyme degradation, reactive oxidative species, or elevated temperatures during inflammation by choosing robustly biocompatible or inert materials. However, SRNs respond to these very stimuli in a positive manner, not only to counteract the potential for damage but to actively utilize them to enhance performance, control and efficiency of drug delivery. In the following subsections, SRNs classified as per major internal stimuli are presented.

### *3.2.1 pH-RESPONSIVE POLYMERIC NANOCARRIERS*

A pervasive stimulus that widely used for the development of smart nanocarriers based drug-delivery application is the physiological pH in vivo (Fig. 3.2). It is considered as an ideal trigger for specific and selective drugs release, particularly for targeting pathologies like tumors or localized inflammation where the local environment presents a lower pH than that of normal tissue (Gerweck and Seetharaman, 1996; Lehner et al., 2012). Intracellular components like endosomes and lysosomes also normally present at pH 6.5 and 5.5, respectively (Manchun et al., 2012). The release of the drug can be triggered through the degradation of the polymer backbone, through the degradation of pH-sensitive polymer–drug linkers (Mellman, 1996; Mrkvan et al., 2005; Torchilin, 2000), or via a combination of both these mechanisms. Overall, this contributes to the specificity of the treatment, as drug release takes place selectively in the areas of interest. Therefore, the rational design of such systems involves constructing a polymeric nanocarrier that is stable at physiological pH but will degrade in tumors or in the acidic intracellular organelles upon cellular uptake or any other areas of inflammation, where an acidification of the microenvironment occurs.

The design of pH-sensitive nanocarriers involves the application of two principal strategies. The first of these is the use of a pH-sensitive polymer, which either undergoes hydrolytic degradation in a pH-sensitive manner or changes key physicochemical properties like polarity, thus leads to the destabilization of the nanocarriers matrix and concomitant drug release from the cargo. Some pioneering work on pH-sensitive polymeric systems was done by Uhrich and coworkers (Erdmann and Uhrich, 2000), with the synthesis and characterization of biodegradable polymeric prodrugs based on salicylic acid (the active component of aspirin) linked to poly(anhydride-ester). This prodrug was found to be stable under acidic conditions (pH 3) but released salicylic acid slowly at normal alkaline pH (pH 7.4) and more quickly in basic conditions (pH 10), owing to the hydrolytic degradation of the polymer backbone. This prodrug showed potential in the management of gastrointestinal disease, where release in the basic environment of the lower intestine is necessary.

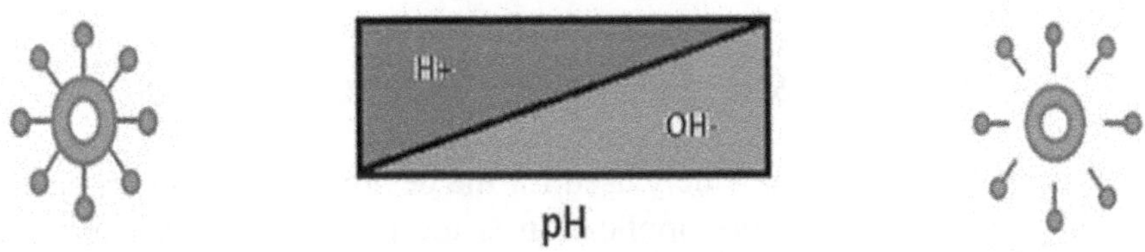


**FIGURE 3.2** Schematic representation of pH-responsive polymeric nanocarriers.

Along the same lines, polyacetals synthesized by the reaction between vinyl ether and alcohols were first described by Heller et al. (1980). These systems can be hydrolyzed under mildly acidic conditions (pH 6.5) but are stable at physiological pH (7.4) (Tomlinson et al., 2002). Additional important class of pH-sensitive systems is the poly(amidoamine)s, which undergo hydrolytic degradation of the amidic bonds of their backbone in aqueous media. These have been used as antineoplastic drug–polymer conjugates for intracellular gene and toxin delivery (Ferruti et al., 2002). Richardson and coworkers published their results with linear poly(acrylic acid) (PAA) that displayed negligible toxicity which is nonspecific in nature, exhibit pH-dependent lysis of the cellular membrane, and transport gene and toxin. In vitro and in vivo studies using these systems have depicted that PAAs are capable of permeabilizing the endocytic vesicular membranes in vivo; this is crucial for their use as nonviral gene delivery systems (Richardson et al., 2010).

Polyesters are also often used as components of pH-sensitive nanocarriers, owing to their pH-dependent degradation properties. A very important example of such polymers is the polylactides (PLA). These are FDA approved for clinical use; they also have demonstrated biodegradability due to hydrolysis under physiological conditions (Jung-Kwon, 2011). As an example, Ahmed and coworkers blended poly(ethylene glycol) with poly(lactic acid) (PEG-*b*-PLA), PEG-*b*-polycaprolactone (PCL), and PEG-*b*-poly(butadiene) (PEG-*b*-PBD) and showed the block of lactic acid gets degraded in acidic media leading to the doxorubicin (DOX) and paclitaxel (PTX) release in a controlled fashion from the encapsulated matrix. This formulation demonstrated shrinkage of rapidly growing tumors in a murine model upon a single intravenous injection (Ahmed et al., 2006).

In contrast to the relatively prolonged release from polyesters, biodegradable cationic polymers based on poly(β-amino esters) (PbAE) used for the design of site-specific drug and gene delivery systems due to their rapid dissolution under acidic conditions. For example, it has been observed that PEO-modified PbAE (PEO-PbAE) nanoparticles loaded with PTX demonstrate enhanced drug accumulation in the tumor tissue upon intravenous administration in human ovarian adenocarcinoma (SKOV-3) xenografts; there was also a significant inhibition of tumor growth in comparison with non-pH-sensitive nanoparticles based on PCL (Devalapally et al., 2007; Shenoy et al., 2005).

Another variation of stimuli-responsive polymers is those capable of sensing minute changes in microenvironmental pH and eliciting a corresponding alteration in physical properties such as morphology or hydrophobicity (Lehner et al., 2012). This behavior is characteristic of micellar systems, where the drug is physically encapsulated in the core of the nanocarrier, and any change in the polymer properties leads to micellar disintegration and subsequent drug release.

Kim et al. (2008) synthesized folic acid and DOX-conjugated pH-sensitive mixed-micelle, which was designed to treat multidrug resistance (MDR) in cancer therapy. The mixed-micelles were composed of poly(histidine)-*co*-phenylalanine-*b*-PEG (His-*co*-Phe) and poly(L-lactic acid) (PLLA)-*b*-PEG-folate. The pH sensitivity of the micelles was conferred by the poly(His-*co*-Phe) core-forming block, owing to the conversion of histidine from a nonionized to hydrophilic state by protonation in acidic (endosomal) media; this, in turn, led to micelle destabilization.

Lee et al. (2008) reported on DOX-loaded polymeric micelles consisting of PLLA-*b*-PEG-*b*-poly(L-His)-TAT (transactivator of transcription) and poly(L-His)-*b*-PEG. The micelle core was pH-sensitive owing to the polyHis block, leading to disintegration and DOX release in the acidic endosomes, while the micelles were actively targeted by the attachment of TAT. The micelles were tested in vivo in several tumor models and demonstrated increased tumor growth inhibition compared to free DOX and minimum weight loss.

Likewise, the synthesis of micelles based on PAA and poly(methacrylic acid) already reported for treatment purpose, owing to their swelling capability in response to pH. In these cases, incorporation of ionizable monomer units into polymer backbones leads to pH-dependent phase transitions and solubility changes (Alexander, 2005; Kyriakides et al., 2002).

A good example of the effective design of pH-responsive nanocarrier is provided by Tomlinson et al. (2003). These researchers have successfully demonstrated a prolonged blood half-life and enhanced tumor accumulation as well as lower liver and spleen uptake for amino pendant polyacetal–doxorubicin (DOX) conjugates when compared to the clinical conjugate *N*-(2-hydroxypropyl) methacrylamide (HPMA) copolymer–DOX.

### 3.2.2 ENZYME-RESPONSIVE POLYMERIC NANOCARRIERS

An emerging arena in the field of stimuli-responsive polymeric nanocarriers is related to nanomaterials that experience macroscopic property changes mediated by enzymatic catalytic activity. The growing importance of this field is still evident by various recent articles (de la Rica et al., 2012; Fleige et al., 2012; Hu et al., 2012). These systems have a distinctive specificity, as enzymes are highly selective biomolecules. Moreover, enzymes are critical components in many biological pathways and are functional under mild conditions which is a crucial feature in vivo. For instance, enzymes are key targets for drug development, since they play a key role in cell regulation. Enzymatically sensitive nanomaterials can be designed to deliver and release their cargos specifically at the target site (Andresen et al., 2005; Minelli et al., 2010) where the enzyme is overexpressed or its activity is increased due to a pathological condition. Taking into account that up- or downregulation of enzyme expression or activity is a key aspect in many diseases, enormous benefit can be obtained by the use of enzyme-responsive nanomaterials as delivery systems in the areas of diagnostics and therapeutics (Fig. 3.3).

To date, two main approaches have been adopted for developing enzyme responsive materials. In the first and most common approach, the nanocarrier itself is sensitive to enzymatic transformation. This is achieved by the use of either an enzymatically degradable polymer or enzymatically sensitive linkers between the drug and the polymer. The second method is based on the surface modification of the nanocarriers with molecules that confer enzymatically triggered changes of the physical properties of the carrier solution. This approach is extremely versatile and has been mainly used in developing inorganic enzyme-responsive nanoparticles (Medintz et al., 2005; Stevens et al., 2004; Zelzer and Ulijn, 2010) that can be act as drug-delivery carrier system has not been so fully developed because of toxicological concerns.

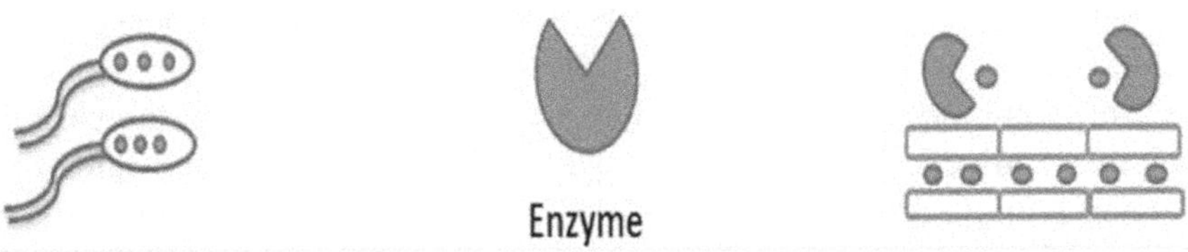


**FIGURE 3.3** Schematic representation of enzyme-responsive polymeric nanocarriers.

In the rest of this section, key examples of enzyme-responsive nanocarriers will be discussed, as well as their applications for enzyme-mediated drug release. This category includes polymeric nanocarriers whose structural scaffold presents a responsive behavior tailored by enzymatic activity, which is achieved by the use of enzymatically degradable polymers. Generally, the advantage of using biodegradable polymers is that they will eventually be degraded into small molecules that can easily be excreted by the body. This is also very useful for the management of various diseases which require prolong administration, for example, tissue regeneration, regenerative, and alternate medicine (Hardwicke et al., 2010; Santamaria et al., 2009; Shaunak et al., 2004).

In the broad category of enzyme degradable polymers, dextrins degradable by amylase are considered to be very promising options. Among these, systems based on polyglutamates susceptible to degradation by cathepsin B are considered to be the best choices. The lysosomal protease cathepsin B is known to take part in various extracellular degradation processes (Mort and Buttle, 1997). This enzyme acts primarily as a carboxypeptidase, essentially as peptidyldipeptidase, as with poly(L-glutamic acid) (PGA). In the case of PGA as a platform, the PGA–PTX conjugate OpaxioTM, developed by Cell Therapeutics Inc., is the most clinically advanced formulation. It reached clinical trials mainly for nonsmall cell lung carcinoma and ovarian carcinoma as well (Lammers et al., 2012), in combination with radiotherapy (Dipetrillo et al., 2006) or cisplatin. In this construct, PTX is linked to PGA through an ester bond shielded during blood circulation to avoid release triggered by plasma esterases. Adopting this approach, the high level of enzyme specificity might permit, in the near future, the use of personalized therapy.

Similarly, another conjugate developed by Cell Therapeutics Inc., CT-2106, is a PGA–camptothecin prodrug based on conjugation through an ester linkage and has shown enhanced anticancer efficacy against B16 melanoma tumors (Bhatt et al., 2003; de Vries et al., 2000, 2001; Singer et al., 2000, 2001). This conjugate is also in phase II of clinical trials (Singer et al., 2001).

Enzymatically sensitive polymers are also widely used in the case of polymeric micelles that physically encapsulate their cargo and release it upon enzymatic degradation of their core polymer. One example of this is the work of Mao and Gan (2009), who synthesized amphiphilic diblock copolymers based on poly(glycidol-*block*-ε-caprolactone) (PG-*b*-PCL) with well-organized manner with pendant hydroxyl groups through the formation of hydrophilic block. These copolymers formed 74–95-nm micelles that demonstrated enzymatically triggered release of the encapsulated dye (pyrene) in the presence of lipase, due to degradation of the PCL block, which resulted in micelle dissociation.

The action of several enzymes can also lead to changes in the morphology of polymeric micelles. This has been exploited in the recent work by Ku et al. (2011). In this case, micelles composed of polymer–peptide block copolymers were prepared, containing substrates either for protein kinase A, for protein phosphatase-1, or for matrix-metalloproteinases (MMP) 2. The phenomenon of the reversible transformation of the micelles morphology via phosphorylation/dephosphorylation cycle resulting in a reversible change of the micellar size was tested. They also studied the peptide sequence-directed alteration in morphology in response to proteolysis. Although this system has not yet been tested as a nanocarrier, it seems to be a good candidate for application in drug delivery, where morphology and the surface chemistry are key aspects in the localizing and pharmacokinetics of nanomaterials.

Although outside the scope of this review, it is worth mentioning that in the field of hydrogels, there has been recent interest in systems that degrade under cellular response. For instance, the work of Aimetti et al. (2009) describes a PEG hydrogel platform with human neutrophil elastase (HNE) degradable cross-links formed using thiolene photo polymerization, which results in a hydrogel degradable at inflammation sites. In this work, rhodamine-labeled bovine serum albumin was entrapped and was successfully released upon treatment with HNE (Aimetti et al., 2009).

### *3.2.3 OXIDATION-RESPONSIVEPOLYMERICNANOCARRIERS*

A very important class of SRNs takes advantage of the differences in redox potential in the extra- and intracellular environment. It has been reported that this difference is between 100 and 1000-fold, with the extracellular space being oxidative, whereas the zone inside the cell is strongly reductive (Lehner et al., 2012; Meng et al., 2009; Roy et al., 2010). The reductive intracellular environment can mainly be attributed to the presence of high concentrations of glutathione (GSH, 0.5–10 mM), in contrast to the low amounts present in the blood and the extracellular matrices (2–20 μM) (Wu et al., 2004). It is therefore obvious that this physiological property can prove very useful for the intracellular release of bioactive molecules. Several polymer therapeutics have been designed using this approach for the intracellular delivery of diverse cargoes such as DNA, siRNA, proteins, and low-molecular-weight drugs (Fig. 3.4). In addition, reduction-sensitive nanocarriers can prove valuable for tumor-specific drug delivery in cancer therapy because of the highly reducing conditions (fourfold higher GSH concentration) in the tumor as compared to normal tissue (Wu et al., 2004).

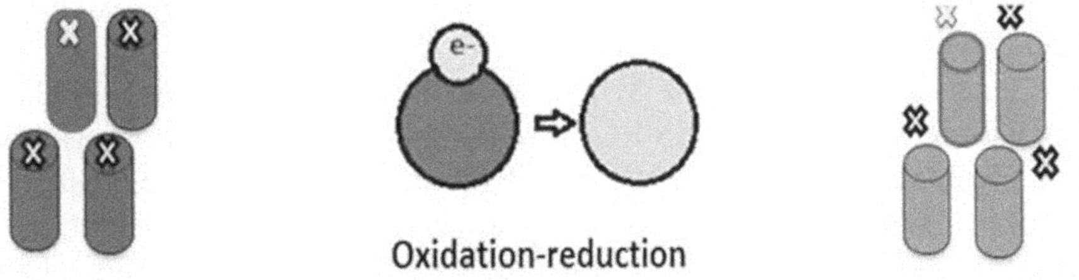


**FIGURE 3.4** Schematic representation of oxidation-responsive polymeric nanocarriers.

The redox sensitivity of nanocarriers for drug delivery is usually modulated by disulfide bridges that are either intentionally incorporated in the polymer synthetically or are already present on the polymer matrix in its native state. These bridges can be cleaved under reducing conditions known as a redox switch or in the presence of other thiols (Roy et al., 2010). This cleavage results either in the destabilization of the carrier, leading to the release of the encapsulated drug, or to the cleavage of the linker between the carrier and the drug, again liberating the therapeutic, resulting overall in a redox switch.

For drug-delivery applications, cross-linking through disulfides can be a useful strategy for increasing the stability of polymeric micelles or other types of nanoparticles in vivo. Wang et al. (2011) synthesized disulfide bridged block copolymer of PCL and poly(ethylethylene phosphate) (PCL-SSPEEP), which self-assembled into 90-nm micellar structures. When these micelles were loaded with DOX, more drug accumulation and retention was observed in MDR cells compared to normal block copolymers, and rapid drug release in the reductive intracellular environment, resulting in higher cytotoxicity. This increased efficacy could be attributed to the intracellular dissociation of the micelles owing to the presence of GSH, leading to DOX release.

Sun et al. (2011) developed reduction-sensitive micelles comprised poly(ethylene oxide)-*b*-poly{*N*-methacryloyl-*N*-(*t*-butyloxycarbonyl) cystamine} (PEO-*b*-PMABC) diblock copolymers that were stable in physiological conditions but dissociated under reducing environments, leading to the rapid release of the encapsulated DOX. These micelles demonstrated higher antitumor efficacy compared to micelles without reduction sensitivity against T24 human bladder cancer cells. Micelles based on thiolated pluronics have also been developed, with disulfide cross-linked cores, loaded with PTX (Abdullah Al et al., 2011). These demonstrate accumulation in the cytoplasm and increased cytotoxicity as compared to controls when incubated with A549 cells.

Another example is the synthesis of the redox-responsive PEG-*b*-poly(lactic acid) (MPEG-SS-PLA) by Song et al. (2011) and its use for the preparation of nanoparticles for PTX delivery. Cytotoxicity assays were performed with A549, MCF-7, and HeLa cells; enhanced antitumor efficacy was observed compared to free PTX while empty nanoparticles showed no cytotoxicity, which was attributed to the release of PTX in a triggered and continuous manner owing to redox sensitivity.

Reduction-responsive drug delivery systems (DDS) have also been proven valuable for the delivery of proteins in the cytosol. In recent work, caspase 3 was encapsulated in a thin, positively charged *N*-(3-aminopropyl) methacrylamide polymer shell interconnected by disulfide cross-links. The redox-responsive nanocapsules formed in this way were able to induce apoptosis in several human cancer cell lines, owing to efficient cellular uptake and protein release in the reductive cytosol (Zhao et al., 2011).

In summary, very important progress has been made with regard to redox responsive polymeric nanocarriers during the past decade. Given their significant advantages, it appears that they will soon overtake the pH- and enzyme-responsive systems, especially when applied to peptide, protein, or gene delivery.

### *3.2.4 SMART NANOCARRIERS RESPONSIVE TO EXTRINSIC STIMULI*

In addition to biomaterials responsive to the most common stimuli described above, various reports describe polymeric nanocarriers which release their payload in response to other exogenous triggers, for example, magnetic field, temperature, light, and ultrasound. These are described briefly in the following subsections.

### *3.2.5 TEMPERATURE-RESPONSIVE POLYMERIC NANOCARRIERS*

One of the most important subclasses of polymeric biomaterials that are sensitive to extrinsic stimuli is the temperature-responsive polymers (Chilkoti et al., 2002; Liu et al., 2009; Talelli and Hennink, 2011). The applicability of the stimuli-responsiveness of these systems is conceptually very different; in most cases, it does not serve to enable drug release but rather to facilitate self-assembly. Thermosensitive polymers are those whose solutions are characterized by a lower critical solution temperature (LCST). Below this temperature, the polymer chains form hydrogen bonds with water and are therefore in a stretched and completely dissolved state, whereas when heated above this temperature, they dehydrate (owing to hydrogen bond cleavage) and become insoluble. This property is extremely useful for the self-assembly of diblock copolymers, made of a thermosensitive block (both hydrophobic and hydrophilic, modulated by temperature) and a permanently hydrophobic or hydrophilic block. Typically, they self-assemble into polymeric micelles, with the thermosensitive blocks aggregating as the micellar core and the nonresponsive blocks assembling a corona, respectively. In addition, when these nanoparticles encapsulate a drug, it is possible to enable drug release by local hypo- or hyperthermia, resulting in a change of the thermosensitive polymer state and dissociation of the micellar structure. However, the latter has not yet been proven, and as such, this review will focus solely on the use of thermosensitivity for polymer self-assembly into nanoscaled carriers.

Among the most commonly used thermosensitive polymers are poly(*N*-isopropylacrylamide) (pNIPAAm) and pluronics (polyethylene glycol-*b*-polypropylene oxide-*b*-polyethylene glycol, PEG-*b*-PPO-*b*-PEG). pNIPAAm, with an LCST of 32°C, has been used as a hydrophobic core for polymeric micelles with a PEG corona, as well as a hydrophilic shell combined with several other hydrophobic blocks. Work on the synthesis and applications of NIPAAm polymers as components for DDS has been extensively described in several reviews over the years (Aoshima and Kanaoka, 2008; Schmaljohann, 2006; Talelli and Hennink, 2011; Wei et al., 2009). Similarly, pluronics are another well-known class of thermosensitive polymers, often used as a micellar component for drug delivery (Batrakova and Kabanov, 2008; Kabanov et al., 2002). These triblock copolymers consist of a poly(propylene oxide) block in the middle flanked by two PEO blocks. Those are self-assembled into the core of the micelles with a PPO block above the LCST of this block. Pluronic micelles loaded with DOX have attained phase II–III clinical trials, where they have demonstrated slower clearance and enhanced antitumor activity than free drug, in advanced MDR adenocarcinoma patients, with an acceptable toxicity profile and antitumor activity (Valle et al., 2011).

Even though they are very promising, these two polymers display disadvantages in that they are nonbiodegradable and also the micelles of these polymers can only disintegrate under local hypo- or hyperthermia to result in controlled drug release, which limits their clinical use. Also, more importantly, they degrade over time under physiological conditions (Soga et al., 2004a). When this pluronic is copolymerized with PEG, the resultant PEG-*b*-pHPMAmLac *n* block copolymers self-assemble in aqueous solutions into small (60 nm) monodisperse micelles; under physiological conditions, these micelles disintegrate over time (owing to HPMAmLac *n* degradation) leading to controlled drug release (Soga et al., 2004b, 2005). Micelles of diblock copolymers of mPEG-*b*-pHPMAm-Lac *n* have been loaded with several anticancer drugs (PTX, DOX) and some formulations have already been tested in vivo in mice, where they have demonstrated prolonged circulation times, as well as increased antitumor efficacy and animal survival as compared to the free drug (Rijcken et al., 2007, 2010; Soga et al., 2005; Talelli et al., 2010).

In summary, thermosensitive polymers can prove very valuable as components of drug-delivery systems. This is mainly attributable to the easy preparation of nanoparticles by simply changing the solution temperature, which in most cases results in spontaneous self-assembly. However, to confer controlled release properties on these systems, either local hypo- or hyperthermia has to be applied or they need to be combined with another stimulus, such as pH or enzyme sensitivity for tailorable and controlled drug release properties (Fig. 3.5).

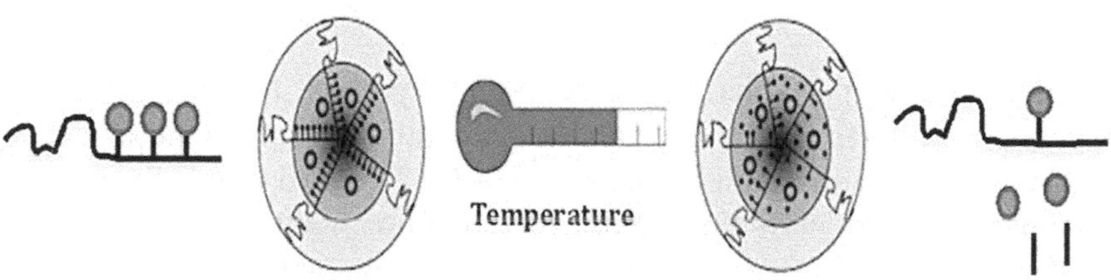


**FIGURE 3.5** Schematic representation of temperature-responsive polymeric nanocarriers.

### 3.2.6 MAGNETIC FIELD-RESPONSIVE NANOCARRIERS

The design of magnetically responsive nanocarriers in most cases involves the use of paramagnetic or superparamagnetic nanoparticles within a polymeric carrier (Fig. 3.6). Of these, superparamagnetic iron oxide nanoparticles (SPIONs) comprise magnetite ($Fe_3O_4$) or maghemite ($Fe_2O_3$) and are sized between 1 and 100 nm (Yigit et al., 2012). Because of their low toxicity and their high relaxation times, they have attracted a lot of attention as magnetic resonance imaging (MRI) contrast agents, particularly as an alternative to gadolinium-based agents. Lately, however, their possible applications in drug delivery have been investigated and a lot of research has been directed toward them (Pankhurst et al., 2003).

Drug-loaded SPIONs have been investigated for magnetically targeted, image-guided drug delivery in which they can be directed to the areas of interest by an external magnetic field, while allowing simultaneous tracking of their in vivo fate by means of MRI (Neuberger et al., 2005). In addition, SPIONs have been shown to produce elevated temperatures when exposed to an alternating magnetic field, making them highly useful for tumor destruction by localized hyperthermia (Kumar and Mohammad, 2011).

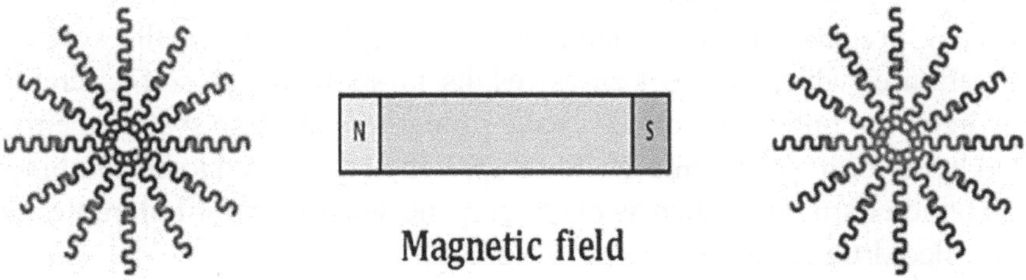


**FIGURE 3.6** Schematic representation of magnetic field-responsive nanocarriers.

This last property has been recently expolited for the design of magnetic field-responsive nanocarriers, in combination with thermosensitive polymers. As an example, Baeza et al. (2012) coated mesoporous silica nanoparticles with pNIPAAm and loaded them with a fluorescent dye and SPIONs. They observed that application of an alternating magnetic field caused release of the dye from the particles, which could be attributed to the collapse of the pNIPAAm coating because of the heating effect of the SPIONs (Baeza et al., 2012). Using this approach, the particles can act as dual therapeutic agents: they cause drug release through the nanocarrier dissociation, combined with killing of cancer cells by hyperthermia. In a similar approach, Hoare et al. (2009) prepared nanocomposite membranes composed of pNIPAAm nanogels and SPIONs loaded with a fluorescent dye; they observed pulsed release of the dye upon the application of magnetic cycles.

### *3.2.7 ULTRASOUND-RESPONSIVE NANOCARRIERS*

Ultrasound is another noninvasive stimulus that may be employed to trigger drug release from nanocarriers (Fig. 3.7). Even though it is very promising, this application has not been extensively exploited to date. Exposure of a part of the body to ultrasound at specific intensities and frequencies results in hyperthermia. Additionally, the formation of high- and low-pressure waves generate oscillating bubbles that eventually collapse, resulting in the disruption of the polymer assemblies around them (Husseini and Pitt, 2009; Marmottant and Hilgenfeldt, 2003; Roy et al., 2010).

In addition, there are reports suggesting that, through the application of ultrasound, the degradation rates of biodegradable polymers can be increased (Kost et al., 1988). As an example, pluronic block copolymer micelles have been tested in combination with ultrasound, and an increased drug release rate was observed (Munshi et al., 1997). Interestingly, ultrasound has also been shown to increase cellular uptake and nucleus internalization of DOX-loaded pluronic micelles (Marin et al., 2001). In other recent work, Du et al. (2011) prepared DOX-loaded PLGAmPEG nanobubbles, which demonstrated drug release only upon application of ultrasound. In addition, when administered in vivo, 85% more tumor inhibitory effect was observed in the group treated with the nanoparticles combined with ultrasound, as compared to the nanoparticles without ultrasound (Du et al., 2011).

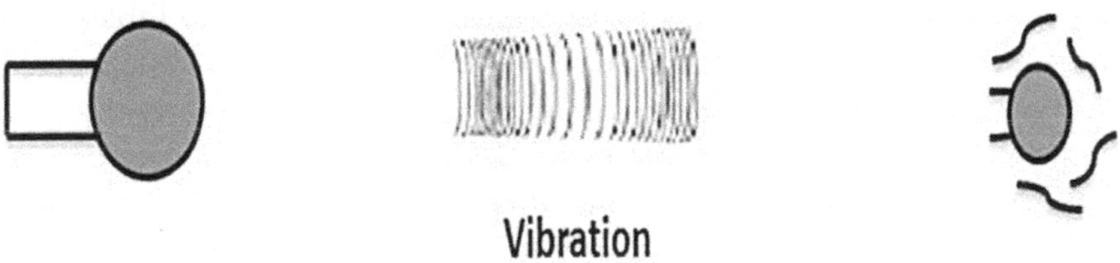


**FIGURE 3.7** Schematic representation of ultrasound-responsive nanocarriers.

### *3.2.8 LIGHT-RESPONSIVE NANOCARRIERS*

Light-responsive DDS are synthesized by incorporating a moiety that is photo-responsive and, upon irradiation, capable of releasing the payload (Fomina et al., 2010; Katz and Burdick, 2010) (Fig. 3.8). Various polymeric systems with such photo-responsiveness have been developed recently (Katz and Burdick, 2010; Zhao, 2012). The advantages of such a stimulus are its noninvasive nature and the potential for highly precise spatial and temporal application (Fomina et al., 2010). However, the patient is required to stay in the dark for some time upon administration to avoid premature release. The most common photosensitive molecules used are azobenzene (Kumar and Neckers, 1989) and *o*-nitrobenzyl (Zhao, 2012) groups. Photosensitive moieties containing copolymers have been used to prepare polymeric micelles and vesicles that enable release upon illumination, owing to changes in polarity (Cabane et al., 2010, 2011; Tong et al., 2005; Wang et al., 2004).

**FIGURE 3.8** Schematic representation of light-responsive nanocarriers.

## 3.3 FUNCTIONALIZATION OF POLYMERIC NANOCARRIERS FOR STIMULI RESPONSIVENESS

Several researchers have studied functionalization of polymeric nanoparticles; some classical examples of which have been reviewed in Table 3.1. A report shows that several drug loaded polymeric nanocarrier systems have received regulatory approval for conducting clinical trials where many of them are being evaluated for safety and efficacy in the marketplace (Zhang et al., 2008). The methods of functionalization for SRN-based therapeutics have to be considered very cautiously so as to improve the desired physiochemical properties without affecting the efficacy or safety profiles of the active ingredients (Huang and Han, 2006; Mammeri et al., 2005; Sandi et al., 2005).

## 3.4 CONCLUSION

Stimuli-responsive, smart nanocarriers are a very important emerging area of drug delivery. Those most commonly studied involve endogenous stimuli (pH, enzymes, and redox potential), which take advantage of specific conditions that exist intracellularly or in the microenvironment of the disease site, enabling selective drug release. Of these, types of carriers offer the most promise in terms of versatility, specificity, and tunability for therapeutic purposes. The responsiveness of carriers can be further enhanced by functionalizing them with the appropriate ligand systems.

**TABLE 3.1** Various Examples of Classical Polymeric Nanocarriers with Diverse Functionalization and Clinical Applications.

| **Nanoparticles** | **Functionalization** | **Drug** | **Applications** | **References** |
|---|---|---|---|---|
| PLLA-*b*-PEG | Folate targeted | Doxorubicin | Solid tumors | Lee et al. (2003) |
| PEG-PE | Lipid conjugated | Paclitaxel | Several cancers | Wang et al. (2005) |
| PEG | Lipid conjugated | Tamoxifen | Lung cancer | Gao et al. (2002) |
| PCL-*b*-trimethylene carbonate-PEG | Serum protein bound | Ellipticin | Anticancer | Liu et al. (2005) |
| PEG | Albumin conjugated | Doxorubicin | Several cancers | Wosikowski et al. (2003) |
| PLGA | Alendronate | Estrogen | Bone-osteoporosis | Choi and Kim (2007) |
| Poly(DEAP-lys)-*b*-PEG-*b*-PLLA | Poly(lysine) | Doxorubicin | pH sensitive tumor | Oh et al. (2009) |
| PLLA-PEG | Biotin | Anticancer | Cancer therapy | Patil et al. (2009) |
| PLA | Aptamer | Anticancer | Prostate cancer | Farokhzad et al. (2004) |
| mPEG/PLGA | Peptidomimetics | Anticancer | Brain cancer | Olivier (2005) |
| PLA | Galactose | Retinoic acid | Hepatocytes | Cho et al. (2001) |
| PLGA | Lectin | Rifampicin | Antitubercular | Sharma et al. (2004) |
| Solid lipid nanoparticles | Compritol | Rifampicin | Antitubercular | Singh et al. (2013) |

Overall, the "smartness" of a stimuli-responsive carrier can be specifically tailored and modulated as per the therapeutic application and the needs for in vivo safety and efficacy.

## 3.5 FUTURE PERSPECTIVES

Other than the endogenous stimuli, there is an growing interest in stimuli that are applied externally. In most cases, these stimuli are applied using techniques that are already employed in clinics for other purposes, such as magnetic field and ultrasound. Even though this approach is very promising, the main challenge is to coordinate drug accumulation with exogenous stimuli application with the input of adequate energy to trigger drug release. For this reason, this approach is commonly combined with imaging techniques to monitor in vivo biodistribution of such stimuli-responsive delivery systems. Most of the work published so far for exogenous stimuli is still at the proof-of-concept level, and the in vivo efficacy and safety of these nanocarriers still has to be assessed. However, the existing literature strongly suggests that these types of carriers have great potential as clinical candidates and will be the therapeutics of the future.

## DISCLOSURES

The authors declare no known conflicts of interest or association with any relevant entities associated with any technology referenced herein at the time of the writing of this chapter.

## ACKNOWLEDGMENTS

The authors are thankful to the Translational Health Science & Technology Institute (an autonomous research institute funded by the Department of Biotechnology, Ministry of Science and Technology, Government of India), for providing the necessary support and also highly thankful to Shoolini University, Solan, Himachal Pradesh, for providing the access of necessary literature resources and essential library facilities (Yogananda Knowledge Centre) for writing this chapter. Acknowledgment also goes to all the authors of all published sources referenced in this chapter.

## KEYWORDS


- **smart polymer**
- **nanocarriers**
- **stimuli responsive**
- **drug delivery**


## REFERENCES


Abdullah Al, N.; Lee, H.; Lee, Y. S.; Lee, K. D; Park, S. Y. Development of Disulfide Core-Crosslinked Pluronic Nanoparticles as an Effective Anticancer Drug Delivery System. *Macromol. Biosci.* **2011,** *11*, 1264–1271.

Adamo, G.; Grimaldi, N.; Campora, S.; Sabatino, M. A.; Dispenza, C.; Ghersi, G. Glutathione-Sensitive Nanogels for Drug Release. *Chem. Eng. Trans.* **2014,** *38*, 457–462.

Ahmed, F.; Pakunlu, R. I.; Srinivas, G.; Brannan, A.; Bates, F.; Klein, M. L.; Minko, T.; Discher, D. E. Shrinkage of a Rapidly Growing Tumor by Drug Loaded Polymersomes: pH-Triggered Release Through Copolymer Degradation. *Mol. Pharm.* **2006,** *3*, 340–350.

Aimetti, A. A.; Machen, A. J.; Anseth, K. S. Poly(Ethylene Glycol) Hydrogels Formed by Thiol–Ene Photopolymerization for Enzyme-Responsive Protein Delivery. *Biomaterials* **2009,** *30*, 6048–6054.

Alexander, C. Stimuli Responsive Polymers for Biomedical Applications. *Chem. Soc. Rev.* **2005,** *34*, 276–285.

Allemann, E.; Gurny, R.; Doelker, E. Drug-Loaded Nanoparticles: Preparation Methods and Drug Targeting Issues. *Eur. J. Pharm. Biopharm.* **1993,** *39*, 173–191.

Andresen, T. L.; Jensen, S. S; Jorgensen, K. Advanced Strategies in Liposomal Cancer Therapy: Problems and Prospects of Active and Tumor Specific Drug Release. *Progr. Lipid Res.* **2005,** *44*, 68–97.

Aoshima, S; Kanaoka, S. Synthesis of Stimuli-Responsive Polymers by Living Polymerization: Poly(*N*-Isopropylacrylamide) and Poly(Vinyl Ether)s. *Adv. Polym. Sci.* **2008,** *210*, 169–208.

Baeza, A.; Guisasola, E.; Ruiz-Hernandez, E.; Vallet-Regi, M. Magnetically Triggered Multidrug Release by Hybrid Mesoporous Silica Nanoparticles. *Chem. Mater.* **2012,** *24*, 517–524.

Batrakova, E. V.; Kabanov, A. V. Pluronic Block Copolymers: Evolution of Drug Delivery Concept from Inert Nanocarriers to Biological Response Modifiers. *J. Control. Release* **2008,** *130*, 98–106.

Bhatt, R. L.; de Vries, P.; Tulinsky, J.; Bellamy, G.; Baker, B.; Singer, J. W.; Klein, P. Synthesis and In Vivo Antitumor Activity of Poly(L-Glutamic Acid) Conjugates of 20(*S*)-Camptothecin. *J. Med. Chem.* **2003,** *46*, 190–193.

Brigger, I.; Dubernet, C.; Couvreur, P. Nanoparticles in Cancer Therapy and Diagnosis. *Adv. Drug Deliv. Rev.* **2002,** *54*, 631–651.

Cabane, E.; Malinova, V.; Meier, W. Synthesis of Photocleavable Amphiphilic Block Copolymers: Toward the Design of Photosensitive Nanocarriers. *Macromol. Chem. Phys.* **2010,** *211*, 1847–1856.

Cabane, E.; Malinova, V.; Menon, S.; Palivan, C. G.; Meier, W. Photoresponsive Polymersomes as Smart, Triggerable Nanocarriers. *Soft Matter* **2011,** *7*, 9167–9176.

Chilkoti, A.; Dreher, M. R.; Meyer, D. E.; Raucher, D. Targeted Drug Delivery by Thermally Responsive Polymers. *Adv. Drug Deliv. Rev.* **2002,** *54*, 613–630.

Cho, C. S.; Cho, K. Y.; Park, I. K.; Kim, S. H.; Sasagawa, T.; Uchiyama, M.; Akaike, T. Receptor-Mediated Delivery of All Trans-Retinoic Acid to Hepatocyte Using Poly(L-Lactic Acid) Nanoparticles Coated with Galactose-Carrying Polystyrene. *J. Control. Release* **2001,** *77*, 7–15.

Choi, S. W.; Kim, J. H. Design of Surface Modified Poly(D,L-Lactide-*co*-Glycolide) Nanoparticles for Targeted Drug Delivery to Bone. *J Control. Release* **2007,** *122*, 24–30.

de la Rica, R.; Aili, D.; Stevens, M. M. Enzyme-Responsive Nanoparticles for Drug Release and Diagnostics. *Adv. Drug Deliv. Rev.* **2012,** *64*, 967–978.

de Vries, P.; Bhatt, R.; Stone, I.; Klein, P.; Singer, J. Optimization of CT-2106: A Water-Soluble Poly-L-Glutanlic Acid (PG)–Camptothecin Conjugate with Enhanced In Vivo Anti-tumor Efficacy. *Clin. Cancer Res.* **2001,** *7*, 3673S.

de Vries, P.; Bhatt, R.; Tulinsky, J.; Heasley, E.; Stone, I.; Klein, P.; Li, C.; Wallace, S.; Lewis, R.; Singer, J. Water-Soluble Poly-(L)-Glutamic Acid (PG)-Camptothecin (CPT) Conjugates Enhance CTP Stability and Efficacy In Vivo. *Clin. Cancer Res.* **2000,** *6*, 4511S.

Devalapally, H.; Shenoy, D.; Little, S.; Langer, R.; Amiji, M. Poly(Ethylene Oxide)-Modified Poly(Beta-Amino Ester) Nanoparticles as a pH-Sensitive System for Tumor-Targeted Delivery of Hydrophobic Drugs: Part 3. Therapeutic Efficacy and Safety Studies in Ovarian Cancer Xenograft Model. *Cancer Chemother. Pharmacol.* **2007,** *59*, 477–484.

Dipetrillo, T.; Milas, L.; Evans, D.; Akerman, P.; Ng, T.; Miner, T.; Cruff, D.; Chauhan, B.; Iannitti, D.; Harrington, D.; Safran, H. Paclitaxel Poliglumex (PPX-Xyotax) and Concurrent Radiation for Esophageal and Gastric Cancer: A Phase I Study. *Am. J. Clin. Oncol: Cancer Clin. Trial* **2006,** *29*, 376–379.

Du, L.; Jin, Y.; Zhou, W.; Zhao, J. Ultrasound-Triggered Drug Release and Enhanced Anticancer Effect of Doxorubicin-Loaded Poly(D,L-Lactide-*co*-Glycolide)-Methoxy-Poly(Ethylene Glycol) Nanodroplets. *Ultrasound Med. Biol.* **2011,** *37*, 1252–1258.

Erdmann, L.; Uhrich, K. E. Synthesis and Degradation Characteristics of Salicylic Acid-Derived Poly(Anhydride-Esters). *Biomaterials* **2000,** *21*, 1941–1946.

Farokhzad, O. C.; Jon, S.; Khademhosseini, A.; Tran, T. T.; LaVan, A.; Langer, R. Nanoparticle-Aptamer Bioconjugates: A New Approach for Targeting Prostate Cancer Cells. *Cancer Res.* **2004,** *64*, 7668–7672.

Ferruti, P.; Marchisio, M. A.; Duncan, R. Poly(Amido-Amine)s: Biomedical Applications. *Macromol. Rapid. Commun.* **2002,** *23*, 332–355.

Fleige, E.; Quadir, M. A.; Haag, R. Stimuli-Responsive Polymeric Nanocarriers for the Controlled Transport of Active Compounds: Concepts and Applications. *Adv. Drug Deliv. Rev.* **2012,** *64*, 866–884.

Fomina, N.; Mcfearin, C.; Sermsakdi, M.; Edigin, O.; Almutairi, A. UV and Near-IR Triggered Release from Polymeric Nanoparticles. *J. Am. Chem. Soc.* **2010,** *132*, 9540–9542.

Gao, Z.; Lukyanov, A.; Singhal, A.; Torchilin, V. Diacyllipid-Polymer Micelles as Nanocarriers for Poorly Soluble Anticancer Drugs. *Nano Lett.* **2002,** *2*, 979–982.

Gerweck, L. E.; Seetharaman, K. Cellular pH Gradient in Tumor versus Normal Tissue: Potential Exploitation for the Treatment of Cancer. *Cancer Res.* **1996,** *56*, 1194–1198.

Hardwicke, J.; Moseley, R.; Stephens, P.; Harding, K.; Duncan, R.; Thomas, D. W. Bioresponsive Dextrin-rhEGF Conjugates: In Vitro Evaluation in Models Relevant to Its Proposed Use as a Treatment for Chronic Wounds. *Mol. Pharm.* **2010,** *7*, 699–707.

Heller, J.; Penhale, D. W. H.; Helwing, R. F. Preparation of Polyacetals by the Reaction of Divinyl Ethers and Polyols. *J. Polym. Sci., C: Polym. Lett.* **1980,** *18*, 5–11.

Hoare, T.; Santamaria, J.; Goya, G. F.; Irusta, S.; Lin, D.; Lau, S.; Padera, R.; Langer, R.; Kohane, D. S. A Magnetically Triggered Composite Membrane for On-demand Drug Delivery. *Nano Lett.* **2009,** *9*, 3651–3657.

Hu, J.; Zhang, G.; Liu, S. Enzyme-Responsive Polymeric Assemblies, Nanoparticles and Hydrogels. *Chem. Soc. Rev.* **2012,** *41*, 5933–5949.

Huang, W. Y.; Han, C. D. Dispersion Characteristics and Rheology of Organoclay Nanocomposites Based on a Segmented Main-Chain Liquid-Crystalline Polymer Having Pendent Pyridyl Group. *Macromolecules* **2006,** *39*, 257–267.

Husseini, G. A.; Pitt, W. G. Ultrasonic-Activated Micellar Drug Delivery for Cancer Treatment. *J. Pharm. Sci.* **2009,** *98*, 795–811.

Jung-Kwon, O. Polylactide (PLA)-Based Amphiphilic Block Copolymers: Synthesis, Self-Assembly, and Biomedical Applications. *Soft Matter* **2011,** *7*, 5096–5108.

Kabanov, A. V.; Batrakova, E. V.; Alakhov, V. Y. Pluronic R Block Copolymers as Novel Polymer Therapeutics for Drug and Gene Delivery. *J. Control. Release* **2002,** *82*, 189–212.

Katz, J. S.; Burdick, J. A. Light-Responsive Biomaterials: Development and Applications. *Macromol. Biosci.* **2010,** *10*, 339–348.

Kawashima, Y. Nanoparticulate Systems for Improved Drug Delivery. *Adv. Drug Deliv. Rev.* **2001,** *47*, 1–2.

Kim, D.; Lee, E. S.; Oh, K. T.; Gao, Z. G.; Bae, Y. H. Doxorubicin-Loaded Polymeric Micelle Overcomes Multidrug Resistance of Cancer by Double-Targeting Folate Receptor and Early Endosomal pH. *Small* **2008,** *4*, 2043–2050.

Kost, J.; Leong, K.; Langer, R. Ultrasonically Controlled Polymeric Drug Delivery. *Makromol. Chem. Macromol. Symp.* **1988,** *19*, 275–285.

Ku, T. H.; Chien, M. P.; Thompson, M. P.; Sinkovits, R. S.; Olson, N. H.; Baker, T. S.; Gianneschi, N. C. Controlling and Switching the Morphology of Micellar Nanoparticles with Enzymes. *J. Am. Chem. Soc.* **2011,** *133*, 8392–8395.

Kumar, C. S. S. R.; Mohammad, F. Magnetic Nanomaterials for Hyperthermia-Based Therapy and Controlled Drug Delivery. *Adv. Drug Deliv. Rev.* **2011,** *63*, 789–808.

Kumar, G. S.; Neckers, D. C. Photochemistry of Azobenzene-Containing Polymers. *Chem. Rev.* **1989,** *89*, 1915–1925.

Kyriakides, T. R.; Cheung, C. Y.; Murthy, N.; Bornstein, P.; Stayton, P. S.; Hoffman, A. S. pH-Sensitive Polymers That Enhance Intracellular Drug Delivery In Vivo. *J. Control. Release* **2002,** *78*, 295–303.

Lammers, T.; Kiessling, F.; Hennink, W. E.; Storm, G. Drug Targeting to Tumors: Principles, Pitfalls and (Pre-)Clinical Progress. *J. Control. Release* **2012,** *161*, 175–187.

Langer, R.; Tirrell, D. A. Designing Materials for Biology and Medicine. *Nature* **2004,** *428*, 487–492.

Lee, E. S.; Gao, Z.; Kim, D.; Park, K.; Kwon, I. C.; Bae, Y. H. Super pH-Sensitive Multifunctional Polymeric Micelle for Tumor pHe Specific TAT Exposure and Multidrug Resistance. *J. Control. Release* **2008,** *129*, 228–236.

Lee, E. S.; Na, K.; Bae, Y. H. Polymeric Micelle for Tumor pH and Folate-Mediated Targeting. *J. Control. Release* **2003,** *91*, 103–113.

Lehner, R.; Wang, X.; Wolf, M.; Hunziker, P. Designing Switchable Nanosystems for Medical Application. *J. Control. Release* **2012,** *161*, 307–316.

Liu, J.; Zeng, F.; Allen, C. Influence of Serum Protein on Polycarbonate-Based Copolymer Micelles as a Delivery System for a Hydrophobic Anti-cancer Agent. *J. Control. Release* **2005,** *103*, 481–497.

Liu, R.; Fraylich, M.; Saunders, B. R. Thermoresponsive Copolymers: From Fundamental Studies to Applications. *Colloid Polym. Sci.* **2009,** *287*, 627–643.

Mammeri, F.; Le Bourhis, E.; Rozes, L.; Sanchez, C. Mechanical Properties of Hybrid Organic–Inorganic Materials. *J. Mater. Chem.* **2005,** *15*, 3787–3811.

Manchun, S.; Dass, C. R.; Sriamornsak, P. Targeted Therapy for Cancer Using pH-Responsive Nanocarrier Systems. *Life Sci.* **2012,** *90*, 381–387.

Mao, J.; Gan, Z. The influence of Pendant Hydroxyl Groups on Enzymatic Degradation and Drug Delivery of Amphiphilic Poly[Glycidol-Block-(Epsiloncaprolactone)] Copolymers. *Macromol. Biosci.* **2009,** *9*, 1080–1089.

Marin, A.; Muniruzzaman, M.; Rapoport, N. Mechanism of the Ultrasonic Activation of Micellar Drug Delivery. *J. Control. Release* **2001,** *75*, 69–81.

Marmottant, P.; Hilgenfeldt, S. Controlled Vesicle Deformation and Lysis by Single Oscillating Bubbles. *Nature* **2003,** *423*, 153–156.

Medintz, I. L.; Uyeda, H. T.; Goldman, E. R.; Mattoussi, H. Quantum Dot Bioconjugates for Imaging, Labelling and Sensing. *Nat. Mater.* **2005,** *4*, 435–446.

Mellman, I. Endocytosis and Molecular Sorting. *Ann. Rev. Cell Dev. Biol.* **1996,** *12*, 575–625.

Meng, F.; Hennink, W. E.; Zhong, Z. Reduction-Sensitive Polymers and Bioconjugates for Biomedical Applications. *Biomaterials* **2009,** *30*, 2180–2198.

Minelli, C.; Lowe, S. B.; Stevens, M. M. Engineering Nanocomposite Materials for Cancer Therapy. *Small* **2010,** *6*, 2336–2357.

Mort, J. S.; Buttle, D. J. Cathepsin, B. *Int. J. Biochem. Cell Biol.* **1997,** *29*, 715–720.

Mrkvan, T.; Sirova, M.; Etrych, T.; Chytil, P.; Strohalm, J.; Plocova, D.; Ulbrich, K.; Rihova, B. Chemotherapy Based on HPMA Copolymer Conjugates with pH-Controlled Release of Doxorubicin Triggers Anti-tumor Immunity. *J. Control. Release* **2005,** *110*, 119–129.

Munshi, N.; Rapoport, N.; Pitt, W. G. Ultrasonic Activated Drug Delivery from Pluronic P-105 Micelles. *Cancer Lett.* **1997,** *118*, 13–19.

Neuberger, T.; Schopf, B.; Hofmann, H.; Hofmann, M.; Von Rechenberg, B. Superparamagnetic Nanoparticles for Biomedical Applications: Possibilities and Limitations of a New Drug Delivery System. *J. Magn. Magn. Mater.* **2005,** *293*, 483–496.

Oh, K. T.; Oh, Y. T.; Oh, N. M.; Kim, K.; Lee, D. H.; Lee, E. S. A Smart Flower-Like Polymeric Micelle for pH-Triggered Anticancer Drug Release. *Int. J. Pharm.* **2009,** *375*, 163–169.

Olivier, J. C. Drug Transport to Brain with Targeted Nanoparticles. *NeuroRx* **2005,** *2*, 108–119.

Pankhurst, Q. A.; Connolly, J.; Jones, S. K.; Dobson, J. Applications of Magnetic Nanoparticles in Biomedicine. *J. Phys. D: Appl. Phys.* **2003,** *36*, R167–R181.

Panyam, J.; Labhasetwar, V. Biodegradable Nanoparticle from Drug and Gene Delivery to Cells and Tissue. *Adv. Drug Deliv. Rev.* **2003,** *55*, 329–347.

Patil, Y. B.; Toti, U. S.; Khadir, A.; Ma, L.; Panyam, J. Single-Step Surface Functionalization of Polymeric Nanoparticles for Targeted Drug Delivery. *Biomaterials* **2009,** *30*, 859–866.

Richardson, S. C. W.; Pattrick, N. G.; Lavignac, N.; Ferruti, P.; Duncan, R. Intracellular Fate of Bioresponsive Poly(Amidoamine)s *In Vitro* and *In Vivo*. *J. Control. Release* **2010,** *142*, 78–88.

Rijcken, C. J.; Snel, C. J.; Schiffelers, R. M.; van Nostrum, C. F.; Hennink, W. E. Hydrolysable Core-Crosslinked Thermosensitive Polymeric Micelles: Synthesis, Characterisation and *In Vivo* Studies. *Biomaterials* **2007,** *28*, 5581–5593.

Rijcken, C. J. F.; Talelli, M.; van Nostrum, C. F.; Storm, G.; Hennink, W. E. Crosslinked Micelles with Transiently Linked Drugs—A Versatile Drug Delivery System. *Eur. J. Nanomed.* **2010,** *3*, 19–24.

Roy, D.; Cambre, J. N.; Sumerlin, B. S. Future Perspectives and Recent Advances in Stimuli-Responsive Materials. *Prog. Polym. Sci.* **2010,** *35*, 278–301.

Sandi, G.; Kizilel, R.; Carrado, K. A.; Fernandez-saavedra, R.; Castagnola, N. Effect of the Silica Precursor on the Conductivity of Hectorite-Derived Polymer Nanocomposites. *Electrochim. Acta* **2005,** *50*, 3891–3896.

Santamaria, B.; Benito-Martin, A.; Conrado Ucero, A.; Stark Aroeira, L.; Reyero, A.; Jesus Vicent, M.; Orzaez, M.; Celdran, A.; Esteban, J.; Selgas, R.; Ruiz-Ortega, M.; Lopez Cabrera, M.; Egido, J.; Perez-Paya, E.; Ortiz, A. A Nanoconjugate Apaf-1 Inhibitor Protects Mesothelial Cells from Cytokine Induced Injury. *PLoS ONE* **2009,** *4*, e6634.

Schmaljohann, D. Thermo- and pH-Responsive Polymers in Drug Delivery. *Adv. Drug Deliv. Rev.* **2006,** *58*, 1655–1670.

Sharma, A.; Sharma, S.; Khuller, G. K. Lectin-Functionalized Poly(Lactide-*co*-Glycolide) Nanoparticles as Oral/Aerosolized Antitubercular Drug Carriers for Treatment of Tuberculosis. *J. Antimicrob. Chemother.* **2004,** *54*, 761–766.

Shaunak, S.; Thomas, S.; Gianasi, E.; Godwin, A.; Jones, E.; Teo, I.; Mireskandari, K.; Luthert, P.; Duncan, R.; Patterson, S.; Khaw, P.; Brocchini, S. Polyvalent Dendrimer Glucosamine Conjugates Prevent Scar Tissue Formation. *Nat. Biotechnol.* **2004,** *22*, 977–984.

Shenoy, D.; Little, S.; Langer, R.; Amiji, M. Poly(Ethylene Oxide)-Modified Poly(Ξ2-Amino Ester) Nanoparticles as a pH-Sensitive System for Tumor-Targeted Delivery of Hydrophobic Drugs: Part 2. In Vivo Distribution and Tumor Localization Studies. *Pharm. Res.* **2005,** *22*, 2107–2114.

Singer, J. W.; Bhatt, R.; Tulinsky, J.; Buhler, K. R.; Heasley, E.; Klein, P.; de Vries, P. Water-Soluble Poly-(L-Glutamic Acid)-Gly-Camptothecin Conjugates Enhance Camptothecin Stability and Efficacy In Vivo. *J. Control. Release* **2001,** *74*, 243–247.

Singer, J. W.; de Vries, P.; Bhatt, R.; Tulinsky, J.; Klein, P.; Li, C.; Milas, L.; Lewis, R. A.; Wallace, S. Conjugation of Camptothecins to Poly(L-Glutamic Acid). *Ann. N.Y. Acad. Sci.* **2000,** *922*, 136–150.

Singh, H.; Bhandari, R.; Kaur, I. P. Encapsulation of Rifampicin in a Solid Lipid Nanoparticulate System to Limit Its Degradation and Interaction with Isoniazid at Acidic pH. *Int. J. Pharm.* **2013,** *446*, 106–111.

Soga, O.; van Nostrum, C. F.; Hennink, W. E. Poly(*N*-(2-Hydroxypropyl) Methacrylamide Mono/Di-Lactate): A New Class of Biodegradable Polymers with Tuneable Thermosensitivity. *Biomacromolecules* **2004a,** *5*, 818–821.

Soga, O.; van Nostrum, C. F.; Hennink, W. E. Thermosensitive and Biodegradable Polymeric Micelles with Transient Stability. *J Control. Release* **2005,** *101*, 383–385.

Soga, O.; van Nostrum, C. F.; Ramzi, A.; Visser, T.; Soulimani, F.; Frederik, P. M.; Bomans, P.; H. H.; Hennink, W. E. Physicochemical Characterization of Degradable Thermosensitive Polymeric Micelles. *Langmuir* **2004b,** *20*, 9388–9395.

Song, N.; Liu, W.; Tu, Q.; Liu, R.; Zhang, Y.; Wang, J. Preparation and In Vitro Properties of Redox-Responsive Polymeric Nanoparticles for Paclitaxel Delivery. *Colloid Surf. B: Biointerfaces* **2011,** *87*, 454–463.

Soppimath, K. S.; Aminabhavi, T. M.; Kulkarni, A. R.; Rudzinski, W. E. Biodegradable Polymeric Nanoparticles as Drug Delivery Devices. *J. Control. Release* **2001,** *70*, 1–20.

Stevens, M. M.; Flynn, N. T.; Wang, C.; Tirrell, D. A.; Langer, R. Coiled-Coil Peptide-Based Assembly of Gold Nanoparticles. *Adv. Mater.* **2004,** *16*, 915–918.

Sun, P.; Zhou, D.; Gan, Z. Novel Reduction-Sensitive Micelles for Triggered Intracellular Drug Release. *J. Control. Release* **2011,** *155*, 96–103.

Talelli, M.; Hennink, W. E. Thermosensitive Polymeric Micelles for Targeted Drug Delivery. *Nanomedicine* **2011,** *6*, 1245–1255.

Talelli, M.; Iman, M.; Varkouhi, A. K.; Rijcken, C. J.; Schiffelers, R. M.; Etrych, T.; Ulbrich, K.; van Nostrum, C. F.; Lammers, T.; Storm, G.; Hennink, W. E. Core-Crosslinked Polymeric Micelles with Controlled Release of Covalently Entrapped Doxorubicin. *Biomaterials* **2010,** *31*, 7797–7804.

Tomlinson, R.; Heller, J.; Brocchini, S.; Duncan, R. Polyacetal-Doxorubicin Conjugates Designed for pH-Dependent Degradation. *Bioconj. Chem.* **2003,** *14*, 1096–1106.

Tomlinson, R.; Klee, M.; Garrett, S.; Heller, J.; Duncan, R.; Brocchini, S. Pendent Chain Functionalized Polyacetals that Display pH-Dependent Degradation: A Platform for the Development of Novel Polymer Therapeutics. *Macromolecules* **2002,** *35*, 473–480.

Tong, X.; Wang, G.; Soldera, A.; Zhao, Y. How Can Azobenzene Block Copolymer Vesicles Be Dissociated and Reformed by Light? *J. Phys. Chem. B* **2005,** *109*, 20281–20287.

Torchilin, V. P. Drug Targeting. *Eur. J. Pharm. Sci.* **2000,** *11* (Suppl. 2), S81–S91.

Valle, J. W.; Armstrong, A.; Newman, C.; Alakhov, V.; Pietrzynski, G.; Brewer, J.; Campbell, S.; Corrie, P.; Rowinsky, E. K.; Ranson, M. A Phase 2 Study of SP1049c, Doxorubicin in P-Glycoprotein-Targeting Pluronics, in Patients with Advanced Adeno-Carcinoma of the Esophagus and Gastroesophageal Junction. *Invest. New Drug* **2011,** *29*, 1029–1037.

Wang, G.; Tong, X.; Zhao, Y. Preparation of Azobenzene-Containing Amphiphilic Diblock Copolymers for Light-Responsive Micellar Aggregates. *Macromolecules* **2004,** *37*, 8911–8917.

Wang, J.; Mongayt, D.; Torchilin, V. P. Polymeric Micelles for Delivery of Poorly Soluble Drugs: Preparation and Anticancer Activity In Vitro of Paclitaxel Incorporated into Mixed Micelles Based on Poly(Ethylene Glycol)–Lipid Conjugate and Positively Charged Lipids. *J. Drug Target.* **2005,** *13*, 73–80.

Wang, Y. C.; Wang, F.; Sun, T. M.; Wang, J. Redox-Responsive Nanoparticles from the Single Disulfi Debond-Bridged Block Copolymer as Drug Carriers for Overcoming Multidrug Resistance in Cancer Cells. *Bioconj. Chem.* **2011,** *22*, 1939–1945.

Wei, H.; Cheng, S. X.; Zhang, X. Z.; Zhuo, R. X. Thermo-Sensitive Polymeric Micelles Based on Poly(*N*-Isopropylacrylamide) as Drug Carriers. *Prog. Polym. Sci. (Oxford)* **2009,** *34*, 893–910.

Wosikowski, K.; Biedermann, E.; Rattel, B.; Breiter, N.; Jank, P.; Loser, R.; Jansen, G.; Peters, G. J. *In Vitro* and *In Vivo* Antitumor Activity of Methotrexate Conjugated to Human Serum Albumin in Human Cancer Cells. *Clin. Cancer Res.* **2003,** *9*, 1917–1926.

Wu, G.; Fang, Y. Z.; Yang, S.; Lupton, J. R.; Turner, N. D. Glutathione Metabolism and Its Implications for Health. *J. Nutr.* **2004,** *134*, 489–492.

Yigit, M. V.; Moore, A.; Medarova, Z. Magnetic Nanoparticles for Cancer Diagnosis and Therapy. *Pharm Res.* **2012,** *29*, 1180–1188.

Zelzer, M.; Ulijn, R. V. Next-Generation Peptide Nanomaterials: Molecular Networks, Interfaces and Supramolecular Functionality. *Chem. Soc. Rev.* **2010,** *39*, 3351–3357.

Zhang, L.; Gu, F. X.; Chan, J. M.; Wang, A. Z.; Langer, R. S.; Farokhzad, O. C. Nanoparticles in Medicine: Therapeutic Applications and Developments. *Clin. Pharmacol. Ther.* **2008,** *83*, 761–769.

Zhao, M.; Biswas, A.; Hu, B.; Joo, K. I.; Wang, P.; Gu, Z.; Tang, Y. Redox Responsive Nanocapsules for Intracellular Protein Delivery. *Biomaterials* **2011,** *32*, 5223–5230.

Zhao, Y. Light-Responsive Block Copolymer Micelles. *Macromolecules* **2012,** *45*, 3647–3657.

CHAPTER 4

# GOLD NANOCONJUGATES FOR SMART DRUG DELIVERY AND TARGETING

AYUOB AGHANEJAD[1†], PARHAM SAHANDI ZANGABAD[1,2†], JALEH BARAR[1,3], and YADOLLAH OMIDI[1,3*]

[1]*Research Center for Pharmaceutical Nanotechnology, Tabriz University of Medical Science, Tabriz, Iran*

[2]*Department of Materials Science and Engineering, Sharif University of Technology, Tehran, Iran*

[3]*Department of Pharmaceutics, Faculty of Pharmacy, Tabriz University of Medical Science, Tabriz, Iran*

*Corresponding author. E-mail: yomidi@tbzmed.ac.ir*

## CONTENTS



†These authors contributed equally to this work.

## ABSTRACT

In the recent decades, advanced nanomaterials have been used for the development of smart multifunctional drug-delivery systems (DDSs) for simultaneous diagnosis and therapy. Of these nanomaterials, gold nanoparticles (GNPs) have been utilized for the production of intelligent nanomedicines, the so-called diapeutics/theranostics. While GNPs are considered as inert nanomaterials, they show great potential for the surface modification and decoration with stimuli-responsive and targeting agents such as temperature and/or pH sensitive linkers, aptamers or antibodies for site-specific imaging and therapy. On account of unique physicochemical properties (e.g., surface multivalency, shape, size, morphology, etc.) of GNPs, they possess great capability as targeted DDSs for imaging and delivering cargo molecules in different diseases, including various types of solid tumors. Based on their in vitro and in vivo applications, GNPs can be listed among the most promising nanomaterials for development of smart multifunctional stimuli-responsive diapeutics/theranostics.

## 4.1 INTRODUCTION

Accurate diagnosis and effective therapy of formidable diseases such as malignancies are still a challenging task. Various attempts have so far been focused on the passive and active targeting of molecular markers associated with the initiation and/or progression diseases. In the case of ominous diseases, in particular solid tumors, the early diagnosis is the key for successful treatment. Given that the nanoscaled drug-delivery systems (DDSs) provide improved means for the transportation of drugs/genes to the diseased cells, various types of nanomedicines in different forms of organic and inorganic nanoparticles (NPs), grafted hybrid nanoconjugates (NCs), and nanosystems (NSs) have been developed. In the recent decades, with the growing advances in nanotechnology as an interdisciplinary filed, nanoscaled platforms have been moved toward simultaneous targeted imaging and therapy. Such seamless nanomedicines (so-called diapeutics/theranostics) are believed to become game-changing tools for the treatment of intimidating diseases. Notwithstanding, the merits of using DDSs are noticeable in disease microenvironment, wherein they are able to provide on-demand site-specific liberation of cargo drug molecules. These nanoscaled DDSs can indeed improve the physicochemical properties of drugs such as solubility as well as their pharmacokinetics and pharmacodynamics behaviors in the target site (Allen and Cullis, 2004; Barar and Omidi, 2013c). Further, multifunctional nanomedicines can be engineered in a way to be responsive upon certain stimuli. For example, in the case of solid tumors, the cancerous cells are able to form a lenient milieu, so-called tumor microenvironment (TME), wherein the pH is dramatically acidified because of aberrant metabolism of glucose (Barar and Omidi, 2013a). The biological elements of TME can be targeted by multimodal nanomedicines (Omidi and Barar, 2014), and these nanoscaled targeted therapies can be pH-responsive in order to liberate the loaded anticancer drug/gene solely within the acidified TME (Gupta et al., 2016; She et al., 2013; Wilson et al., 2013). In fact, the conventional cancer therapies often encounter with several limitations, including nonspecific toxicity and side effects (Oberoi et al., 2013), as well as emergence of drug-resistance and multidrug resistance (MDR) to chemotherapeutics (Kibria et al., 2014; Wang et al., 2014b). As a result, specific delivery of anticancer agents by targeted NSs can warrant profound impacts of the cytotoxic agents solely on the diseases cells. In some other diseases, wherein a certain biological barrier such as blood–brain barrier (BBB) and blood–ocular barrier (BOB) may hinder the entrance of drug molecules, utilization of multimodal nanomedicines can be clinically beneficial. For example, in the case of neurodegenerative diseases such as stroke, Alzheimer's (AD), and Parkinson's disease (PD), crossing the BBB is crucial for efficient delivery of designated drugs even though the BBB is partially impaired (Omidi and Barar, 2012; Omidi et al., 2016; Saraiva et al., 2016). Ocular barriers can also imped the entrance of locally or systemically administered drugs into the eye, while such biological hindrance can be circumvented using nanoscaled DDSs (Barar et al., 2008, 2016).

In the case of solid tumors, nanoscaled untargeted DDSs provide an opportunity for drug to be accumulated within the TME by a passive targeting via an enhanced permeability and retention (EPR) effect, while the targeted NSs combine the EPR effect with the active targeting possibility imposing maximal impacts specifically on the malignant cells. In the Research Center for Pharmaceutical Nanotechnology, we have developed and examined various types of DDSs and found that multifunctional NSs armed with targeting agents provide substantially improved drug delivery to the designated cells (Azhdarzadeh et al., 2016; Heidari Majd et al., 2013a, 2013b; Matthaiou et al., 2014; Rahmanian et al., 2016; Same et al., 2016). Figure 4.1 depicts the active and passive targeting mechanisms schematically.

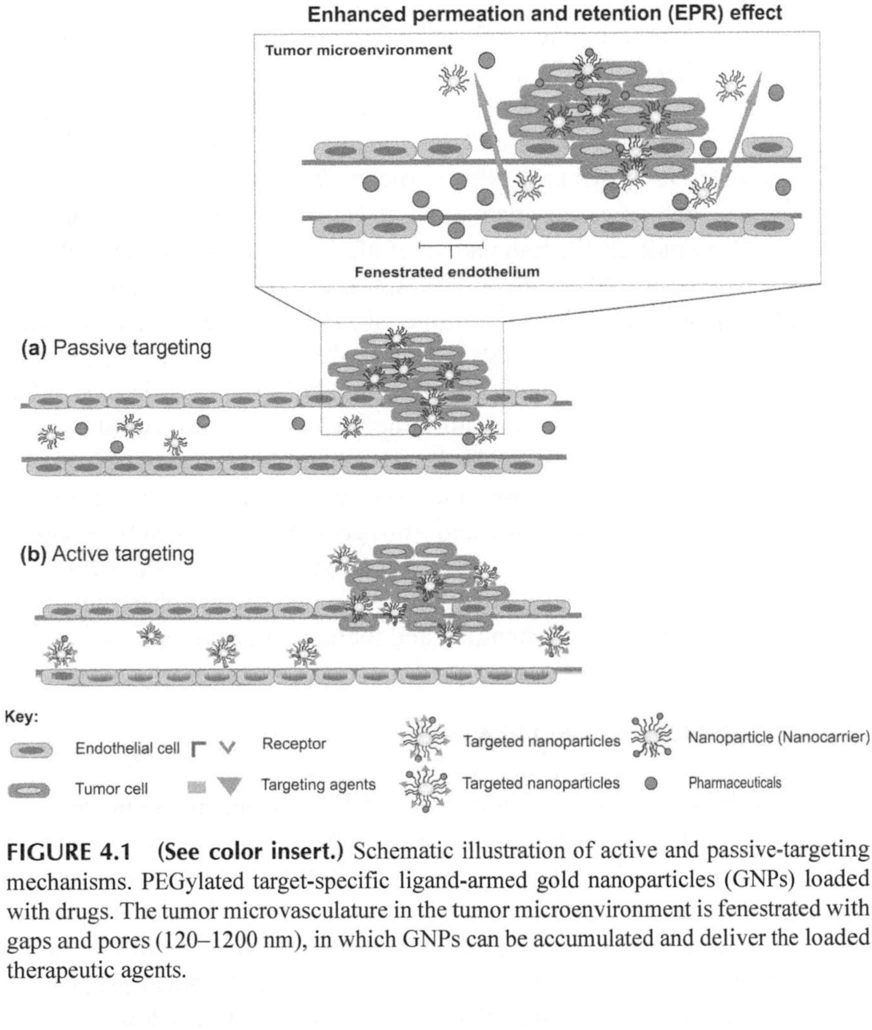


**FIGURE 4.1 (See color insert.)** Schematic illustration of active and passive-targeting mechanisms. PEGylated target-specific ligand-armed gold nanoparticles (GNPs) loaded with drugs. The tumor microvasculature in the tumor microenvironment is fenestrated with gaps and pores (120–1200 nm), in which GNPs can be accumulated and deliver the loaded therapeutic agents.

Nanoscaled DDSs can be categorized into two main inorganic and organic NSs. Depending on the end-point applications and objectives, they can be armed with different entities to meet passive and/or active targeting purposes while their decoration with imaging agents results in the production of diapeutics/theranostics. In the case of NSs with inorganic core, dual imaging can be plausible though the conjugation of a radionuclide or a fluorophore

(Same et al., 2016). No matter what the nature of NSs is, they need to be biologically compatible and show capability of circumventing the immune clearance function (Inturi et al., 2015; Moghimi and Szebeni, 2003). Various inorganic materials have been used as the core in formulation of NSs, in particular gold and iron oxide (Heidari Majd et al., 2013a, 2013b; Same et al., 2016). Further, a wide range of advanced materials has been utilized in production of nanoscaled DDSs such as solid-lipid-based NPs (Dolatabadi and Omidi, 2016), albumin-based NPs (Langiu et al., 2014; Li et al., 2014), stimuli-responsive polymeric NPs (Crucho, 2015; Fleige et al., 2012), nanocarbons such as carbon nanotubes (CNTs) and nanoscaled graphene oxide (GO) (Omidi, 2011; Rahmanian et al., 2016), bacterial-derived NPs (Mokhtarzadeh et al., 2016), metal NPs (Cai et al., 2016; Retnakumari et al., 2010), mesoporous silica NPs (Karimi et al., 2016b; Qu et al., 2015a; Xiao et al., 2014), and radiolabeled NPs (Black et al., 2015; Same et al., 2016).

Among nanoscaled inorganic materials, gold (Au) nanoparticles (GNPs) offer suitable characteristics including appreciable biocompatibility with no or negligible cytotoxicity and high biocompatibility, controllable physicochemical properties via altering their size and shape. GNPs appear to be promising agents for bioimaging and drug/gene delivery as well as photo-activated treatment strategies. In the current study, we provide a comprehensive overview on GNPs' applications in biomedical and pharmaceutical sciences with main focus on imaging and therapy of cancer.

## 4.2 GOLD NANOCONJUGATES

Owing to the enhanced emission rate and fluorescent quantum yield of GNPs, they can be considered for various optical applications including bioimaging. In fact, GNPs having fewer than 300 Au atoms possess unique optical and electronic characteristics. The thiolation of GNPs grants greater stability and also provides possibility for their surface modification to exhibit desired reactivities, polyfunctionalization, and optical properties. They offer large stokes-shifted *fluorescent* emission with a lower photobleaching and size-dependent fluorescence tenability, which make them favorable fluorescent agents for the biological probing—necessary for the theranostic applications in medicine (Dixit et al., 2015). Besides, GNPs show appropriate water solubility, great photostability, and high surface-to-volume ratio and hence high capacity for the surface functionalization

(Guo et al., 2012; Qu et al., 2015b; Tian et al., 2012). As the nanocarriers, they can be loaded with various cargo molecules such as anticancer drugs, proteins, and nucleic acids (Deng et al., 2016) and also conjugated with a wide variety of materials through covalent binding via thiol/amine groups, generating gold nanoconjugates (GNCs/AuNCs). GNCs have successfully been used in photodynamic therapy (PDT), photothermal therapy (PTT), photo-controlled delivery of drugs/genes, and radiotherapy as a sensitizing agent (Huang et al., 2016). Taken all, GNCs have been listed among the most efficient multifunctional DDSs/gene-delivery systems (GDSs) for simultaneous detection and therapy of formidable diseases, in particular solid tumors. Therefore, the main focus of this study is based on the applications of GNPs in cancer diagnosis and therapy.

## 4.3 GNPs SYNTHESIS METHODS

### *4.3.1 THE TURKEVITCH METHOD*

In 1951, Turkevitch developed one of the most popular methods for the synthesis of GNPs with a size range around 20 nm through the reduction of $HAuCl_4$ in water in the presence of citrate (Turkevich et al., 1951). Further studies on the ratios between the stabilizing/reducing agents and gold salt resulted in development of improved approaches to produce size-controllable GNPs (Frens, 1973; Kimling et al., 2006; Tyagi et al., 2016; Wuithschick et al., 2015). Following protocols represent a simplified Turkevitch method for the engineering of GNPs:

1. Dissolve 1 g (1 eq) of hydrogen tetrachloroaurate(III) hydrate ($HAuCl_4$) in 100 mL of ultrapure water.
2. Heat the solution at 90°C and mix at maximum speed.
3. Dissolve 10 eq of sodium tricitrate in 5 mL of ultrapure water and heat at 90°C.
4. Quickly add preheated sodium tricitrate solution to the gold solution.
5. Vigorously stir the reaction mixture at 90°C for 30 min.
6. Cool the reaction mixture to room temperature.
7. Centrifuge the reaction mixture at 5000 *g* for 10 min at 25°C.
8. Remove the aqueous phase and purify the GNPs by washing (3–5 times) with ethanol.

### 4.3.2 THE BRUST–SCHIFFRIN METHOD

The thiol-stabilized GNPs have already been introduced (Giersig and Mulvaney, 1993). Later on, some comprehensive studies by Brust et al. in 1994 showed the possibility of using thiols in different chain lengths to generate GNPs with controlled size (Brust et al., 1994). In this method, the surfactant is utilized for transferring gold(III) salt from aqueous phase to organic phase; then, the reaction mixture is reduced using an appropriate reducing agent such as sodium borohydride in the presence of thiol derivatives. To obtain functional thiol-capped GNPs, the influence of various factors has been examined, including different types of thiol derivatives, reaction conditions, reduction rate, thiol/gold ratio, and temperature (Liz-Marzan, 2013). Some methods have been used for the synthesis of the sulfur-containing ligands such as thiols and sulfides derivatives used for capping of GNPs, while surface modifications can be accomplished through linkers such as alkylthiosulfates (Kraft et al., 2014), alkanethioacetates (Zhang et al., 2009), thermally stable alkanethiols (Comeau and Meli, 2012), and arenethiol (Yan et al., 2010). Such modifications result in the synthesis of mono- and/or multilayer thiol-stabilized GNPs. Following protocols represent simplified methods of Brust–Schiffrin.

*One-phase Brust–Schiffrin method*

1. Add the aqueous solution of 1 g (1 eq) $HAuCl_4$ to 100 mL of tetrahydrofuran.
2. Add thiol-functionalized stabilizer (3 eq) to gold solution under rapidly stirring.
3. Dissolve the 20 eq of sodium borohydride ($NaBH_4$) in 10 mL of $H_2O$.
4. Quickly add $NaBH_4$ solution to the reaction mixture.
5. Stir the reaction mixture at room temperature for 12 h.
6. Evaporate the majority of the solvent under a reduced pressure.
7. Purify the GNPs by washing (3–5 times) with ethanol.

*Two-phase Brust–Schiffrin method:*

1. Dissolve 1 g (1 eq) of $HAuCl_4$ in 100 mL of ultrapure water (solution should be clear yellow).

2. Dissolve 1.2 eq of tetra-*n*-octylammonium bromide (TOAB) in 100 mL of toluene.
3. Add TOAB solution to gold solution.
4. Mix at maximum speed and allow to react at room temperature for 30 min.
5. Add dropwise thiol-functionalized stabilizer (3 eq) to the mixture under rapidly stirring.
6. Dissolve 20 eq of sodium borohydride ($NaBH_4$) in 10 mL of ultra-pure water.
7. Quickly add $NaBH_4$ solution to the reaction mixture.
8. Stir the reaction mixture at room temperature for 12 h.
9. Separate aqueous phase from the organic solution.
10. Vaporize the solvent under a reduced pressure.
11. Purify the GNPs by washing (3–5 times) with ethanol.

Figure 4.2 shows the methods developed by Turkevitch and Brust–Schiffrin methods.

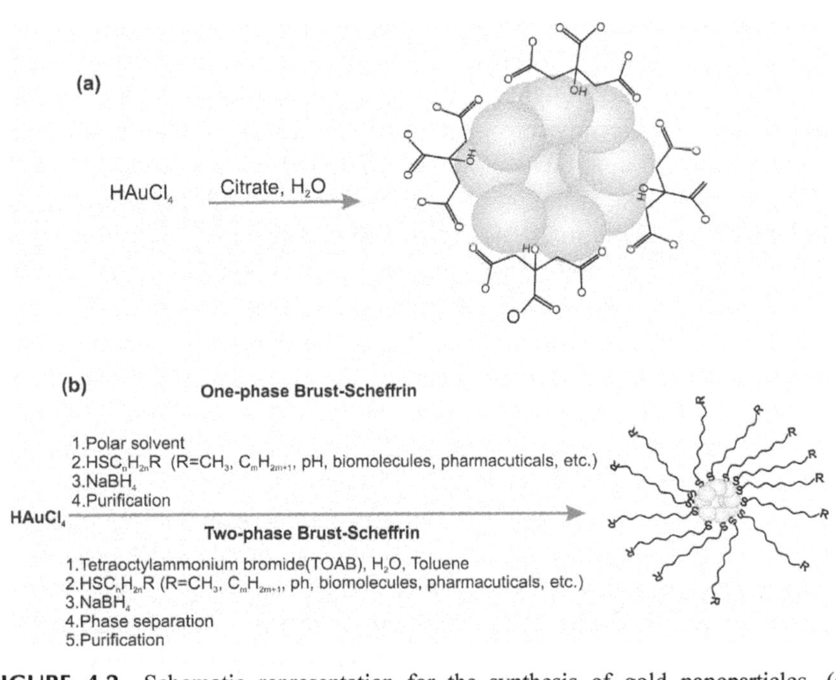


**FIGURE 4.2** Schematic representation for the synthesis of gold nanoparticles. (a) Turkevitch method and (b) Burst–Scheffrin method.

## 4.4 THE IMPACT OF GNPs ON CANCER THERAPY

Solid tumors are associated with some important alterations in the epigenetic patterns (e.g., methylation and acetylation) as well as genetic patterns including up-/downregulation of arrays of genes and transcription factors and irregular activation of different signaling pathways. The aberrant metabolism pathways, in particular on glucose and L-tryptophan (Trp) metabolisms, result in creation of permissive milieu, i.e., TME. Anomalous production of acidic substances through glycolysis acidifies the extracellular fluid, deforms the extracellular matrix, and compels a new co-op setting among cancerous and stromal cells. Within such abnormal environment, Trp is metabolized to kynurenine that is able to occupy certain receptors of lymphocytes. As a result, the immune system fails to impose an intact immunosurveillance function. The cancerous cells, in order to go through unchecked cell division and metastasis, need to restructure themselves toward some type of mesenchymal niches, the so-called cancer stem cells. The TME is such a hospitable bed wherein arrays of biomarkers of cancer and stromal cells are tied up in various important processes such as proliferation and differentiation of cancer cells. Such activity, in return, demands continued supply of nutrients. As a result, irregular angiogenesis with highly fenestrated vasculatures can emerge, which in return can enhance the oncotic pressure of the interstitial fluid within the TME. It should be noted that anticancer agents fail to enter into the core of tumor, in large part because of the pressurized interstitial fluid (Barar and Omidi, 2013a; Omidi and Barar, 2014). Smart stimuli-responsive NSs may be the solution for such problem. In the recent years, conducted attempts in the area of innovative structures have given a vivid rise toward development of materials that can show feedback and respond to the environmental changes including external or internal stimuli such as pH alterations, enzyme activity, intracellular glutathione (GSH) level, light irradiation, thermal changes, electromagnetic fields, ultrasound waves, mechanical forces, etc. Such smart materials have found many applications in multifunctional nanomedicine as intelligent stimuli-responsive micro-/nanocarriers for the delivery of diagnostic and/or therapeutic agents. Smart micro-/nanocarriers, such as GNCs, are able to accurately target specific cells/tissues and deliver the cargo molecules (e.g., drugs, genes, proteins, diagnostics agents, etc.) to the target sites. Such strategy results in enhanced site-specific therapeutic effects with reduced doses in a controlled-release manner and hence decreased systemic toxicity of highly cytotoxic drugs (Barar and Omidi, 2014). Among NPs used as smart nanoscale DDSs, GNCs seem to be one of the most appropriate NSs for the simultaneous imaging and therapy of solid tumors through PTT/PDT and stimuli-responsive delivery of cytotoxic agents with little undesired impacts on the healthy normal cells (Ahn and Jung, 2015).

## 4.5 SMART STIMULI-RESPONSIVE GOLD NANOSTRUCTURE

Advanced stimuli-responsive NSs as ground-breaking class of materials have revolutionized the field of nanomedicines and theranostics. In fact, smart materials that can respond to the exogenous (e.g., light, temperature, magnetic/electrical fields, mechanical force) and/or endogenous (e.g., redox activity, pH changes, biomolecules like certain enzymes, GSH and ATP) stimuli have found a wide range of applications in biosensing, tissue engineering, imaging, and targeted therapy (Mendes, 2008). Accordingly, various stimuli-responsive materials have been developed and used as nanocarriers, substrates, surfaces and films, and bulk gels (Cao and Wang, 2016). Stimuli-sensitive moieties can be triggered through different mechanisms. For example, the light responsiveness occurs for azobenzene and spiropyran derivatives through a reversible *cis–trans* configuration isomerization (Li and Keller, 2009; Son et al., 2014; Thambi et al., 2016), for coumarin via bond cleavage (Ji et al., 2013), for diselenide linkers via photo-oxidative bond cleavage (Zhang et al., 2016a), and for cinnamoyl by photo-dimerization (Zhang and Kim, 2016). The best paradigms for the temperature responsive materials include poly(*N*-isopropylacrylamide) (pNIPAAm) and poly(*N*-acryloyl glycinamide) (Karimi et al., 2016a). The pH-responsiveness properties can be attained by capitalizing on alkali-swellable carboxyl groups, polyacetals (polyoxymethylene), acid-swellable pyridine groups, redox groups, disulfides, diselenides, nitroxides, ferrocene, etc.

Various types of smart NSs have been developed to serve as advanced stimuli-responsive DDSs, which show different degrees of responsiveness depending on the nature of materials used. Such smart stimuli-responsive DDSs are formulated using different materials, including polymeric moieties, noble metal NPs such as Au, Ag, Pt, and Pd (Zhang et al., 2015a), semiconductor NPs (Huang et al., 2015b), lanthanide ions (e.g., $La^{3+}$, $Tm^{3+}$, $Er^{3+}$, and $Ho^{3+}$), doped upconversion NPs (Dong et al.,

2015; Hodak et al., 2016; Ruggiero et al., 2016; Yang et al., 2015), CNTs such as single-walled CNTs (Haruyuki and Noritaka, 2016), GO (Tran et al., 2015), fluorescent carbon NPs (Wang et al., 2014a), and self-assembled colloid particles (Zhang, 2015). It should be noted that integrating stimuli-responsive moieties and NPs into nanocarriers can provide site-specific and controlled cargo delivery and release (Kwon et al., 2015) and also synergize the impacts of treatment modality through combining different effects such as PTT/PDT with chemotherapy, immunotherapy, and radiotherapy (Li et al., 2016).

In fact, the surface decoration of GNPs results in different types of stimuli-responsive GNCs specific to exogenous and/or endogenous stimuli. Targeted GNCs can be devised to release the loaded cargo molecules, on demand, in a controlled manner once the NSs reached the target site (Fathi et al., 2015).

### 4.5.1 LIGHT IRRADIATION-RESPONSIVE GNPs

Photo-irradiation can be exploited for the activation of drug release by various mechanisms. The main mechanisms include two photon-sensitive deprotection strategy (Mahmoodi et al., 2016), two photon-responsive bond cleavage, photo-oxidative bond cleavage, photo-dimerization, photothermal reactions (Ahmad et al., 2016; Jabeen et al., 2014), photodynamic reactions (Chu et al., 2016; Zhang et al., 2015b), and upconversion mechanisms. Photo-responsive DDSs offer many advantages, including simultaneous imaging and light-triggered delivery of cargo molecules as well as PTT/PDT. Among various light-sensitive inorganic metal NSs, some noble metal NPs show photo-excited localized thermal energy production through absorption of irradiated photons. These metals include Au, silver (Ag), and platinum (Pt), whose NPs provide robust means for the production of light-responsive and photo-induced heat generating NSs (Fomina et al., 2012). It should be pointed out that the emission of the GNPs (e.g., as gold nanoclusters, nanospheres, and nanocages) occurs mostly in the near-infrared (NIR) region, which can be used for the detection of metastatic cells without any interference with the autofluorescence of human cells. This favors precise detection of cancer cells and liberation of chemotherapy agents as well as induction of PTT/PDT on target cells per se. Recent studies on the light-activated GNPs/GNCs have mostly been capitalized on the photo-triggered temperature-responsive bond-cleavage release ability, PTT, PDT as well as radiotherapy. We will discuss these applications in the following sections.

### *4.5.2 PHOTOTHERMAL THERAPY AND PHOTODYNAMIC THERAPY BY GNPs*

The PTT and PDT are de novo modalities that can be used as photo-reactive DDSs for treatment of various diseases in particular cancers (Kharkwal et al., 2011). Mechanistically, in PTT, materials with high light absorption such as noble metals (Au, Ag, and Pt) are used to convert the light irradiation into the thermal energy. Owing to the surface plasmon resonance (SPR) potential of these metals, their activation is based on the resonant dipole oscillation of the free electrons in the conduction band of the metal along an electromagnetic field activated by high absorption of an irradiated external source of energy. They generate strong SPR effect, through which PTT/PDT can be performed (Zhao and Karp, 2009).

The ultraviolet (UV) is a high-energy light—hazardous to the human cells/tissues. Thus, the NIR ray is the preferred light in PTT modality, in large part due to its safety with minimal photodamage to the biological specimens. In comparison with the UV/visible light, the NIR ray can penetrate into deeper tissues (Cao and Wang, 2016; Jayakumar et al., 2012; Yavuz et al., 2009). However, the NIR light lacks strong photon energies and quickly scatters. Hence, to solve such shortcomings, some novel approaches have been developed including upconversion and two-photon absorption (Olejniczak et al., 2015). In fact, regarding the PTT applications, it should be noted that the NPs with high plasmonic absorption have absorption range mostly at the NIR region. Likewise, by reason of the great absorption cross-section and remarkable photothermal conversion efficiency of GNPs, they show great potential to be used for PTT (Hwang et al., 2014). Photothermal applications of GNPs have been well reviewed previously (Abadeer and Murphy, 2016).

Mechanistically, the PDT photosensitizing moieties absorb photon energy from a light irradiating source with appropriate wavelength. The best paradigms with such effects include chlorin e6 (Ce6) phthalocyanines, porphyrins, phenothiazinium, boron-dipyrromethene, squaraine, and riboflavin curcumin (Abrahamse and Hamblin, 2016). The excited photosensitizers (PSs) transfer the photon energy to the surrounding oxygen molecules; thus, reactive oxygen species (ROS) such as hydroxyl radicals and single oxygens are produced. ROS is cytotoxic and can attack the neighboring cells within the solid tumor and cause tumor ablation via different mechanisms including (1) direct damage to cells, (2) activating immune system response against cancerous cells, and (3) shutdown of tumor vasculature. However, the absorption spectra of the PSs are in the visible light range with a low tissue penetration. Therefore, developing PSs with absorption in the NIR range is preferred, in large part because of deeper tissue penetration. This visible-to-NIR shift can be obtained through chemically modifying PSs or employing approaches such as the fluorescence resonance energy transfer with upconversion and two-photon absorption mechanisms (Yu et al., 2016b). In addition, the poor water solubility and cellular uptake of PSs together with their nonspecific biological impacts rationalize the use of targeted nanocarriers as a promising solution.

Further, the combination of PTT and PDT can lead to the enhanced therapeutic effects. For example, the localized surface plasmon resonance (LSPR) effect of GNPs integrated with ROS-generating PSs can result in combined PTT and PDT. Matching of the excitation wavelength of PSs with the excitation wavelength of LSPR of GNPs can enhance the efficiency of PDT through their strong near-field LSPR distribution. Accordingly, an interesting nanoplatform has recently been developed for the simultaneous PTT and PDT through LSPR activation (Chu et al., 2016). Theses researchers synthesized gold nanorings loaded with PSs, sulfonated aluminum phthalocyanines (AlPcS), and showed that a femtosecond 1064-nm laser irradiation (i.e., as the LSPR source) was able to induce generation of the photothermal heat. In addition, two-photon absorption of the AlPcS was able to generate singlet oxygens, in which the use of gold nanoring loaded with AlPcS could substantially enhance the two-photon absorption excitation as compared to the gold nanoring alone. This system was reported to diminish the threshold intensity needed for the inactivation of cancer cells.

### *4.5.3 PHOTO-TRIGGERED THERMO-RESPONSIVE LIBERATION OF DRUG FROM GNPs*

Among light-responsive mechanisms used in advanced GNP-based DDSs, the photo-cleavage of linking moieties is considered as one of the most common strategies for the triggered cargo release. A novel system was engineered by the conjugation of 5-fluorouracil molecules onto the GNPs via a photo-cleavable *o*-nitrobenzyl linkage (Agasti et al., 2009). They capitalized on a protecting cage to provide a nontoxic condition, in which the liberation of payload molecules was triggered through a UV light irradiation. It should be stated that photothermal and photodynamic mechanisms can be used to trigger the release of the loaded cargos from NSs. To obtain a photo-activated thermo-stimuli release, photo-thermal responsive agents can be exploited, in which the release of payloads can be triggered through the absorption of the irradiated light and its conversion to a localized heat (Kim et al., 2016b). In fact, nanomaterials such as Au, germanium (Ge), Ag, and Pt have been proposed for the photothermal liberation of payloads. The photodynamic mechanism is based on the use of photo-oxidizable moieties, wherein a photo-induced ROS generation triggers the photo-cleavage leading to the release of payloads.

### *4.5.4 GOLD NANOCAGES AS STIMULI-RESPONSIVE DDS*

Gold nanocages (GNCgs) are defined as miniscule structures with porous walls and hollow interiors. They offer unique photoacoustic (PA) properties and light-responsiveness to the NIR electromagnetic spectrum (700–900 nm). In a study, Yavuz et al. (2009) conjugated the thermo-sensitive pNIPAAm polymer onto the GNCgs via thiolate bonds. These researchers assumed that the conformation of the coated polymer chains could be altered by means of temperature, induced by the NIR irradiation. They showed that an appropriate NIR laser irradiation could induce photothermal effect on the GNCgs though their LSPR absorption. As a result, increased temperature above the lower critical solution temperature of the pNIPAAm (i.e., 39°C) could lead to rapid inevitable conformational changes of the polymer chains, resulting in accessibility of the pores of the GNCgs and hence liberation of payloads. Once the NIR irradiation turned off, however, the temperature lowered and the polymer chains restored their extended form with hydrophilicity and literally terminating the payload release. Taken all, these DDSs appear to be a remotely controlled turn-off/on drug release system with high spatiotemporal resolutions (Fig. 4.3).

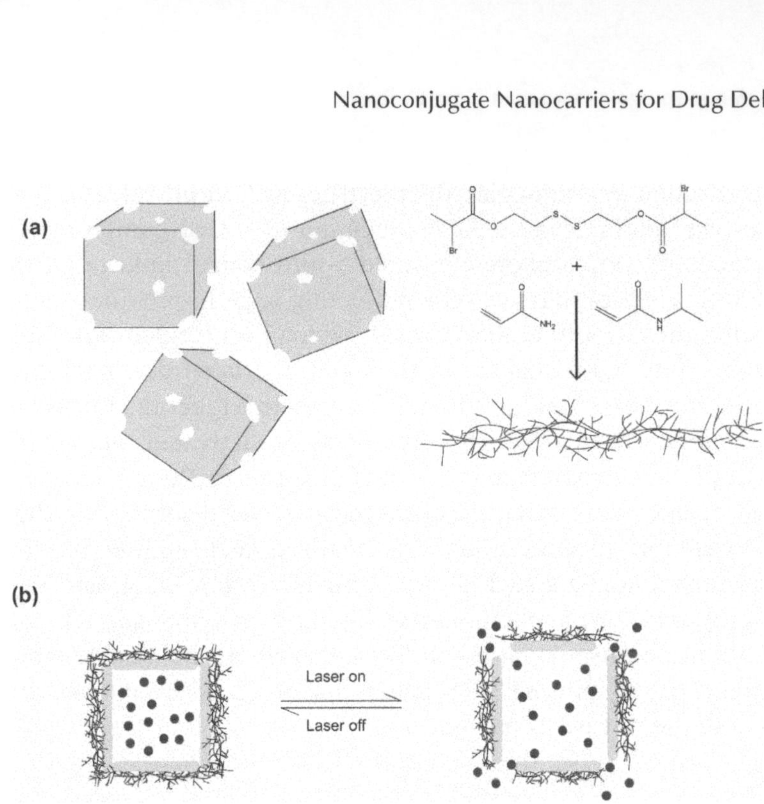


**FIGURE 4.3** Schematic representation of gold nanocages. (a) Au nanocage is used as smart delivery system. (b) Based on NIR laser irradiation, increased temperature above the LCST of the polymer can lead to conformational changes and thus the liberation of the loaded drug molecules.

It should be noted that GNPs can be functionalized with targeting agents such as antibodies (Abs)/aptamers (Aps) and armed with stimuli-responsive moieties, which results in the production of targeted DDS form which the liberation of payload drug can be remotely controlled (Fig. 4.4).

In a study, GNCs were engineered through conjugation of Ap/hairpin DNA (hpDNA) onto the GNPs loaded with cargo biomolecules (Luo et al., 2011). In this study, GNPs' surface was modified with hpDNA that increased the GNPs physiological stability. The NS was loaded with doxorubicin (DOX) and further decorated with the sgc8c DNA Ap. The reversible conjugation of DOX molecules to hpDNA was obtained via proximate base pairs interaction. Given that the sgc8c Ap recognizes the transmembrane receptor protein tyrosine kinase 7 (expressed in T-cell acute lymphoblastic leukemia) with the binding affinity ($K_d$) of 1.0 nM, the engineered system was able to selectively interact with the target cells and induce a specific enhanced cytotoxicity. In a recent study, a plasmon resonant continuous-wave light irradiation (532 nm and 2.0 W/cm$^2$ power) was shown to trigger a photo-induced thermally activated controlled liberation of the loaded DOX molecules, through which nonspecific side effects in the healthy cells/tissues were substantially decreased (Yang et al., 2016). In fact, a light-responsive system was designed for targeted delivery of DOX as a photo-chemotherapy platform. A single-stranded DNA (ssDNA) was hybridized and masked the sgc8 Ap conjugated onto the GNPs or SWCNTs. Once the NIR laser irradiation was spatially/temporally focused on the tumor site, the ssDNAs were dehybridized via a localized photo-thermal effect of the nanocarriers and the Aps were unmasked and recognized the cancer cells by restoring their cell-binding affinity. As a result, the engineered NSs were able to induce profound selective toxicity against cancer cells through specific targeting and on-demand hyperthermia-based liberation of DOX molecules.

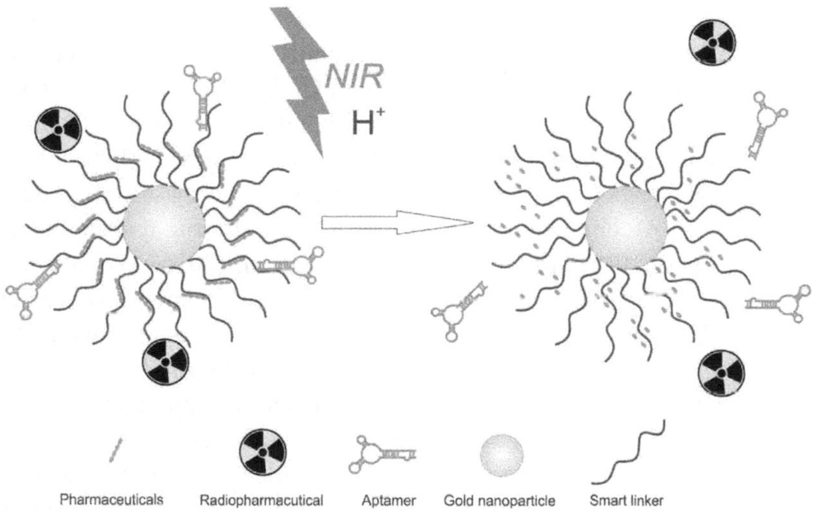


**FIGURE 4.4 (See color insert.)** Schematic illustration of dual NIR-light- and pH stimuli-triggered GNPs.

### *4.5.5 GOLD NANORODS AS STIMULI-RESPONSIVE DDS*

Like GNPs, gold nanorods (GNRs) provide a distinct nanoplatform as smart DDS for targeted delivery and stimuli-responsive controlled release of cargo drugs/genes. In an interesting study, Arunkumar et al. developed DOX-conjugated GNRs using poly(sodium 4-styrene sulfonate) (PSS) to serve as a controlled-release NS. These researchers conjugated DOX molecules onto the GNRs through PSS (DOX@PSS-GNRs) that resulted in an enhanced anticancer effect on the Dalton lymphoma ascites in Swiss albino mice with reduced accumulation of DOX molecules in myocardium (i.e., fourfold) and hence lowered cardiotoxicity (Arunkumar et al., 2015). In addition, an increased hyperthermia by laser irradiation with 0.1 $W/cm^2$ power has been reported. Recently, Ko et al. fabricated a stable NS composed of an amphiphilic polymer polyethylene glycol (PEG)-block-polycaprolactone (PEG-*b*-PCL)-functionalized GNRs loaded with DOX molecules. The PEG-*b*-PCL polymer, which possessed a disulfide bond and prepared using a facile synthetic route via copper(I)-free click chemistry, was covalently conjugated onto the GNRs by Au–S bonds, and DOX molecules were loaded on the hydrophobic PCL domain of the NS. The release of DOX molecules was negligible at physiological conditions. However, upon introduction of the NIR irradiation (808 nm), the NS displayed a light-induced plasmon resonance adsorption and generated heat. Such impacts changed the crystalline PCL blocks to amorphous state as a thermal transition and enhanced the liberation of DOX molecules in a time-dependent manner (Ko et al., 2016). Similarly, a dual-stimuli (pH and NIR ray) responsive DDS drug delivery has been recently developed using DNA-conjugated GNRs to fight the multidrug resistant cancer (Zhang et al., 2016b). In another study, Song et al. (2015) capitalized on reduced graphene oxide (rGO)-loaded ultrasmall plasmonic GNR vesicles (rGO-GNRVe) with a size rage of 65 nm, which were loaded with DOX molecules. The NS showed remarkable PA and photothermal effects. Given that the hybrid NS possessed cavity and large surface area, they displayed high DOX-loading capacity. While the NS showed high tumor accumulation through passive targeting, it released the loaded DOX molecules under the NIR laser irradiation (808 nm and 0.25 $W/cm^2$) and revealed effective inhibition of tumor growth.

### 4.5.6 IMPACTS OF NANOCOMPOSITES AND NANOPLATES

In comparison with the chemotherapy or PTT alone, the noble metal-based NSs with both PTT/PDT and chemotherapy abilities together with capability of photo-responsive liberation of cargo molecules have led to profound synergistic effects in cancer therapy. For example, Shi et al. coated the surface of palladium (Pd) nanosheets with gold to get core–shell

Pd@Au nanoplates that were PEGylated and loaded with platinum(IV) prodrug, that is, *c,c,t*-$Pt(NH_3)_2Cl_2(OOCCH_2CH_2COOH)_2$. The engineered nanoplate (Pd@Au-PEG-Pt) showed high loading efficiency and stability in physiological conditions. The exposure of the nanoplates to physiological reducing biomolecules such as GSH or ascorbic acid was found to convert platinum(IV) to cytotoxic and hydrophilic platinum(II) that could be released from the nanocomposites, while the NIR irradiation facilitated the Pt(II) specie's liberation and also imposed marked photothermal effect. As shown in Figure 4.5, the Pd@Au-PEG-Pt nanoplates were highly accumulated in tumor site (29% ID/g) through a passive targeting mechanism and imposed a combined chemotherapy and PTT effects on the target site resulting in profound eradication of the tumor tissue under relatively low power densities (Shi et al., 2016).

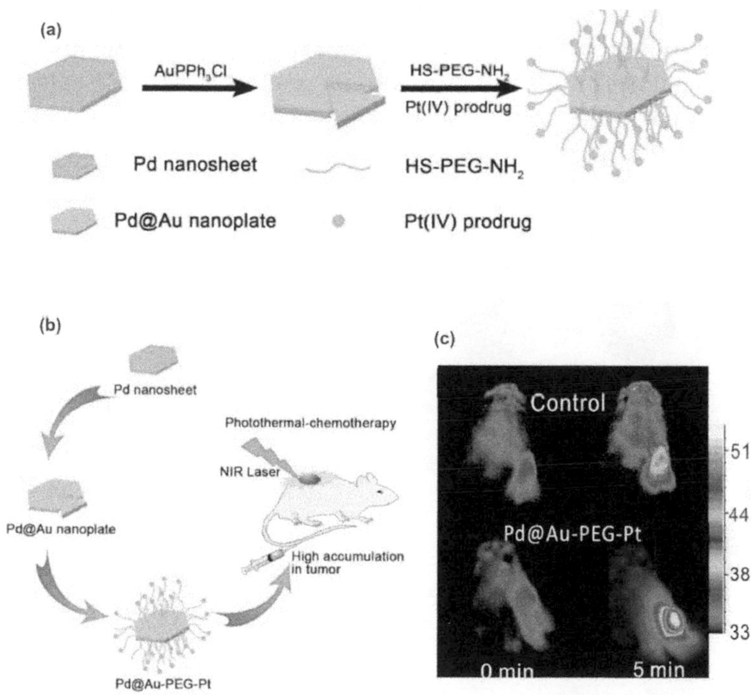


**FIGURE 4.5 (See color insert.)** Gold nanoplates as theranostics. (a) Process for preparation of Pd@Au–PEG–Pt nanoplate. (b) Prodrug conjugated Pd@Au nanoparticle for photothermal therapy and chemotherapy. (c) NIR-thermal images of tumor-bearing mice after Pd@Au–PEG–Pt injection at different times. (Data adapted with permission from Shi et al. © 2016, Royal Society of Chemistry.)

### 4.5.7 HOLLOW GOLD NANOSPHERES

On the basis that the light-triggered drug release as a controlled chemotherapy approach can be integrated with combined PTT/PDT, Liu et al. showed that anti-epidermal growth factor receptor (EGFR) Ab-conjugated hollow gold nanospheres (HGNSs) were able to enhance the radiocytotoxic targeting of cervical cancer by megavoltage radiation energies (Liu et al., 2015). Similarly, anti-TROP2 Ab-conjugated HGNSs were developed for the targeted photothermal destruction of cervical cancer cells (Liu et al., 2014). Deng et al. engineered HGNSs grafted with a thermoresponsive polymer, poly(oligo(ethylene oxide) methacrylate-*co*-2-(2-methoxyethoxy)ethyl methacrylate) (p(OEGMA-*co*-MEMA)). The HGNSs were loaded with DOX molecules and Ce6 as the photosenitizer. By exposure to a 650-nm laser irradiation, PTT and PDT were activated simultaneously, upon which the controlled release of the cargo molecules occurred (Deng et al., 2016). Such impacts were shown to enhance the antitumor activity and eradication of cancer cells as compared to the negligible toxicity observed in the control group with no laser irradiation (Fig. 4.6).

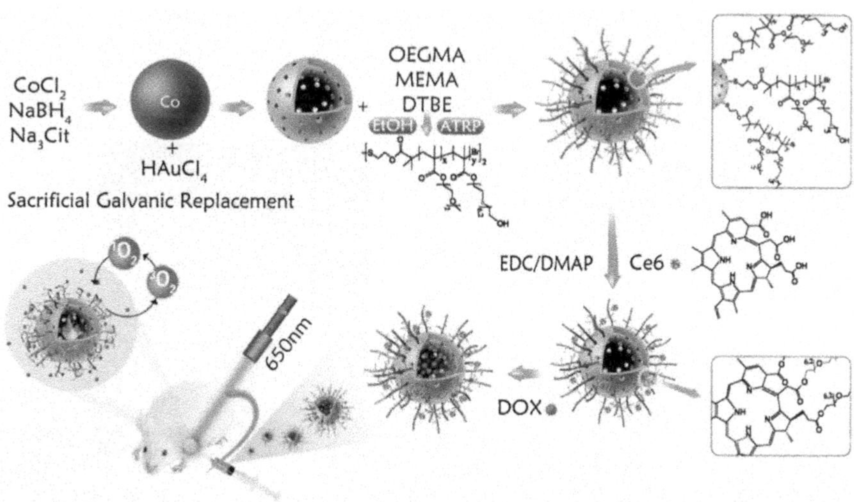


**FIGURE 4.6** Schematic representation of the synthesis process and mechanism of temperature-sensitive smart gold nanoconjugates. (Data adapted with permission from Deng et al. © 2016, Royal Chemical Society.)

### 4.5.8 LOGIC-GATED GNCs

Logic operations seem to pave a new avenue to develop the next generation of molecular devices (Pan et al., 2016; Romieu, 2015). Of these devices, the logic-gated systems can be an innovative approach for the stimuli-responsive controlled liberation of drugs. For example, smart NSs with photothermal-encoded logic-gated spatiotemporal drug release ability have recently been developed using smart GNCgs for controlled intracellular release of drugs (Shi et al., 2014). Two different GNCgs, conjugated with photothermal-responsive polymeric shell, were designed with different LSPR. It should be noted that the laser irradiation of GNCgs with the LSPR can heat them locally by the photothermal effect. Thus, the NS was able to absorb the NIR laser and convert it into a localized heat, resulting in the collapse of the surface-attached polymer chains and drug release. In the turn-off state of NIR laser, the polymer chains were able to restore their initial extended conformation with no release of payloads. Two different GNCgs coated with smart copolymer were loaded with enzyme and prodrug, and the logic gates including "AND," "OR," and "INHIBIT" were operated. Only when the two types of GNCgs opened by simultaneously applying two different NIR laser irradiations (i.e., 670 and 808 nm) as inputs, enzymatic reactions could be occurred. As a result, the AND logic gate was operated, and a fluorescent signal was produced as output, through which active prodrug molecule release could be regulated by the two NIR lights. Despite the AND logic gate, loading with an isoenzyme or an enzyme inhibitor resulted in OR or INHIBIT logic gate, enabling manually and site-specific dosage regulations. Taken all, an orthogonally triggered release, controllable prodrug activation process, and dosage regulation of active drug by defined different logic operations can be achieved by such NSs (Fig. 4.7).

### 4.5.9 GNPs AS RADIOSENSITIZERS FOR X-RAY IRRADIATION

Recently, miscellaneous NPs such as gold and selenium have been considered as radiosensitizing agents to improve the radiotherapy effect. Irradiated X-ray can be absorbed by GNPs, wherein the generation of ROSs is triggered, giving rise to more effective toxicity to the under irradiation cells and tissues, particularly cancerous cells. GNPs have widely

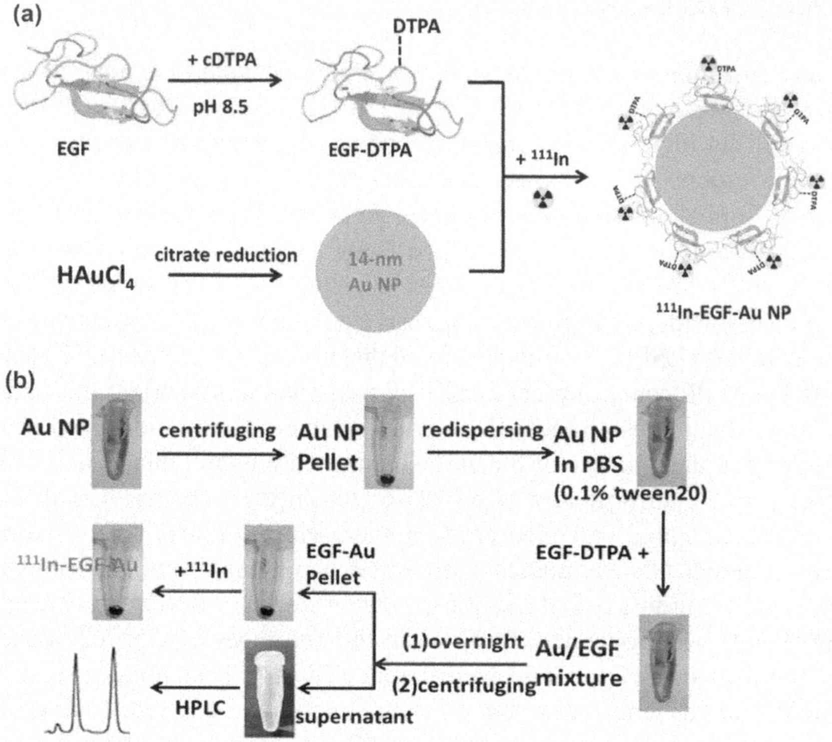


**FIGURE 4.7** Synthetic scheme of $^{111}$In-EGF-GNP theranostics. (a) GNPs are synthesized via the citrate reduction procedures and radiolabeled with $^{111}$In via DTPA, and then conjugated with chelating ligand DTPA-coupled EGF to form EGF-Au NP. (b) The NS is purified and then radiolabeled with $^{111}$In to produce $^{111}$In-EGF-Au NP. (Data adapted with permission from Song et al., 2016, https://creativecommons.org/licenses/by/4.0/.)

been investigated as radiosensitizers for the X-ray-mediated radiotherapy (Garnica-Garza, 2013). GNPs are able to convert low linear energy transfer (LET) X-rays to high energy LET X-rays by having interaction with the surface of GNPs. Most of the recently developed NSs, particularly gold-based NSs, are simultaneously equipped with targeting moieties and radio-sensitizing capability. In a study conducted by Vallis group, epidermal growth factor (EGF) disulfide bond-grafted GNPs were radiolabeled with indium-111 ($^{111}$In) via $^{111}InCl_3$ ($^{111}$In-EGF-GNP) (Song et al., 2016). It was shown that the total radioactivity of the NPs internalized by MCF-7 and MDA-MB-468 cells was 1.3% and 15%, respectively. Therefore, after 4-h incubation, the MDA-MB-468 cells endured more radiotoxicity than the MCF-7 cells (i.e., survival fraction of 17.1% and 89.8%, respectively). Hence, $^{111}$In-EGF-GNP NS indicated high loading efficiency of $^{111}$In, specific targeting of EGFR-positive cancer cells, and enhanced radiotoxicity. Figure 4.6 illustrates the schematic synthesis process of $^{111}$In-EGF-GNP.

Cai et al. developed an innovative auger electron-emitting radiation NS based on PEGylated $^{111}$In-labeled trastuzumab (TZB)-modified GNP (TZB-GNP-$^{111}$In) for targeting and treatment of HER2-positive breast cancer cells via intratumoral injection. The TZB-GNP-$^{111}$In nanocarrier showed increased cellular internalization and also enhanced cytotoxicity in the SK-BR-3 and MDA-MB-361 cells as compared to GNP-$^{111}$In (Cai et al., 2016).

## *4.5.10 TARGETED GOLD NANOCARRIERS*

Recently, an anti-EGFR Ab-conjugated HGNSs with a radiosensitizing function were developed to selectively target the cervical cancer HeLa cells. As a result, high radiocytotoxicity-induced apoptosis was achieved through a megavoltage radiotherapy (i.e., 6-MV X-rays). As compared to the naked gold nanospheres, the Ab-armed gold nanospheres resulted in higher cellular uptake of the NC and noticeable tumor growth inhibition. Under the X-ray irradiation, the Ab-armed gold nanospheres down-regulated Bcl-2 expression and upregulated expression of Bax, Bad, and caspase 3, inducing apoptosis (Liu et al., 2015). In another study performed by Li et al., arginine-glycine-aspartate (RGD) peptide-conjugated GNRs (RGD-GNR) were exploited for the NIR laser irradiated PTT. An enhanced radiotherapy efficiency was shown based on the targeting of integrin αvβ3 in the melanoma cancer A-375 cells. The engineered RGD-GNR conjugates were present on both cell surface and cytoplasm of A-375 cells owning to the high expression of integrin αvβ3. NIR laser irradiation (808 nm) of RGD-GNRs with various concentrations resulted in an accelerated photothermal heating. Thus, the photothermal-assisted radiotherapy by RGD-GNRs synergistically enhanced the radiosensitivity in the A-375 cells, mainly because of the active targeting of the integrin αvβ3. Such actively targeting by the RGD-GNRs showed great effectiveness as radiosensitizers by both low and high energy X-rays as well as the NIR-activated local photothermal treatment, leading to synergistic effects through integration of radiotherapy and thermotherapy and hence giving a resulting in the induction of the apoptosis in the cancer cells (Li et al., 2015b).

## 4.6 GNPs RESPONSIVE TO ENDOGENOUS STIMULI

Various biological environments exhibit specific parameters (e.g., pH, functional expression of biomarkers, etc.) that are altered in pathological conditions. Of these, pH variation can be exploited for designing smart pH-responsive DDSs. For instance, pH values vary within different sections of the gastrointestinal tract, or the pH of wound sites becomes acidic. In TME, while the intracellular pH of cancer cells is increased (pH ~7.4), the extracellular pH is decreased (pH ~6.6–6.9). Different microorganisms are able to change the pH of their surrounding milieu, in large part by release of biomolecules like enzymes (Alvarez-Lorenzo and Concheiro, 2014). Further, various intracellular compartments have acidic pH, including endosomes (pH ~5.5–6), lysosomes (pH ~4.5–5), and Golgi apparatus (pH ~6.4) (Song et al., 2013). Such pH differences have made this parameter as an appropriate stimulus, whose potential can be used for development of pH-responsive DDSs. The release of the encapsulated cargo molecules from the pH-responsive nanocarriers (e.g., smart gold-based NSs) can occur in response to the alterations of pH through mechanisms such as polymeric network swelling and/or dissolution. In 2003, Liu and Balasubramanian reported on the development of a proton-fueled DNA (PF-DNA) NS that was a ssDNA consisting of four stretches of cytosine (C)-rich sequences, so-called i-motif domain (Liu and Balasubramanian, 2003). The PF-DNA showed fast and greatly reversible conformational shifts in a four-stranded i-motif arrangement at the weakly acidic condition with pH values lower than pH 6.0, while it revealed a random coil structure at the higher pHs (>6.4). Later, Song et al. capitalized on the PF-DNA system and developed a pH-sensitive nanocarrier, in which multiple i-motif strands were conjugated onto the GNPs through a gold-thiol assembly. This nanocarrier indicated uniform size distribution, biocompatibility, water solubility, stability, and high drug loading. Herein, pH sensitivity was achieved by employing the intracellular acidic milieus of endosomes/lysosomes, which led to pH-activated intracellular release of intercalated DOX molecules in cancer cells, and hence markedly enhanced cytotoxicity. In addition, the PEGylation of the nanocarrier prolonged its blood circulation while reduced its nonspecific interactions (e.g., serum protein adsorption), resulting in an enhanced targeted delivery of DOX molecules to the diseased cells (Song et al., 2013).

## 4.7 DUAL STIMULI-RESPONSIVE GNPs

In DDSs, integration of multiple stimuli-responsive moieties as one single NS can lead to noticeable results, including (1) maximized site-specific delivery and accumulation of drug molecules, (2) minimized nonspecific delivery of drug molecules and hence reduced undesired side effects, (3) on-demand liberation of drug molecules in the target site. Dual stimuli-responsive gold NCs have been also studied. Zhang et al. reported a dual stimuli-responsive DNA-GNR NCs loaded with DOX molecules intercalated within a double-stranded i-motif assembly (i.e., dsM1/MC2), in which the M1 and MC2, respectively, represent the thiolated i-motif strand and the complementary strand with two deliberately presented mismatches to adjust the duplex stability at the normal physiological pH (Zhang et al., 2016b). The NS was built based on a biotin-PEG passivated i-motif conjugate of DNA-GNR in the form of Biotin-PEG-GNR-DNA/DOX (BPGDD). The NS, sensitive to pH and NIR light, was used for targeted inhibition of MDR in MCF-7/ADR cancer cells, while it showed enhanced stability, longer blood circulation, and decreased undesired immunogenicity and cytotoxicity. Within slight acidic microenvironments, the M1 strand was able to form a stable i-motif structure that could dehybridize the MC2 strand and result in the liberation of DOX molecules. The NIR laser irradiation could photothermally affect the GNR structure and simultaneously increase the release of DOX molecules. Taken all, this multimodal nanomedicine was shown to substantially enhance the cellular uptake with decreased efflux of drug molecules; hence, it was proposed as a robust NS for the treatment of drug-resistance human breast adenocarcinoma.

## 4.8 GOLD NANOSTRUCTURES AS THERANOSTICS

For a large number of scientists, simultaneous imaging and therapy of the diseased cells is a great desire, which is now plausible through development of seamless multifunctional NSs. Ideally, one single NS should be able to precisely detect and monitor the diseased cells and treat them on demand through combined mechanisms such as PTT, PDT, and chemotherapy. Such platform is composed of different parts, including (1) vehicle such as organic and inorganic delivery systems; (2) targeting agent such as Ab, Ap, and ligand specific to designated molecular marker; (3) cargo drug molecules loaded, conjugated or encapsulated into the vehicle; and (4) imaging agent. To have the PTT/PDT effects, it is necessary to include a system with PA properties such as noble metals (e.g., Au and Pt). Of these, because of the unique light-to-temperature conversion, gold nanostructures have been used for PTT/PDT and simultaneous imaging (Alkilany and Murphy, 2010). Furthermore, the simultaneous imaging and PTT application of gold nanostructures have been shown together with a PS agent conjugated to GNPs to enhance their therapeutic impacts. Liang et al. developed nanoprobes composed of gold nanostars armed with anti-CD44v6 Ab for targeted imaging and therapy of gastric cancer stem cells (GCSCs). These nanoprobes showed profoundly high affinity toward GCSCs. Once irradiated by a 790-nm-NIR laser treatment with a low power density (1.5 W/cm$^2$ for 5 min), the gold nanostars resulted in marked eradication of GCSCs through PTT effect (Liang et al., 2015). In a study, Wu et al. developed Ag–Au shell–core nanostructures that were further labeled with Rhodamine 6G (Rh6G). The Rh6G-labeled Ag–Au nanocomposites were then conjugated with a thiol-capped aptamer (5′-HS-(CH2)6-GGT TGC ATG CCG TGG GGA GGG GGG TGG GTT TTA TAG CGT ACT CAG-3′) specific to adenocarcinomic human alveolar epithelial A549 cells. The Ap-armed Rh6G-labeled Ag–Au nanocomposites was found to be capable of specific detection of A549 cells as analyzed using surface-enhanced Raman scattering (SERS) assay through Rh6G (Wu et al., 2013). Herein, conjugation of the nanocomposites with Ap (a 45-base oligonucleotide) provided high affinity and targeting ability and specificity toward A549 cells distinguished from other cancer cells such as HeLa and MCF-7 cells as well as other subtypes of lung cancer cells such as NCI-H157, NCI-H446, NCI-H1299, and NCI-H520. Once conducted a low NIR laser irradiation (i.e., power density of 0.2 cm$^{-2}$), PTT effects were observed specifically on the target cells with no damage to surrounding healthy cells and tissues. In addition, high SERS signaling of Rh6G molecules made specific detection and monitoring of the target cells possible, at which treatment process of cancer cells was applicable even at very low concentration of about 10 cells/mL. In another attempt for integration of cancer targeting of GNCgs with their diagnostic ability, Wang et al. studied in vivo pharmacokinetics of heterofunctional PEG-conjugated GNCgs (with diameters of 30 and 55 nm) via Au–S linking (Wang et al., 2012). They showed that the PEGylated GNCgs with 30 nm size displayed an improved pharmacokinetics, and a better blood circulation time with lower uptake by the reticuloendothelial system. Then, the engineered nanocages were labeled with $^{64}Cu$ and undergone positron emission tomography (PET), which indicated a strong theranostics potential, as shown in Figure 4.8.

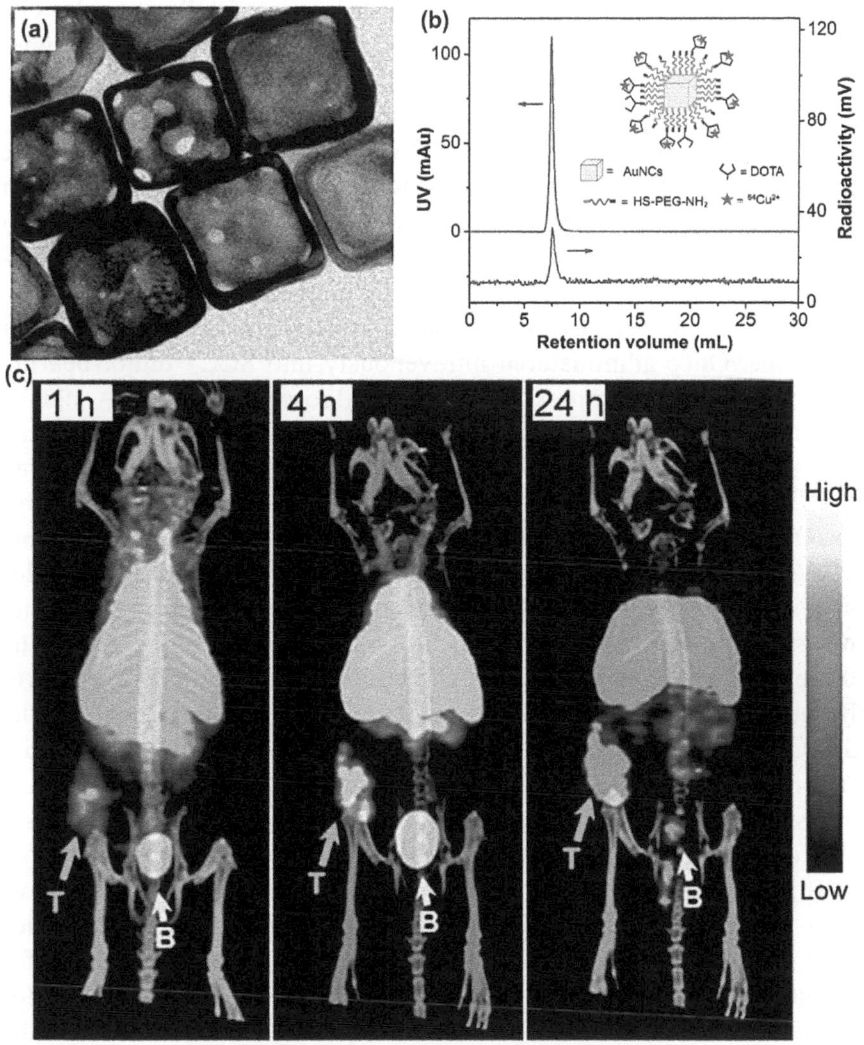


**FIGURE 4.8 (See color insert.)** A $^{64}$Cu-DOTA-PEG-Au nanocage. (a) Transmission electron microscopy micrograph of $^{64}$Cu-DOTA-PEG-Au nanocage. (b) Radio-HPLC analysis of the $^{64}$Cu-DOTA-PEG-GNP. (c) PET/CT images of the $^{64}$Cu-DOTA-PEG-GNP in a mouse bearing an EMT-6 tumor at different times. *B*, bladder; *T*, tumor. (Data adapted with permission from Wang et al., 2012, © 2012, American Chemical Society.)

As previously stated, diagnostic ability of gold nanostructures is also possible to be integrated with PTT and PDT. Gao et al. reported a hybrid Au–GO nanostructure with simultaneous PTT and PDT ability integrated with diagnostic potential. They developed a fluorescent/PA image-guided PTT nanocomplex through ligating GNPs onto the GO nanosheets. The Au–GO (GA) nanocomplex was further conjugated with the NIR dye (Cy5.5) labeled matrix metalloproteinase-14 (MMP-14) substrate (i.e., Gly-Ala-Thr-Arg-eu-Phe-Gly-Ile-Arg-Gly) peptide (CP) through amide bond to form the tumor targeting diapeutic/theranostic probe (CPGA). In the engineered nanocomplex, the Cy5.5 fluorescent signal is quenched by the SPR capacity from GA nanocomplex, while it can boost strong signals of fluorescence upon degradation by MMP-14 that is a key endopeptidase overexpressed on the surface of tumor cells. The GA nanocomplex exhibited higher photothermal effect than Au or GO alone. Once administered intravenously into SCC7 tumor-bearing mice, CPGA nanocomplex showed high fluorescence and PA signals within the tumor site. Upon irradiation of treated tumor with a NIR laser, substantial inhibition of tumor was seen with no relapse (Gao et al., 2016). Jang et al. conjugated a PS, pyropheophorbide-a (PPa), onto the surface of GNRs via a protease-cleavable GPLGVRG peptide (-Gly-Pro-Leu-Gly-Val-Arg-Gly) linker, a substrate of matrix metalloproteinase-2 (MMP2), which is important for the tumor progression and metastasis. In the engineered NSs, the conjugated PSs were inactive due to the NGR quenching effect. As shown in Figure 4.9, once exposed to the protease matrix metalloprotease-2 (MMP2)-rich environment, the GPLGVRG peptide liker is cleaved by the enzyme overexpressed by tumor cells such as HT1080. Such phenomenon can results in marked emission of fluorescence that can be used for imaging and phototoxic PDT (Jang and Choi, 2012).

In radiotherapy, also, various NSs with theranostic ability have been reported. Butterworth et al. evaluated a gold-conjugated dithiolated diethylenetriamine pentaacetic acid NPs as a theranostic agent for simultaneous radiosensitization and CT-image contrast enhancement for prostate cancer. Radio-irradiation when exposing the cancer cells to the NPs significantly decreased the cancer survival ratio than the radiation only; thus, tumor growth delay was extended from 16.9 days to 38.3 days. Also, the CT enhancement of 10% was reported (Butterworth et al., 2016).

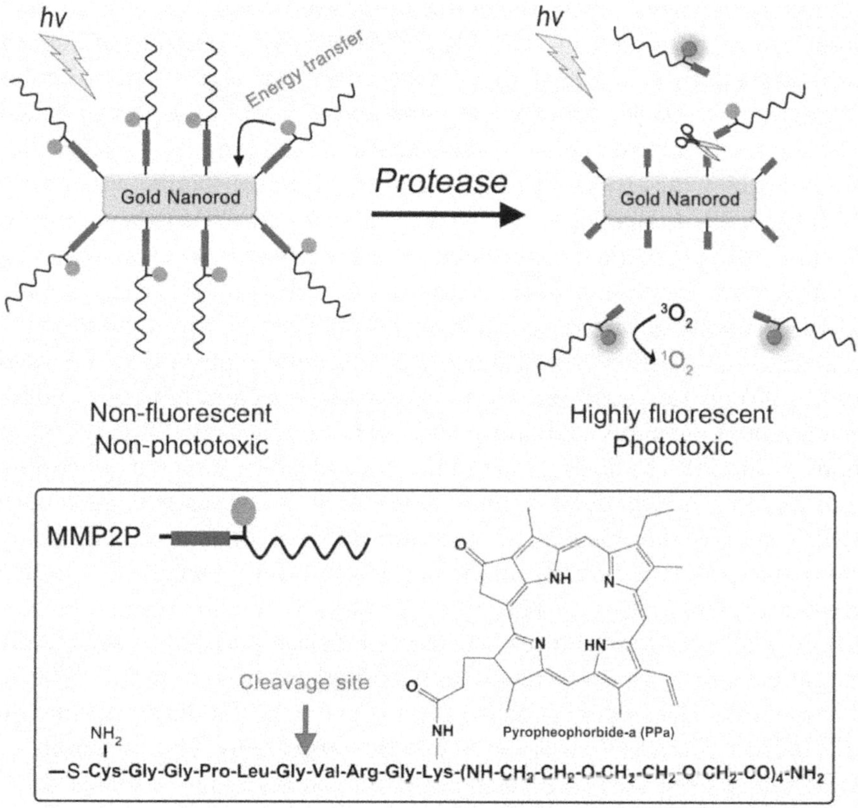


**FIGURE 4.9** Schematic illustration of an enzyme activatable fluorescence imaging and photodynamic therapy. Photosensitizer (PPa) was conjugated onto the gold nanorod (GNR) through a peptide linker that is substrate to matrix metalloproteinase-2 (MMP2) to form MMP2P-GNRs. The GNR-conjugated PPa molecules are quenched and inactive. Upon exposure to MMP2, the peptide linker is cleaved and the PPa is activated resulting in fluorescence imaging and PDT. Box shows the structure of the MMP2P containing MMP2-cleavable peptide sequence and photosensitizer. Arrow indicates cleavable site by MMP2 activity. (Image was adapted from Jang and Choi, 2012, per Creative Commons License.)

## 4.9 GNPs FOR GENE DELIVERY

In gene therapy, the main challenge is to specifically deliver the gene-based medicines to the related pathological sites for genetic correction and modification (Chandler and Venditti, 2016; Naldini, 2015). Various types of nucleic acids can be employed for the gene therapy approaches, including antisense oligonucleotides, plasmid DNA (pDNA), small interfering RNA (siRNA), miRNA, and Ap. Delivery of genes of interest into target cells is a crucial step, however, several significant obstacles act against it that hinder efficient gene delivery and transport of naked nucleic acids into the cells. In this regard, developing new approaches for enhancing the efficiency of gene delivery is of substantial importance (Foldvari, 2014; Hendel et al., 2015; Negishi et al., 2015). Accordingly, several efficient methods have been developed in recent years, including viral vectors (Kotterman et al., 2015) and nonviral vectors (Karimi et al., 2013), as well as breakthrough innovations such as the technology of clustered regularly interspaced short palindromic repeats-Cas9 (Hendel et al., 2015). However, the viral and nonviral vectors may pose undesired immunogenicity and nonspecific genotoxicity, respectively (Barar and Omidi, 2013b). In fact, due to the several safety concerns regarding exploitation of viral vectors, their gene-delivery applications have been restricted. On the other hand, nonviral delivery of genes owns many favorable properties including (a) negligible safety concerns, (b) much higher capability to transport a wide range of cargos (biomolecules, drugs, nucleic acids), and (c) site-specific delivery ability to selectively target cells through using functionalized molecular recognition ligands. In nonviral GDSs, a micro-/nanocarrier is exploited for the encapsulation of nucleic acids and their delivery into the target cells. Thus, a variety of GDSs have been developed using organic and inorganic particles (Li et al., 2015a; Loh et al., 2016; Urbiola et al., 2015). GDSs with capability of stimuli-responsive gene delivery have also been developed (Kim et al., 2016a).

Inorganic noble metal nanostructures (e.g., GNPs, GNCs, and GNRs) can be a promising nonviral GDSs for the delivery of nucleic acids. Recently, Yin et al. developed a light-activated GNR coated with polyelectrolyte polymers, poly(styrene sulfate). The engineered GNRs were conjugated with DOX molecules and electrostatically loaded with K-Ras siRNA molecules. The NS functioned as a combined delivery system, in which DOX and K-Ras siRNA molecules served as chemotherapy and K-Ras gene silencing agents, respectively. It was successfully used for the treatment of pancreatic adenocarcinoma Panc-1 cells. Thus, the codelivery of DOX and siRNA incorporated with a light treatment with 665 nm wavelength led to a synergistic antitumor effect where light irradiation induced marked release of DOX molecules. These researchers showed about 80% gene silencing efficiency by highly effective K-Ras gene downregulation, 75% proliferation inhibition, and 90% in vivo tumor growth inhibition with at least 25 days tumor growth suppression (Yin et al., 2015). In another study, a NIR-triggered gold NS with mesoporous silica (MSN) shell capped with covalently attached ODN (i.e., DNA duplexes) was employed for DOX and siRNA intracellular release. Upon 808-nm continuous-wave laser irradiation, photothermal energy conversion induced the NS, by which DNA duplex biogates were dehybridized, leading to release of DOX molecules. Alternatively, use of green fluorescent protein (GFP)-interfering siRNA, conjugated to the pores of MSN as biogate in lieu of DNA duplexes followed by a sequential laser ON–OFF irradiation mode, resulted in silencing of the GFP expression in GFP-expressing human epithelial carcinoma HeLa cells as well as the liberation of DOX and induction of cytotoxicity in A549 cells (Chang et al., 2012). Yu et al. constructed a GSH-responsive GDS using gold cysteamine (AuCM), pDNA, cell-penetrating polypeptide derived from the transactivator of transcription (pTAT), and hyaluronic acid (HA) to form AuCM/pDNA/pTAT/HA nanocomplexes. The NS was able to completely protect pDNA against enzymatic degradation, while showing effective cellular uptake in CD44 receptors overexpressed HepG 2 cells and resulting in high transfection efficiency in the presence of GSH (Yu et al., 2016a). Further, RGD-decorated dendrimer-entrapped GNPs were shown to be highly efficient and specific targeted GDS in gene therapy of stem cells (Kong et al., 2015). It is envisioned that the applications of gold nanostructures continue to grow rapidly as one of the most efficient nonviral GDSs.

## 4.10 CYTOTOXIC EFFECTS OF GNPs

Like any other pharmaceutical formulations, the gold nanostructures may impose undesired biological impacts (Jain et al., 2014). To evaluate these effects, in vitro and/or in vivo toxicological studies have been carried out on various types of gold nanostructures in cells/animal models (Browning et al., 2013; Siddiqi et al., 2012; Uchiyama et al., 2014).

The overall interpretation from these studies is that the GNPs are biologically inert when the core size is larger than 2 nm, below which show some degrees of toxicity perhaps as a consequence of its surface reactivity and catalytic effects. Because of negligible surface reactivity, the optimum size of gold-based theranostics was shown to be about 10 nm (Chithrani et al., 2006; Yang et al., 2014). It has been reported that the cellular response to GNPs is largely size dependent. GNPs with 1.4 nm in diameter were shown to induce primarily fast cell death through necrosis within 12 h period of time, however closely related GNPs with 1.2 nm in diameter predominantly caused programmed cell death by apoptosis (Pan et al., 2007). GNPs (60 nm in diameter) internalization by the murine macrophage cells were shown to induce no cytotoxicity or any detectable proinflammatory mediators (Zhang et al., 2011). While pure GNPs appear to induce negligible cytotoxicity in the human alveolar type II epithelial cells, the particles with an excess of sodium citrate residue on their surface could be toxic without affecting their cellular uptake (Uboldi et al., 2009). Further, it has been shown that the aggregation of GNPs results in markedly higher cytotoxicity than that of the well-dispersed ones (Huang et al., 2015a). The surface modification (e.g., PEGylation and citrate-stabilization) of gold nanostructures can profoundly affect the cellular toxicity of these NSs (Tlotleng et al., 2016).

Taken all, various toxicological evaluations of GNPs can be performed on culture cells, including viability tests, cell morphology assays, gene expression analyses, cellular uptake examinations, ROS analyses, and micro-motility and electric cell-substrate impedance sensing analyses. In addition to the impacts of GNPs' shape and size, the results of some investigations have shown that the toxicity and in vitro responses of GNPs may be arisen from the core composition, type of stabilizing agents, and purity and/or its degradation products. In other words, the cellular uptake of gold nanostructures can be affected by the surface functionality, charge, shape, size, concentration of NPs, incubation conditions, and type of the cell and even culture media (Jain et al., 2014; Shao et al., 2015; Tan et al., 2016; Yang et al., 2011). Thus, appropriate optimization steps need to be conducted prior to any interventions by the gold nanostructures.

## 4.11 CONCLUSION

In this chapter, we summarized recent developments on the production of the multifunctional GNSs with the capability of treating various diseases. Gold nanostructures distinctive properties (e.g., size, composition, physical, and optical specificities) facilitate their utilization for diapeutics/theranostics applications. Therefore, with surface engineering techniques, gold nanostructures could be used for the passive and/or active targeting, PTT, PDT, drug and gene delivery, imaging, and other diagnostic/therapeutic applications. Despite many advancements, the great challenges in the translation of multifunctional gold nanostructures seem to be the development of gold NSs with high in vivo stability, no immune-response, and low/no toxicity.

## KEYWORDS

- **gold nanoparticles**
- **gold nanoconjugates**
- **smart drug-delivery systems**
- **drug targeting**
- **multifunctional nanosystems**
- **theranostics**

## REFERENCES

Abadeer, N. S.; Murphy, C. J. Recent Progress in Cancer Thermal Therapy Using Gold Nanoparticles. *J. Phys. Chem. C* **2016,** *120*, 4691–4716.

Abrahamse, H.; Hamblin, M. R. New Photosensitizers for Photodynamic Therapy. *Biochem. J.* **2016,** *473*, 347–364.

Agasti, S. S.; Chompoosor, A.; You, C. C.; Ghosh, P.; Kim, C. K.; Rotello, V. M. Photo-Regulated Release of Caged Anticancer Drugs from Gold Nanoparticles. *J. Am. Chem. Soc.* **2009,** *131*, 5728–5729.

Ahmad, R.; Fu, J.; He, N.; Li, S. Advanced Gold Nanomaterials for Photothermal Therapy of Cancer. *J. Nanosci. Nanotechnol.* **2016,** *16*, 67–80.

Ahn, J.; Jung, J. H. pH-Driven Assembly and Disassembly Behaviors of DNA-Modified Au and $Fe_3O_4$@$SiO_2$ Nanoparticles. *Bull. Korean Chem. Soc.* **2015,** *36*, 1922–1925.

Alkilany, A. M.; Murphy, C. J. Toxicity and Cellular Uptake of Gold Nanoparticles: What We Have Learned So Far? *J. Nanopart. Res.* **2010,** *12*, 2313–2333.

Allen, T. M.; Cullis, P. R. Drug Delivery Systems: Entering the Mainstream. *Science* **2004,** *303*, 1818–1822.

Alvarez-Lorenzo, C.; Concheiro, A. Smart Drug Delivery Systems: From Fundamentals to the Clinic. *Chem. Commun.* **2014,** *50*, 7743–7765.

Arunkumar, P.; Raju, B.; Vasantharaja, R.; Vijayaraghavan, S.; Preetham Kumar, B.; Jeganathan, K.; Premkumar, K. Near Infra-Red Laser Mediated Photothermal and Antitumor

Efficacy of Doxorubicin Conjugated Gold Nanorods with Reduced Cardiotoxicity in Swiss Albino Mice. *Nanomedicine* **2015,** *11*, 1435–1444.

Azhdarzadeh, M.; Atyabi, F.; Saei, A. A.; Varnamkhasti, B. S.; Omidi, Y.; Fateh, M.; Ghavami, M.; Shanehsazzadeh, S.; Dinarvand, R. Theranostic MUC-1 Aptamer Targeted Gold Coated Superparamagnetic Iron Oxide Nanoparticles for Magnetic Resonance Imaging and Photothermal Therapy of Colon Cancer. *Colloids Surf. B: Biointerfaces* **2016,** *143*, 224–232.

Barar, J.; Aghanejad, A.; Fathi, M.; Omidi, Y. Advanced Drug Delivery and Targeting Technologies for the Ocular Diseases. *Bioimpacts* **2016,** *6*, 49–67.

Barar, J.; Javadzadeh, A. R.; Omidi, Y. Ocular Novel Drug Delivery: Impacts of Membranes and Barriers. *Expert Opin. Drug Deliv.* **2008,** *5*, 567–581.

Barar, J.; Omidi, Y. Dysregulated pH in Tumor Microenvironment Checkmates Cancer Therapy. *Bioimpacts* **2013a,** *3*, 149–162.

Barar, J.; Omidi, Y. Intrinsic Bio-signature of Gene Delivery Nanocarriers May Impair Gene Therapy Goals. *Bioimpacts* **2013b,** *3*, 105–109.

Barar, J.; Omidi, Y. Nanoparticles for Ocular Drug Delivery. In *Nanomedicine in Drug Delivery*; Kumar, A., Mansour, H. M., Friedman, A., Blough, E. R., Eds.; CRC Press: Boca Raton, FL, 2013c; pp 287–335.

Barar, J.; Omidi, Y. Surface Modified Multifunctional Nanomedicines for Simultaneous Imaging and Therapy of Cancer. *BioImpacts* **2014,** *4*, 3.

Black, K. C.; Akers, W. J.; Sudlow, G.; Xu, B.; Laforest, R.; Achilefu, S. Dual-Radiolabeled Nanoparticle SPECT Probes for Bioimaging. *Nanoscale* **2015,** *7*, 440–444.

Browning, L. M.; Huang, T.; Xu, X. H. Real-Time In Vivo Imaging of Size-Dependent Transport and Toxicity of Gold Nanoparticles in Zebrafish Embryos Using Single Nanoparticle Plasmonic Spectroscopy. *Interface Focus* **2013,** *3*, 20120098.

Brust, M.; Walker, M.; Bethell, D.; Schiffrin, D. J.; Whyman, R. Synthesis of Thiol-Derivatised Gold Nanoparticles in a Two-Phase Liquid–Liquid System. *J. Chem. Soc, Chem. Commun.* **1994,** *7*, 801–802.

Butterworth, K. T.; Nicol, J. R.; Ghita, M.; Rosa, S.; Chaudhary, P.; McGarry, C. K.; McCarthy, H. O.; Jimenez-Sanchez, G.; Bazzi, R.; Roux, S.; Tillement, O.; Coulter, J. A.; Prise, K. M. Preclinical Evaluation of Gold–DTDTPA Nanoparticles as Theranostic Agents in Prostate Cancer Radiotherapy. *Nanomedicine* **2016**, *11*, 2035–2047.

Cai, Z.; Chattopadhyay, N.; Yang, K.; Kwon, Y. L.; Yook, S.; Pignol, J. P.; Reilly, R. M. $^{111}$In-Labeled Trastuzumab-Modified Gold Nanoparticles Are Cytotoxic In Vitro to HER2-Positive Breast Cancer Cells and Arrest Tumor Growth In Vivo in Athymic Mice After Intratumoral Injection. *Nucl. Med. Biol.* **2016,** *43*, 818–826.

Cao, Z. Q.; Wang, G. J. Multi-Stimuli-Responsive Polymer Materials: Particles, Films, and Bulk Gels. *Chem. Rec.* **2016,** *16*, 1398–1435.

Chandler, R. J.; Venditti, C. P. Gene Therapy for Metabolic Diseases. *Transl. Sci. Rare Dis.* **2016,** *1*, 73–89.

Chang, Y. T.; Liao, P. Y.; Sheu, H. S.; Tseng, Y. J.; Cheng, F. Y.; Yeh, C. S. Near-Infrared Light-Responsive Intracellular Drug and siRNA Release Using Au Nanoensembles with Oligonucleotide-Capped Silica Shell. *Adv. Mater.* **2012,** *24*, 3309–3314.

Chithrani, B. D.; Ghazani, A. A.; Chan, W. C. Determining the Size and Shape Dependence of Gold Nanoparticle Uptake into Mammalian Cells. *Nano Lett.* **2006,** *6*, 662–668.

Chu, C. K.; Tu, Y. C.; Hsiao, J. H.; Yu, J. H.; Yu, C. K.; Chen, S. Y.; Tseng, P. H.; Chen, S.; Kiang, Y. W.; Yang, C. C. Combination of Photothermal and Photodynamic Inactivation of Cancer Cells Through Surface Plasmon Resonance of a Gold Nanoring. *Nanotechnology* **2016,** *27*, 115102.

Comeau, K. D.; Meli, M. V. Effect of Alkanethiol Chain Length on Gold Nanoparticle Monolayers at the Air–Water Interface. *Langmuir* **2012,** *28*, 377–381.

Crucho, C. I. Stimuli-Responsive Polymeric Nanoparticles for Nanomedicine. *ChemMedChem* **2015,** *10*, 24–38.

Deng, X.; Chen, Y.; Cheng, Z.; Deng, K.; Ma, P.; Hou, Z.; Liu, B.; Huang, S.; Jin, D.; Lin, J. Rational Design of a Comprehensive Cancer Therapy Platform Using Temperature-Sensitive Polymer Grafted Hollow Gold Nanospheres: Simultaneous Chemo/Photothermal/Photodynamic Therapy Triggered by a 650 nm Laser with Enhanced Anti-Tumor Efficacy. *Nanoscale* **2016,** *8*, 6837–6850.

Dixit, S.; Novak, T.; Miller, K.; Zhu, Y.; Kenney, M. E.; Broome, A. M. Transferrin Receptor-Targeted Theranostic Gold Nanoparticles for Photosensitizer Delivery in Brain Tumors. *Nanoscale* **2015,** *7*, 1782–1790.

Dolatabadi, J. E. N.; Omidi, Y. Solid Lipid-Based Nanocarriers as Efficient Targeted Drug and Gene Delivery Systems. *TrAC—Trends Anal. Chem.* **2016,** *77*, 100–108.

Dong, H.; Du, S. R.; Zheng, X. Y.; Lyu, G. M.; Sun, L. D.; Li, L. D.; Zhang, P. Z.; Zhang, C.; Yan, C. H. Lanthanide Nanoparticles: From Design Toward Bioimaging and Therapy. *Chem. Rev.* **2015,** *115*, 10725–10815.

Fathi, M.; Barar, J.; Aghanejad, A.; Omidi, Y. Hydrogels for Ocular Drug Delivery and Tissue Engineering. *Bioimpacts* **2015,** *5*, 159–164.

Fleige, E.; Quadir, M. A.; Haag, R. Stimuli-Responsive Polymeric Nanocarriers for the Controlled Transport of Active Compounds: Concepts and Applications. *Adv. Drug Deliv. Rev.* **2012,** *64*, 866–884.

Foldvari, M. Nanopharmaceutics Innovations in Gene Therapy: Moving Towards Non-Viral and Non-Invasive Delivery Methods. *Nanomed. Biother. Discov*. **2014,** *4* (2), 1000e135.

Fomina, N.; Sankaranarayanan, J.; Almutairi, A. Photochemical Mechanisms of Light-Triggered Release from Nanocarriers. *Adv. Drug Deliv. Rev.* **2012,** *64*, 1005–1020.

Frens, G. Controlled Nucleation for the Regulation of the Particle Size in Monodisperse Gold Suspensions. *Nature* **1973,** *241*, 20–22.

Gao, S.; Zhang, L.; Wang, G.; Yang, K.; Chen, M.; Tian, R.; Ma, Q.; Zhu, L. Hybrid Graphene/Au Activatable Theranostic Agent for Multimodalities Imaging Guided Enhanced Photothermal Therapy. *Biomaterials* **2016,** *79*, 36–45.

Garnica-Garza, H. M. Microdosimetry of X-Ray-Irradiated Gold Nanoparticles. *Radiat. Prot. Dosimetry* **2013,** *155*, 59–63.

Giersig, M.; Mulvaney, P. Preparation of Ordered Colloid Monolayers by Electrophoretic Deposition. *Langmuir*, **1993,** *9*, 3408–3413.

Guo, Y.; Wang, Z.; Shao, H.; Jiang, X. Stable Fluorescent Gold Nanoparticles for Detection of Cu2+ with Good Sensitivity and Selectivity. *Analyst* **2012,** *137*, 301–304.

Gupta, J.; Mohapatra, J.; Bhargava, P.; Bahadur, D. A pH-Responsive Folate Conjugated Magnetic Nanoparticle for Targeted Chemo-thermal Therapy and MRI Diagnosis. *Dalton Trans.* **2016,** *45*, 2454–2461.

Heidari Majd, M.; Asgari, D.; Barar, J.; Valizadeh, H.; Kafil, V.; Abadpour, A.; Moumivand, E.; Mojarrad, J. S.; Rashidi, M. R.; Coukos, G.; Omidi, Y. Tamoxifen Loaded Folic

Acid Armed PEGylated Magnetic Nanoparticles for Targeted Imaging and Therapy of Cancer. *Colloids Surf. B: Biointerfaces* **2013a,** *106*, 117–125.

Heidari Majd, M.; Asgari, D.; Barar, J.; Valizadeh, H.; Kafil, V.; Coukos, G.; Omidi, Y. Specific Targeting of Cancer Cells by Multifunctional Mitoxantrone-Conjugated Magnetic Nanoparticles. *J. Drug Target.* **2013b,** *21*, 328–340.

Hendel, A.; Bak, R. O.; Clark, J. T.; Kennedy, A. B.; Ryan, D. E.; Roy, S.; Steinfeld, I.; Lunstad, B. D.; Kaiser, R. J.; Wilkens, A. B.; Bacchetta, R.; Tsalenko, A.; Dellinger, D.; Bruhn, L.; Porteus, M. H. Chemically Modified Guide RNAs Enhance CRISPR–Cas Genome Editing in Human Primary Cells. *Nat. Biotechnol.* **2015,** *33*, 985–989.

Hodak, J.; Chen, Z.; Wu, S.; Etchenique, R. Multiphoton Excitation of Upconverting Nanoparticles in Pulsed Regime. *Anal. Chem.* **2016,** *88*, 1468–1475.

Huang, D.; Zhou, H.; Liu, H.; Gao, J. The Cytotoxicity of Gold Nanoparticles is Dispersity-Dependent. *Dalton Trans.* **2015a,** *44*, 17911–17915.

Huang, S.; Duan, S.; Wang, J.; Bao, S.; Qiu, X.; Li, C.; Liu, Y.; Yan, L.; Zhang, Z.; Hu, Y. Folic-Acid-Mediated Functionalized Gold Nanocages for Targeted Delivery of Anti-miR-181b in Combination of Gene Therapy and Photothermal Therapy Against Hepatocellular Carcinoma. *Adv. Funct. Mater.* **2016,** *26*, 2532–2544.

Huang, S.; Liu, J.; He, Q.; Chen, H.; Cui, J.; Xu, S.; Zhao, Y.; Chen, C.; Wang, L. Smart Cu1. 75S Nanocapsules with High and Stable Photothermal Efficiency for NIR Photo-Triggered Drug Release. *Nano Res.* **2015b,** *8*, 4038–4047.

Hwang, S.; Nam, J.; Jung, S.; Song, J.; Doh, H.; Kim, S. Gold Nanoparticle-Mediated Photothermal Therapy: Current Status and Future Perspective. *Nanomedicine* **2014,** *9*, 2003–2022.

Inturi, S.; Wang, G.; Chen, F.; Banda, N. K.; Holers, V. M.; Wu, L.; Moghimi, S. M.; Simberg, D. Modulatory Role of Surface Coating of Superparamagnetic Iron Oxide Nanoworms in Complement Opsonization and Leukocyte Uptake. *ACS Nano* **2015,** *9*, 10758–10768.

Jabeen, F.; Najam-ul-Haq, M.; Javeed, R.; Huck, C. W.; Bonn, G. K. Au-Nanomaterials as a Superior Choice for Near-Infrared Photothermal Therapy. *Molecules* **2014,** *19*, 20580–20593.

Jain, S.; Coulter, J. A.; Butterworth, K. T.; Hounsell, A. R.; McMahon, S. J.; Hyland, W. B.; Muir, M. F.; Dickson, G. R.; Prise, K. M.; Currell, F. J.; Hirst, D. G.; O'Sullivan, J. M. Gold Nanoparticle Cellular Uptake, Toxicity and Radiosensitisation in Hypoxic Conditions. *Radiother. Oncol.* **2014,** *110*, 342–347.

Jang, B.; Choi, Y. Photosensitizer-Conjugated Gold Nanorods for Enzyme-Activatable Fluorescence Imaging and Photodynamic Therapy. *Theranostics* **2012,** *2*, 190–197.

Jayakumar, M. K.; Idris, N. M.; Zhang, Y. Remote Activation of Biomolecules in Deep Tissues Using Near-Infrared-to-UV Upconversion Nanotransducers. *Proc. Natl. Acad. Sci. U.S.A.* **2012,** *109*, 8483–8488.

Ji, W.; Li, N.; Chen, D.; Qi, X.; Sha, W.; Jiao, Y.; Xu, Q.; Lu, J. Coumarin-Containing Photo-Responsive Nanocomposites for NIR Light-Triggered Controlled Drug Release via a Two-Photon Process. *J. Mater. Chem. B* **2013,** *1*, 5942–5949.

Karimi, M.; Avci, P.; Mobasseri, R.; Hamblin, M. R.; Naderi-Manesh, H. The Novel Albumin-Chitosan Core–Shell Nanoparticles for Gene Delivery: Preparation, Optimization and Cell Uptake Investigation. *J. Nanopart. Res.* **2013,** *15*, 1651.

Karimi, M.; Ghasemi, A.; Sahandi Zangabad, P.; Rahighi, R.; Moosavi Basri, S. M.; Mirshekari, H.; Amiri, M.; Shafaei Pishabad, Z.; Aslani, A.; Bozorgomid, M.; Ghosh, D.;

Beyzavi, A.; Vaseghi, A.; Aref, A. R.; Haghani, L.; Bahrami, S.; Hamblin, M. R. Smart Micro/Nanoparticles in Stimulus-Responsive Drug/Gene Delivery Systems. *Chem. Soc. Rev.* **2016a,** *45* (5), 1457–1501.

Karimi, M.; Mirshekari, H.; Aliakbari, M.; Sahandi-Zangabad, P.; Hamblin, M. R. Smart Mesoporous Silica Nanoparticles for Controlled-Release Drug Delivery. *Nanotechnol. Rev.* **2016b,** *5* (2), 195–207.

Kharkwal, G. B.; Sharma, S. K.; Huang, Y. Y.; Dai, T.; Hamblin, M. R. Photodynamic Therapy for Infections: Clinical Applications. *Lasers Surg. Med.* **2011**, *43*, 755–767.

Kibria, G.; Hatakeyama, H.; Harashima, H. Cancer Multidrug Resistance: Mechanisms Involved and Strategies for Circumvention Using a Drug Delivery System. *Arch. Pharm. Res.* **2014,** *37*, 4–15.

Kim, H.; Kim, J.; Lee, M.; Choi, H. C.; Kim, W. J. Stimuli-Regulated Enzymatically Degradable Smart Graphene-Oxide-Polymer Nanocarrier Facilitating Photothermal Gene Delivery. *Adv. Healthc. Mater.* **2016a,** *5*, 1918–1930.

Kim, J.; Kim, J.; Jeong, C.; Kim, W. J. Synergistic Nanomedicine by Combined Gene and Photothermal Therapy. *Adv. Drug Deliv. Rev.* **2016b,** *98*, 99–112.

Kimling, J.; Maier, M.; Okenve, B.; Kotaidis, V.; Ballot, H.; Plech, A.; Turkevich Method for Gold Nanoparticle Synthesis Revisited. *J. Phys. Chem. B* **2006,** *110*, 15700–15707.

Ko, H.; Son, S.; Bae, S.; Kim, J. H.; Yi, G. R.; Park, J. H. Near-Infrared Light-Triggered Thermochemotherapy of Cancer Using a Polymer–Gold Nanorod Conjugate. *Nanotechnology* **2016,** *27*, 175102.

Kong, L.; Alves, C. S.; Hou, W.; Qiu, J.; Mohwald, H.; Tomas, H.; Shi, X. RGD Peptide-Modified Dendrimer-Entrapped Gold Nanoparticles Enable Highly Efficient and Specific Gene Delivery to Stem Cells. *ACS Appl. Mater. Interfaces* **2015,** *7*, 4833–4843.

Kotterman, M. A.; Yin, L.; Strazzeri, J. M.; Flannery, J. G.; Merigan, W. H.; Schaffer, D. V. Antibody Neutralization Poses a Barrier to Intravitreal Adeno-Associated Viral Vector Gene Delivery to Non-Human Primates. *Gene Ther.* **2015,** *22*, 116–126.

Kraft, M.; Adamczyk, S.; Polywka, A.; Zilberberg, K.; Weijtens, C.; Meyer, J.; Gorrn, P.; Riedl, T.; Scherf, U. Polyanionic, Alkylthiosulfate-Based Thiol Precursors for Conjugated Polymer Self-Assembly onto Gold and Silver. *ACS Appl. Mater. Interfaces* **2014,** *6*, 11758–11765.

Kwon, H. J.; Byeon, Y.; Jeon, H. N.; Cho, S. H.; Han, H. D.; Shin, B. C. Gold Cluster-Labeled Thermosensitive Liposmes Enhance Triggered Drug Release in the Tumor Microenvironment by a Photothermal Effect. *J. Control. Release* **2015,** *216*, 132–139.

Langiu, M.; Dadparvar, M.; Kreuter, J.; Ruonala, M. O. Human Serum Albumin-Based Nanoparticle-Mediated In Vitro Gene Delivery. *PLoS ONE* **2014,** *9*, e107603.

Li, C.; Li, Y.; Gao, Y.; Wei, N.; Zhao, X.; Wang, C.; Li, Y.; Xiu, X.; Cui, J. Direct Comparison of Two Albumin-Based Paclitaxel-Loaded Nanoparticle Formulations: Is the Cross-linked Version More Advantageous? *Int. J. Pharm.* **2014,** *468*, 15–25.

Li, L.; Wei, Y.; Gong, C. Polymeric Nanocarriers for Non-Viral Gene Delivery. *J. Biomed. Nanotechnol.* **2015a,** *11*, 739–770.

Li, M.-H.; Keller, P. Stimuli-Responsive Polymer Vesicles. *Soft Matter* **2009,** *5*, 927–937.

Li, P.; Shi, Y. W.; Li, B. X.; Xu, W. C.; Shi, Z. L.; Zhou, C.; Fu, S. Photo-Thermal Effect Enhances the Efficiency of Radiotherapy Using Arg-Gly-Asp Peptides-Conjugated Gold Nanorods That Target αvβ3 in Melanoma Cancer Cells. *J. Nanobiotechnol.* **2015b,** *13*, 52.

Li, Z.; Huang, H.; Tang, S.; Li, Y.; Yu, X. F.; Wang, H.; Li, P.; Sun, Z.; Zhang, H.; Liu, C.; Chu, P. K. Small Gold Nanorods Laden Macrophages for Enhanced Tumor Coverage in Photothermal Therapy. *Biomaterials* **2016,** *74*, 144–154.

Liang, S.; Li, C.; Zhang, C.; Chen, Y.; Xu, L.; Bao, C.; Wang, X.; Liu, G.; Zhang, F.; Cui, D. CD44v6 Monoclonal Antibody-Conjugated Gold Nanostars for Targeted Photoacoustic Imaging and Plasmonic Photothermal Therapy of Gastric Cancer Stem-like Cells. *Theranostics* **2015,** *5*, 970–984.

Liu, D.; Balasubramanian, S. A Proton-Fuelled DNA Nanomachine. *Angew. Chem. Int. Ed. Engl.* **2003,** *42*, 5734–5736.

Liu, J.; Liang, Y.; Liu, T.; Li, D.; Yang, X. Anti-EGFR-Conjugated Hollow Gold Nanospheres Enhance Radiocytotoxic Targeting of Cervical Cancer at Megavoltage Radiation Energies. *Nanoscale Res. Lett.* **2015,** *10*, 218.

Liu, T.; Tian, J.; Chen, Z.; Liang, Y.; Liu, J.; Liu, S.; Li, H.; Zhan, J.; Yang, X. Anti-TROP2 Conjugated Hollow Gold Nanospheres as a Novel Nanostructure for Targeted Photothermal Destruction of Cervical Cancer Cells. *Nanotechnology* **2014,** *25*, 345103.

Liz-Marzan, L. M. Gold Nanoparticle Research Before and After the Brust–Schiffrin Method. *Chem. Commun. (Camb.)* **2013,** *49*, 16–18.

Loh, X. J.; Lee, T. C.; Dou, Q.; Deen, G. R. Utilising Inorganic Nanocarriers for Gene Delivery. *Biomater. Sci.* **2016,** *4*, 70–86.

Luo, Y. L.; Shiao, Y. S.; Huang, Y. F. Release of Photoactivatable Drugs from Plasmonic Nanoparticles for Targeted Cancer Therapy. *ACS Nano* **2011,** *5*, 7796–7804.

Mahmoodi, M. M.; Abate-Pella, D.; Pundsack, T. J.; Palsuledesai, C. C.; Goff, P. C.; Blank, D. A.; Distefano, M. D. Nitrodibenzofuran: A One- and Two-Photon Sensitive Protecting Group That Is Superior to Brominated Hydroxycoumarin for Thiol Caging in Peptides. *J. Am. Chem. Soc.* **2016,** *138*, 5848–5859.

Matthaiou, E. I.; Barar, J.; Sandaltzopoulos, R.; Li, C.; Coukos, G.; Omidi, Y. Shikonin-Loaded Antibody-Armed Nanoparticles for Targeted Therapy of Ovarian Cancer. *Int. J. Nanomed.* **2014,** *9*, 1855–1870.

Mendes, P. M. Stimuli-Responsive Surfaces for Bio-Applications. *Chem. Soc. Rev.* **2008,** *37*, 2512–2529.

Moghimi, S. M.; Szebeni, J. Stealth Liposomes and Long Circulating Nanoparticles: Critical Issues in Pharmacokinetics, Opsonization and Protein-Binding Properties. *Progr. Lipid Res.* **2003,** *42*, 463–478.

Mokhtarzadeh, A.; Alibakhshi, A.; Hejazi, M.; Omidi, Y.; Dolatabadi, J. E. N. Bacterial-Derived Biopolymers: Advanced Natural Nanomaterials for Drug Delivery and Tissue Engineering. *TrAC—Trends Anal. Chem.* **2016,** *82*, 367–384.

Naldini, L. Gene Therapy Returns to Centre Stage. *Nature* **2015,** *526*, 351–360.

Negishi, Y.; Yamane, M.; Kurihara, N.; Endo-Takahashi, Y.; Sashida, S.; Takagi, N.; Suzuki, R.; Maruyama, K. Enhancement of Blood–Brain Barrier Permeability and Delivery of Antisense Oligonucleotides or Plasmid DNA to the Brain by the Combination of Bubble Liposomes and High-Intensity Focused Ultrasound. *Pharmaceutics* **2015,** *7*, 344–362.

Oberoi, H. S.; Nukolova, N. V.; Kabanov, A. V.; Bronich, T. K. Nanocarriers for Delivery of Platinum Anticancer Drugs. *Adv. Drug Deliv. Rev.* **2013,** *65*, 1667–1685.

Olejniczak, J.; Carling, C. J.; Almutairi, A. Photocontrolled Release Using One-Photon Absorption of Visible or NIR Light. *J Control. Release* **2015,** *219*, 18–30.

Omidi, Y. CNT Nanobombs for Specific Eradication of Cancer Cells: A New Concept in Cancer Theranostics. *Bioimpacts* **2011,** *1*, 199–201.

Omidi, Y.; Barar, J. Impacts of Blood–Brain Barrier in Drug Delivery and Targeting of Brain Tumors. *Bioimpacts* **2012,** *2*, 5–22.

Omidi, Y.; Barar, J. Targeting Tumor Microenvironment: Crossing Tumor Interstitial Fluid by Multifunctional Nanomedicines. *Bioimpacts* **2014,** *4*, 55–67.

Omidi, Y.; Gumbleton, M.; Barar, J.; Pourseif, M. M.; Rafi, M. A. Blood-Brain Barrier Transport Machineries and Targeted Therapy of Brain Diseases. *Bioimpacts* **2016,** *6* (4), 225–248.

Pan, Y.; Neuss, S.; Leifert, A.; Fischler, M.; Wen, F.; Simon, U.; Schmid, G.; Brandau, W.; Jahnen-Dechent, W. Size-Dependent Cytotoxicity of Gold Nanoparticles. *Small* **2007,** *3*, 1941–1949.

Pan, Y.; Shi, Y.; Chen, Z.; Chen, J.; Hou, M.; Chen, Z.; Li, C. W.; Yi, C. Design of Multiple Logic Gates Based on Chemically Triggered Fluorescence Switching of Functionalized Polyethylenimine. *ACS Appl. Mater. Interfaces* **2016,** *8*, 9472–9482.

Qu, Q.; Ma, X.; Zhao, Y. Targeted Delivery of Doxorubicin to Mitochondria Using Mesoporous Silica Nanoparticle Nanocarriers. *Nanoscale* **2015a,** *7*, 16677–16686.

Qu, X.; Li, Y.; Li, L.; Wang, Y.; Liang, J.; Liang, J. Fluorescent Gold Nanoclusters: Synthesis and Recent Biological Application. *J. Nanomater.* **2015b,** *2015*, Article ID 784097, 23 pages.

Rahmanian, N.; Eskandani, M.; Barar, J.; Omidi, Y. Recent Trends in Targeted Therapy of Cancer Using Graphene Oxide-Modified Multifunctional Nanomedicines. *J. Drug Target.* **2017,** *25* (3), 202–215.

Retnakumari, A.; Setua, S.; Menon, D.; Ravindran, P.; Muhammed, H.; Pradeep, T.; Nair, S.; Koyakutty, M. Molecular-Receptor-Specific, Non-Toxic, Near-Infrared-Emitting Au Cluster–Protein Nanoconjugates for Targeted Cancer Imaging. *Nanotechnology* **2010,** *21*, 055103_1–055103_12.

Romieu, A. "AND" Luminescent "Reactive" Molecular Logic Gates: A Gateway to Multianalyte Bioimaging and Biosensing. *Org. Biomol. Chem.* **2015,** *13*, 1294–1306.

Ruggiero, E.; Garino, C.; Mareque-Rivas, J. C.; Habtemariam, A.; Salassa, L. Upconverting Nanoparticles Prompt Remote Near-Infrared Photoactivation of Ru(II)–Arene Complexes. *Chemistry* **2016,** *22*, 2801–2811.

Saito, H.; Kato, N. Polyelectrolyte/Carbon Nanotube Composite Microcapsules and Drug Release Triggered by Laser Irradiation. *Jpn. J. Appl. Phys.* **2016,** *55*, 03DF06.

Same, S.; Aghanejad, A.; Akbari Nakhjavani, S.; Barar, J.; Omidi, Y. Radiolabeled Theranostics: Magnetic and Gold Nanoparticles. *Bioimpacts* **2016,** *6*, 169–181.

Saraiva, C.; Praca, C.; Ferreira, R.; Santos, T.; Ferreira, L.; Bernardino, L. Nanoparticle-Mediated Brain Drug Delivery: Overcoming Blood–Brain Barrier to Treat Neurodegenerative Diseases. *J. Control. Release* **2016,** *235*, 34–47.

Shao, X.; Tian, C.; Wei, Q.; Wang, X. D. Cellular Uptakes of Novel Gold Nanoparticle Conjugates via Radioactivity Analysis and Photoacoustic Imaging. *J. Label. Comp. Radiopharm.* **2015,** *58*, S321–S321.

She, W.; Li, N.; Luo, K.; Guo, C.; Wang, G.; Geng, Y.; Gu, Z. Dendronized Heparin–Doxorubicin Conjugate Based Nanoparticle as pH-Responsive Drug Delivery System for Cancer Therapy. *Biomaterials* **2013,** *34*, 2252–2264.

Shi, P.; Ju, E.; Ren, J.; Qu, X. Near-Infrared Light-Encoded Orthogonally Triggered and Logical Intracellular Release Using Gold Nanocage@ Smart Polymer Shell. *Adv. Funct. Mater.* **2014,** *24*, 826–834.

Shi, S.; Chen, X.; Wei, J.; Huang, Y.; Weng, J.; Zheng, N. Platinum(IV) Prodrug Conjugated Pd@Au Nanoplates for Chemotherapy and Photothermal Therapy. *Nanoscale* **2016,** *8*, 5706–5713.

Siddiqi, N. J.; Abdelhalim, M. A.; El-Ansary, A. K.; Alhomida, A. S.; Ong, W. Y. Identification of Potential Biomarkers of Gold Nanoparticle Toxicity in Rat Brains. *J. Neuroinflamm.* **2012,** *9*, 123.

Son, S.; Shin, E.; Kim, B. S. Light-Responsive Micelles of Spiropyran Initiated Hyperbranched Polyglycerol for Smart Drug Delivery. *Biomacromolecules* **2014,** *15*, 628–634.

Song, J.; Yang, X.; Jacobson, O.; Lin, L.; Huang, P.; Niu, G.; Ma, Q.; Chen, X. Sequential Drug Release and Enhanced Photothermal and Photoacoustic Effect of Hybrid Reduced Graphene Oxide-Loaded Ultrasmall Gold Nanorod Vesicles for Cancer Therapy. *ACS Nano* **2015,** *9*, 9199–9209.

Song, L.; Falzone, N.; Vallis, K. A. EGF-Coated Gold Nanoparticles Provide an Efficient Nano-Scale Delivery System for the Molecular Radiotherapy of EGFR-Positive Cancer. *Int. J. Radiat. Biol.* **2016,** *92*, 716–723.

Song, L.; Ho, V. H.; Chen, C.; Yang, Z.; Liu, D.; Chen, R.; Zhou, D. Efficient, pH-Triggered Drug Delivery Using a pH-Responsive DNA-Conjugated Gold Nanoparticle. *Adv. Healthc. Mater.* **2013,** *2*, 275–280.

Tan, G.; Kantner, K.; Zhang, Q.; Soliman, M. G.; Del Pino, P.; Parak, W. J.; Onur, M. A.; Valdeperez, D.; Rejman, J.; Pelaz, B. Cellular Uptake and Bioactivity of Antibody–Gold Nanoparticle Bioconjugates. *J. Biotechnol.* **2016,** *231*, S29–S29.

Thambi, T.; Park, J. H.; Lee, D. S. Stimuli-Responsive Polymersomes for Cancer Therapy. *Biomater. Sci.* **2016,** *4*, 55–69.

Tian, D.; Qian, Z.; Xia, Y.; Zhu, C. Gold Nanocluster-Based Fluorescent Probes for Near-Infrared and Turn-On Sensing of Glutathione in Living Cells. *Langmuir* **2012,** *28*, 3945–3951.

Tlotleng, N.; Vetten, M. A.; Keter, F. K.; Skepu, A.; Tshikhudo, R.; Gulumian, M. Cytotoxicity, Intracellular Localization and Exocytosis of Citrate Capped and PEG Functionalized Gold Nanoparticles in Human Hepatocyte and Kidney Cells. *Cell Biol. Toxicol.* **2016,** *32*, 305–321.

Tran, T. H.; Nguyen, H. T.; Pham, T. T.; Choi, J. Y.; Choi, H. G.; Yong, C. S.; Kim, J. O. Development of a Graphene Oxide Nanocarrier for Dual-Drug Chemo-Phototherapy to Overcome Drug Resistance in Cancer. *ACS Appl. Mater. Interfaces* **2015,** *7*, 28647–28655.

Turkevich, J.; Stevenson, P. C.; Hillier, J. A Study of the Nucleation and Growth Processes in the Synthesis of Colloidal Gold. *Discuss. Faraday Soc.* **1951,** *11*, 55–75.

Tyagi, H.; Kushwaha, A.; Kumar, A.; Aslam, M. A Facile pH Controlled Citrate-Based Reduction Method for Gold Nanoparticle Synthesis at Room Temperature. *Nanoscale Res. Lett.* **2016,** *11*, 362.

Uboldi, C.; Bonacchi, D.; Lorenzi, G.; Hermanns, M. I.; Pohl, C.; Baldi, G.; Unger, R. E.; Kirkpatrick, C. J. Gold Nanoparticles Induce Cytotoxicity in the Alveolar Type-II Cell Lines A549 and NCIH441. *Part Fibre Toxicol.* **2009,** *6*, 18.

Uchiyama, M. K.; Deda, D. K.; Rodrigues, S. F.; Drewes, C. C.; Bolonheis, S. M.; Kiyohara, P. K.; Toledo, S. P.; Colli, W.; Araki, K.; Farsky, S. H. In Vivo and In Vitro Toxicity and Anti-Inflammatory Properties of Gold Nanoparticle Bioconjugates to the Vascular System. *Toxicol Sci.* **2014,** *142*, 497–507.

Urbiola, K.; Blanco-Fernandez, L.; de Ilarduya, C. T. Nanoparticulated Polymeric Systems for Gene Delivery. *Curr. Pharm. Des.* **2015,** *21*, 4193–4200.

Wang, H.; Ke, F.; Mararenko, A.; Wei, Z.; Banerjee, P.; Zhou, S. Responsive Polymer–Fluorescent Carbon Nanoparticle Hybrid nanogels for Optical Temperature Sensing, Near-Infrared Light-Responsive Drug Release, and Tumor Cell Imaging. *Nanoscale* **2014a,** *6*, 7443–7452.

Wang, Y.; Liu, Y.; Luehmann, H.; Xia, X.; Brown, P.; Jarreau, C.; Welch, M.; Xia, Y. Evaluating the Pharmacokinetics and In Vivo Cancer Targeting Capability of Au Nanocages by Positron Emission Tomography Imaging. *ACS Nano* **2012,** *6*, 5880–5888.

Wang, Z.; Wang, Z.; Liu, D.; Yan, X.; Wang, F.; Niu, G.; Yang, M.; Chen, X. Biomimetic RNA-Silencing Nanocomplexes: Overcoming Multidrug Resistance in Cancer Cells. *Angew. Chem. Int. Ed. Engl.* **2014b,** *53*, 1997–2001.

Wilson, J. T.; Keller, S.; Manganiello, M. J.; Cheng, C.; Lee, C. C.; Opara, C.; Convertine, A.; Stayton, P. S. pH-Responsive Nanoparticle Vaccines for Dual-Delivery of Antigens and Immunostimulatory Oligonucleotides. *ACS Nano* **2013,** *7*, 3912–3925.

Wu, P.; Gao, Y.; Lu, Y.; Zhang, H.; Cai, C. High Specific Detection and Near-Infrared Photothermal Therapy of Lung Cancer Cells with High SERS Active Aptamer-Silver-Gold Shell–Core Nanostructures. *Analyst* **2013,** *138*, 6501–6510.

Wuithschick, M.; Birnbaum, A.; Witte, S.; Sztucki, M.; Vainio, U.; Pinna, N.; Rademann, K.; Emmerling, F.; Kraehnert, R.; Polte, J. Turkevich in New Robes: Key Questions Answered for the Most Common Gold Nanoparticle Synthesis. *ACS Nano* **2015,** *9*, 7052–7071.

Xiao, D.; Jia, H. Z.; Zhang, J.; Liu, C. W.; Zhuo, R. X.; Zhang, X. Z. A Dual-Responsive Mesoporous Silica Nanoparticle for Tumor-Triggered Targeting Drug Delivery. *Small* **2014,** *10*, 591–598.

Yan, H.; Wong, C.; Chianese, A. R.; Luo, J.; Wang, L. Y.; Yin, J.; Zhong, C. J. Dendritic Arenethiol-Based Capping Strategy for Engineering Size and Surface Reactivity of Gold Nanoparticles. *Chem. Mater.* **2010,** *22*, 5918–5928.

Yang, D.; Ma, P.; Hou, Z.; Cheng, Z.; Li, C.; Lin, J. Current Advances in Lanthanide Ion (Ln(3+))-Based Upconversion Nanomaterials for Drug Delivery. *Chem. Soc. Rev.* **2015,** *44*, 1416–1448.

Yang, H.; Du, L. B.; Tian, X.; Fan, Z. L.; Sun, C. J.; Liu, Y.; Keelan, J. A.; Nie, G. J. Effects of Nanoparticle Size and Gestational Age on Maternal Biodistribution and Toxicity of Gold Nanoparticles in Pregnant Mice. *Toxicol. Lett.* **2014,** *230*, 10–18.

Yang, H.; Fung, S. Y.; Liu, M. Y. Programming the Cellular Uptake of Physiologically Stable Peptide-Gold Nanoparticle Hybrids with Single Amino Acids. *Angew. Chem. Int. Ed. Engl.* **2011,** *50*, 9643–9646.

Yang, Y.; Liu, J.; Sun, X.; Feng, L.; Zhu, W.; Liu, Z.; Chen, M. Near-Infrared Light-Activated Cancer Cell Targeting and Drug Delivery with Aptamer-Modified Nanostructures. *Nano Res.* **2016,** *9*, 139–148.

Yavuz, M. S.; Cheng, Y.; Chen, J.; Cobley, C. M.; Zhang, Q.; Rycenga, M.; Xie, J.; Kim, C.; Song, K. H.; Schwartz, A. G.; Wang, L. V.; Xia, Y. Gold Nanocages Covered by

Smart Polymers for Controlled Release with Near-Infrared Light. *Nat. Mater.* **2009,** *8*, 935–939.

Yin, F.; Yang, C.; Wang, Q.; Zeng, S.; Hu, R.; Lin, G.; Tian, J.; Hu, S.; Lan, R. F.; Yoon, H. S.; Lu, F.; Wang, K.; Yong, K. T. A Light-Driven Therapy of Pancreatic Adenocarcinoma Using Gold Nanorods-Based Nanocarriers for Co-Delivery of Doxorubicin and siRNA. *Theranostics* **2015,** *5*, 818–833.

Yu, F.; Huang, J.; Yu, Y.; Lu, Y.; Chen, Y.; Zhang, H.; Zhou, G.; Sun, Z.; Liu, J.; Sun, D.; Zhang, G.; Zou, H.; Zhong, Y. Glutathione-Responsive Multilayer Coated Gold Nanoparticles for Targeted Gene Delivery. *J. Biomed. Nanotechnol.* **2016a,** *12*, 503–515.

Yu, Z.; Zeng, L.; Wu, A. Near-infrared Light Responsive Upconversion Nanoparticles for Imaging, Drug Delivery and Therapy of Cancers. *Curr. Nanosci.* **2016b,** *12*, 18–32.

Zhang, C.; Chen, B.-Q.; Li, Z.-Y.; Xia, Y.; Chen, Y.-G. Surface Plasmon Resonance in Bimetallic Core–Shell Nanoparticles. *J. Phys. Chem. C* **2015a,** *119*, 16836–16845.

Zhang, C.; Li, C.; Liu, Y.; Zhang, J.; Bao, C.; Liang, S.; Wang, Q.; Yang, Y.; Fu, H.; Wang, K. Gold Nanoclusters-Based Nanoprobes for Simultaneous Fluorescence Imaging and Targeted Photodynamic Therapy with Superior Penetration and Retention Behavior in Tumors. *Adv. Funct. Mater.* **2015b,** *25*, 1314–1325.

Zhang, H.; Kim, J. C. Spray-Dried Microparticles Composed of Cinnamoyl Poly(*N*-Isopropylacrylamide-*co*-Hydroxyethylacrylate) and Gold Nanoparticle. *J. Appl. Polym. Sci.* **2016,** *133* (43), 44141.

Zhang, Q.; Hitchins, V. M.; Schrand, A. M.; Hussain, S. M.; Goering, P. L. Uptake of Gold Nanoparticles in Murine Macrophage Cells without Cytotoxicity or Production of Pro-inflammatory Mediators. *Nanotoxicology* **2011,** *5*, 284–295.

Zhang, S.; Leem, G.; Lee, T. R. Monolayer-Protected Gold Nanoparticles Prepared Using Long-Chain Alkanethioacetates. *Langmuir* **2009,** *25*, 13855–13860.

Zhang, W.; Lin, W. H.; Pei, Q.; Hu, X. L.; Xie, Z. G.; Jing, X. B. Redox-Hypersensitive Organic Nanoparticles for Selective Treatment of Cancer Cells. *Chem. Mater.* **2016a,** *28*, 4440–4446.

Zhang, W.; Wang, F.; Wang, Y.; Wang, J.; Yu, Y.; Guo, S.; Chen, R.; Zhou, D. pH and Near-Infrared Light Dual-Stimuli Responsive Drug Delivery Using DNA-Conjugated Gold Nanorods for Effective Treatment of Multidrug Resistant Cancer Cells. *J. Control. Release* **2016b,** *232*, 9–19.

Zhang, X. Gold Nanoparticles: Recent Advances in the Biomedical Applications. *Cell Biochem. Biophys.* **2015,** *72*, 771–775.

Zhao, W.; Karp, J. M. Tumour Targeting: Nanoantennas Heat Up. *Nat. Mater.* **2009,** *8*, 453–454.

# PART II

# Vesicle-Based Drug Carriers

CHAPTER 5

# VESICULAR DRUG CARRIERS AS DELIVERY SYSTEMS

SANJA PETROVIĆ, SNEŽANA ILIĆ-STOJANOVIĆ*, ANA TAČIĆ, LJUBIŠA NIKOLIĆ, and VESNA NIKOLIĆ

*University of Niš, Faculty of Technology, Bulevar oslobodjenja 124, 16000 Leskovac, Republic of Serbia*

*Corresponding author. E-mail: ilic.s.snezana@gmail.com*

## CONTENTS

## ABSTRACT

This chapter provides an overview of the delivery systems that represent multifunction carriers of active substances for controlled, delayed, and targeted drug substance delivery, with a special review to the vesicular drug carriers. Drug carrier formulations are designed to connect a number of different possible actions depending on the needs of future purposes for which they are processed. The use of nanomaterials has a lot of preferences and is becoming more pronounced in recent years. Vesicular systems are one of the nanosystems for the drug substances delivery that has a lot of advantages. Considering that, vesicles as liposomes, sphingosomes, ethosomes, transferosomes, niosomes, discomes, bilosomes, aquasomes, enzymosomes, exosomes, and virosome will be comprehensively described. One of the most popular vesicular nanocarriers are liposomes that are classified to the organic, that is, natural materials. Thanks to the structure, liposomes can serve as carriers of the both hydrophilic and lipophilic substances. The process of encapsulation of the active compounds in liposomes is a challenge by itself and knowledge of liposomes character and structure, as well as the appropriate methods for liposome characterizing, are of a crucial importance. Another challenge regarding the use of liposomes is the question of their long-term stability. In addition to the listed advantages and disadvantages of such systems, more accurate knowledge of mechanism behavior of active components incorporated into the nanocarrier under the influence of many internal or external factors, as well as its optimal concentration and stability in the potential formulations, is of crucial importance for further research and use.


## 5.1 INTRODUCTION

Traditional treatments and conventional pharmaceutical dosage forms are mainly based on the immediate release of pharmaceutical active substances. Functional ingredients and innovative delivery systems/carriers lead to the development of a new era of modern products. High degree of sophistication of this new treatment approach is based on the remarkable manipulation. Place of drug release, the starting moment of its release, the interval time during which will be released as well as the amount of drug that will be dismissed over the time mostly depend on the carrier type. Today, consumers are looking for the products of personal care that offer multiple advantages over conventional preparations. Product users also expecting that the innovative technological processes provide safe and effective products. Faced with these trends, pharmaceutical and cosmetic industry strives to develop highly differentiated multifunctional products that will meet needs in quality treatment, care, and esthetics. Processes of drugs delivery based on carrier use have been developed in several basic directions and can be divided into:

- controlled,
- delayed (infected), and
- target drug substance delivery.

The way of control over the drug carrier usually refers to the precise modification of the surface characteristics, which involves the binding of ligands capable of specifically binding to the appropriate, targeted groups of receptors. Controlled-release drug-delivery system refers to the possible control of drug concentration at the target site (Martindale, 2005). An alternative way of directing the carrier with the drug at the target site refers to the ability to manipulate with the application of external factors as electric, magnetic, ultrasound field, etc. In that case, the formulation of the active substance with a suitable carrier must have characteristics which make it sensitive to external stimulants to achieve its movement and the possibility of monitoring (Ilić-Stojanović et al., 2016; O'Hagan et al., 1991; Park et al., 2010). This method of active substances delivery is currently in focus of many well-known companies because of the possibility of manufacturing targeted delivery products (at the cellular and subcellular level). In this way, it is possible to achieve higher specificity of treatment and by that better effect. The increased specificity of treatment is particularly important in the case of antibiotics therapies (Briones et al., 2008) and the anticancer medication (Ferrari, 2010; Kim et al., 2005; Prakash et al., 2010). Generally, systems with delayed drugs release are applied to maintain the concentration of the drug in the systemic circulation in sufficient therapeutic dose at appropriate time. The drug delivery in that case is performed by first-order kinetics. Kinetics of zero-order release is a feature of the formulations in which the drug release takes place regardless of the amount of drug in the delivery system, that is, release is carried out at a constant speed of delivery. Systems for target delivery of therapeutic/active substances include the carrier with incorporated active substance that protects the medicinal active substance of any adverse environmental impact that the active substance experiencing due to the passage to the target. The most commonly used systems for this purpose are liposomes, micro, and nanoparticles. The differences between the carriers are in size, formation, structure, etc. A large number of different drug carriers have been developed in accordance to potential application needs. Their constant improvement follows the upward trend of requests that define the convenience of modern materials for biomedical application. Overall, so far, developed carriers of drugs can be divided into three groups:

- synthetic,
- natural, and
- the cell carriers.

Multifunction carriers of active substances include formations as liposomes, micelles, nanoemulsions, polymeric particles, and many other carrier formulations that are designed in a specific way to connect a number of different possible actions depending on the needs of future purposes for which they are processed. Some of the most popular delivery nanocarriers are presented at Figure 5.1.

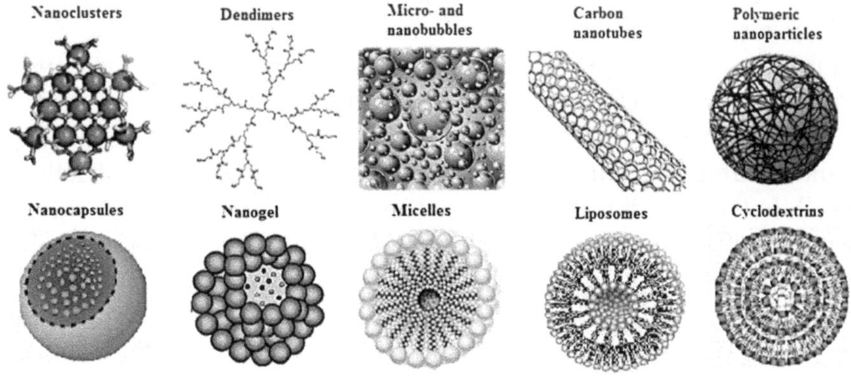


**FIGURE 5.1** Common type of delivery nanocarriers.

Many nanocarriers such as virosomes, niosomes, and ethosomes are not presented in Figure 5.1. The so-called virosome possesses attractive characteristics as a vaccine carrier and adjuvant system (Felnerova et al., 2004). As a class of molecular cluster formed by self-association of nonionic surfactants in an aqueous phase, niosomes represent a unique effective structure of drug-delivery system (Marianecci et al., 2014; Moghassemi and Afra, 2014; Pardakhty, 2016). Similar to liposomes, ethosomes are vesicular carriers composed of phospholipid, water, and ethanol where ethanol has a fluidizing effect on the phospholipid bilayer as well as intercellular pathway in the skin (Ainbinder et al., 2016; Saudagar et al., 2016; Sujatha et al., 2016).

## 5.2 NANOPARTICLES

Nanoparticle systems include nanospheres and nanocapsules (presented at Fig. 5.1) that can be defined as a submicron colloidal systems with a median diameter of 0.003–1 μm. Nanocapsules are reservoir type of systems, while nanospheres are matrix type. The active ingredient in nanocapsules and nanospheres may be incorporated in various ways: dissolved in the matrix nanospheres, adsorbed onto the surface of the nanospheres, dissolved in the liquid phase of nanocapsules, adsorbed onto the surface of the nanocapsules, etc. In the case when the active substance is absorbed onto the support surface, liberation of the active substance from the polymeric nanoparticles is biphasic with an initial burst phase followed by its sustained release (Kwangjae et al., 2008; Patravale and Mandawgade, 2008). Nanoparticle systems can be obtained using a variety of materials such as lipids—liposomes, polymers—polymeric nanoparticles, micelles, dendrimers, viruses—viral nanoparticles, and even organometallic compounds—nanotubes (Kwangjae et al., 2008; Yogeshkumar et al., 2009). Nanoparticles are very stable and have a high affinity for *stratum corneum*, hence high bioavailability of encapsulated active substances to the skin, making them strongly favors precisely in the field of cosmetology among other applications. The nanoparticle penetrate the upper layers of the *stratum corneum* where it merged with the lipids of the skin where the active agents release. Compared to the liposome, nanoparticle can be used for effective delivery of lipophilic substances. Solid-lipid nanoparticles (SLN) may be a promising compound for hydrating cosmetic formulations. Because of their excellent physical stability and compatibility with other additives, they can be applied in many cosmetic formulations without any particular problems.

## 5.3 VESICULAR DRUG CARRIERS

Various lipid and nonlipid vesicles are classified as vesicular drug carriers. Vesicular carriers are essentially lipid or nonlipid particles organized into concentric bilayer particles separated by a layer of water as shown in Figure 5.2. They are all formed with the same purpose to emphasize the controlled drug delivery. The main purpose of these systems is to improve active substance solubility, the therapeutic index, the stability, and rapid release of active drug molecule. The structure of all vesicular drug carriers is primary similar, but still there are differences that have designated them to the liposomes, sphingosomes, ethosomes, niosomes, discomes, transferosomes, aquasomes, bilosomes, etc. Most of them are presented in Figure 5.2.

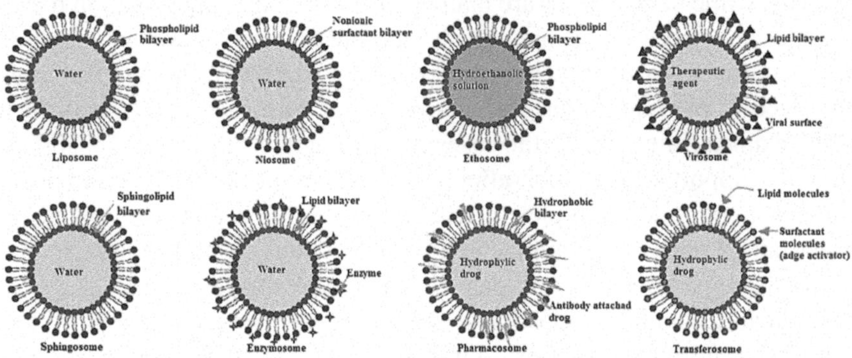


**FIGURE 5.2** **(See color insert.)** Structure differences between some vesicular drug carriers.

### *5.3.1 LIPOSOMES*

Liposomes are lipid vesicles that contain water inside the oval lipid bilayer structure made of phospholipids. They can serve as oral or topical drug carriers. Different types of phospholipids can influence the properties and stability of prepared liposomes, but of all, PC lipids—phosphatidylcholine (PC), are the most commonly used. Liposomes are widely used and as the most widespread vesicles, they will be described in the following. Liposomal formulations can be roughly divided to the rigid vesicles (liposomes and niosomes) and elastic vesicles (transferosomes and ethosomes) (Rajan et al., 2011).

### 5.3.2 SPHINGOSOMES

Unlike liposomes, lipid bilayer of sphingosomes consists of sphingolipids instead of phospholipids. Sphingolipids have a sphingoid base that contains aliphatic amino alcohols including sphingosine. The main advantage of sphingolipids is that they are less liable to oxidation reactions compared to phospholipids. Due to similarity between sphingosomes and biological membranes, sphingosomes can be used as carriers for different cosmetic and pharmaceutical substances in antifungal, cancer, and gene therapy as well as for enzyme delivery. Prostaglandins, amphoterecin B, methotrexate, cisplatin, vincristine, vinblastine, doxorubicin, camphothecin, ciprofloxacin, progesterone, and topotecan are some of active substances incorporated in sphingosomes (Ashok et al., 2013; Lankaldapalli and Damuluri, 2012; Webb et al., 1998).

### 5.3.3 ETHOSOMES

Ethosomes are suitable carriers for transport of substances which have poor skin penetration. Similar to the liposomes, ethosomal carriers are also able to provide an effective intracellular delivery of hydrophilic, as well as lipophilic compounds (Bhalaria et al., 2009; Elsayed et al., 2006; Jain et al., 2004, 2007). Ethosomes represent lipid type of vesicles. Ethanol increases the penetration capacity of ethosomal vesicle. Because of that, ethosomes are suitable carriers for cosmetic active substances. Ethanol content in the ethosomal vesicles varies from 20% to 50%. The ethosome size depends on ethanol content; higher amount of ethanol causes the smaller vesicle size (Kamboj et al., 2013).

### 5.3.4 TRANSFEROSOMES

Transferosomes are the most deformable lipid vesicular carriers. Transferosomes, with the size of 500 nm, are mainly composed of phospholipids, surfactants, alcohol, and hydrating medium (Pirvu et al., 2010; Rajan et al., 2011). They can easily pass through pores that are 5–10 times smaller than the transferosomes size, because of high elasticity which provides good skin penetration without any losses (Kamboj et al., 2013; Rajan et al., 2011). The concentration of surfactant is of crucial importance for transferosomes properties because high surfactant concentration may lead to the vesicles destructions. They provide a good transfer efficiency of high molecular weight compounds and have a better entrapment efficiency of lipophilic compounds. They can be used as carriers of various proteins (Paul et al., 1995), anticancer drugs (Alvi et al., 2011), antifungal drugs (Abdallah, 2013), analgesic and antiinflammatory drugs (Cevc and Blume, 2001), anesthetics (Planas et al., 1992), corticosteroids (Cevc et al., 1997), and hormones (Cevc et al., 1998).

### *5.3.5 NIOSOMES*

Niosomes are lipid vesicles used for the transport of hydrophilic as well as lipophilic molecules. They are a cheaper nonbiological alternative to liposomes, suitable for topical and transdermal application (Coviello et al., 2015; Pardakhty, 2016; Sezgin-Bayindir et al., 2015). Contrary to the phospholipid liposomes bilayer, niosomes have bilayers made of surfactants. For their preparation, besides amphiphile and water, stabilizers are also used to prevent aggregation of vesicles. For the same reason, cholesterol is also commonly used (Kamboj et al., 2013).

### *5.3.6 DISCOMES*

Discomes are very similar to the niosomes. Their size is 12–60 μm. As niosomes, they are derived from nonionic surfactants. Discomes are usually used for ocular drug delivery, as the large size of vesicles does not allow drainage (Kaur et al., 2004).

### *5.3.7 BILOSOMES*

The incorporation of deoxycholic acid in niosomes produces vesicles classified as bilosomes. Bilosomes are more stable compared to niosomes because of bile salts in their structure. They are usually used in therapy of gastrointestinal tract, as vesicles of high bioavailability that can be rapidly absorbed through intestine to the portal circulation (Al-Mahallawi et al., 2015; Chilkawar et al., 2015; Guan et al., 2016).

### 5.3.8 AQUASOMES

Aquasomes are nonlipid vesicles which consist of solid nanocrystalline core coated with three layers of carbohydrates. Active substances are adsorbed on the carbohydrate surface. Nanocrystalline core provides structural stability until carbohydrate coating protects active molecules from dehydration. Aquasome size is usually around 60–300 nm (Goyal et al., 2008; Shahabade et al., 2009; Umashankar et al., 2010). Self-aggregation of vesicles is achieved by noncovalent or labile van der Waals forces.

### 5.3.9 ENZYMOSOMES

Bilayers in which enzymes are covalently immobilized or coupled to the surface are known as enzymosomes. Enzymes can be directly connected to the surface of phospholipid bilayer or through the polyethyleneglycol chain (Vale et al., 2006). Enzymes such as alkaline phosphatase, carboxypeptidase, superoxide dismutase, β-glucosidase, and β-lactamase are the most common enzymes immobilized at liposomes to be used in therapy (Fonseca et al., 1998; Hundekar et al., 2015; Vale et al., 2006). By enzymosomes use, longer circulation half-life and accumulation at the action site can be managed.

### 5.3.10 VIROSOME

Virosomes are lipid type of vesicles, with the size of 150 nm, which compared to the liposomes contain viral envelope proteins incorporated in the phospholipid bilayer. They can be used as carriers of vaccines and adjuvants as well as nucleic acids and active substances (Dong et al., 2016; Felnerova et al., 2004; Swain et al., 2016).

### 5.3.11 EXOSOMES

Exosomes are lipid carriers with the size of up to 100 nm (Simons and Raposo, 2009). They can be secreted by most cells of the human organism and isolated from several extracellular fluids, such as blood, saliva, urine, amniotic fluid, and cerebrospinal fluid. The main difference between liposomes and exosomes is the surface composition complexity. The presence of different proteins and specific lipids in the membrane of exosomes vesicles facilitates targeted drug delivery and drug uptake by the cells. The most abundant proteins in the exosomes membranes are tetraspanins, heat shock proteins, lysosomal proteins, and fusion proteins (Conde-Vancells et al., 2008). Besides proteins, exosomes contain different lipids, such as cholesterol, sphingomyelin, ceramide, phosphoglyceride, and saturated fatty acids (Qin and Xu, 2014). Exosomes as drug-delivery systems are biocompatible, nontoxic, with a long circulation half-life. RNA, curcumin, and doxorubicin are usually incorporated in exosomes (Ha et al., 2016).

Genosomes, vesosomes, photosomes, proteosomes, layersomes, cryptosomes, emulsomes, ufosomes, cubosomes, herbosomes, and colloidosomes are carriers which are not only presented within this title but also represent a vesicular type of carriers.

## 5.4 DELIVERY SYSTEMS—BIOLOGICAL MODEL MEMBRANES

One of the ideas that have commissioned delivery systems is a need to bring them closer to the biological membranes. Biological membranes are lipoprotein composition, that is, they are composed of a lipid bilayer, in which are housed the proteins, until the standard lipid matrix is mainly phospholipid composition. All biological membranes have a similar structural organization, as well as many other common traits.

The major classes of lipids that constitute the phospholipid membranes are sphingolipids, cholesterol, as well as glycolipids and cardiolipins, in small quantities. Different membranes require different containing amounts of the individual lipid class. Lipids affect the physical properties of the membrane such as viscosity and permeability (Gennis, 2013; Milenkovic et al., 2013; Petrović et al., 2014). The most commonly used models of biomembranes—which are formed in an aqueous medium—are micelles, monolayers, bilayers, liposomes, etc. (Fig. 5.3).

Micelles are stable structures due to the interaction of polar groups with water and the forces of attraction between the hydrocarbon chains in the interior of the micelle (Wang et al., 2012). Micelle formation occurs only at the appropriate concentration of lipid which is defined as the critical micelle concentration (CMC) (Kataoka et al., 2001; Mukerjee et al., 1971). On the other hand, large lipid molecules in aqueous solutions can be organized in a two-layer structure: polar "heads" outward and hydrophobic "tails" to the interior. These two-layer structures can be created and closed to the spherical shape, as shown in Figure 5.3, which completely separates the outside from the interior space, known as liposomes (Gregoriadis, 2007, 1984; Petrović, 2016).

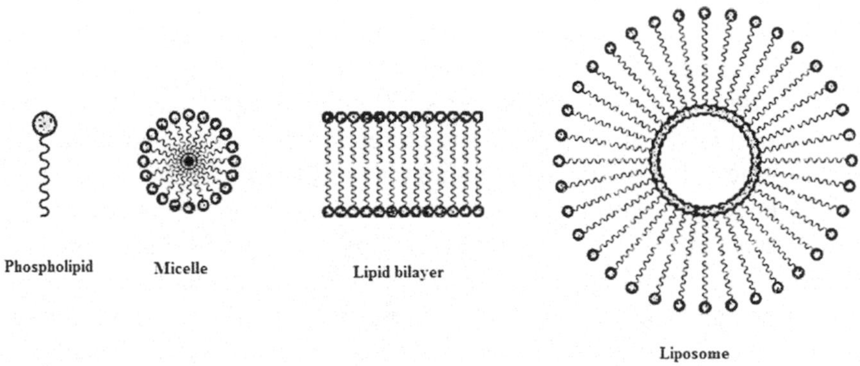


**FIGURE 5.3** Organized lipid structures.

## 5.5 LIPOSOMES, CHARACTER, AND STRUCTURE

Liposomes are microscopic spherical particles whose membrane, which consists of one or more lipid bilayers, "closes" a part of the aqueous phase in which the liposomes are suspended (Fig. 5.3). English scientist Alec D. Bangham in 1960s made the first formulation based on liposomes. The lipid bilayers of liposomes are composed of lipids whose hydrophilic parts are directed to the aquatic environment (heads), while the lipophilic moieties are turning to each other to form a hydrophobic interior of liposomes (Vemuri and Rhodes, 1994). Liposomes are commonly used as carriers for the delivery of a drug substances, which can be incorporated not only into the lipid bilayer (if lipophilic) but also in an aqueous "core" (if hydrophilic) (Petrović, 2016; Vuleta et al., 2003). Namely, the encapsulation of therapeutic bioactive substances in liposomes can increase its penetration into the deeper layers of the skin, which may cause the accumulation and formation of a "depot" of the drug substance that provides a sustained release while reducing the systemic absorption and side effects (Kee-Yeo and Lee, 2006; Meybeck, 1992; Nastruzzi et al., 1993; Yang et al., 2009). Encapsulation of pharmaceutical substances in liposomes also increases the stability of the active substance, which prolongs their shelf-life (Stuchlík and Žák, 2001).

### 5.5.1 LIPIDS—THE BASIC LIPOSOME INGREDIENTS

Phospholipids are natural constituents of cell membranes which make them biodegradable, nontoxic, and nonimmunogenic. They consider as an ideal carriers of bioactive substances. Phospholipids are one of the main groups of lipids in animal, plant, and bacterial membranes. Phosphate part of the structure is polar (hydrophilic) part of phospholipids—polar head of the molecule, until the long-chain unsaturated and/or saturated fatty acids represent nonpolar (hydrophobic) part of phospholipid molecule (Fig. 5.4). Whether the synthesis of phospholipid molecules involved glycerol or sphingosine, phospholipids are divided into glycerophospholipids or sphingophospholipids (Blume, 1992; Borst et al., 2000). Phospholipids are soluble in organic solvents or their mixtures. Due to their amphiphilic nature, they are forming the aggregates in an aqueous medium above the CMC of $10^{-8}$ M. The aggregate structure depends on the chemical structure of the polar head, length, and degree of fatty acids chain saturation, as well as pH and ionic strength of aqueous solution.

The aggregates usually form lamellar structures, but under certain conditions can form the cubic and hexagonal structures (Blume, 1992). Lamellar structure can occur in two phases, gel phase (lipid molecules are organized in the form of quasi-two-dimensional crystal lattice) or liquid-crystal phase (lamellar lipid arrangement of molecules). The phase state of the membrane affects the physical properties of the liposome. The permeability and the fluidity of the liquid crystal layers of the lipid are greater than the permeability and fluidity of the layers in the gel state (Chapman, 1975; Ilya and Veatch, 2016). The transition from one phase to another is reversible and dependent on temperature. Above the critical phase transition temperature ($T_m$), lipids are present in a liquid crystal form. The nature of the polar head of phospholipids, origin of lipids, and the ramified-chain fatty acid chain branching determinate a reduction of $T_m$. Phospholipids whose polar heads make intermolecular H-bonds with neighboring molecules have higher $T_m$. For phospholipids isolated from egg yolk, $T_m$ is placed from −15 to −7°C.

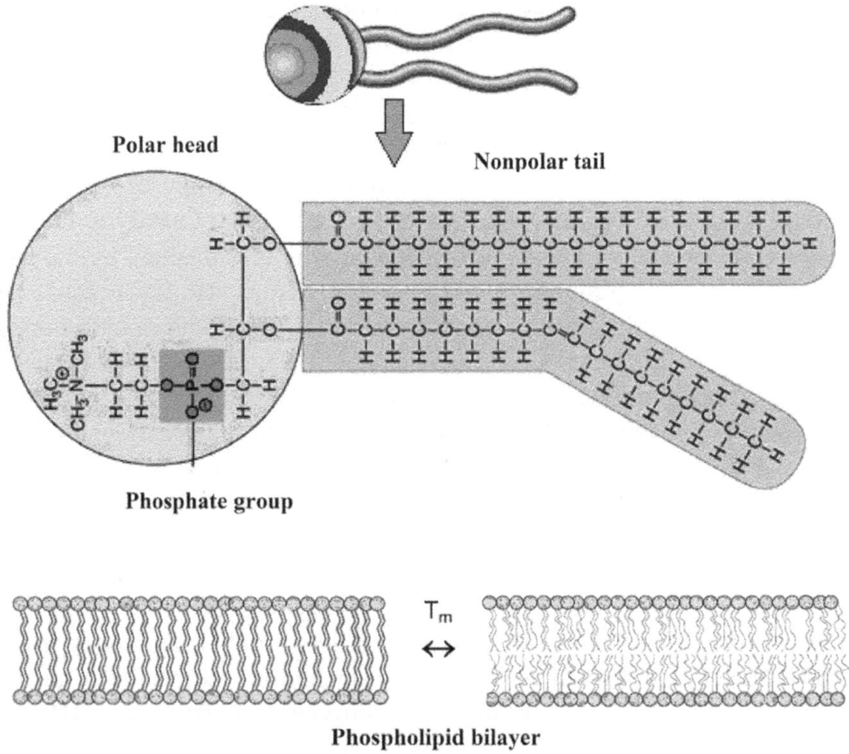


**FIGURE 5.4** Schematic presentation of phospholipid structure and lipid bilayer phase behavior, a gel state (left) and liquid crystal state (right). $T_m$ is a critical phase transition temperature.

Liquid-crystalline bilayers have higher permeability and fluidity than the bilayers in the gel state. At $T_m$, a liquid-crystalline and gel phases are simultaneously present in different parts of the liposome membrane. Due to the presence of both phases at the interfaces of unevenly packaging of phospholipids can be form leaking place from where substances from liposomes can be lost. Thickness of the membrane in the gel phase is greater than the liquid-crystalline phase, so desired physical and chemical properties or selectivity of the target can be achieved by modifying the structure of fatty acid chains or modifications of the polar head of phospholipids.

The most important type of phospholipids that is of importance for the liposomes formation is PC. PC can be obtained from a vegetable raw materials (seeds of soybean, corn, beets), or animal origin (animal liver and egg yolk). Esterified fatty acids differ in the number of carbon atoms and the degree of unsaturation. The most common fatty acids, which are part of the phospholipids, are palmitic acid (16:0), stearic (18:0), oleic (18:1), linoleic (18:2), linolenic (18:3), and arachidonic acid (20:4) (Ioannis et al., 2010; Lingwood and Simons, 2009; London, 2002; Rawicz et al., 2000). In addition to the unsaturated phospholipids, for the liposomes formation are also used saturated phospholipids obtained by catalytic hydrogenation of phospholipids, as well as those who are obtained by natural sources isolation. These phospholipids have greater stability because of the reduced possibility of peroxidation since there are no double bonds in their structures (Dragičević-Ćurić et al., 2005, 2009). Usually, used saturated phospholipids are dipalmitoyl phosphatidylcholine (DPPC), dipalmitoyl phosphoserine (DPPS), dipalmitoyl phosphatidylethanolamine, dipalmitoyl phosphatidic acid, and of unsaturated phospholipids, dioleyl phosphatidylcholine, and dioleoyl phosphatidylgycerol, with one unsaturated bond in the fatty acid segments. Saturated phospholipids with long fatty acid chains are used to form rigid liposomes, nonpermeable membranes. Not only the stability but also a way of incorporation of active components depends on saturation, of the used lipids. Sometimes, even lipids of a similar structures make differences that are primarily manifested through differences in membrane curvature of DPPC and DMPC lipid liposomes, for example, (Milenkovic et al., 2012a, 2012b; Petrović et al., 2014; Tugulea et al., 2013). Depending on the length of phospholipid branches, active components can be inserted into the membrane by different profundity.

### 5.5.2 LIPOSOME CLASSIFICATION

Liposomes can be classified according to the preparation method technology, by layers number in their structure and by size. Most commonly liposomes are classified according to their layered structure and size, while classification by obtained preparation method was very rarely used. The most common ways of classification of liposomes are shown in Table 5.1. There are various methods of liposomes preparation of different sizes and lamellarity. Preparation of the liposomes that include use of ultrasound or extrusion methods is common methods for the unilamellar vesicles preparation (Dua et al., 2012). High-energy sonication is generally used for the synthesis of small liposomes, so-called small unilamellar vesicles (SUV), where the size of the obtained liposomes is in a function of frequency and power of ultrasound and sonication time (Cheng and Gang, 2011). The so-called extrusion method that implies the use of the extruder can be used to generate the SUVs or large unilamellar vesicles (LUV). According to these methods, the liposomes preparation starts by formation a dry lipid film that is hydrated by aqueous solution (most common phosphate buffer solution) and subjected to successive freezing and thawing procedures until the formation of multilamellar vesicles (MLV) (Akbarzadeh et al., 2013; Cheng and Gang, 2011). The suspension is then exposed to the high-frequency ultrasound or extruded through polycarbonate membranes with defined nanopores (Cheng and Gang, 2011; Olson et al., 1979; Yogita and Jadhav, 2014, 2015). Methods for the liposome preparation may also be modified based on the properties, hydrophobicity, and the desired particle size. According to the method of application, liposomes as drug carriers can be used for external (topical) or internal use. When liposomes are used for the transport of bioactive substances, then their size is an important factor for the pharmacokinetics and biodistribution properties, for example may affect the efficiency of the so-called delivery specific types of bioactive substances, like photosensitizers in the photodynamic method or photodynamic therapy (PDT)[1]. One of the photosensitizers that can be used is chlorophyll (Allison and Sibat, 2010; Markovic and Zvezdanovic, 2012; Moser, 1998; Petrović, 2016). Size control of formed liposomes is an important factor in the pharmaceutical formulation and should be optimized to improve the pharmacological properties of pharmaceuticals. SUV liposomes are different from other types of liposomes by a large curvature, which resulted in large membrane potential and therefore increased internal permeability and thermodynamic instability of the system. On the other hand, that kind of system has a tendency toward aggregation and fusion, especially below the phase transition temperatures. SUV liposomes exhibit other disadvantages, such as necessity of applying a high energy (high frequency) for the preparation which can lead to oxidation and hydrolysis of the lipids (Milenkovic et al., 2013; Petrović et al., 2014, 2016). SUVs have a great permeability for hydrophilic molecules. The advantage of small size SUV liposomes is better distribution in the body after intravenous administration.

LUV liposomes have a negligible membrane voltage, which results in a greater stability compared to SUV liposomes. Intermediate unilamellar vesicles are classified between SUV and LUV liposomes. Because of the large

**TABLE 5.1** Nomenclature and the Approximate Size of Different Types of Liposomes.

| | Classification | Abbreviation | Size (μm) | Schematic view |
|---|---|---|---|---|
| By size | Small unilamellar vesicles | SUV | 0.025–0.05 | 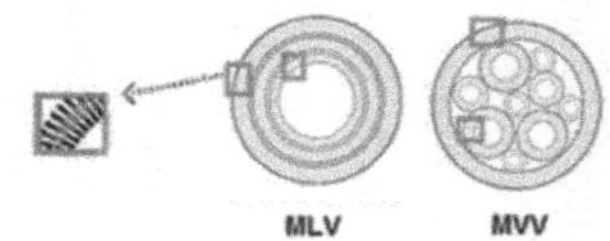  |
| | Large unilamellar vesicles | LUV | >0.1 | |
| | Giant unilamellar vesicles | GUV | >1 | |
| By lamellarity | Multilamellar vesicles | MLV | 0.05–10 | 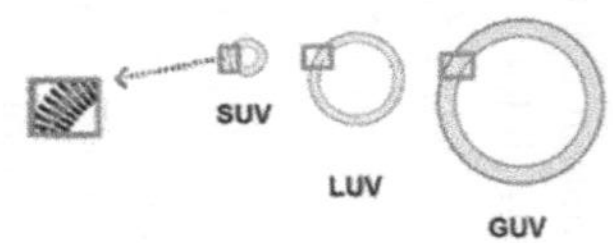  |
| | Unilamellar vesicles | ULV | 0.0225–0.1 | |
| | Multivesicular vesicles | MVV | ~0.4–10 | |
| By preparation method | Dehydration rehydration method | DRV | – | No preview |
| | Reverse phase evaporation method | REV | – | |
| | Vesicles prepared by extrusion technique | VET | – | |
| | Frozen and thawed | FAT | – | |

internal volume, they have a high encapsulation efficiency of hydrophilic substances that can be reached using the method of reverse-phase evaporation and the double emulsification. Due to the presence of only one of the lipid bilayer, their mechanical stability and the ability to retain the encapsulated substance are not as big as multilamellar vesicles (MLV) or a multilamellar large vesicle obtained by reverse-phase evaporation (REV-MLV Reverse-Evaporation Vesicles).

The liposomes, in which the two or more vesicles located next to each other are encapsulated in a larger vesicle, are known as multivesicular vesicles (MVV). The structure of these liposomes is explained by Bangham and his associates (1965). These structures are frequently encountered when creating MLV liposomes. Diameter of these liposomes is usually greater than 1 μm. MLV liposomes have sizes in a range of 100 nm to several microns and have a large number of concentric bilayers. The characteristic of these liposomes is sustained releases of encapsulated hydrophilic substances due to the large number of lamellas, which can cause the "depot effect" (Mravec et al., 2011). Thanks to a large number of lipid membranes, these are ideal carriers for lipophilic substances, while, on the other hand, the entrapment efficiency of the hydrophilic substance is low when conventional methods of dispersion are utilized.

Surface charge of the liposomes, determined by used phospholipids properties, is a very important parameter that defines the aggregation and sedimentation of liposomes as well as the cells–liposomes interactions (fusion or endocytosis). Surface charge of particles is determinate by zeta-potential and is used as a criterion for refusal between the particles. The greater zeta-potential leads to greater forces of repulsion and stability of the dispersion.

### 5.5.3 THE ENCAPSULATION OF SUBSTANCES IN LIPOSOMES

Thanks to the structure, liposomes can serve as carriers of the both hydrophilic and lipophilic substances. Depending on the solubility of the bioactive substance and its distribution coefficient within the lipid bilayer, the localization of the active substances is different. For consequence, the encapsulation as well as degree of efficiency and release rate of the active substances may be variable. Liposome encapsulation of the active substance means transportation way to not only hydrophilic bioactive substances but also permeation barrier, which can have a "depot effect" as a result. Permeation rate depends on the lipid properties of liposomal membrane and on the other side of the molecular size and lipophilicity of the active ingredient (De Gier, 1993; De Gier et al., 1968; Mahrhauser et al., 2015).

Bioactive substances are classified according to the partitioned characteristics to the hydrophilic, lipophilic, amphiphilic (good biphasic solubility between the inner aqueous and lipid phase), and those that are completely biphasic insoluble (this candidates are most problematic for incorporation into liposomes). Of all, lipophilic substances are the best candidates for incorporation into liposomes concerning the stability, cost, and usefulness. Therefore, a large number of drug substances of different solubility are structurally altered to obtain a lipophilic compound and the optimal liposome formulation. Lipophilic interior of a two-layer liposomes membrane allows the incorporation of the lipophilic substances, whereby the concentration ratio of the phospholipid and a lipophilic substance will depend on the properties of the substance and phospholipid, as well as the liposomes preparation procedure. The encapsulation efficiency of the lipophilic substances often goes to 100% if the substance is not used in a concentration greater than its solubility in liposomal lipids. The efficiency of encapsulation depends on liposome lamellarity, that is, MLV liposomes have a significantly higher capacity for lipophilic substances compared to unilamellar (Giuseppina and Molinari, 2015; Luciani et al., 2016; Vijaykumar and Kalepu, 2015). With regard to the amphiphilic substances, they tend to be adsorbed on the surface of the liposome.

### *5.5.4 LIPOSOME STABILITY*

Physical, chemical, and microbiological stability determinates the overall liposome stability. Physical stability is reflected by leakage of active substances that should be kept to a minimum or should not be present at all. Leaking of substances depends on the nature of the lipid bilayer and the characteristics of the active substance. The polar water soluble substances of high molecular weight show less leakage predispositions compared to the amphiphilic low-molecular weight substances. Aggregation processes should also be avoided, that is, flocculation of liposomes. Physical stability of liposomes is depended on lipid composition and preparation methods (e.g., liposomes obtained by detergents dialysis are the most stable and the least stable methods are sonication and microfluidization). Also, charged liposomes are more stable than neutral liposomes, because the reduced fusion and aggregation (Pippa et al., 2013; Tan et al., 2013). Chemical stability of liposomes is followed by the degree of hydrolysis of the ester bonds and oxidation of the unsaturated fatty acids of phospholipids in the lipid bilayer. Oxidation process is reflected in the following reactions: separation of hydrogen atoms of the $CH_2$ groups of the lipid chain (*initiation*), the double bond migration to form conjugated diene and triene, hydroperoxide formation, and hydroperoxide decomposition to aldehydes which results in shortening of the fatty acids chain. Oxidative stress, which occurs in the cell membrane, or can be induced in biomembrane models as liposomes are, can be as one of the possible consequences of a so-called lipid peroxidation (LP), that is, oxidation of the lipid component of biomembranes which lead to the formation of the peroxidic structures. The initiators and promoters of LP in vivo and in vitro are usually free radical ($O_2^-$, OH·) or non-ROS species ($^1O_2$, $O_3$), whether they are internally or externally present in observed system. LP can be initiated photochemically by some photosensitizer, such as chlorophyll (Petrović, 2016) or benzophenone that has great selectivity both in solution (Marković and Patterson, 1989), as well as organized systems, micelles (Marković and Patterson, 1993; Marković et al., 1990), or compressed monolayers (Marković, 2001; Moser, 1998). The content of LP can be quantitatively evaluated through appropriate tests, such as conjugated diene and TBA–MDA test. The conjugated dienes test is a spectrophotometric method that is based on the absorption of ultra violet light in the area of 230–235 nm, that is characteristic of conjugated diene structures with alternating arrangement of single and double bonds (–C=C–C=C–) (Halliwell and Chirico, 1993; Wheatley, 2000). Some authors are using the absorbance ratio at 234 and 215 nm, the so-called peroxidation index (Gregoriadis, 2007, 1984; Torchilin and Weissig, 2003). General lack of the method is weakly expressed peak at 234 at low diene concentrations (Mimica-Dukić, 1997). On the other hand, TBA–MDA test is rapid, simple, and sensitive and is used for measuring the amount of malondialdehyde (MDA) present in the sample. MDA is a secondary product, formed as a degradation product of peroxide fatty acids having at least three unsaturated double bonds between which is a methylene group. The result of reaction is pink colored complex whose absorption maximum lies in the range of 532–535 and 245–305 nm. The oxidation that occurs in liposomes is also possible in the case of saturated lipids liposomes, but compared with unsaturated lipids liposomes is being far smaller (Petrović, 2016).

To prevent oxidation, saturated synthetic phospholipids should be used as well as nonoxigen solvents. The liposome preparation methods that don't require high temperatures and sonication are also preferred. It is also preferred that storage of liposomes is in a stream of inert gas. The addition of antioxidants (vitamins C and E) also increases stability as well as the addition of disodium EDTA to remove heavy metals that catalyze the oxidation. Microbiological stability of liposomes which actually means the sterility of liposomes can be achieved by bacteriological filtration, by γ-radiation (that could destabilize the liposomes) (Nicolosi et al., 2015). Sterilization under high pressure, ethylene oxide, or by autoclaving is also in use. Preservatives sterilization by parabens can be standing when it embedded in the bilayer membrane (Ioele et al., 2016) and this is expressed when the hydrophilic compounds are encapsulated and when the liposome membranes are not stabilized by the addition of vitamin E and phytosterols.

### *5.5.5 METHODS FOR LIPOSOME CHARACTERIZATION*

The objective of each formulation, weather pharmaceutical or cosmetic, is the sustainability, stability, and quality. The stability of liposomes is important for the stability of liposome-based products. The stability and the liposome quality is determined by following parameters: liposome lamellarity, the homogeneity (size and size distribution), the internal volume and encapsulation efficiency, the release of the encapsulated substance, the liposome fluidity (liposome membrane thermodynamic state), the surface charge of the liposome as well as microbiological and toxicological tests. The lamellarity determination implies the number of lamellae (phospholipid bilayers in a liposome), which usually obtained by electron microscopy method (Chiba et al., 2014; Hoppe et al., 2013; Klang et al., 2013) and confocal laser scanning microscopy (Pozzi et al., 2014). To determine the liposome size, the dynamic dissipation of laser light, optical microscopy, electron microscopy, fragmentation under the influence field, and turbidimetric method are commonly used. The internal liposome volume can be measured directly by determination the amount of water inside the liposomes using the NMR method where the external water media must be replaced by spectroscopically inert liquid, or indirectly by applying water-soluble markers $^{14}C$ and $^{3}H$–$^{3}H$-glucose or sucrose which mark the encapsulated aqueous phase to get the percentage value of encapsulated substance. Increasing the encapsulation efficiency of water-soluble substances is achieved by increasing the liposomes size at a constant lipids concentration. The substances release, that is, leaking of substances from liposomes in pharmaceutical preparations, can be followed using the detection methods and special markers as glucose or carboxyfluorescein (Liu and Ben, 2013). The surface charge of the liposomes is depended on many factors like type of phospholipids that form a bilayer membrane and the molecule type that enters into the composition of the phospholipids polar heads. Phospholipids, that is, liposomes, can be neutral, negatively, and positively charged. Surface charge is characterized by zeta potential. Higher zeta potential means the greater repulsion between liposomes and therefor the liposome stability is greater (Műller, 1996). Zeta potential is important parameter for the in vivo liposome administration monitoring because it affects the distribution in the body and the clearance from the circulation (Murata et al., 2014; Soheyla and Zahir, 2013). Zeta potential depends on the ionization degree of phospholipids molecules which depends on the dispersion pH value (Műller, 1996).

### *5.5.6 LIPOSOME USES IN DIFFERENT INDUSTRIES*

The tendency of consumers has oriented the recent scientific research toward organic or natural products. All cosmetic products placed on EU market must meet the requirements of the EU Regulation 1223/2009 on cosmetic products. According to the latest EU regulations, all the cosmetic or pharmaceutical products must satisfy the Cosmos regulative—Cosmetics Organic and Natural Standard (Cosmos-Standard, 2011). In addition, Organic and Sustainable Industry Standards (OASIS) provide similar regulations. All the formulation ingredients must meet those regulations. Liposomes are nanoparticle systems that unlike other nanoparticles can be applied in so-called products that are classified as organic or natural. That is another advantage of liposome use when it comes to narrowing development.

Due to their specific characteristics, liposomes are widely used in different areas of industry, including pharmaceutical, cosmetic, and food industry. In pharmaceutical industry, liposomes are used as drug-delivery systems for systemic or local application. They can be used in form of liquid, semi-solid, or solid preparations, which can be applied topical or parenteral. The role of liposomes as drug carriers can be the protection of encapsulated drugs from the enzymatic degradation, increasing the solubility of lipophilic and amphiphilic drugs, improvement of the drug penetration into cells, presentation of the antigen on their surface, and targeted drug delivery (Laouini et al., 2012). Different pharmaceutical active substances can be incorporated into liposomes, such as antibiotics (Gaspar et al., 2008), antiparasitic agents (Lopes, 2012), antineoplastics (Hirai et al., 2010; Zhang et al., 2005), glucocorticoids (Bhardwaj and Burgess, 2010), and others. An overview of some commercial drug-delivery systems based on liposomes is presented in Table 5.2.

Liposomes can be used for incorporation and topical delivery of cosmetic active substances. Preparations based on liposomes represent the new generation of products designed for skin antiaging and protection. Lipids present in liposomes hydrate the skin and reduce the dryness of the skin, which is the primary cause of skin aging. Betz et al. (2005) studied *in vivo* the effect of liposomes prepared from different lipid components on skin elasticity, water content, and barrier function. It has been shown that liposomes prepared from egg phospholipids have significantly higher moisturizing effect than liposomes prepared from soya phospholipids. Reduction of phospholipids content in liposome formulations decreased the moisturizing effect of preparation. An overview of some cosmetic products based on liposomes is presented in Table 5.2.

Different active substances can be incorporated into liposomes, such as antioxidants, vitamins, proteins, and enzymes which protect skin from the negative effect of UV radiation and atmospheric pollution. Coenzyme Q10 is a vitamin-like compound which is often used in skin care products, due to its antioxidant activity which protects skin cells from harmful effect of free radicals and aging (Fuller et al., 2006). It has been shown that topical application of coenzyme Q10 during 6 months decreases wrinkle depth (Hoppe et al., 1999). However, bioavailability of coenzyme Q10 after topical application is limited by its liphophilicity and thermolability. For that reason, several preparations based on lipids have been formulated for encapsulation of coenzyme Q10, including nanoliposomes (Xia et al., 2006), nanoemulsions and nanostructured lipid carriers (Junyaprasert et al., 2009), and liposomes (Lee and Tsai, 2010). Stege et al. (2000) have shown that topical application of egzogenic photolyase represents a very efficient method for skin protection from harmful UVB radiation effect and for photodamage reparation. Hofer et al. (2011) have

**TABLE 5.2** An Overview of Some Commercial Drug Delivery Systems and Cosmetic Products Based on Liposomes.

| Drug-delivery systems based on liposomes | | | |
|---|---|---|---|
| **Drug name** | **Active ingredient** | **Approved indication** | **Manufacturer** |
| Ambisome | Amphotericin B | Severe fungal infections | Gilead Sciences, Inc. |
| DaunoXome | Daunorubicin | HIV-related Kaposi's sarcoma | Galen Limited |
| | | Other blood tumors | |
| Doxil | Doxorubicin | Ovarian cancer | Janssen Products, LP |
| | | AIDS-related Kaposi's sarcoma | |
| | | Multiple myeloma | |
| Lipodox | Doxorubicin | Ovarian cancer | Sun Pharmaceuticals Industries Limited |
| | | AIDS-related Kaposi's sarcoma | |
| | | Multiple myeloma | |
| Myocet | Doxorubicin | In combination with cyclophosphamide for treatment of metastatic breast cancer in adult women | Teva UK Limited |
| Visudyne | Verteporfin | Age-related macular degeneration, pathologic myopia, and ocular histoplasmosis | Novartis Pharmacauticals AG |
| Depocyt | Cytarabine | The intrathecal treatment of lymphomatous meningitis | Sigma-Tau Pharmaceuticals, Inc. |
| DepoDur | Morphine sulfate | For the treatment of pain following major surgery | Pacira Pharmaceuticals, Inc. |
| Marqibo | Vincristine sulfate | Philadelphia chromosome–negative acute lymphoblastic leukemia | Spectrum Pharmaceuticals, Inc. |

**TABLE 5.2** *(Continued)*

| Cosmetic products based on liposomes | | | |
|---|---|---|---|
| **Product** | **Key ingredients** | **Use** | **Manufacturer** |
| Capture | Liposomes in gel | Antiaging | Christian Dior |
| Effect du Soleil | Tanning agents in liposomes | Skin tanning | L'Oréal |
| Advanced Night Repair® | Liposome | Antiaging | Estée Lauder |
| Royal Jelly Lift Concentrate | Liposome and cellspan complex | Antiwrinkle | Royal Jelly |
| Moisture Liposome Serum | Liposome | Hydration | Decorté |
| Moisture Liposome Treatment Liquid | Liposome | Hydration | Decorté |
| Moisture Liposome Eye Cream | Liposome and polyglutamic acid | Hydration | Decorté |
| Moisture Liposome Face Cream | Liposome | Hydration and antiaging | Decorté |
| Liposome MultiActive Chamomille | Multilamellar liposome with evening primrose oil, jojoba oil, vitamins A and E, provitamin B5, hyaluronic acid, and chamomile | Moisturizing | Dr. Baumann |
| Liposome MultiActive Vitamin E + C | Multilamellar liposome with vitamins C and E | UV protection | Dr. Baumann |
| Liposome MultiActive Ceramid | Multilamellar liposome with ceramide | Skin barrier function improvement | Dr. Baumann |
| NutraEffects | Deep penetrating moisture complex with liposomes and vitamins A, C, E, β-carotene, and chia seeds | Hydration/moisturizing | Avon |

shown that application of after-sun lotion based on liposomes with incorporated DNA-repair enzymes (photolyase from *Anacystis nidulans* and *Micrococcus luteus* extract with endonuclease activity) can effectively prevent polymorphic light eruption. Liposomes can be used for encapsulation of these compounds, as they protect them from degradation, increase their bioavailability, and enable their sustained release.

Even though liposomes are primarily used for incorporation of different pharmaceutical and cosmetic active substances, during last few decades, interest for the application of liposomes in food industry has increased. Liposomes are nontoxic, biodegradable, and biocompatible systems which can be used in food production, ingredients stabilization, and delivery of food functional components, such as proteins, enzymes, vitamins, antioxidants, and flavor agents. Sustained release of incorporated enzymes from liposomes can be useful in different fermentation processes (Kirby, 1990; Law and King, 1985), as it decreases fermentation procedure and increases product quality. Marsanasco et al. (2011) have applied liposomes as carriers for vitamin C and vitamin E, which are used for orange juice fortification. Juice organoleptic properties were not changed, while the antioxidant activity of incorporated vitamins has been kept, even after heat treatment. Liposomes can be used as carriers for essential oil compounds and increase solubility and chemical stability of incorporated compounds. Sebaaly et al. (2015) have studied liposomes prepared from hydrogenized and nonhydrogenized soya phospholipids as carriers for clove essential oil and its main component, eugenol. Eugenol incorporated in liposomes is protected from degradation by UV radiation, while simultaneously keeping its antioxidant activity. Additionally, prepared liposomes have shown sufficient stability during testing period.

## 5.6 CONCLUSIONS

The use of nanomaterials has a lot of advantages and is becoming more pronounced in recent years. Vesicular nanosystems are renowned and commonly used as a transfer systems for targeted and controlled active substance delivery. A lipid vesicles-liposomes, as nanomaterials that meet the criteria of the latest standards, since belongs to the organic, that is, natural materials, will be more exploited in the future. The process of encapsulation of the active compounds in liposomes is a challenge by itself, since a number of recent studies do not provide sufficient information about the actually effective amount of encapsulated active ingredient. Another challenge regarding the use of liposomes is the question of their long-term stability. In addition to the listed advantages and disadvantages of such systems, based on the above facts, it can be concluded that the more accurate knowledge of mechanism behavior of active components incorporated into the nanocarrier under the influence of many internal or external factors, as well as its optimal concentration and stability in the potential formulations, is of crucial importance for further research and use.

## ACKNOWLEDGMENT

Financial support provided by the Ministry of Education, Science and Technological Development, the Republic of Serbia (projects no. TR 34012) is gratefully acknowledged.

## KEYWORDS


- **vesicles**
- **liposomes**
- **drug-delivery systems**
- **nanocarriers**
- **photodynamic therapy**
- **biological membranes**


## REFERENCES

Abdallah, M. H. Transfersomes as a Transdermal Drug Delivery System for Enhancement the Antifungal Activity of Nystatin. *Int. J. Pharm. Pharm. Sci.* **2013,** *5* (4), 560–567.

Ainbinder, D.; Godin, B.; Touitou, E. Ethosomes: Enhanced Delivery of Drugs to and Across the Skin. In *Percutaneous Penetration Enhancers Chemical Methods in Penetration Enhancement*; Springer: Berlin Heidelberg, DE, 2016; pp 61–75.

Akbarzadeh, A.; Rezaei-Sadabady, R.; Davaran, S.; Joo, S. W.; Zarghami, N.; Hanifehpour, Y.; Samiei, M.; Kouhi, M.; Nejati-Koshki, K. Liposome: Classification, Preparation, and Applications. *Nanoscale Res. Lett.* **2013,** *8* (1), 1.

Allison, R. R.; Sibat, C. H. Oncologic Photodynamic Therapy Photosensitizers: A Clinical Review. *Photodiagn. Photodyn. Ther.* **2010,** *7*, 61–75.

Al-Mahallawi, A. M.; Abdelbary, A. A.; Aburahma, M. H. Investigating the Potential of Employing Bilosomes as a Novel Vesicular Carrier for Transdermal Delivery of Tenoxicam. *Int. J. Pharm.* **2015,** *485* (1), 329–340.

Alvi, I. A.; Madan, J.; Kaushik, D.; Sardana, S.; Pandey, R. S.; Ali, A. Comparative Study of Transfersomes, Liposomes, and Niosomes for Topical Delivery of 5-Fluorouracil to Skin Cancer Cells: Preparation, Characterization, *In-Vitro* Release, and Cytotoxicity Analysis. *Anticancer Drugs* **2011,** *22* (8), 774–782.

Ashok, K.; Kumar, A. R.; Nama, S.; Brahmaiah, B.; Desu, P. K.; Rao, C. B. Sphingosomes: A Novel Vesicular Drug Delivery System. *Int. J. Pharm. Res. Biosci.* **2013,** *2* (2), 305–312.

Bangham, A. D.; Standish, M. M.; Watkins, J. C. Diffusion of Univalent Ions Across the Lamellae of Swollen Phospholipids. *J. Mol. Biol.* **1965,** *13*, 238–252.

Betz, G.; Aeppli, A.; Menshutina, N.; Leuenberger, H. In Vivo Comparison of Various Liposome Formulations for Cosmetic Application. *Int. J. Pharm.* **2005,** *296*, 44–54.

Bhalaria, M. K.; Naik, S.; Misra, A. N. Ethosomes: A Novel Delivery System for Antifungal Drugs in the Treatment of Topical Fungal Diseases. *Indian J. Exp. Biol.* **2009,** *47* (5), 368.

Bhardwaj, U.; Burgess, D. J. Physicochemical Properties of Extruded and Non-extruded Liposomes Containing the Hydrophobic Drug Dexamethasone. *Int. J. Pharm.* **2010,** *388*, 181–189.

Blume, A. Phospholipids as Basic Ingredients. In *Liposome Dermatics*; Springer: Berlin Heidelberg, 1992; pp 29–37.

Borst, J. W.; Visser, N. V.; Kouptsova, O.; Visser, A. J. W. G. Oxidation of Unsaturated Phospholipids in Membrane Bilayer Mixtures Is Accompanied by Membrane Fluidity Changes. *Biochim. Biophys. Acta* **2000,** *1487*, 61–73.

Briones, E.; Colino, C. I.; Lanao, J. M. Delivery Systems to Increase the Selectivity of Antibiotics in Phagocytic Cells. *J. Control. Release* **2008,** *125* (3), 210–227.

Cevc, G.; Blume, G. New, Highly Efficient Formulation of Diclofenac for the Topical, Transdermal Administration in Ultra-deformable Drug Carriers, Transfersomes. *Biochem. Biophys. Acta* **2001,** *1514*, 191–205.

Cevc, G.; Blume, G.; Schätzlein, A. Transfersomes-Mediated Transepidermal Delivery Improves the Regio-specificity and Biological Activity of Corticosteroids *In Vivo*. *J. Control. Release* **1997,** *45* (3), 211–226.

Cevc, G.; Gebauer, D.; Stieber, J.; Schätzlein, A.; Blume, G. Ultraflexible vesicles, Transfersomes, Have an Extremely Low Pore Penetration Resistance and Transport Therapeutic Amounts of Insulin Across the Intact Mammalian Skin. *BBA—Biomembranes* **1998,** *1368* (2), 201–215.

Chapman, D. Phase Transitions and Fluidity Characteristics of Lipids and Cell Membranes. *Q. Rev. Biophys.* **1975,** *8* (2), 185–235.

Cheng, S. J.; Gang, Z. Liposomal Nanostructures for Photosensitizer Delivery. *Laser Surg. Med.* **2011,** *43*, 734–748.

Chiba, M.; Miyazaki, M.; Ishiwata, S. I. Quantitative Analysis of the Lamellarity of Giant Liposomes Prepared by the Inverted Emulsion Method. *Biophys. J.* **2014,** *107* (2), 346–354.

Chilkawar, R. N.; Nanjwade, B. K.; Nwaji, M. S.; Idris, S. M.; Mohamied, A. S. Bilosomes Based Drug Delivery System. *J. Chem. Appl.* **2015,** *2* (1), 1–5.

Conde-Vancells, J.; Rodriguez-Suarez, E.; Embade, N.; Gil, D.; Matthiesen, R.; Valle, M.; et al. Characterization and Comprehensive Proteome Profiling of Exosomes Secreted by Hepatocytes. *J. Proteome Res.* **2008,** *7* (12), 5157–5166.

Cosmos-Standard. *Cosmetics Organic and Natural Standard.* European Cosmetics Standards Working Group: Brussels, 2011. http://www.cosmos-standard.org/docs/COSMOS-standard-final-jan-10.pdf (accessed Nov 2013).

Coviello, T.; Trotta, A. M.; Marianecci, C.; Carafa, M.; Di Marzio, L.; Rinaldi, F.; et al. Gel-Embedded Niosomes: Preparation, Characterization and Release Studies of a New System for Topical Drug Delivery. *Colloids Surf. B: Biointerfaces* **2015,** *125*, 291–299.

De Gier, J. Osmotic Behaviour and Permeability Properties of Liposomes. *Chem. Phys. Lipids* **1993,** *64* (1–3), 187–196.

De Gier, J.; Mandersloot, J. G.; Van Deenen, L. L. M. Lipid Composition and Permeability of Liposomes. *BBA—Biomembranes* **1968,** *150* (4), 666–675.

Dong, W.; Rijken, P.; Ugwoke, M. Method for Preparing Virosomes. Patent Appl. EP 3,017,040 A1, July 2, 2013.

Dragičević-Ćurić, N.; Scheglmann, D.; Albrecht, V.; Fahr, A. Development of Liposomes Containing Ethanol for Skin Delivery of Temoporfin: Characterization and In Vitro Penetration Studies. *Colloids Surf. B: Biointerfaces* **2009,** *74*, 114–122.

Dragičević-Ćurić, N.; Stupar, M.; Milić, J.; Zorić, T.; Krajišnik, D.; Vasiljević, D. Hydrophilic Gels Containing Chlorophyllin-Loaded Liposomes: Development and Stability Evaluation. *Pharmazie* **2005,** *60* (8), 588–592.

Dua, J. S.; Rana, A. C.; Bhandari, A. K. Liposome: Methods of Preparation and Applications. *Int. J. Pharm. Stud. Res.* **2012,** *3*, 14–20.

Elsayed, M. M.; Abdallah, O. Y.; Naggar, V. F.; Khalafallah, N. M. Deformable Liposomes and Ethosomes: Mechanism of Enhanced Skin Delivery. *Int. J. Pharm.* **2006,** *322* (1), 60–66.

Felnerova, D.; Viret, J. F.; Glück, R.; Moser, C. Liposomes and Virosomes as Delivery Systems for Antigens, Nucleic Acids and Drugs. *Curr. Opin. Biotechnol.* **2004,** *15* (6), 518–529.

Ferrari, M. Frontiers in Cancer Nanomedicine: Directing Mass Transport Through Biological Barriers. *Trends Biotechnol.* **2010,** *28* (4), 181–188.

Fonseca, M. J.; Vingerhoeds, M. H.; Haisma, H. J.; Crommelin, D. J. A.; Storm, G. Development of an Optimized Immuno-Enzymosome Formulation for Application in Cancer Therapy. In *Future Strategies for Drug Delivery with Particulate Systems*; Diederichs, J. E., Muller, R. H., Eds.; CRC Press, Medpharm GmbH Scient. Publ.: Stuttgart, DE, 1998; pp 152–157.

Fuller, B.; Smith, D.; Howerton, A.; Kern, D. Anti-inflammatory Effects of CoQ10 and Colorless Carotenoids. *J. Cosmet. Dermatol.* **2006,** *5*, 30–38.

Gaspar, M. M.; Cruz, A.; Penha, A. F.; Reymão, J.; Sousa, A. C.; Eleutério, C. V.; Domingues, S. A.; Fraga, A. G.; Longatto Filho, A.; Cruz, M. E. M.; Pedrosa, J. Rifabutin Encapsulated in Liposomes Exhibits Increased Therapeutic Activity in a Model of Disseminated Tuberculosis. *Int. J. Antimicrob. Ag.* **2008,** *31* (1), 37–45.

Gennis, R. B. *Biomembranes: Molecular Structure and Function*; Springer Science Business Media: Berlin, 2013.

Giuseppina, B.; Molinari, A. Liposomes as Nanomedical Devices. *Int. J. Nanomed.* **2015,** *10* (1), 975–999.

Goyal, A. K.; Khatri, K.; Mishra, N.; Mehta, A.; Vaidya, B.; Tiwari, S.; Vyas, S. P. Aquasomes—A Nanoparticulate Approach for the Delivery of Antigen. *Drug Dev. Ind. Pharm.* **2008,** *34* (12), 1297–1305.

Gregoriadis, G. *Liposome Technology*: Liposome Preparation and Related Techniques, 3rd ed.; Informa Healthcare USA, Inc.: New York, 2007.

Gregoriadis, G. *Liposome Technology, Volume I*: *Preparation of Liposomes*; CRC Press Inc.: Boca Raton, FL, 1984.

Guan, P.; Lu, Y.; Qi, J.; Wu, W. Readily Restoring Freeze-Dried Probilosomes as Potential Nanocarriers for Enhancing Oral Delivery of Cyclosporine A. *Colloids Surf. B: Biointerfaces* **2016,** *144*, 143–151.

Ha, D.; Yang, N.; Nadithe, V. Exosomes as Therapeutic Drug Carriers and Delivery Vehicles Across Biological Membranes: Current Perspectives and Future Challenges. *Acta Pharm. Sin. B* **2016,** *6* (4), 287–296.

Halliwell, B.; Chirico, S. Lipid Peroxidation: Its Mechanism, Measurement, and Significance. *Am. J. Clin. Nutr.* **1993,** *57* (5), 715–725.

Hirai, M.; Minematsu, H.; Hiramatsu, Y.; Kitagawa, H.; Otani, T.; Iwashita, S.; Kudoh, T.; Chen, L.; Li, Y.; Okada, M.; Salomon, DS.; Igarashi, K.; Chikuma, M.; Seno, M. Novel and Simple Loading Procedure of Cisplatin into Liposomes and Targeting Tumor Endothelial Cells. *Int. J. Pharm.* **2010,** *391* (1–2), 274–283.

Hofer, A.; Legat, F. J.; Gruber-Wackernagel, A.; Quehenberger, F.; Wolf, P. Topical Liposomal DNA-Repair Enzymes in Polymorphic Light Eruption. *Photochem. Photobiol. Sci.* **2011,** *10* (7), 1118–1128.

Hoppe, S. M.; Sasaki, D. Y.; Kinghorn, A. N.; Hattar, K. In-Situ Transmission Electron Microscopy of Liposomes in an Aqueous Environment. *Langmuir* **2013,** *29* (32), 9958–9961.

Hoppe, U.; Bergemann, J.; Diembeck, W.; Ennen, J.; Gohla, S.; Harris, I.; et al. Coenzyme Q10, a Cutaneous Antioxidant and Energizer. *Biofactors* **1999,** *9*, 371–378.

Hundekar, Y. R.; Nanjwade, B. K.; Mohamied, A. S.; Idris, N. F.; Srichana, T.; Shafioul, A. S. M. Nanomedicine to Tumor by Enzymosomes. *J. Nanotechnol. Nanomed. Nanobiotechnol.* **2015,** *2* (1), 1–5.

Ilić-Stojanović, S. S.; Nikolić, L. B.; Nikolić, V. D.; Petrović, S. D. Smart Hydrogels for Pharmaceutical Applications Smart Hydrogels for Pharmaceutical Applications. In *Novel Approaches for Drug Delivery*; Keservani, R. K., Sharma, A. K., Kesharwani, R. K.; IGI Global: Hershey, PA, 2016.

Ilya, L.; Veatch, L. S. The Continuing Mystery of Lipid Rafts. *J. Mol. Biol.* **2016,** *428* (24), 4749–4764.

Ioannis, S. A.; Theodoros, H. V.; Sotirios, K.; Athanasios, E. L. Lipids, Fats and Oils. In *Advances in Food Biochemistry*; Fatih, Y., Eds.; CRC Press: Boca Raton, FL, 2010; pp 131–203.

Ioele, G.; Tavano, L.; Muzzalupo, R.; De Luca, M.; Ragno, G. Stability-Indicating Methods for NSAIDs in Topical Formulations and Photoprotection in Host–Guest Matrices. *Mini-Rev. Med. Chem.* **2016,** *16* (8), 676–682.

Jain, S.; Tiwary, A. K.; Sapra, B; Jain, N. K. Formulation and Evaluation of Ethosomes for Transdermal Delivery of Lamivudine. *AAPS Pharm. Sci. Technol.* **2007,** *8* (4), 249–257.

Jain, S.; Umamaheshwari, R. B.; Bhadra, D.; Jain, N. K. Ethosomes: A Novel Vesicular Carrier for Enhanced Transdermal Delivery of an anti-HIV Agent. *Indian J. Pharm. Sci.* **2004,** *66* (1), 72–81.

Junyaprasert, V. B.; Teeranachaideekul, V.; Souto, E. B.; Boonme, P.; Muller, R. H. Q10-Loaded NLC Versus Nanoemulsions: Stability, Rheology and In Vitro Skin Permeation. *Int. J. Pharm.* **2009,** *377*, 207–214.

Kamboj, S.; Saini, V.; Maggon, N.; Bala, S.; Jhawat, V. Vesicular Drug Delivery Systems: A Novel Approach for Drug Targeting. *Int. J. Drug Delivery* **2013,** *5* (2), 121.

Kataoka, K.; Harada, A.; Nagasaki, A. Block Copolymer Micelles for Drug Delivery: Design, Characterization and Biological Significance. *Adv. Drug Deliv. Rev.* **2001,** *47*, 113–131.

Kaur, I. P.; Garg, A.; Singla, A. K.; Aggarwal, D. Vesicular Systems in Ocular Drug Delivery: An Overview. *Int. J. Pharm.* **2004,** *269* (1), 1–14.

Kee-Yeo, K.; Lee, C. Development and Prospect of Emulsion Technology in Cosmetics. *J. Soc. Cosmet. Chem.* **2006,** *32* (4), 209–217.

Kim, S. H.; Jeong, J. H.; Chun, K. W.; Park, R. T. G. Target-Specific Cellular Uptake of PLGA Nanoparticles Coated with Poly(L-Lysone)–Poly(Ethylene-Glycol)–Folate Conjugate, *Langmuir* **2005,** *21*, 8852–8857.

Kirby, C. Delivery Systems for Enzymes. *Chem. Br.* **1990,** *26*, 847–850.

Klang, V.; Claudia, V.; Matsko, N. B. Electron Microscopy of Pharmaceutical Systems. *Micron* **2013,** *44*, 45–74.

Kwangjae, C.; Xu, W.; Shuming, N.; Zhuo, C.; Dong, M. S. Therapeutic Nanoparticles for Drug Delivery in Cancer. *Clin. Cancer Res.* **2008,** *14* (5), 1310–1316.

Lankaldapalli, S.; Damuluri, M. Sphingosomes: Applications in Targeted Drug Delivery. *Int. J. Pharm. Chem. Biol. Sci.* **2012,** *2* (4), 507–516.

Laouini, A.; Jaafar-Maalej, C.; Limayem-Blouza, I.; Sfar, S.; Charcosset, C.; Fessi, H. Preparation, Characterization and Applications of Liposomes: State of the Art. *J. Colloid Sci. Biotechnology* **2012,** *1*, 147–168.

Law, B. A.; King, J. S. Use of Liposomes for Proteinase Addition to Cheddar Cheese. *J. Diary Res.* **1985,** *52*, 183–188.

Lee, W.; Tsai, T. Preparation and Characterization of Liposomal Coenzyme Q10 for In Vivo Topical Application. *Int. J. Pharm.* **2010,** *395*, 78–83.

Lingwood, D.; Simons, K. Lipid Rafts as a Membrane-Organizing Principle. *Science* **2009,** *327*, 46–50.

Liu, Q.; Ben, J. B. Liposomes in Biosensors. *Analyst* **2013,** *138* (2), 391–409.

London, E. Insights into Lipid Rafts Structure and Formation from Experiments in Model Membranes. *Curr. Opin. Struct. Biol.* **2002,** *12*, 480–486.

Lopes, R.; Eleutério, C. V.; Gonçalves, L. M. D.; Cruz, M. E. M.; Almeida, A. J. Lipid Nanoparticles Containing Oryzalin for the Treatment of Leishmaniasis. *Eur. J. Pharm. Sci.* **2012,** *45* (4), 442–450.

Luciani, P.; Debora, B.; Piero, B. Liposomes and Biomacromolecules. In *Colloids in Drug Delivery*; Monzer, F., Ed.; CRC Press: Boca Raton, FL, 2016; pp 150–395.

Mahrhauser, D. S.; Reznicek, G.; Kotisch, H.; Brandstetter, M.; Nagelreiter, C.; Kwizda, K; Valenta, C. Semi-Solid Fluorinated-DPPC Liposomes: Morphological, Rheological and Thermic Properties as well as Examination of the Influence of a Model Drug on Their Skin Permeation. *Int. J. Pharm.* **2015,** *486* (1), 350–355.

Marianecci, C.; Di Marzio, L.; Rinaldi, F; Celia, C.; Paolino, D.; Alhaique, F.; Carafa, M. Niosomes from 80s to Present: The State of the Art. *Adv. Colloid Interface* **2014,** *205*, 187–206.

Markovic, D. Z.; Zvezdanovic, J. B. Impact of Molecular Organization on UV-Irradiation Effects to Chlorophyll Stability: A Base to Understand Biomedical Applications. In *Chlorophyll: Structure, Production and Medicinal Uses*; Lee, H., Salcedo, E., Eds.; Nova Science Publishers Inc., New York, NY, 2012; pp 1–42.

Marković, D. Z. Benzophenone-Sensitized Peroxidation in Compressed Lipid Monolayers at Airwater Interface. *Collect. Czech. Chem. Commun.* **2001,** *66*, 1603–1614.

Marković, D. Z.; Durand, T.; Patterson, L. K. Hydrogen Abstraction from Lipids by Triplet States of Derivatized Benzophenone Photosenzitizers. *Photochem. Photobiol.* **1990,** *51*, 389–394.

Marković, D. Z.; Patterson, L. K. Benzophenone-Sensitized Lipid Peroxidation in Linoleate Micelles, *Photochem. Photobiol.* **1993,** *58*, 329–334.

Marković, D. Z; Patterson, L. K. Radical Processes in Lipids. Selectivity of Hydrogen Abstraction from Lipids by Benzophenone Triplet. *Photochem. Photobiol.* **1989,** *49* (5), 531–535.

Marsanasco, M.; Márquez, A. L.; Wagner, J. R.; del Alonso, S. V.; Chiaramoni, N. S. Liposomes as Vehicles for Vitamins E and C: An Alternative to Fortify Orange Juice and Offer Vitamin C Protection after Heat Treatment. *Food Res. Int.* **2011,** *44*, 3039–3046.

Martindale. *The Complete Drug Reference*, 34th ed.; Pharmaceutical Press: London, UK, 2005; pp 1056–1058, 1576–1582.

Meybeck, A. Past, Present and Future of Liposome Cosmetics. In *Liposome Dermatics*; Springer: Berlin Heidelberg, DE, 1992; pp 341–345.

Milenkovic, S.; Barbanta-Patrascu, M.; Baranga, G; Markovic, D.; Tugulea, L. Comparative Spectroscopic Studies on Liposomes Containing Chlorophyll *a* and Chlorophyllide *a*. *Gen. Physiol. Biophys.* **2013,** *32*, 559–567.

Milenkovic, S.; Tugulea, L.; Barbanta-Patrascu, M.; Baranga, G.; Markovic, D. Spectroscopic Studies on Liposome Containing Chlorophyll *a* and Chlorophyllide *a*. In *2012 Annual Scientific Conference*, Bucharest, Romania, 2012a, 3.17; p 64.

Milenkovic, S.; Tugulea, L.; Barbanta-Patrascu, M.; Baranga, G.; Markovic, D. Quercetin Effects on Model Lipid Membranes Monitored by Chlorophyll Derivatives. In *12th Eurasia Conference on Chemical Sciences*, Corfu, Greece, 2012b, S2-OP11; p 45.

Mimica-Dukić, N. *In Vivo* i *In Vitro* ispitivanja antioksidantnih svojstava biljnih ekstrakata. *Arch. Pharm.* **1997,** *5* (7), 475–493.

Moghassemi, S.; Afra, H. Nano-Niosomes as Nanoscale Drug Delivery Systems: An Illustrated Review. *J. Control. Release* **2014,** *185*, 22–36.

Moser, J. G. Definitions and General Properties of 2nd and 3rd Generation Photosensitizers. In *Photodynamic Tumor Therapy: 2nd and 3rd Generation Photosensitizers*; Moser, J. G., Ed.; Harwood Academic Publishers: Netherlands, 1998; pp 3–7.

Mravec, F.; Klucakova, M.; Pekar, M. Fluorescence Spectroscopy Study of Hyaluronan–Phospholipid Interactions. In: *Advances in Planar Lipid Bilayers and Liposomes*; Ales, I., Ed.; Academic Press, Cambridge, MA, 2011; pp 235–254.

Mukerjee, P.; Mysels, K. J. Critical Micelle Concentrations of Aqueous Surfactant Systems (No. NSRDS-NBS-36). National Standard Reference Data System, 1971.

Műller, R. H. Zetapotential-Theorie. In *Zetapotential und Partikelladung in der Laberpraxis*; Műller, R. H., Ed.; Wissenschaftliche Verlagsgesellschaft mbH: Stuttgart, DE, 1996; pp 19–99.

Murata, M.; Kohei, T.; Hirofumi, T. Real-Time In Vivo Imaging of Surface-Modified Liposomes to Evaluate Their Behavior after Pulmonary Administration. *Eur. J. Pharm. Biopharm.* **2014,** *86* (1), 115–119.

Nastruzzi, C.; Esposito, E.; Menegatti, E.; Walde, P. Use and Stability of Liposomes in Dermatological Preparations. *J. Appl. Cosmetol.* **1993,** *11*, 77–91.

Nicolosi, D.; Cupri, S.; Genovese, C.; Tempera, G.; Mattina, R.; Pignatello, R. Nanotechnology Approaches for Antibacterial Drug Delivery: Preparation and Microbiological Evaluation of Fusogenic Liposomes Carrying Fusidic Acid. *Int. J. Antimicrob. Ag.* **2015,** *45* (6), 622–626.

O'Hagan, D. T.; Rahman, D.; McGee, J. P.; Jeffery, H.; Davies, M. C.; Williams, P.; Davis, S. S.; Challacombe, S. J. Biodegradable Microparticles as Controlled Release Antigen Delivery Systems. *Immunology* **1991,** *73* (2), 239–242.

Olson, F.; Hunt, C. A.; Szoka, F. C.; Vail, W. J. Papahadjopoulos, D. Preparation of Liposomes of Defined Size Distribution by Extrusion Through Polycarbonate Membranes. *BBA—Biomembranes* **1979,** *557* (1), 9–23.

Pardakhty, A. Non-Ionic Surfactant Vesicles (Niosomes) as New Drug Delivery Systems. In *Novel Approaches for Drug Delivery*; Keservani, R. K., Sharma, A. K., Kesharwani, R. K., Eds.; IGI Global: Hershey, PA, 2016; pp 89–119.

Park, J. H.; Saravanakumar, G.; Kim, K.; Kwon, I. C. Target Drug Delivery of Low Molecular Drugs Using Chitosan and Its Derivatives. *Adv. Drug Deliv. Rev.* **2010,** *62* (1), 28–41.

Patravale, V. B.; Mandawgade, S. D. Novel Cosmetic Delivery Systems: An Application Update. *Int. J. Cosmet. Sci.* **2008,** *30* (1), 19–33.

Paul, A.; Cevc, G.; Bachhawat, B. K. Transdermal Immunization with Large Proteins by Means of Ultradeformable Drug Carriers. *Eur. J. Immunol.* **1995,** *25* (12), 3521–3524.

Petrović, S. Stability of Chlorophyll on Oxidative Stress in Water Medium and in Liposomes. *Ph. D. Thesis*; Faculty of Technology, University of Niš: Serbia, March 2016.

Petrović, S.; Savić, S.; Zvezdanović, J.; Cvetković, D.; Marković, D. A Lipids Microenvironment Impact on Liposomes with Incorporated Pigments. In *International Conference on Radiation and Applications in Various Fields of Research (RAD 2016)*; Ristić, G., Ed.; Serbia, 2016; p 71.

Petrović, S.; Tugulea, L; Marković, D; Barbinta-Patrascu, M. Chlorophyll *a* and Chlorophyllide *a* inside Liposomes Made of Saturated and Unsaturated Lipids: A Possible Impact of the Lipids Microenvironment, *Acta Period Technol.* **2014,** *45*, 215–227.

Pippa, N.; Psarommati, F.; Pispas, S.; Demetzos, C. The Shape/Morphology Balance: A Study of Stealth Liposomes *via* Fractal Analysis and Drug Encapsulation. *Pharm. Res.-Dordr.* **2013,** *30* (9), 2385–2395.

Pirvu, C. D.; Hlevca, C.; Ortan, A.; Prisada, R. Ă. Z. V. A. N. Elastic Vesicles as Drugs Carriers Through the Skin. *Farmacia* **2010,** *58* (2), 128.

Planas, M. E.; Gonzalez, P.; Rodriguez, L.; Sanchez, S.; Cevc, G. Noninvasive Percutaneous Induction of Topical Analgesia by a New Type of Drug Carrier, and Prolongation of Local Pain Insensitivity by Anesthetic Liposomes. *Anesth. Anal.* **1992,** *75* (4), 615–621.

Pozzi, D.; Colapicchioni, V.; Caracciolo, G.; Piovesana, S.; Capriotti, A. L.; Palchetti, S.; De Grossi, S.; Riccioli, A.; Amenitsch, H.; Laganà, A. Effect of Polyethyleneglycol (PEG) Chain Length on the Bio–Nano-Interactions Between PEGylated Lipid Nanoparticles and Biological Fluids: From Nanostructure to Uptake in Cancer Cells. *Nanoscale* **2014,** *6* (5), 2782–2792.

Prakash, J.; De Jong, E.; Post, E.; Gouw, A. S.; Beljaars, L.; Poelstra, K. A Novel Approach to Deliver Anticancer Drugs to Key Cell Types in Tumors Using a PDGF Receptor-Binding Cyclic Peptide Containing Carrier. *J. Control. Release* **2010,** *145*, 91–101.

Qin, J.; Xu, Q. Functions and Applications of Exosomes. *Acta Pol. Pharm.—Drug Res.* **2014,** *71* (4), 537–543.

Rajan, R.; Jose, S.; Mukund, V. B.; Vasudevan, D. T. Transferosomes—A Vesicular Transdermal Delivery System for Enhanced Drug Permeation. *J. Adv. Pharm. Technol. Res.* **2011,** *2* (3), 138.

Rawicz, W.; Oldrich, K. C.; McIntosh, T.; Needham, D.; Evans, E. Effects of Chain Length and Unsaturation on Elasticity of Lipid Bilayers. *Biophys. J.* **2000,** *79*, 328–339.

Saudagar, R. B.; Samuel, S. Ethosomes: Novel Noninvasive Carrier for Transdermal Drug Delivery. *Annu. Rev. Biophys. Biomol. Struct.* **2016,** *6* (2), 135–138.

Sebaaly, C.; Jraij, A.; Fessi, H.; Charcosset, C.; Greige-Gerges, H. Preparation and Characterization of Clove Essential Oil-Loaded Liposomes. *Food Chem.* **2015,** *178*, 52–62.

Sezgin-Bayindir, Z.; Antep, M. N.; Yuksel, N. Development and Characterization of Mixed Niosomes for Oral Delivery Using Candesartan Cilexetil as a Model Poorly Water-Soluble Drug. *AAPS Pharm. Sci. Tech.* **2015,** *16* (1), 108–117.

Shahabade, G. S.; Bhosale, A. V.; Mutha, S. S.; Bhosale, N. R.; Khade, P. H. An Overview on Nanocarrier Technology—Aquasomes. *J. Pharm. Res.* **2009,** *2* (7), 1174–1177.

Simons, M.; Raposo, G. Exosomes–Vesicular Carriers for Intercellular Communication. *Curr. Opin. Chem. Biol.* **2009,** *21* (4), 575–581.

Soheyla, H.; Zahir, F. Effect of Zeta Potential on the Properties of Nano-Drug Delivery Systems—A Review (Part 1). *Trop. J. Pharm. Res.* **2013,** *12* (2), 255–264.

Stege, H.; Roza, L.; Vink, A. A.; Grewe, M.; Ruzicka, T.; Grether-Beck, S.; Krutmann, J. Enzyme Plus Light Therapy to Repair DNA Damage in Ultraviolet-B-Irradiated Human Skin. *Proc. Natl. Acad. Sci. U.S.A.* **2000,** *97* (4), 1790–1795.

Stuchlík, M.; Žák, S. Lipid-Based Vehicle for Oral Drug Delivery. *Biomed. Pap.* **2001,** *145* (2), 17–26.

Sujatha, S.; Sowmya, G.; Chaitanya, M.; Reddy, V. K.; Monica, M.; Kumar, K. K. Preparation, Characterization and Evaluation of Finasteride Ethosomes. *Int. J. Drug Deliv.* **2016,** *8* (1), 01–11.

Swain, S.; Beg, S.; Babu, SM. Liposheres as a Novel Carrier for Lipid Based Drug Delivery: Current and Future Directions. *Rec. Pat. Drug Deliv. Formul.* **2016,** *10* (1), 59–71.

Tan, C.; Xia, S.; Xue, J.; Xie, J.; Feng, B.; Zhang, X. Liposomes as Vehicles for Lutein: Preparation, Stability, Liposomal Membrane Dynamics, and Structure. *J. Agric. Food Chem.* **2013,** *61* (34), 8175–8184.

Torchilin, V.; Weissig, V. *Liposomes: A Practical Approach*, 2nd ed.; Oxford University Press Inc.: New York, 2003.

Tugulea, L.; Barbinta-Patrascu, M.; Baranga, G.; Milenkovic, S.; Markovic, D. Spectroscopic Studies on Liposomes Containing Chlorophyll *a* and Chlorophyllide *a*. *IUPAC 2013 Solar Energy for World Peace*; International Union of Pure and Applied Chemistry: Istanbul, 2013, PC-O-09.

Umashankar, M. S.; Sachdeva, R. K.; Gulati, M. Aquasomes: A Promising Carrier for Peptides and Protein Delivery. *Nanomedicine* **2010,** *6* (3), 419–426.

Vale, C. A.; Corvo, M. L.; Martins, L. C. D.; Marques, C. R.; Storm, G.; Cruz, M. E. M.; Martins, M. B. F. Construction of Enzymosomes: Optimization of Coupling Parameters. *NSTI-Nanotechnol.* **2006,** *2*, 396–397.

Vemuri, S.; Rhodes, C. T. Development and Characterization of a Liposome Preparation by a pH-Gradient Method. *J. Pharm. Pharmacol.* **1994,** *46*, 778–783.

Vijaykumar, N.; Kalepu, S. Recent Advances in Liposomal Drug Delivery: A Review. *Pharm. Nanotechnol.* **2015,** *3* (1), 35–55.

Vuleta, G.; Milić, J.; Cekić, N. Savremeni kozmetički proizvodi za negu kože- formulacije i zahtevi za kvalitet. *Hem. Ind.* **2003,** *57* (10), 463–470.

Wang, W.; Cheng, D.; Gong, F.; Miao, X.; Shuai, X. Design of Multifunctional Micelle for Tumor-Targeted Intracellular Drug Release and Fluorescent Imaging. *Adv. Mater.* **2012,** *24*, 115–120.

Webb, M. S.; Bally, M. B.; Mayer, L. D.; Miller, J. J.; Tardi, P. G. Sphingosomes for Enhanced Drug Delivery. U.S. Patent No. 5,814,335. September 29, 1998.

Wheatley, R. Some Recent Trends in the Analytical Chemistry of Lipid Peroxidation. *TrAC—Trend Anal. Chem.* **2000,** *19* (10), 617–628.

Xia, S.; Xu, S.; Zhang, X. Optimization in the Preparation of Coenzyme Q10 Nanoliposomes. *J. Agric. Food Chem.* **2006,** *54*, 6358–6366.

Yang, G. M.; Gu, J. M.; Song, L. Y. The Application of Liposome Technology in Cosmetic Preparation. *Chinese J. Aesth. Med.* **2009,** *3*, 059.

Yogeshkumar, M.; Marilena, L.; Seifalian, A. Liposomes and Nanoparticles: Nanosized Vehicles for Drug Delivery in Cancer. *Trends Pharmacol. Sci.* **2009,** *30* (11), 592–599.

Yogita, P.; Jadhav, S. Novel Methods for Liposome Preparation. *Chem. Phys. Lipids* **2014,** *177*, 8–18.

Yogita, P.; Jadhav, S. Preparation of Liposomes for Drug Delivery Applications by Extrusion of Giant Unilamellar Vesicles. In *Nanoscale and Microscale Phenomena*; Springer, India, 2015; pp 17–29.

Zhang, J. A.; Anyarambhatla, G.; Ma, L.; Ugwu, S.; Xuan, T.; Sardone, T.; Ahmad, I. Development and Characterisation of a Novel Cremophor EL Free Liposome Based Paclitaxel (LEP-ETU) Formulation. *Eur. J. Pharm. Biopharm.* **2005,** *59* (1), 177–187.

# CHAPTER 6

# siRNA DELIVERY WITH LIPOSOMES AS PLATFORM TECHNOLOGY

PREETHI NAIK and MANGAL S. NAGARSENKER*

*Department of Pharmaceutics, Bombay College of Pharmacy, Mumbai 400098, India*

**Corresponding author. E-mail: mangal.nagarsenker@gmail.com*

## CONTENTS

## ABSTRACT

Over the past decade, research in the field of gene silencing mediated through the RNA interference (RNAi)-driven small interfering ribonucleic acid (siRNA)-based therapies has grown exponentially. Though siRNA has shown the potential to address a wide array of diseases, it has failed to deliver on that promise because of the challenge of intracellular delivery. Nanocarrier systems, suitably designed to form supramolecular assemblies with siRNA, can aid in achieving safe and effective delivery of siRNA to the cytosol of target cells. This chapter discusses the phenomenon of RNAi, the different attempts for improving intracellular delivery of siRNA with special focus on excipient driven liposomal nanocarrier systems, clinical relevance of siRNA delivery vehicles, and the toxicity concerns associated with such carrier systems.

## 6.1 INTRODUCTION

Gene therapy has intrigued researchers for several years. It refers to the transfer of nucleic acids into target cells for treatment. In this approach of disease management, regulation of gene expression or alteration of genes forms the treatment modality. It is important for the readers to understand the subtle difference between gene silencing and gene knockdown. Gene silencing describes regulation of gene expression wherein the cell is prevented from translating proteins. This is generally brought about by degrading the mRNA at transcriptional or posttranscriptional stage of protein synthesis. On the other hand, gene knockdown refers to the genetic alteration by complete deletion of some genes from the genome of the organism. While gene silencing is a temporary phenomenon and is reversible, gene knockout is permanent. The present chapter relates to gene silencing phenomenon by exploring RNA interference (RNAi) process with small interfering RNA (siRNA) in the spotlight. Over the last two and half decades, siRNA has established exponential rise in research interest by academicians and industrial research groups in understanding and presenting an attractive and a powerful tool for silencing specific phenotypes. The promise of protein downregulation by degradation of specific RNA has fascinated researchers. This can be brought about by one of the four strategies, namely, administration of

1. antisense oligonucleotides: short single-stranded oligonucleotides that hybridize with complementary mRNA and results in RNA cleavage;
2. ribozymes: three-dimensional catalytic RNA composed of three helices that catalyze cleavage of RNAs;
3. miRNA: microRNA; and
4. siRNA: small interfering RNA.

Amongst all, siRNA seems to be the most stable for intracellular delivery, more effective, and many folds potent than the others (Bertrand et al., 2002; Reischl and Zimmer, 2009). miRNA and siRNA operate through RNAi machinery. RNAi refers to a specific mechanism wherein double-stranded RNA (dsRNA) self-assembles into a multicomponent nuclease called RNA-induced silencing complex (RISC) that binds to complementary mRNA and initiates cascade of events resulting in the enzymatic degradation of mRNA.

The stepping stone for the discovery of the revolutionary phenomenon of RNAi was laid back in 1998 by Fire and Melo with experimentation of regulation of gene expression using dsRNA on nematode, *Caenorhabditis elegans*, which fetched them the Nobel Prize in Physiology and Medicine in 2006 (Bruno, 2011; Montgomery, 1998). Over the following years, the intermediates in RNAi machinery were identified by different groups of researchers (Sen and Blau, 2006). Tuschl and coworkers demonstrated the proof of principle that synthetic siRNA could achieve sequence-specific gene silencing in HeLa cell line (Elbashir et al., 2001). Shortly thereafter, the first successful in vivo gene silencing using siRNA was achieved for hepatitis C target in mice (Sen and Blau, 2006). In this experiment, siRNA was injected in large volume via tail vein and it was observed that siRNA was delivered to the liver and could protect it against the virus for several days (McCaffrey et al., 2002). Though therapeutic efficacy of siRNA therapy could be established, this technique remained limited to rodent administration. Persistent efforts by scientists in overcoming challenges of siRNA delivery eventually resulted in the first siRNA therapy in humans via intravitreal administration against wet age related macular degeneration in the year 2004 (Bruno, 2011). Since then, the arena of siRNA-based therapy has seen advancement with increasing number of candidates entering clinical trials against plethora of broad spectrum diseases in the recent times.

This chapter attempts to discuss the potentials and the challenges of siRNA delivery while briefly explaining its molecular mechanism of action. The chapter deliberates upon the discovery of RNAi phenomenon, the success and failures of naked siRNA, different attempts to improve siRNA delivery through chemical modifications, vector-based delivery, and innovative, excipient-driven nanocarrier delivery. There is abundance of literature with regards to the subject of siRNA-based therapeutics and it is a herculean task to discuss each one of the continuously growing, wide variety of delivery approaches. This chapter covers discussion of liposome-based siRNA delivery shedding light on some of the general considerations for formulation development with special emphasis on its application in cancer. Also, a brief discussion on the clinical phase of such therapeutics has been recorded.

Let us begin with understanding of phenomenon of RNAi and its operation before answering the question of what is siRNA.

### 6.1.1 PHENOMENON OF RNAi

Cellular functioning is ensured by expression of specific proteins. Protein expression is governed by two processes, namely, transcription and translation. Information or instructions to cellular machinery are acquired from DNA by transcription into mRNA within the nucleus. The formed mRNA interacts with ribosomal complex in the cytosol and translates into polymer of amino acids called proteins. Expression of proteins can be manipulated by gene regulation via RNA silencing, either by suppressing transcription or by inducing posttranscriptional degradation of mRNA (McManus and Sharp, 2002; Zimmermann et al., 2006). RNAi is one means of achieving posttranscriptional protein suppression. RNAi refers to a biological mechanism, wherein a dsRNA suppresses the expression of specific gene. This endogenous pathway resulting in posttranscriptional silencing of genes can be triggered by cytoplasmic presence of small fragments of RNA known as siRNA (Davidson and McCray, 2011).

### 6.1.2 RISC MECHANISM

The heart of RNAi machinery lies in formation of multicomponent ribonucleoprotein complex called RNA-induced silencing complex (RISC).

The siRNA solely cannot result in mRNA degradation but it must self-assemble into cytoplasmic complex of RISC to exert its function. RISC is not a single-step process but involves multiple steps, functioning in an orderly manner to eventually result in mRNA degradation (Kawamata and Tomari, 2010). Briefly, RISC is composed of two crucial proteins, namely, Dicer enzyme, the initiator, and Argonaute enzyme, the effector (Pratt and MacRae, 2009). In the presence of RNA substrate, the Dicer enzyme generates siRNA (Tijsterman and Plasterk, 2004). The fragments of siRNA are loaded onto Argonaute enzyme which unwinds and incorporates the antisense/guide strand into activated RISC. The sense strand/passenger strand of siRNA is degraded. The antisense strand selectively seeks out for mRNA that is complementary to it and Argonaute enzyme slices the mRNA, thus degrading it (Hutvagner and Simard, 2008). Interestingly, during this catalytic cleavage, siRNA molecule is recovered and can degrade several mRNA resulting in amplified gene silencing (Whitehead et al., 2009). The detailed description of gene-silencing machinery of RISC is explained in the mini review (Pratt and MacRae, 2009).

From gene regulation stand point, two types of small RNAs operate through RISC portal, namely, siRNA and miRNA (micro-RNA).

siRNA is a short, 20–25 nucleotide base pair in length, double-stranded, straight-chain RNA, derived from precursor long dsRNA or hairpin RNA by the action of the catalytic enzyme called Dicer in the cytoplasm. Gene silencing resulting from this siRNA–mRNA complexation is highly selective and specific owing to high degree of complementarity. siRNA can be generated endogenously by the enzymatic action of Dicer on long dsRNA or short hairpin (shRNA). shRNA consists of two complementary RNA sequences held together by a loop of 4–11 nucleotides. shRNA is synthesized in the nucleus or can be transfected into cells via DNA vectors or plasmids that encode for shRNA. The shRNA transcribed in the nucleus is transported to the cytosol wherein short RNA fragments are generated by the actions of Dicer which are further incorporated into RISC to elucidate silencing.

miRNA was discovered 5 years before the discovery of siRNA by Ambros and coworkers and was viewed as regulators of endogenous genes. These RNA molecules are transcribed endogenously from the genome in the nucleus by the action of RNA polymerase II, as primary miRNA (pri-miRNA). pri-miRNA comprises irregularities in double-strand stem in the form of loops and flanking segments. The catalytic Drosha–Pasha complex in the nucleus trims the pri-miRNA to pre-miRNA that is transported to cytosol. The pre-miRNA is incorporated into miRISC which involves different enzymes from Dicer and Argonaute families but eventually results in mRNA degradation. The loops irregularities in the nucleotide sequence result in mismatch and imperfect complementarity with mRNA, as opposed to total complementarity of siRNA. Elaborated details of similarities and differences between the origin and functioning of siRNA and miRNA can be read in the review (Carthew and Sontheimer, 2009; Khatri et al., 2012; Rao et al., 2009). Though the functional outcomes of siRNA, miRNA, and shRNA are similar, each represents intrinsically different molecular entities and therefore their molecular mechanism of action, RNAi pathways, advantages, and disadvantages are distinct to each class (Carthew and Sontheimer, 2009; Rao et al., 2009; Tang, 2005).

In this chapter, cellular delivery of siRNA has been focused.

## 6.2 CHALLENGES WITH SIRNA DELIVERY

For RNAi to function, it is imperative that siRNA molecules are delivered into the cytosol of target cell and be incorporated into RNAi machinery. Enlisted below are some of the barriers to intracellular delivery of naked siRNA.

### *6.2.1 PHYSICOCHEMICAL BARRIERS*

#### *6.2.1.1 SIZE AND HYDROPHILICITY*

siRNA molecules are too large (12 kDa) and too hydrophilic owing to the phosphodiester bonds that diffusion across cell membranes is difficult (Aliabadi et al., 2012; Xia et al., 2016).

#### *6.2.1.2 SURFACE CHARGE*

The polyphosphate (~40/molecule) backbone of siRNA imparts negative charge to the molecule. Since the cell membrane surface charge is negative, internalization of negatively charged siRNA is prevented by electrostatic repulsion (Raemdonck et al., 2008).

## 6.2.2 PHYSIOLOGICAL BARRIERS

### 6.2.2.1 DEGRADATION BY NUCLEASES

siRNA is highly susceptible to serum endonucleases and rapidly undergoes enzymatic degradation resulting in short plasma half-life <30 min on systemic administration (Aliabadi et al., 2012; Raemdonck et al., 2008).

### 6.2.2.2 RENAL CLEARANCE

The renal glomeruli molecular weight cutoff is 60 kDa, much larger than molecular weight of 12 kDa of siRNA resulting in rapid renal clearance (Dykxhoorn et al., 2006; Aliabadi et al., 2012). Forty-fold increased accumulation of radiolabeled siRNA was observed in the kidneys after 1 h of low-pressure intravenous administration (Bruno, 2011).

### 6.2.2.3 POOR PHARMACOKINETICS

High susceptibility to nuclease degradation coupled with rapid renal clearance is responsible for poor pharmacokinetic profile of intravenously administered naked siRNA (Hatanaka et al., 2010; Musacchio and Torchilin, 2012).

### 6.2.2.4 OPSONIZATION

siRNA accumulates in the organs of reticuloendothelial system like liver and spleen where the macrophage cells like Kupffer cells readily clear them (Whitehead et al., 2009). siRNA can stimulate innate immune system by activating interferon response. These oligonucleotides can associate nonspecifically with serum proteins that are phagocytozed by macrophages and monocytes (Kanasty et al., 2013).

### 6.2.2.5 DEGRADATION IN THE ENDOSOME

siRNA is taken up through endocytosis in the form of endocytic vesicles that in due course mature and merge into lysosomal vesicles. The low pH and high enzymatic activity in the milieu of late endosome and lysosome degrades siRNA (Dominska and Dykxhoorn, 2010).

## 6.3 siRNA DELIVERY APPROACHES

It is evident that naked siRNA cannot be delivered alone and that the potentials of RNAi mechanism cannot be harvested unless siRNA is efficiently delivered into the host cells.

### *6.3.1 CHEMICAL MODIFICATIONS*

One of the preliminary strategies employed was chemical modification of the oligonucleotide backbone to generate nuclease-resistant siRNA. Several modifications were proposed and practically studied. Few examples have been enlisted below.

#### *6.3.1.1 CHANGES IN SUGAR MOIETY*

It was realized that 2-OH position on the ribose was not an important structural requirement for activity. Therefore, this position was expansively modified to generate versatile molecules with improved nuclease stability and better potency. *O*-alkylation, especially methylation, worked well. The best results were obtained with fluoro-substitution which was inspired from development of DNA mimics. Changes at the ring oxygen position have also been attempted. These modifications can be on either of the sense or antisense stands or both. The most common and well-tolerated site appears to be the end terminal 3′ overhangs of the antisense strand.

#### *6.3.1.2 PHOSPHATE LINKAGE MODIFICATION*

Replacement of the phosphodiester group with phosphorothioate or boranophosphate linkage has been reported. However, this strategy did not significantly alter siRNA biodistribution and kinetics, on the other hand resulted in reduced potency and higher cytotoxicity in few cases.

#### 6.3.1.3 MODIFICATIONS IN THE BASE STRUCTURE

Alterations in the adenine and uracil structure by diaminopurine and halogenated uracil/thiouracil have been reported.

#### 6.3.1.4 MODIFICATIONS AT TERMINI

Reports suggest that 21 nucleotides, double-stranded chain with 2 nucleotide overhangs at 3′ end is the ideal siRNA structure. It is a consensus that the 5′ phosphate group on the antisense strand is an important structure feature for RNAi and modifications at this position may result in loss of activity. Tethering of fluorescent dyes at the termini has enabled biophysical and biodistribution analysis to be performed.

#### 6.3.1.5 MODIFICATION ON SENSE STRAND

Another strategy employed was conjugation of small molecules like polyethylene glycol or cholesterol or peptides like cell penetrating peptides to the sense strand. Cholesterol-conjugated siRNA could achieve *in vivo* silencing of endogenous Apolipoprotein B (Apo B) mRNA in liver and lowering of plasma cholesterol levels in transgenic mice (Soutschek et al., 2004). In an attempt to device alternative lipid–siRNA conjugates, it was found that bile acid derivatives and long-chain ($C > 18$) fatty acid conjugates could effectively elucidate RNAi via HDL receptors in hepatocytes of mice (Wolfrum et al., 2007). In another study, DiFiglia et al. (2007) detailed the ability of a cholesterol-modified siRNA to silence mutant gene associated with Huntington's disease by a single intrastiatal injection which attenuated the striatal and cortical neuropathological and behavioral deficits observed in a rapid-onset mouse model.

Though chemical modification strategy offers some advantages over naked siRNA, not all modifications are effective and well tolerated. Judicious choice of backbone modifications and siRNA sequences must be done to maintain functionality of siRNA while empowering serum stability. Activation of immune response and off-target mRNA silencing are some of the other challenges that may be encountered.

### 6.3.2 VECTOR-BASED SIRNA DELIVERY

The next strategy employed was the development of siRNA vectors, the two categories being viral and nonviral carrier systems.

#### 6.3.2.1 VIRAL VECTORS

The process of insertion of viral material into host cell in course of natural replication cycle of a virus was exploited in development of viral vectors that evolved as tools for gene therapy. Retrovirus, lentivirus, adenovirus, and adeno-associated virus are some of the viral vectors reported for RNAi (Davidson and McCray, 2011; Heilbronn and Weger, 2010). Major differences among these vectors are in their packaging capacity, immunogenicity, and duration of expression. For example, lentivirus allows packaging of larger inserts up to 15 kb, less immunogenic, and exhibit long-lasting expression; adeno-associated vector packs ~4.5 kb, mildly immunogenic and imparts transient expression and adenovirus can pack <10 kb, is highly immunogenic, and imparts transient expression. shRNA carrying lentivirus has shown successful regulation of specific brain targets following local administration (Dreyer, 2010). HIV suppression by retrovirus-shRNA in T-cell line in vitro was showed in a recent study. Also, hematopoietic stem cells that were transduced ex vivo with lentivirus–shRNA and reinfused into HIV patients have been of clinical significance (Morris and Rossi, 2006). In general, expression of RNAi systems is achieved by delivery of plasmids into viral vectors. Following host cell internalization of the vectors, RNAi triggers like shRNA or miRNA are transcribed from the plasmid and processed via RISC mechanism to target mRNA. Since the RNAi triggers are incorporated into plasmids that remain episomal (not part of chromosomal DNA), transient expression of RNAi can be achieved in dividing cells like epithelial tissue or in nondividing cells like neuronal tissue thus avoiding insertional mutagenesis. Following cell division, the episomal plasmid is lost. Tissue targeting can also be achieved by manipulation of capsid genes since the structure of the viral capsid (protein coat of virus) influences tissue tropism, that is, the virus may infect diverse cells or specific cells within tissue.

Though viral vectors have been one of the earliest carrier systems employed, their utility is limited by the disadvantages of high immunogenicity, ineffective delivery, handling difficulty with high processing cost.

Expression of viral genes by the vectors can induce immune response in the host cell. These viral vectors may infect as multiples copies in one cell, resulting in amplified dosing of RNAi triggers, inefficient delivery, and toxicity which is undesirable.

#### *6.3.2.2 NONVIRAL VECTORS*

Bacterial vectors were also attempted as an alternative to viral vectors. This innovative platform was based on the findings that bacteria like *Escherichia coli* with recombinant plasmid expressing shRNA could effectively elucidate RNAi after in vivo administration in mice (Xiang et al., 2006). Research also showed that therapeutic bacteria could enter and deliver shRNA in tumor cells of cancer patient (MacDiarmid et al., 2007, 2009).

#### *6.3.2.3 DELIVERY SYSTEMS*

The limitations of biological carriers for RNAi machinery opened new opportunities for delivery systems. By their design and characteristics, these delivery systems would interact with siRNA and self-assemble into supramolecular complexes whose goal is to overcome the shortcomings of naked siRNA. Some of the desirable properties of such carriers would be imparting stability against enzymatic degradation, ensuring efficient intracellular trafficking and release of siRNA into target cells while maintaining the functional efficacy of the cargo.

Literature is abundant with reports of many carrier systems. We would refer the readers to Aliabadi et al. (2012), Bruno (2011), Dominska and Dykxhoorn (2010), Foged (2012), and Kanasty et al. (2013) that extensively describe the different siRNA carrier systems. In the following chapter, liposome-based siRNA carriers would be discussed with special emphasis on the excipients that impart siRNA-carrying property.

## 6.4 NEED FOR LIPOSOMAL DELIVERY

Many siRNA carriers have been reported for *in vitro* applications but are rendered inappropriate for *in vivo* utility owing to poor safety profile and unsatisfactory tissue delivery. Most siRNA-based therapies that have entered clinical trials are limited to local delivery like intravitreal administration of siRNA against macular degeneration. However, systemic delivery of siRNA remains an unmet goal.

We know that cell membrane is composed of phospholipids and in nature intra- and intercellular transport of material is facilitated by the flexibility, mobility, and self-assembly of these lipids. This observation prompted the development of phospholipid-based delivery systems that imparted biocompatibility along with effective cellular internalization (Mendes et al., 2013). Liposomes are vesicular carrier systems wherein self-assembly of amphiphilic phospholipids generates dichotomous hydrophilic and hydrophobic domains that house both lipophilic and hydrophilic cargo. Liposomes offer the essential property of biocompatibility, biodegradability, nonimmunogenicity, commercial viability along with the expected properties of stability, intracellular internalization while providing robust gene silencing for siRNA-based therapies (Kanasty et al., 2013; Nagarsenker and Marwah, 2016; Xia et al., 2016; Wu and McMillan, 2009). Thus, liposomes became an obvious choice for siRNA delivery. However, it was soon realized that liposomes composed of neutral lipids themselves could not load siRNA and that stronger interactive mechanism was necessary (Xia et al., 2016). Covalent conjugation of siRNA with lipid molecules like cholesterol and short-chain fatty acids was sought. This enabled siRNA loading into liposomes with appreciable in vivo administration success (Jeong et al., 2008). The limitations of chemically modified siRNA have been discussed in the previous section. Hence, alternative approach involving physical interaction of lipids and siRNA was sought.

A closer look at structure and physicochemical properties of polynucleotides reveals that the polyphosphate backbone imparts slightly negative charge to the polynucleotide. Hence, electrostatic interaction between the negatively charged phosphate and positively charged aminolipid can yield stable lipoplex. This opened the doors for development of array of cationic carrier systems wherein siRNA was electrostatically bound to charged excipient and efficiently loaded into the carrier (Xia et al., 2016). These noncovalent complexes are spontaneously formed by simple mixing and incubation. This supramolecular assembly consists of multilamellar structures with polynucleotide chain sandwiched between cationic membrane into lamellar phase or inverted hexagonal phase depending on the lipid composition (Barreleiro et al., 2003; Oberle et al., 2000). The need for liposomal delivery system for carrying siRNA has been depicted in Figure 6.1.

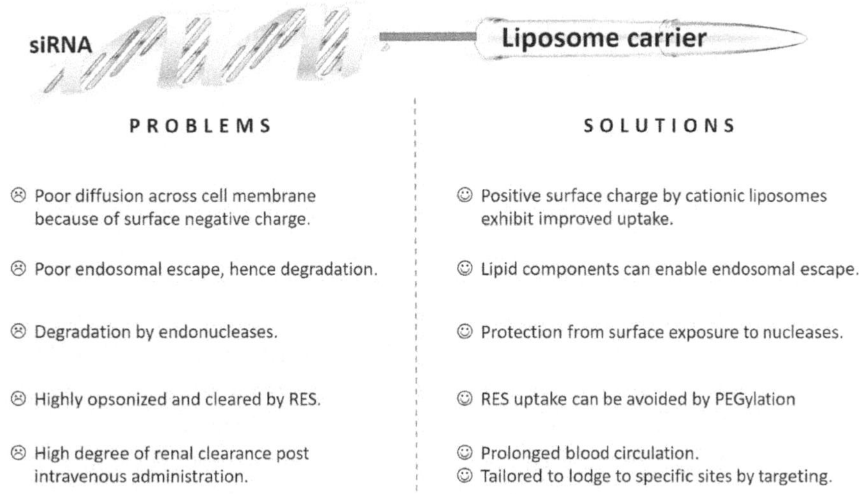


**FIGURE 6.1** Need for liposomal delivery vehicle for siRNA.

It is well documented that nanoparticles employ various internalization pathways like clathrin-mediated or caveolae-mediated endocytosis, micropinocytosis, and others (ur Rehman et al., 2013). Uptake may virtually occur in every cell which interact with the particles. However, every entry does not necessarily translate to efficacy of the cargo. Intracellular transfection of siRNA is governed by particle uptake, stability, and ability to escape endosome. The true challenge lies in not just delivering siRNA to the site but also ensuring its effective release (like endosomal escape, avoidance of accumulation in lysosomes) for functional efficacy.

## 6.5 EXCIPIENT-DRIVEN SIRNA DELIVERY

In the following section of the chapter, we shall focus on the excipient-driven siRNA interactions to tailor efficient delivery.

### *6.5.1 CATIONIC LIPIDS*

The basis for siRNA carrier systems grew from the knowledge of DNA delivery systems. In pursuit of development of nonviral DNA carriers, Felgner and collaborators synthesized the first cationic agent, DOTMA for DNA delivery following which many popular and commercially viable cationic agents evolved (Pisani et al., 2011). The different cationic lipids like DOTAP, DDAB, DOSPA, Dc-Chol, and others employed for siRNA delivery has been detailed in reference (Koynova and Tenchov, 2014; Singh et al., 2015; ur Rehman et al., 2013). The common structural feature of a cationic lipid is one or more positively charged quaternary nitrogen group attached via a linker to a hydrophobic core composed of alkyl or acyl hydrocarbon chain (ur Rehman et al., 2013). This unique structure enables these entities to self-assemble or be readily incorporated into liposomal vesicles. The significance of this entity in the carrier systems lies in not only complexing the cargo but also in attributing overall positive charge to the carrier that ensures easier biological membrane interaction with anionic cell surface molecules, sulfated proteoglycans, sialic acid, and others resulting in rapid fusion or endocytosis and intracellular internalization. In some cases, the cationic lipid may induce transient membrane perturbations due to lipid intermingling resulting pore-like structures across the cell membrane that facilitates siRNA delivery (Foged, 2012; Kanasty et al., 2013; MacLachlan, 2007). Two parameters that govern such lipoplex performance are membrane charge density (Average charge per unit area of membrane) and charge ratio (siRNA:charge).

### *6.5.2 HELPER LIPIDS*

Helper lipids are usually included to aid the cationic agent to provide particle stability, biocompatibility, and enhance cell membrane interaction. They improve siRNA delivery to the cytosol by different mechanisms. DOPE is one helper lipid that causes pH-dependent phase transition of liposomal membrane to nonlamellar structures. The ability of DOPE to control the structure of the cationic lipoplexes also depends on the molecular shape of the cationic lipid which is determined by the ratio of head group area over hydrophobic area (Smisterová et al., 2001). Helper lipid property of DOPE is attributed to its unique structure wherein the surface area of the hydrophilic head is smaller than the hydrophobic tail resulting in inverted cone formation of the lipid and these transient conformational disorders causes the lipid membrane to assume an overall hexagonal phase. The lamellar to nonlamellar transitions is instrumental in initiating lipid mixing between liposomes and target cell membrane (Cheng and Lee, 2016; Mochizuki et al., 2013). Though many research experiments over structure-efficiency relationships have been studied the conclusions are not consistent especially for siRNA delivery. While most experiments support role of DOPE in H phase transition and resultant transfection in vitro, few report that DOPE reduces transfection in vivo (Cheng and Lee, 2016; Foged, 2012).

Cholesterol is another example of helper lipid that operates by lowering the phase transition temperature causing the lipoplex membrane to become more flexible and increases the propensity for cell membrane fusion. Wolfrum et al. (2007) formulated liposomes containing high concentration of cholesterol that exhibited satisfactory siRNA delivery, owing to membrane destabilization. Interestingly, Rao et al. reported some studies wherein cationic lipid has high affinity for cholesterol and elucidate transfection in presence of cholesterol and not DOPE and vice versa. Such observation was explained by the differential influence of each of the lipids on the surface charge density and distribution (Dabkowska et al., 2012). Since cholesterol is endogenous to the body and cholesterol-containing particles are readily uptaken by many internalization pathway, lipoplex-containing cholesterol performs better in vivo (Rao, 2010).

The phosphatidylcholines form stable lamellar bilayered membranes that is influenced by the degree of unsaturation and chain length of the hydrophobic tail. These lipids form liposomes with better stability that lack phase transition to nonlamellar phases. Hence, these lipids represent a compromise between stability and transfection efficiency at cellular level (Cheng and Lee, 2016).

### *6.5.3 IONIZABLE LIPIDS*

Though cationic lipids result in effective condensation of siRNA, they are marred by high toxicity, attraction to RES uptake, interaction with plasma proteins, and activation of immune response owing to high positive charge (Xue et al., 2014). A balance of minimal toxicity with high efficacy can be achieved with ionizable lipids wherein the p*Ka* of lipid can be tailored such that it should remain unprotonated during circulation but should become protonated in either the early or late endosome (Kanasty et al., 2013). The distinct structural difference between permanently cationic and ionizable cationic lipid is the presence of tertiary amino group that is unionized at physiologic pH but readily ionizes at low pH. This causes repulsion between the ionized groups, increasing the surface charge and charge distribution along the lipid membrane which further impart endosomal membrane interactions in a fashion similar to cationic lipids (ur Rehman et al., 2013). Jayaraman et al. (2012) evaluated several ionizable lipids to maximize potency of siRNA carrying lipid nanoparticles by altering p*K*a of lipids and reported that p*K*a range of 5.6–6.9 was the most efficacious. Moreover, the overall p*K*a of liposome dominates individual lipids and therefore coformulation of lipids within the optimal range exhibited higher fold silencing than individual lipids. With a view to understand mechanism by which maximum gene silencing potency can be achieved by ionizable lipids, Lin and collaborators investigated lipid nanoparticles containing four ionizable lipids that were differing marginally in their structures. They inferred that the relative potencies of each of the studied lipids were attributed to the p*K*a and their stability against degradation by endogenous lipase. Hence, gene silencing is a function of lipoplex uptake, distribution, intracellular processing, and ability to escape endocytic pathways (Lin et al., 2013). Similar study by Yu et al. (2014) showed that cationic lipid like DOTAP showed satisfactory cellular uptake of siRNA but low silencing efficacy in comparison to DODMA, an ionizable lipid which conferred better endosomal escape property.

### *6.5.4 PEGYLATED LIPIDS*

PEGylation serves many purposes like particle stabilization with respect to size and shape, prevents particle aggregation, increases circulation times, and reduces unintended off-target uptake like opsonization/RES uptake or nanoparticles (Bao et al., 2013). Literature is abundant with reports wherein PEG incorporation has improved delivery efficiency of nanoparticles and the most common shielding lipids are those containing different molecular weights of PEG chains anchored to the lipid. In an independent study, Kolli et al. (2013) synthesized pH-responsive liposomes composed of oxime-linked PEGylated lipid wherein oxime linker was stable at physiologic pH 7.4 but degraded at pH 5.5 of the endosome. It showed improved delivery, release, and siRNA efficacy against hepatitis virus in liver cells both in vitro and in vivo. Another approach of pH-sensitive nanoparticle was developed by conjugating PEG with agents to incorporate groups like imidazole and methacrylic acid residues which protonated at low pH in the endosome. The PEG coat interacted with endosomal membrane, diffused from the carrier, thus enabling release of siRNA into cytosol (Lin et al., 2012). On similar strategy, sheddable coatings of PEG-ceramides on Apo B siRNA-loaded liposomes showed 1000-fold increase in potency in comparison to cholesterol-conjugated Apo B siRNA following a single intravenous administration (Foged, 2012).

### *6.5.5 TARGETING LIPIDS*

To ensure target-specific accumulation of lipoplex, several endogenous or exogenous targeting ligands have been conjugated with lipids and incorporated into carriers. The ligand can be preinserted in the liposomal vesicle before condensation with siRNA to form lipoplex. This approach has the disadvantage that surface exposure of ligand may be compromised by siRNA condensation. By the postinsertion strategy, wherein the ligand is incubated with the lipoplex, ligand-loading efficiency may be compromised (Foged, 2012). Some of the examples of targeting moieties include protein ligands like antibodies, vitamins like folate, glycolipids, and others (Ozpolat et al., 2014).

Figure 6.2 is a pictorial depiction of the role of excipients in designing liposomal carrier for siRNA.

## 6.6 FORMULATION ASPECTS

Physicochemical properties of the liposomal carrier/lipoplex are an equal contributor as the excipients forming them, for effectively delivering siRNA. Like the optimization parameters practiced for chemically synthesized small molecules, properties of size, shape, surface charge, surface engineering, and others may influence biodistribution of lipoplex.

### *6.6.1 PARTICLE SIZE*

There are varied reports on the optimal particle size required for effective cellular uptake. Size heterogeneity is a factor difficult to control and factors like incubation media, time of incubation, and charge ratio influence the size distribution. Study by Ross and Hui (1999) reported that particle size of lipoplex does influence internalization and was a major determinant for in vitro transfection efficiency, while Spagnou et al. (2004) reported that siRNA

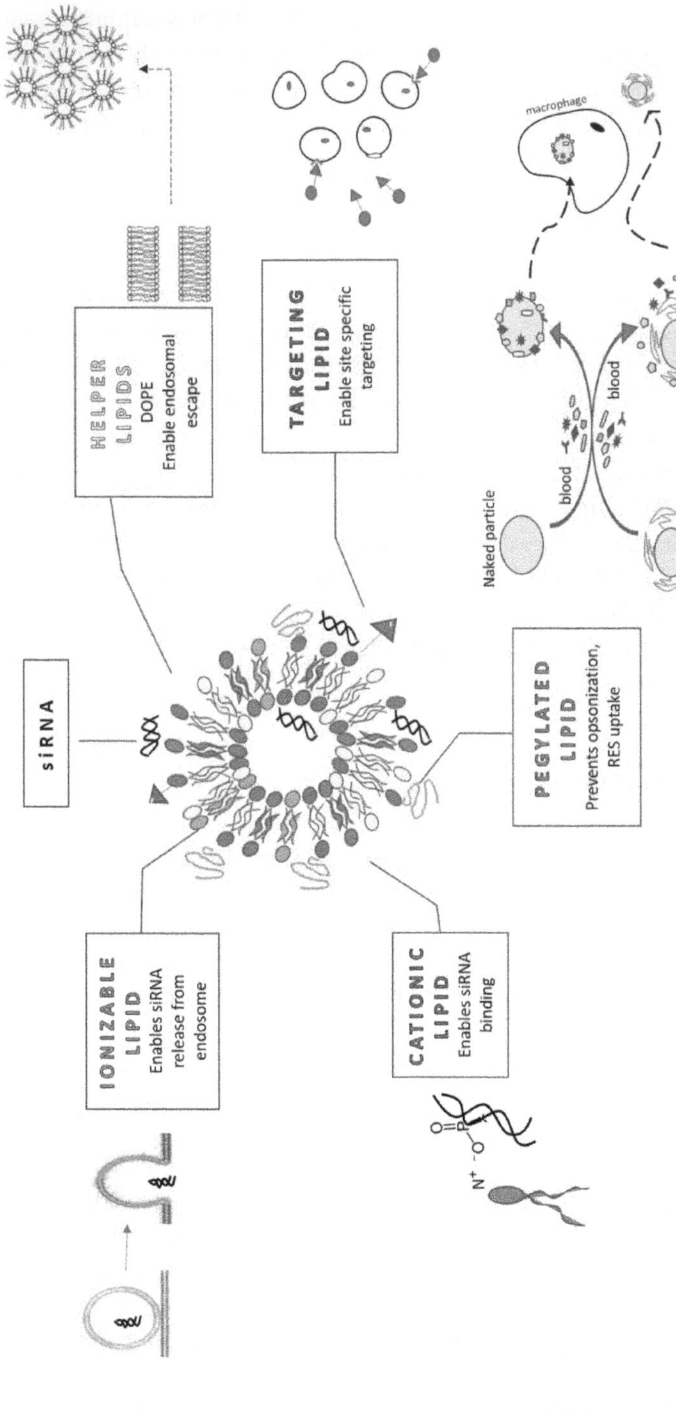


**FIGURE 6.2** **(See color insert.)** Role of excipient in liposomal siRNA delivery vehicle.

carrying lipoplex size over the range of 60–400 nm did not significantly influence gene silencing. Ramezani and collaborators screened different liposomal formulations to investigate the effect of cationic liposome size and observed that lipoplex with size less than 100 nm resulted in high transfection (Ramezani et al., 2009). Szoka et al. suggested that for systemic gene delivery, a size cutoff of 100 nm is highly critical for overcoming in vivo barriers like interactions with blood components, RES uptake, tumor access, and extracellular matrix (Li and Szoka, 2007). Considering the enhanced permeation and retention effect observed with tumor tissues, a size range of 100–500 nm is desirable for internalization by passive transport mechanisms like endocytosis (Reischl and Zimmer, 2009; Torchilin, 2011).

### 6.6.2 SURFACE CHARGE

Positively charge on the liposomes is essential for effective siRNA condensation and cellular uptake. Since the negative charge bearing glycocalyx in the outer cell membrane appreciably interacts with cationic particles, it is imperative that the overall charge on the lipoplex remains positive. However, the high surface charge may compromise safety and efficacy. The use of ionizable lipids has now led to development of neutral/slightly positive charged particles at physiologic pH but assume functionally effective positive surface charge at low pH of tumor or the endosome.

### 6.6.3 LAMELLARITY

The significance of this property is with regards to interaction with cell membrane and membrane fusion. Multilamellar vesicular structures possessed ability to better internalize and translate gene silencing even though the size of liposomes was large (Ramezani et al., 2009).

### 6.6.4 N/P RATIO

This is a parameter that defines the molar ratio of cationic lipid nitrogen (N) to siRNA phosphate (P). This ratio can vary from 1 to 10, or higher, and is named the nitrogen/phosphate ratio (N/P ratio). The exact ratio is cell line and experiment specific and requires optimization. Different ratios should be used in trials to determine the most effective formulation based on the application. N/P ratio can be determined by taking nanomole of nitrogen in the lipid, divided by the nanomole of phosphate in the RNA. N/P ratio influences the final characteristics of the lipoplexes, such as size, surface zeta potential, and reproducibility, thereby reflecting their efficiency following transfection.

## 6.7 siRNA FOR CANCER THERAPY

Cancer remains leading cause of death worldwide and cancer treatment has been a formidable challenge facing the scientific community for decades. Tremendous amount of efforts and resources have been employed in the years of research toward development of safe and effective treatment modalities. Traditionally, cancer treatment involved a macrolevel approach of surgery, radiation, and chemotherapy that resulted in extensive adverse effects and significant chances of relapse. Today, with the advancement in technology, increasing scientific knowledge of molecular mechanism of disease progression, the fight against cancer is more targeted. It is well established that cancer is caused by specific gene mutations and many important genetic information associated with different cancers have been discovered with precise identification of the mutations and characterization of the different pathways through which they act. RNAi phenomenon with siRNA therapy presents an opportunity as innovative platform with novel approach in the gene silencing for diseases like cancer which were once believed to be incurable. Cancer treatment using siRNA-based therapeutics is determined by the fundamental principle of loss of function activity of siRNA. siRNA can be designed to downregulate genes associated with overexpressed oncogenes like cyclin-dependent kinases, malfunctional oncogenes like insulin growth factor, viral oncogenes, tumor–host interactions like vascular endothelial growth factor, tumor resistance to chemo or radiotherapy, and antiapoptotic/antiproliferative effects like Bcl-2 proteins (Kim et al., 2016; Young et al., 2016).

## 6.8 TOXICITY CONCERNS OF CATIONIC LIPOSOMES

Toxicity of cationic agents is the foremost concern limiting their application as major candidates for systemic delivery. Activation of complement system, stimulation of immune response via Toll-like receptor, induction of reactive oxygen species, disruption of cell membrane, DNA damage, and hepatotoxicity are some of commonly encountered toxicity issues. Agents with multivalence of the charge group elicit pronounced toxicity in comparison to monovalent counterparts. Owing to the nanometric size of these systems, the toxicity issues get significantly amplified due to larger surface area for interaction with biological system. In cases of repeat dosing, the risk of accumulation of toxins within the body and risk of cumulative toxicity can be anticipated. Comprehensive review on the nanotoxicity concerning siRNA delivery vehicles can be referred by the readers in Xue et al. (2014). Efficient siRNA delivery nanocarriers with minimal toxicity concerns need to be developed by optimizing the physical properties like surface exposure reduction by PEGylation, optimizing the composition with regards to cationic agent and neutral lipids, designing and synthesizing newer lipids with low toxicity, reducing off-target carrier lodging by incorporation of targeting moieties, exploring alternate routes of administration, developing hybrid particles, and adopting codelivery or combination therapy.

## 6.9 CLINICAL RELEVANCE OF SIRNA DELIVERY VEHICLES

Over the recent years, the number of siRNA carrying delivery vehicles entering clinical phase following successful efficacy in preclinical rodent models has seen a significant rise. References like Bruno (2011), Kanasty et al. (2013), Kim et al. (2016), and Ozcan et al. (2015) enlist the different siRNA-based therapeutics in clinical phase. Tenacious efforts by few pharmaceutical companies have launched siRNA-loaded products onto clinical trials but none have seen the light to the marketplace. While some products fail to pass safety specifications, others fail to deliver the primary objective of efficacy.

The application of siRNA therapy can be either by localized delivery to the intended viscera or by intravenous administration for systemic delivery. Local delivery modality is explored for organs that are easily accessible for administration like eyes, respiratory organs, or vagina. Some of the products in the different phases of clinical trials have been enlisted in Table 6.1.

**TABLE 6.1** Summary of Clinical Studies Based on siRNA Lipid Nanoparticles.

| Sr. no. | Product | Initiator | Target | Route of administration | Clinical phase status |
|---|---|---|---|---|---|
| **Local administration** | | | | | |
| 1 | Bevasiranib (VEGF siRNA) | Opko Health Inc. | Retinal neovascularization in age-related macular degeneration | Intravitreal | Phase III terminated in 2009 |
| 2 | Bamosiran SYL040012 | Sylentis, SA | β2-adrenergic receptor in ocular hypertension or open angle glaucoma | Ocular | Phase I |
| 3 | QPI-1007 | Quark Pharmaceuticals | Caspase-2 enzyme in glaucoma | Intravitreal | Phase II |
| 4 | ALNRSV01 | Alnylam Pharmaceuticals | Respiratory syncyial viral infection postlung transplant | Intranasal | Phase II |
| 5 | TD-101 | | Keratin 6A N171K mutant in the rare skin disorder pachyonychia congenita | Intralesional | Phase I terminated |
| **Systemic administration** | | | | | |
| 6 | siRNA-EphA2-DOPC | MD Anderson Cancer Center | Advanced cancer<br>Solid tumor | Intravenous | Phase I |
| 7 | Atu027 | Silence Therapeutics | Cancer<br>Protein kinase N3 | Intravenous | Phase II |
| 8 | TKM-80301 | Tekmira Pharmaceuticals | Advanced solid tumors<br>Polo-like kinase 1 | Intravenous | Phase II |
| 9 | DCR-MYC siRNA | Dicerna Pharmaceuticals | Hepatocellular carcinoma | Intravenous | Phase I |
| 10 | ALN-TTRsc siRNA-GalNac conjugate | Alnylam Pharmaceuticals | Transtyretin-mediated amyloidosis cardiopathy | Subcutaneous | Phase II |

## 6.10 CONCLUSION

siRNA-based therapeutics presents a powerful and promising tool for treatment of many diseases, especially cancer. It can be viewed as a treatment modality for targets that were once believed to be undruggable. With knowledge of the specific genes/proteins involved in disease progression, specific siRNA can be designed rationally to silence practically any gene. Researchers today have gathered the know-how of overcoming the challenges to siRNA delivery, yet, most of the table-top research with cell culture in laboratories and effective in vivo efficacy in preclinical models have failed to translate to bedside therapy. Concerted efforts toward development of safe, effective, and commercially viable siRNA carrier vehicles are the need of the hour. Since there is limited information regarding the potential toxicity of nanoparticles, the design of the delivery vehicle should be judiciously made after considerable deliberations. Also, quality standards, techniques, and tools for evaluation of toxicity concerning nanoparticles should be developed which can help in faster and economical screening of nanocarriers with improved clinical relevance.

## KEYWORDS

- **siRNA**
- **cationic liposomes**
- **helper lipids**
- **ionizable lipids**
- **RNAi phenomenon**
- **lipoplex**

## REFERENCES

Aliabadi, H. M.; Landry, B.; Sun, C.; Tang, T.; Uludağ, H. Supramolecular Assemblies in Functional siRNA Delivery: Where Do We Stand? *Biomaterials* **2012,** *33* (8), 2546–2569.

Bao, Y.; Jin, Y.; Chivukula, P.; Zhang, J.; Liu, Y.; Liu, J.; Clamme, J.-P.; Mahato, R. I.; Ng, D. Ying, W. Effect of PEGylation on Biodistribution and Gene Silencing of siRNA/Lipid Nanoparticle Complexes. *Pharm. Res.* **2013,** *30* (2), 342–351.

Barreleiro, P. C.; May, R. P.; Lindman, B. Mechanism of Formation of DNA–Cationic Vesicle Complexes. *Faraday Dis.* **2003,** *122*, 191–201.

Bertrand, J.-R.; Pottier, M.; Vekris, A.; Opolon, P.; Maksimenko, A.; Malvy, C. Comparison of Antisense Oligonucleotides and siRNAs in Cell Culture and In Vivo. *Biochem. Biophys. Res. Commun.* **2002,** *296* (4), 1000–1004.

Bruno, K. Using Drug-Excipient Interactions for siRNA Delivery. *Adv. Drug Deliv. Rev.* **2011,** *63* (13), 1210–1226.

Carthew, R. W.; Sontheimer, E. J. Origins and Mechanisms of miRNAs and siRNAs. *Cell* **2009,** *136* (4), 642–655.

Cheng, X.; Lee, R. J. The Role of Helper Lipids in Lipid Nanoparticles (LNPs) Designed for Oligonucleotide Delivery. *Adv. Drug Deliv. Rev.* **2016,** *99*, 129–137.

Dabkowska, A. P.; Barlow, D. J.; Hughes, A. V.; Campbell, R. A.; Quinn, P. J.; Lawrence, M. J. The Effect of Neutral Helper Lipids on the Structure of Cationic Lipid Monolayers. *J. R. Soc. Interface* **2012,** *9* (68), 548–561.

Davidson, B. L.; McCray, P. B. Current Prospects for RNA Interference-Based Therapies. *Nat. Rev. Genet.* **2011,** *12* (5), 329–340.

DiFiglia, M.; Sena-Esteves, M.; Chase, K.; Sapp, E.; Pfister, E.; Sass, M.; Yoder, J.; Reeves, P.; Pandey, R. K.; Rajeev, K. G. Therapeutic Silencing of Mutant Huntingtin with siRNA Attenuates Striatal and Cortical Neuropathology and Behavioral Deficits. *Proc. Natl. Acad. Sci.* **2007,** *104* (43), 17204–17209.

Dominska, M.; Dykxhoorn, D. M. Breaking Down the Barriers: siRNA Delivery and Endosome Escape. *J. Cell. Sci.* **2010,** *123* (8), 1183–1189.

Dreyer, J. L. Lentiviral Vector-Mediated Gene Transfer and RNA Silencing Technology in Neuronal Dysfunctions. *Methods Mol. Biol.* **2010,** *614*, 3–35.

Dykxhoorn, D.; Palliser, D.; Lieberman, J. The Silent Treatment: siRNAs as Small Molecule Drugs. *Gene Ther.* **2006,** *13* (6), 541–552.

Elbashir, S. M.; Harborth, J.; Lendeckel, W.; Yalcin, A.; Weber, K.; Tuschl, T. Duplexes of 21-Nucleotide RNAs Mediate RNA Interference in Cultured Mammalian Cells. *Nature* **2001,** *411* (6836), 494–498.

Foged, C. siRNA Delivery with Lipid-Based Systems: Promises and Pitfalls. *Curr. Top. Med. Chem.* **2012,** *12* (2), 97–107.

Hatanaka, K.; Asai, T.; Koide, H.; Kenjo, E.; Tsuzuku, T.; Harada, N.; Tsukada, H.; Oku, N. Development of Double-Stranded siRNA Labeling Method Using Positron Emitter and Its In Vivo Trafficking Analyzed by Positron Emission Tomography. *Bioconj. Chem.* **2010,** *21* (4), 756–763.

Heilbronn, R.; Weger, S. Viral Vectors for Gene Transfer: Current Status of Gene Therapeutics. In *Drug Delivery: Handbook of Experimental Pharmacology*; Springer: Berlin-Heidelberg, 2010; pp 143–170.

Hutvagner, G.; Simard, M. J. Argonaute Proteins: Key Players in RNA Silencing. *Nat. Rev. Mol. Cell Biol.* **2008,** *9* (1), 22–32.

Jayaraman, M. S.; Ansell, M.; Mui, B. L.; Tam, Y. K.; Chen, J.; Du, X.; Butler, D.; Eltepu, L.; Matsuda, S.; Narayanannair, J. K. Maximizing the Potency of siRNA Lipid Nanoparticles for Hepatic Gene Silencing In Vivo. *Angew. Chem. Int. Ed.* **2012,** *51* (34), 8529–8533.

Jeong, J. H.; Mok, H.; Oh, Y.-K.; Park, T. G. siRNA Conjugate Delivery Systems. *Bioconj. Chem.* **2008,** *20* (1), 5–14.

Kanasty, R.; Dorkin, J. R.; Vegas, A.; Anderson, D. Delivery Materials for siRNA Therapeutics. *Nat. Mater.* **2013,** *12* (11), 967–977.

Kawamata, T.; Tomari, Y. Making RISC. *Trends Biochem. Sci.* **2010,** *35* (7), 368–376.

Khatri, N. I.; Rathi, M. N.; Baradia, D. P.; Trehan, S.; Misra, A. In Vivo Delivery Aspects of miRNA, shRNA and siRNA. *Crit. Rev. Ther. Drug Carrier Syst.* **2012,** *29* (6), 487–527.

Kim, H. J.; Kim, A.; Miyata, K.; Kataoka, K. Recent Progress in Development of siRNA Delivery Vehicles for Cancer Therapy. *Adv. Drug Deliv. Rev.* **2016,** *104*, 61–77.

Kolli, S.; Wong, S.-P.; Harbottle, R.; Johnston, B.; Thanou, M.; Miller, A. D. pH-Triggered Nanoparticle Mediated Delivery of siRNA to Liver Cells In Vitro and In Vivo. *Bioconj. Chem.* **2013,** *24* (3), 314–332.

Koynova, R.; Tenchov, B. Enhancing Nucleic Acid Delivery, Insights from the Cationic Phospholipid Carriers. *Curr. Pharmcol. Biotechnol.* **2014,** *15* (9), 806–813.

Li, W.; Szoka, Jr., F. C. Lipid-Based Nanoparticles for Nucleic Acid Delivery. *Pharm. Res.* **2007,** *24* (3), 438–449.

Lin, P. J. Tam,; Y. Y. C.; Hafez, I.; Sandhu, A.; Chen, S.; Ciufolini, M. A.; Nabi, I. R.; Cullis, P. R. Influence of Cationic Lipid Composition on Uptake and Intracellular Processing of Lipid Nanoparticle Formulations of siRNA. *Nanomedicine* **2013,** *9* (2), 233–246.

Lin, S.-Y.; Zhao, W.-Y.; Tsai, H.-C.; Hsu, W.-H.; Lo, C.-L.; Hsiue, G.-H. Sterically Polymer-Based Liposomal Complexes with Dual-Shell Structure for Enhancing the siRNA Delivery. *Biomacromolecules* **2012,** *13* (3), 664–675.

MacDiarmid, J. A.; Amaro-Mugridge, N. B.; Madrid-Weiss, J.; Sedliarou, I.; Wetzel, S.; Kochar, K.; Brahmbhatt, V. N.; Phillips, L.; Pattison, S. T.; Petti, C. Sequential Treatment of Drug-Resistant Tumors with Targeted Minicells Containing siRNA or a Cytotoxic Drug. *Nat. Biotechnol.* **2009,** *27* (7), 643–651.

MacDiarmid, J. A.; Mugridge, N. B.; Weiss, J. C.; Phillips, L.; Burn, A. L.; Paulin, R. P.; Haasdyk, J. E.; Dickson, K.-A.; Brahmbhatt, V. N.; Pattison, S. T. Bacterially Derived 400 nm Particles for Encapsulation and Cancer Cell Targeting of Chemotherapeutics. *Cancer Cell* **2007,** *11* (5), 431–445.

MacLachlan, I. Liposomal Formulations for Nucleic Acid Delivery. In *Antisense Drug Technology: Principles, Strategies, and Applications*. CRC Press Taylor & Francis Group: Boca Raton, FL, 2007; vol. 2, pp 237–270.

McCaffrey, A. P.; Meuse, L.; Pham, T.-T. T.; Conklin, D. S.; Hannon, G. J.; Kay, M. A. Gene expression: RNA Interference in Adult Mice. *Nature* **2002,** *418* (6893), 38–39.

McManus, M. T.; Sharp, P. A. Gene Silencing in Mammals by Small Interfering RNAs. *Nat. Rev. Genet.* **2002,** *3* (10), 737–747.

Mendes, A. C.; Baran, E. T.; Reis, R. L.; Azevedo, H. S. Self-Assembly in Nature: Using the Principles of Nature to Create Complex Nanobiomaterials. *Wiley Interdiscip. Rev. Nanomed. Nanobiotechnol.* **2013,** *5* (6), 582–612.

Mochizuki, S.; Kanegae, N.; Nishina, K.; Kamikawa, Y.; Koiwai, K.; Masunaga, H.; Sakurai, K. The Role of the Helper Lipid Dioleoylphosphatidylethanolamine (DOPE) for DNA Transfection Cooperating with a Cationic Lipid Bearing Ethylenediamine. *Biochim. Biophys. Acta—Biomembranes* **2013,** *1828* (2), 412–418.

Montgomery, M. Potent and Specific Genetic Interference by Double-Stranded RNA in *Caenorhabditis elegans*. *Nature* **1998,** *391*, 806–811.

Morris, K.; Rossi, J. Lentiviral-Mediated Delivery of siRNAs for Antiviral Therapy. *Gene Ther*. **2006,** *13* (6), 553–558.

Musacchio, T.; Torchilin, V. P. siRNA Delivery: From Basics to Therapeutic Applications. *Front. Biosci. (Landm. Ed.)* **2012,** *18*, 58–79.

Nagarsenker, M. S.; Marwah, M. S. Liposomes: Concept and Therapeutic Applications. In *Novel Approaches for Drug Delivery India*; Keservani, R. K., Sharma, A. K., Kesharwani, R. K., Eds.; Medical Information Science Reference, an Imprint of IGI Global: Hershey, PA, 2017; pp 52–82.

Oberle, V.; Bakowsky; Zuhorn, I. S.; Hoekstra, D. Lipoplex Formation under Equilibrium Conditions Reveals a Three-Step Mechanism. *Biophys. J.* **2002,** *79* (3), 1447–1454.

Ozcan, G.; Ozpolat, B.; Coleman, R. L.; Sood, A. K.; Lopez-Berestein, G. Preclinical and Clinical Development of siRNA-Based Therapeutics. *Adv. Drug Deliv. Rev.* **2015,** *87*, 108–119.

Ozpolat, B.; Sood, A. K.; Lopez-Berestein, G. Liposomal siRNA Nanocarriers for Cancer Therapy. *Adv. Drug Deliv. Rev.* **2014,** *66*, 110–116.

Pisani, M.; Mobbili, G.; Bruni, P. Neutral Liposomes and DNA Transfection. *Nonviral Gene Therapy*; INTECH Open Access Publisher, Europe, 2011; pp 320–346.

Pratt, A. J.; MacRae, I. J. The RNA-Induced Silencing Complex: A Versatile Gene-Silencing Machine. *J. Biol. Chem.* **2009,** *284* (27), 17897–17901.

Raemdonck, K.; Vandenbroucke, R. E.; Demeester, J.; Sanders, N. N.; De Smedt, S. C. Maintaining the Silence: Reflections on Long-Term RNAi. *Drug Discov. Today* **2008,** *13* (21), 917–931.

Ramezani, M.; Khoshhamdam, M.; Dehshahri, A.; Malaekeh-Nikouei, B. The Influence of Size, Lipid Composition and Bilayer Fluidity of Cationic Liposomes on the Transfection Efficiency of Nanolipoplexes. *Colloids Surf., B: Biointerfaces* **2009,** *72* (1), 1–5.

Rao, D. D.; Vorhies, J. S.; Senzer, N.; Nemunaitis, J. siRNA vs. shRNA: Similarities and Differences. *Adv. Drug Deliv. Rev.* **2009,** *61* (9), 746–759.

Rao, N. M. Cationic Lipid-Mediated Nucleic Acid Delivery: Beyond Being Cationic. *Chem. Phys. Lipids* **2010,** *163* (3), 245–252.

Reischl, D.; Zimmer, A. Drug Delivery of siRNA Therapeutics: Potentials and Limits of Nanosystems. *Nanomedicine* **2009,** *5* (1), 8–20.

Ross, P.; Hui, S. Lipoplex Size Is a Major Determinant of In Vitro Lipofection Efficiency. *Gene Ther.* **1999,** *6* (4), 651–659.

Sen, G. L.; Blau, H. M. A Brief History of RNAi: The Silence of the Genes. *FASEB J.* **2006,** *20* (9), 1293–1299.

Singh, Y.; Tomar, S.; Khan, S.; Meher, J. G.; Pawar, V. K.; Raval, K.; Sharma, K.; Singh, P. K.; Chaurasia, M.; Reddy, B. S. Bridging Small Interfering RNA with Giant Therapeutic Outcomes Using Nanometric Liposomes. *J. Control. Release* **2014,** *220*, 368–387.

Smisterová, J.; Wagenaar, A.; Stuart, M. C.; Polushkin, E.; ten Brinke, G.; Hulst, R.; Engberts, J. B.; Hoekstra, D. Molecular Shape of the Cationic Lipid Controls the Structure of Cationic Lipid/Dioleylphosphatidylethanolamine–DNA Complexes and the Efficiency of Gene Delivery. *J. Biol. Chem.* **2001,** *276* (50), 47615–47622.

Soutschek, J.; Akinc, A.; Bramlage, B.; Charisse, K.; Constien, R.; Donoghue, M.; Elbashir, S.; Geick, A.; Hadwiger, P.; Harborth, J.; John, M.; Kesavan, V.; Lavine, G.; Pandey, R. K.; Racie, T.; Rajeev, K. G.; Rohl, I.; Toudjarska, I.; Wang, G.; Wuschko, S.; Bumcrot, D.; Koteliansky, V.; Limmer, S.; Manoharan, M.; Vornlocher, H.-P. Therapeutic Silencing of an Endogenous Gene by Systemic Administration of Modified siRNAs. *Nature* **2004,** *432* (7014), 173–178.

Spagnou, S.; Miller, A. D.; Keller, M. Lipidic Carriers of siRNA: Differences in the Formulation, Cellular Uptake, and Delivery with Plasmid DNA. *Biochemistry* **2004,** *43* (42), 13348–13356.

Tang, G. siRNA and miRNA: An Insight into RISCs. *Trends Biochem. Sci.* **2005,** *30* (2), 106–114.

Tijsterman, M.; Plasterk, R. H. Dicers at RISC: The Mechanism of RNAi. *Cell* **2004,** *117* (1), 1–3.

Torchilin, V. Tumor Delivery of Macromolecular Drugs Based on the EPR Effect. *Adv. Drug Deliv. Rev.* **2011,** *63* (3), 131–135.

ur Rehman, Z.; Zuhorn, I. S.; Hoekstra, D. How Cationic Lipids Transfer Nucleic Acids into Cells and Across Cellular Membranes: Recent Advances. *J. Control. Release* **2013,** *166* (1), 46–56.

Whitehead, K. A.; Langer, R.; Anderson, D. G. Knocking down Barriers: Advances in siRNA Delivery. *Nat. Rev. Drug Discov.* **2009,** *8* (2), 129–138.

Wolfrum, C.; Shi, S.; Jayaprakash, K. N.; Jayaraman, M.; Wang, G.; Pandey, R. K.; Rajeev, K. G.; Nakayama, T.; Charrise, K.; Ndungo, E. M. Mechanisms and Optimization of In Vivo Delivery of Lipophilic siRNAs. *Nat. Biotechnol.* **2007,** *25* (10), 1149–1157.

Wu, S. Y.; McMillan, N. A. Lipidic Systems for In Vivo siRNA Delivery. *AAPS J.* **2009,** *11* (4), 639–652.

Xia, Y.; Tian, J.; Chen, X. Effect of Surface Properties on Liposomal siRNA Delivery. *Biomaterials* **2016,** *79*, 56–68.

Xiang, S.; Fruehauf, J.; Li, C. J. Short Hairpin RNA-Expressing Bacteria Elicit RNA Interference in Mammals. *Nat. Biotechnol.* **2006,** *24* (6), 697–702.

Xue, H. Y.; Liu, S.; Wong, H. L. Nanotoxicity: A Key Obstacle to Clinical Translation of siRNA-Based Nanomedicine. *Nanomedicine* **2014,** *9* (2), 295–312.

Young, S. W. S.; Stenzel, M.; Jia-Lin, Y. Nanoparticle–siRNA: A Potential Cancer Therapy? *Crit. Rev. Oncol. Hematol.* **2016,** *98*, 159–169.

Yu, B.; Wang, X.; Zhou, C.; Teng, L.; Ren, W.; Yang, Z.; Shih, C.-H.; Wang, T.; Lee, R. J.; Tang, S. Insight into Mechanisms of Cellular Uptake of Lipid Nanoparticles and Intracellular Release of Small RNAs. *Pharm. Res.* **2014,** *31* (10), 2685–2695.

Zimmermann, T. S.; Lee, A. C.; Akinc, A.; Bramlage, B.; Bumcrot, D.; Fedoruk, M. N.; Harborth, J.; Heyes, J. A.; Jeffs, L. B.; John, M. RNAi-Mediated Gene Silencing in Nonhuman Primates. *Nature* **2006,** *441* (7089), 111–114.

## CHAPTER 7

# THERANOSTIC APPLICATION OF INDOCYANINE GREEN LIPOSOMES

SABYASACHI MAITI* and SHALMOLI SETH[2]

*Department of Pharmaceutics, Gupta College of Technological Sciences, Ashram More, G.T. Road, Asansol 713301, West Bengal, India*

**Corresponding author. E-mail: sabya245@rediffmail.com*

## CONTENTS

## ABSTRACT

The ability of liposomes to encapsulate both diagnostic and therapeutic agents has drawn much attention of the research scientists in utilizing liposomes as nanocarriers for theranostic applications. Because of their unique structure, liposomes can improve the stability of these agents in biological environments and control their delivery to the intended sites, especially to cancer cells. Moreover, the well-known biocompatible characteristics of liposomes can provide better pharmacokinetics and biodistribution profiles of theranostic agents than other nanocarriers in clinical studies. Near-infrared (NIR) fluorescence imaging has attracted significant attention due to the low absorption and scattering of photons within the NIR region by living tissue. The deeper penetration of NIR fluorescence imaging compared to visible wavelengths widens the possibility of deeper tissue imaging in vivo, as well as long-term imaging without off-target effects of excitation. One potential NIR fluorescence probe is indocyanine green (ICG) that has been approved for clinical use by the United States Food and Drug Administration. Recently, attempts have been made to explore the potential theranostic applications of ICG-liposome nanocarriers. This chapter highlighted recent progress in developing modified liposomes for either diagnostic or therapeutic, as well as theranostic applications.

## 7.1 INTRODUCTION

The small vesicles consist of single or multiple concentric lipid bilayers surrounding an aqueous internal space are called liposomes. Only the polar hydrophilic groups are in contact with the external and internal aqueous solutions forming a bilayer with thickness of about 4–7 nm (Balgavý et al., 2001). The bilayers are usually composed of phospholipids (phosphatidylcholine or phosphatidylethanoamine) and cholesterol to encapsulate various active water-soluble drug in the hydrophilic core and/or insoluble drug in the hydrophobic membrane for targeted and controlled delivery of therapeutic agents. Cholesterol is commonly added to confer stability to the liposomes (Kirby et al., 1980; Ulrich, 2002). Cholesterol is located in the free space between the alkyl chains and it increases bilayer rigidity and decreases lateral diffusion in the lipid bilayer. The liposomal encapsulation could improve pharmacokinetics and therapeutic efficacy with fewer adverse effects (Muthu and Feng, 2013).

Liposomes are biocompatible, because their structure mimics the lipid bilayers found in the body. This enables drug delivery with relatively low immune response and toxicity. In addition to drug encapsulation, the large aqueous core offers possibilities to encapsulate small nanoparticles or other agents in the liposomes (Ulrich, 2002). Liposomes can be designed to extend the half-life of the drug in blood circulation and improve their delivery to target tissues and cells. The size and surface properties of liposomes can be easily modified by careful selection of components and preparation methods. Due to their versatility, liposomal formulation became the first nanomedicine approved by the United States Food and Drug Administration (US FDA) for clinical application in 1995.

Depending on the method of preparation, the diameter of spherical phospholipid vesicles ranged from 40 nm to a few micrometers. Liposomes larger than 100 nm are mostly retained at the injection site whereas small liposomes (40–70 nm) are distributed to the lymphatic system and blood circulation. Intestinal lipases render per oral delivery impossible for the liposomes, though a limited degree of success have been achieved after polymer coating of the liposomes (Takeuchi et al., 2005a, 2005b). However, the liposomes can be administered via topical routes such as the skin or eyes, or as injections into the blood circulation or vitreous cavity, and inhalation to the respiratory tract, although distinct advantages and challenges exist with these routes. In most studies, the liposomes are given as injections (e.g., intravenous, subcutaneous). Intravenous injections enable rapid systemic blood circulation and fast distribution of the drug to easily accessible tissues (Moghimi and Szebeni, 2003; Torchilin, 2005). On contrary, subcutaneous injections can be used for prolonged local effect or preferable targeting of the lymphatic system (Oussoren et al., 1997).

Like most nanoparticles, liposomes are most commonly internalized into the cells through endocytosis pathways (Straubinger et al., 1983). After cellular internalization, liposomes are entrapped within the endosomes (Medina-Kauwe et al., 2005). Often, the endocytosis leads the cargo into lysosomes that may break down the liposomes and release the drug cargo by enzymatic activity and acidic environment (Ulrich, 2002). Some drugs may tolerate the lysosomal conditions, but compounds that are more sensitive should escape from the endosomes before trafficking to the lysosomes and degradation. Passive diffusion is the simplest form of drug release from liposomes. This process is driven by the concentration gradient between the liposome core and outside medium. The lipid bilayer properties (i.e., rigidity and charge) control the rate of passive drug release. Reduction in the alkyl chain length and increased unsaturation render the bilayer more permeable. Likewise, the properties of the drug affect the passive permeation through the bilayer. Generally, relatively small and lipophilic compounds can pass through the liposomal bilayers faster than larger molecules that need disruption in the ordered bilayer for their escape from the liposome (Szoka Jr. and Papahadjopoulos, 1980). Charged compounds have notably reduced permeation from the liposomes compared to neutral molecules of the same size.

Despite safety and biocompatibility of liposomes, the major drawbacks of conventional liposomes are instability, insufficient drug loading, faster drug release, and shorter circulation time in the blood. When liposomes reach the blood circulation, serum opsonins selectively bind to their surface (Harashima et al., 1994; Koo et al., 2005). The extent of binding depends on the size and lipid composition of the liposomes. Mononuclear phagocyte system recognizes the opsonized liposomes and removes them from the bloodstream. In general, larger and negatively charged liposomes are eliminated from the blood circulation more quickly than smaller and neutral liposomes, respectively (Chonn et al., 1992; Senior and Gregoriadis, 1982), but this view has been contested by some reports (Immordino et al., 2006). Irrespective of the charge and size, the reticuloendothelial system, primarily in the liver and spleen, removes the uncoated liposomes efficiently from the systemic circulation within a few hours. Henceforth, the liposomes are coated with polyethylene glycol (PEG) or other hydrophilic polymer that sterically protects the liposomes from opsonization and prolongs their circulation (Allen et al., 2002). It has been suggested that by its excluded volume effect, the long PEG chain coated on the liposome surface prevents the adsorption of plasma protein onto the liposome surface and, as a result, effectively reduces the liposome aggregation in plasma (Muthu et al., 2011).

The sterically stabilized liposomes can accumulate in the tumors by enhanced permeability and retention (EPR) effect (Chonn and Cullis, 1998). The prolonged circulation (~24–48 h) can help to achieve better targeting effect of the drug carriers allowing more time for their interaction with the target (Torchilin, 1996). The leaky vasculature of tumor tissues and the lack of normal lymphatic drainage facilitate the EPR effect, nanoparticle accumulation, and, consequently, the delivery of drug in those areas.

The liposome surface can be further modified with ligands, such as antibodies, peptides, and other molecules that recognize and bind to the target cells (Park, 2002). The active targeting enables preferential accumulation in pathological tissues in comparison to the healthy tissues, thereby protecting the patient from adverse drug reactions and enhancing bioavailability. The targeting ligands can be bound covalently to the liposome surface or tethered at the end of the PEG chain (Maruyama et al., 1997). Cancer cell targeting with antibodies, transferrin, folate, and other ligands has especially been widely studied (Lee and Huang, 1996; Noble et al., 2014). In principle, the PEG tethered ligand should be readily available for binding at the target cell surface, but in practice, the ligand may be masked within the PEG layer (Lehtinen et al., 2012).

Even though an optimal liposomal size and active targeting may increase the uptake of the liposomes to the target cells, the therapeutic effect might remain unsatisfactory due to the inefficient endosomal escape and drug release from the liposomes (Shum et al., 2001). One of the most common methods for increased contents release within the cells is to formulate pH-sensitive liposomes. After being endocytosed in the stable form, these liposomes break down as a result of the lower pH inside the endosomes (Chu et al., 1990). In addition to pH-sensitive formulations, enzymatic degradation has been utilized as a release mechanism. Higher concentration of proteases in cancer tissue has been used to enhance drug release from liposomes (Banerjee et al., 2009). The thermosensitive liposomes can be triggered by external heating of the target. The most sophisticated thermosensitive liposome formulations are stable at body temperature but release the contents rapidly after an increase in temperature over 40°C (Tagami et al., 2011). Exogenous triggering of drug release is an attractive option, because it allows precise and time-dependent control of drug release. In addition to the temperature, pH, and enzymes, light has also been explored as an intriguing stimulus to remotely trigger cargo release. The light as a trigger is advantageous in that various parameters, such as exposure time, wavelength, beam diameter, and laser intensity, can be readily controlled to adapt to different purposes. The light must induce some structural changes in the system that leads to the drug release (Fomina et al., 2012). Ultraviolet and visible light can be used to trigger only superficial tissues, because the wavelengths under 700 nm do not reach tissues deeper than 1 cm due to high scattering and absorption by tissue components (Juzenas et al., 2002; Weissleder and Ntziachristos, 2003; Klohs et al., 2008).

On the other hand, near-infrared (NIR) light (700–900 nm) has minimal absorbance to hemoglobin (<650 nm), water (>900 nm), and lipids, and it can penetrate to the tissues several centimeters depending on the light intensity (Weissleder, 2001). Moreover, light in the NIR region does not cause significant heating in the application area and is well tolerated in ocular and other tissues (Delori et al., 2007; Mochizuki-Oda et al., 2002; Nagasaki and Shinkai, 2007).

Some photothermal molecules produce heat upon light irradiation. One of the most interesting options is indocyanine green (ICG), a phototriggering agent which converts IR light into heat, which is transferred to the bilayer to induce membrane permeability of liposomes and can be used to trigger drug release from thermosensitive formulations (Sawa et al., 2004; Ma et al., 2013). ICG is the only NIR molecule approved by US FDA and European Medicines Agency for clinical use (Pauli et al., 2010; Summerton et al., 2012). ICG is widely used in the clinic for determination of cardiac output, hepatic function and liver blood flow, inspection of retinal and choroidal vessels (Dzurinko et al., 2004), and diagnosis of burn depth (Still et al., 2001).

It is an amphiphilic, inert (nonionizing), and nontoxic compound having a molecular weight of 751.4 Da (Alander et al., 2012). The tricarbocyanine dye is composed of two hydrophobic polycyclic parts connected to a carbon chain. Each polycyclic part is attached to a sulfate group, which results in hydrophilic properties (Desmettre et al., 2000). Its wide acceptance is due to its low toxicity (after intravenous administration LD50 of 50–80 mg/kg for animal subjects) (Costa et al., 2001), the fast binding with plasma proteins, and quick excretion by the liver into bile juice. This dye fixes rapidly and intensely to serum proteins in vivo after intravenous injection, producing fluorescence signal (Shimada et al., 2015) without alteration of protein structures (Kochubey et al., 2005).

Currently, ICG is widely employed for diagnostic and therapeutic applications by virtue of its fluorescent properties. ICG has been used in photodynamic therapy (PDT) and photothermal therapy (PTT) owing to its strong absorption band that allows deeper tissue penetration causing significant heating (Skřivanová et al., 2006). In PDT, light absorption generates reactive oxygen species (ROS) and causes necrosis and apoptosis of tumor cells (Abels, 2004; Engel et al., 2008). In PTT, the photoresponsive agents generate local heating after cellular uptake and the malignant tissue is destroyed (Miao et al., 2015). Toyota et al. (2014) synthesized ICG-C18 derivative by substituting one of the sulfonate groups of ICG with an alkyl chain and used as NIR fluorescence probe for noninvasive visualization of sentinel lymph nodes following liposomal incorporation. Kraft and Ho (2014) found that ICG was incorporated into the lipid membrane of liposomes completely and stably.

Despite several advantages, the clinical use of ICG is still limited. ICG nonspecifically binds to plasma proteins and, consequently, exhibits fast body clearance (plasmatic half-life of 2–4 min) and marked instability in vivo, with a lack of target specificity. Further, this contrast agent is unstable in water solutions and aggregates because of its physicochemical properties, inducing self-quenching and low quantum yield (Sheng et al., 2013). Saxena et al. (2003) reported that in vitro degradation of ICG in water solutions is strongly accelerated by high temperatures and light exposure. Due to these limitations, ICG itself is not always suitable for cancer detection (Shemesh et al., 2014). For this reason, long-lasting ICG-loaded probes need to be developed to allow efficient and prolonged imaging. Furthermore, the peripheral distribution of ICG after systemic administration is limited by hepatobiliary elimination, which restricts its tumor accumulation for specific treatment (Ott, 1998). In addition, PDT-induced DNA damage may be minimized in normal tissue, localizing the therapeutic effect to tumor cells by using targeting strategies (Shemesh et al., 2014).

To address the several issues, the encapsulation of the dye into numerous carriers has been proposed. Experimental studies demonstrated that this approach could improve ICG stability, prolong its circulation in the blood, and target the dye to a specific site, making ICG an ideal fluorescence marker (Saxena et al., 2006). For these reasons, the ICG loading into delivery systems makes its application in cancer diagnosis and therapy as well as in cancer theranostics more valuable, combining imaging and therapy in a single platform to get therapeutic protocols that are more specific to individuals (Xie et al., 2010).

Theranostics describes the codelivery of therapeutic and imaging agents in a single formulation that is used for simultaneous diagnosis and treatment (Warner, 2004). This term defines "an integrated nanotherapeutic system" which can diagnose, deliver targeted therapy, and monitor the response to therapy (Sumer and Gao, 2008).

Codelivery enables noninvasive, real-time visualization of drug fate, including drug pharmacokinetics and biodistribution profiles and intratumoral accumulation. Nanocarriers are advantageous for theranostics as their size and versatility enables integration of multiple functional components in a single platform (Charron et al., 2015). The diseases that are currently the main cause of morbidity and/or mortality are in great need of quick diagnosis and treatment. Cancer is one of these diseases and initial research into theranostics is oriented toward oncology (Ahmed et al., 2012). Theranostic liposomes are generally designed to facilitate simultaneous diagnosis and therapy of cancer (Choi et al., 2012).

The liposome vesicles have the capacity for covalent and noncovalent encapsulation of both hydrophobic and hydrophilic cargo. Thus, the vesicles are well suited to codeliver therapeutic and diagnostic agents. The nanosized imaging agents such as organic dyes, iron oxide, quantum dots, and gold nanoparticles can be entrapped within the hydrophobic core or linked covalently to the surface of the theranostic liposomes and the anticancer agent can be either encapsulated in the core or embedded in the lipophilic bilayer shell. There are multiple examples of liposomal formulations that incorporate targeting, therapeutic, and imaging functionalities (Al-Jamal et al., 2009; Chen et al., 2004; Erdogan and Torchilin, 2010; Ponce et al., 2006; Strijkers et al., 2010).

These theranostic liposomes can then be further conjugated with molecular probe for targeting. Such multifunctional liposomes may circulate for prolonged periods in the blood, evading host defenses, and gradually release the drug at target site and simultaneously facilitate in vitro or in vivo imaging. This chapter focuses on the latest developments on the theranostic applications of liposomal nanocarriers.

## 7.2 THERANOSTIC APPLICATIONS OF ICG-LIPOSOMES

The theranostic approach in the field of cancer is mostly oriented toward diagnosis and drug delivery. As development progresses, the concept opens up a wide range of diagnostic applications of ICG-liposomes (Table 7.1).

In addition to diagnostic applications, ICG liposomes have also been studied for efficient, light-triggered release of both small (calcein) and large molecular (dextran, 20 kDa) contents (Lajunen et al., 2016). After exposure to 1-W laser light for 15 s, a fluorescent dye calcein was released from the liposomes and higher ICG content accelerated the release rate.

**TABLE 7.1** ICG-liposomes and Their Preclinical Investigations for Diagnostic Applications.

| Cell line/animals | Results of investigation | Sources |
|---|---|---|
| Lewis lung carcinoma tumor bearing mice | Fluorescence imaging showed selective accumulation in tumor site after intravenous injection | Jing et al. (2015) |
| | Efficient photothermal ablation of tumor through a one-time NIR laser irradiation | |
| Nude mice were intraperitoneally injected with gastric cancer cells | ICG derivatives with 18-carbon alkyl chain can effectively target peritoneal disseminated tumors and can be easily detected by a NIR imaging system | Hoshino et al. (2015) |
| Mice | Following subcutaneous administration, lipid-bound ICG (250:1) in liposomes caused an enhanced lymph node and lymphatic vessel visualization | Kraft and Ho (2014) |
| Tumor-bearing mice | Fluorescent signal from targeted scVEGF-Lip/ICG grew slower than untargeted Lip/ICG reaching maximum at 30-min postinjection and declined much slower | Zanganeh et al. (2013) |
| | Higher scVEGF-Lip/ICG tumor accumulation was confirmed by the analysis of fluorescence on cryosections of tumors | |
| RAW 264.7 macrophage cell, normal mouse model | M-LP-ICG demonstrated a high uptake in macrophage cell because the high density of mannose | Jeong et al. (2013) |
| | M-LP-ICG can be used as an optical contrast agent for imaging not only SLN but also the draining lymph nodes | |
| Tumor-bearing mice | Liposomes with dextran and polyethylene glycol demonstrated similar thermal release properties; however, in vitro macrophage uptake was greater with dextran | Turner et al. (2012) |
| | Noninvasive in vivo NIR imaging showed tumor accumulation of liposomes with both coatings | |
| | Ex vivo NIR imaging correlated well with actual ICG concentrations in various organs of healthy mice | |

**TABLE 7.1** *(Continued)*

| Cell line/animals | Results of investigation | Sources |
|---|---|---|
| Mice, mice bearing B16 luciferase-expressing melanomas | Sequential near-IR imaging of intradermally injected LP-ICG enabled quantification of lymphatic flow | Proulx et al. (2010) |
| | Increased flow through draining lymph nodes was observed in mice bearing VEGF-C-expressing tumors without metastases, whereas a decreased flow pattern was seen in mice with a higher lymph node tumor burden | |
| | Potential for imaging of lymphedema or improved sentinel lymph detection in cancer | |
| C57Bl/6 mice | ICG fluorescence accumulated only in active CNV lesions from cationic liposomes | Hua et al. (2012) |
| | Noninvasive scanning laser ophthalmoscope imaging technique provided indication for use of current intraocular anti-angiogenic drugs in neovascular eye diseases | |
| Monkey | Blood velocity in the selected choroidal vessel segments was measured | Peyman et al. (1996) |
| Hamster skin flap model | Laser-induced release of liposome-encapsulated dye for a real-time quantification of thermal damage of blood vessels | Mordon et al. (1996) |
| Mouse xenograft model, MCF-7 cells | HAS-coated liposomes were efficiently uptaken by MCF-7 cells through endocytosis after 0.5 h incubation | Chen et al. (2016) |
| | Tumor was clearly detected and the signal lasted at least up to 24 h after injection | |
| | Nanoprobe could be an ideal contrast agent for in vivo NIR fluorescence imaging | |
| Mice | C18-ICG-liposomes accumulated only in popliteal lymph node 1 h after injection into the footpad, whereas unmodified ICG accumulated in popliteal lymph node and other organs like liver | Toyota et al. (2014) |
| | LP-ICG-C18 is a useful NIR-fluorescence probe for noninvasive in vivo bioimaging, especially for the sentinel lymph node | |

**TABLE 7.1** *(Continued)*

| Cell line/animals | Results of investigation | Sources |
|---|---|---|
| Mice bearing A549 tumors (adenocarcinomic human alveolar basal epithelial cells) | Covalently conjugated ICG with DOPE was incorporated into liposome bilayers and used as NIR photoactivating probe<br>NIR fluorescence signal intensities of the tumors gradually increased as time passed compared with the normal regions | Suganami et al. (2012) |
| Murine squamous cell carcinoma SCCVII tumor cells, CH3/He mice | LP-ICG-C18 accumulated in the tumor through the EPR effect, and exerted good antitumor effect and induced cell apoptosis in vivo<br>Effective for the diagnosis and treatment of early esophageal, oral, and pharyngeal cancer | Maruyama et al. (2015) |

CNV, choroidal neovascularization; DOPE, dioleoylglycerophosphoethanolamine; M-LP-ICG, mannosylated liposome-encapsulated indocyanine green; scVEGF, single-chain VEGF; SLN, sentinel lymph nodes.

Only 1:25 (ICG/lipid) formulation significantly caused passive leakage of calcein without light induction from the liposomes (Fig. 7.1A). At 1-W laser power, the release of calcein steadily increased with increasing light exposure time (Fig. 7.1B). Almost 94% calcein was released in 4 min, and about half of the calcein (46%) was released in 1 min of light activation. Increasing the laser power to 3 W caused significantly faster calcein release from the liposomes (91% calcein release after light exposure of 15 s). Calcein release from the control samples at the same temperature (37°C) without light activation showed only minimal calcein release (0–2%). Similarly, the negative control liposomes without ICG released <1% calcein after laser light exposure (Fig. 7.1B). Almost all laser-irradiated liposomes released FITC-dextran more than the corresponding control samples without laser exposure. The release of FITC-dextrans (4, 10, and 20 kDa) from ICG liposomes became slower with 1 than 3-W laser exposure. Further, the release of 4 kDa FITC-dextran was slower than that of 10 and 20 kDa FITC-dextrans. For instance, the triggering with 3-W laser for 15 s caused 47%, 84%, and 95% release of 4, 10, and 20 kDa FITC-dextrans from the liposomes, respectively.

There has been a significant advancement toward simultaneous diagnosis and therapy using ICG-liposomes. However, it is not possible to cover all the attempts in this chapter; henceforth, a brief overview of latest theranostic applications of ICG-liposomes is presented herein.

Colorectal cancer is one of the most common cancers with relatively poor prognosis. An early diagnosis is the best way to improve overall survival. Portnoy and coworkers (2011) developed liposome-based NIR probe that combines both imaging and targeting abilities. ICG and cetuximab monoclonal antibody for epidermal growth factor receptor (EGFR) were attached to liposomes by passive adsorption. Cetuximab-adsorbed fluorescent liposomes effectively internalized and selectively bound to A431 colon carcinoma cells overexpressing EGFR. The binding of cetuximab-targeted fluorescent liposomes to A431 compared with normal IEC-6 colon cells (normal enterocytes expressing physiological EGFR levels) was larger by a factor of 3.5. Further, it was larger than nonspecific binding of liposomes without the targeting moiety by 40%. Due to relatively high ICG fluorescence and specific tumor cell-recognizing ability, the liposomes hold tremendous potential as diagnostic imaging pharmaceutical tools and promise for the development of multimodality agents with both imaging and therapeutic capabilities.

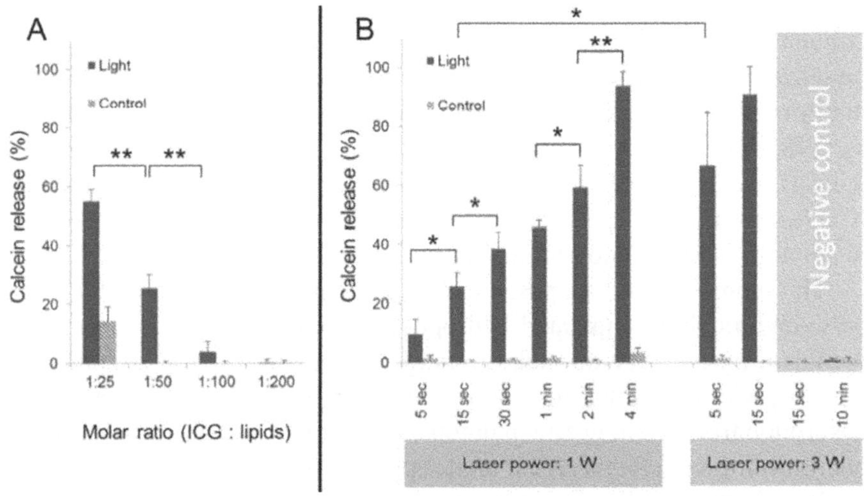


**FIGURE 7.1** (A) Calcein release from the liposomes with various ICG/lipid molar ratios. The samples were heated to +37°C and light-triggered samples were exposed to 808-nm 1-W laser light for 15 s. (B) Calcein release from the liposomes with 1:50 ICG/lipid molar ratio and negative control liposomes without ICG. Exposure times and 808-nm laser powers of light are shown on the horizontal axis. Error bars indicate the standard deviations ($n = 3$). Statistical analysis between the light-triggered samples: *$p < 0.05$; **$p < 0.005$.

[Reprinted (adapted) with permission from Lajunen, T.; Kontturi, L.-S.; Viitala, L.; Manna, M.; Cramariuc, O.; Róg, T.; Bunker, A.; Laaksonen, T.; Viitala, T.; Murtomäki, L.; Urtti, A. Indocyanine Green-Loaded Liposomes for Light-Triggered Drug Release. *Mol. Pharmaceutics* **2016,** *13*, 2095–2107, Copyright (2016) American Chemical Society.]

Turner et al. (2012) developed temperature-sensitive PEG- or dextran-coated liposomes to study the biodistribution of liposomes in tumor-bearing mice using ICG. The optimized formulation had a transition temperature around 42°C and was composed of L-α-phosphatidylcholine, cholesterol, 1,2-dipalmitoyl-*sn*-glycero-3-phosphocholine (dipalmitoyl phosphatidylcholine [DPPC]), and DSPE-PEG 2000. At 37°C, liposomal formulation was stable but a burst release was observed at 42°C. Afterward, Shemesh et al. (2014) focused their study on in vivo PDT of breast cancer using theranostic liposomal system. The photodynamic effect caused by ICG stopped the growth of TNBC (triple negative breast cancer) cells in vitro. In another study (Shemesh et al., 2015), a significant loss of cell viability was reported following photoactivation of liposomal ICG via NIR irradiation. Quantitative pharmacokinetic profile and fluorescence imaging-based biodistribution patterns were obtained using the human TNBC xenograft model in nude mice. A significant increase in systemic distribution and circulation half-life of liposomal ICG was noted, and NIR fluorescence imaging demonstrated enhanced accumulation of liposomes within the tumor region. Tumor growth in mice treated with liposomal ICG followed by NIR irradiation was significantly reduced compared to those treated with free ICG, saline, and irradiation alone. In vivo PDT using liposomal ICG demonstrated targeted biodistribution and superior antitumor efficacy in a human TNBC xenograft model compared to free ICG.

The submicron-sized liposomes (ssLips) have also been modified with polyvinyl alcohol-conjugated hydrophobic anchors (PVA-R) to improve peptide drug delivery through the lung (Murata et al., 2012). Compared with peptide drug solutions, the modified liposomes provided sustained absorption from the lung after pulmonary administration and prolonged the pharmacological effects of the model peptide drug. However, the effects of PVA-R on the behavior of pulmonary liposome systems in the whole body require in vivo characterization. NIR optical imaging is a powerful tool for in vivo observations of dynamic liposome behaviors because it is a minimally invasive, nonionizing method that permits sensitive deep tissue imaging (Licha and Olbrich, 2005). Therefore, the mechanisms of enhanced peptide drug absorption, the effects of PVA-R modification on the pharmacodynamics, and retention of liposomes in the lung were evaluated using in vivo imaging in rats by Murata et al. (2014). ICG, which has a fluorescence emission wavelength of approximately 820 nm, was used as a liposome tracer in rat bodies.

To monitor pulmonary delivery of liposomes in rats, the fluorescence of ICG-labeled multilamellar vesicles (MLVs) with or without surface modification was observed in the whole body at specific time points. Fluorescence intensities in the lungs did not differ greatly between animals treated with unmodified and PVA-R-modified MLVs, disappearing within 6 h postadministration in both cases. Hence, PVA-R modification of MLVs had negligible effects. Earlier, it has been demonstrated that the uptake of micron-sized particles by alveolar macrophages was significantly greater than that of submicron-sized particles (Chono et al., 2010). Therefore, it was assumed that MLVs with or without surface modification are rapidly cleared by alveolar macrophages, potentially owing to their large particle sizes (883.4 nm for unmodified and 920.3 nm for modified MLVs). PVA-R-modified ssLips emitted sustained fluorescence in rats and retained at higher extent in the rat lungs after intratracheal administration than unmodified ssLips. Perhaps, the liposomal surface modification improved the stability and retention of liposomes in the lung owing to the thick and flexible layer (ca. 20–30 nm) of PVA-R on liposome surfaces. This hypothesis was confirmed by in vitro cellular association studies using rat alveolar macrophage cell line. PVA-R-modified liposomes significantly decreased cellular association compared with unmodified liposomes. Therefore, it was ensured that a thick and flexible surface layer of PVA-R produced steric hindrance on ssLips surfaces and reduced recognition and uptake by alveolar macrophages (Takeuchi et al., 2000; Nakano et al., 2008).

Zhao et al. (2015) reported a noninvasive NIR-driven, doxorubicin (DOX)/ICG-loaded temperature-sensitive liposomes. The theranostic system exhibited an enhanced response to NIR laser. The laser irradiation causes threefold increase in drug release. DOX/ICG was used to treat cancer by chemo/PTT. Furthermore, DOX/ICG liposomes showed laser-controlled release of DOX in tumor, enhanced ICG, and DOX retention by 7 times and 4 times compared with free drugs. Thermo-sensitive liposomes manifested high efficiency to promote cell apoptosis and completely eradicated tumor without side effect. After endocytosis by MCF-7 breast adenocarcinoma cells, the liposomes in cellular endosomes can cause hyperthermia through laser irradiation, and then endosomes are disrupted and DOX is released simultaneously for increased cytotoxicity and high antitumor performance. The mechanism DOX release and cytotoxicity in tumor for combinatorial cancer therapy is schematically shown in Figure 7.2.

Kono et al. (2015) functionalized PEGylated liposomes with thermo-sensitive poly[2-(2-ethoxy) ethoxyethyl vinyl ether-*block*-octadecyl vinyl ether] (EOEOVE-ODVE) and trastuzumab antibody (Herceptin, HER) for targeting human epidermal growth factor-2. ICG was incorporated for NIR fluorescence imaging. The liposomes (<125 nm) retained DOX in the interior below physiological temperature but released the same molecules immediately at temperature >40°C.

Compared to HER-free liposomes, the liposomes exhibited about 2–5 times higher association with their target cells and greatly internalized into the target cells, overexpressing Her-2, such as SK-OV3 (human ovarian carcinoma), SK-BR3 (human breast adenocarcinoma) cells, and killed these cells when heated at 45°C for 5 min.

Intravenous administration to mice bearing SK-OV3 tumor caused efficient accumulation of liposomes in the tumor than those without HER. The tumor treated with HER-conjugated liposomes displayed 2.3 times stronger fluorescence than the tumor treated with HER-free liposomes. The higher accumulation of HER-conjugated liposomes could be attributed to the specific interaction between HER and Her-2 of the target tumor cells, which might have enabled their retention at the tumor site. As was evident from NIR fluorescence imaging, the HER-conjugated liposomes stayed there for more than 48 h and the extent of accumulation reached plateau about 7 h after administration. As the mouse tumor site was heated to 44°C for 10 min at 7 h after administration of DOX-loaded liposomes, the tumor growth was suppressed strongly thereafter. Indeed, this enhancement of the tumor suppressive effect might be due to the higher release DOX molecules from the liposomes and efficient accumulation at the tumor tissues. HER-conjugated liposomes produced more efficient tumor suppressive effects.

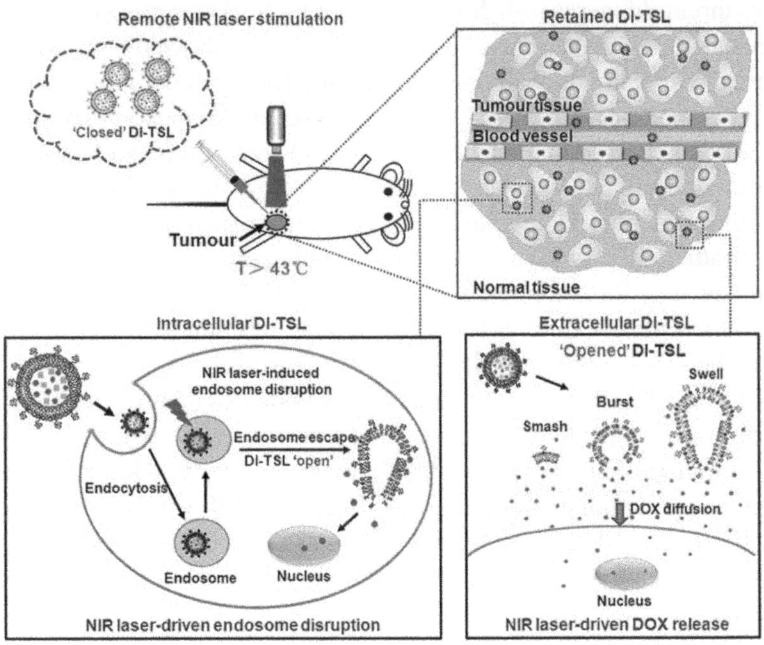


**FIGURE 7.2** A scheme showing the mechanism of remote NIR-triggered DOX release and further cytotoxicity in tumor. The DI-TSL are treated with remote NIR laser (808 nm, 0.5 W/cm$^2$, 5 min) after the injection and kill cancer cells in different ways: (1) Intracellular DI-TSL escape from cell endosomes by NIR laser induced endosomal disruption, and DOX release from "opened" DI-TSL and enter the cytosol after cellular uptake. (2) Extracellular DI-TSL immediate released DOX through smash, burst, and swell, and DOX diffuses into the tumor along a high-concentration gradient, attacking tumor cells.

(Reproduced from Zhao, P.; et al. NIR-Driven Smart Theranostic Nanomedicine for On-demand Drug Release and Synergistic Antitumor Therapy. *Sci. Rep.* **2015,** *5*, 14258. DOI:10.1038/srep14258, with permission from Nature Publishing Group.)

Multispectral optoacoustic tomography (MSOT) is a powerful modality that allows high-resolution imaging of photoabsorbers deep within tissue, beyond the classical depth and resolution limitations of conventional optical imaging. Recently, a powerful in vivo MSOT contrast agent has been developed by Beziere et al. (2015). They incorporated a strong photo-absorbing probe ICG into PEGylated liposomes that demonstrated an enhanced optoacoustic imaging characteristics compared to gold nanorods. The lipid layer was composed of hydrogenated L-α-phosphatidylcholine, cholesterol, and 1,2-distearoyl-*sn*-glycero-3-phosphoethanolamine-*N*-[methoxy(polyethylene glycol)-2000] (DSPE-PEG2000) having different concentrations of ICG (25–75 μM). The tissue distribution profile of ICG-liposomes in albino mice was determined and compared to free ICG using whole-body NIR fluorescence imaging. Strong fluorescent signals were obtained from the animals 5 min after injection; however, no specific tissue was confidently identified using the imaging system. The emitted fluorescent signal decreased within 5 h and was almost completely lost after 24 h. To verify the tissue distribution of the contrast agents, all animals were euthanized at 24 h postinjection and major organs were harvested and imaged. The injections of ICG liposomes (50–75 μM) resulted in a strong fluorescent signal from the liver showing higher optical stability of liposome-ICG in vivo. Images of free ICG revealed complete loss of signal from all organs. They further indicated that ICG and liposomal ICG as contrast agents did not cause any acute adverse events because there were no signs of tissue necrosis, fibrosis, or inflammation in any excised organs such as liver, kidney, and heart. Using MSOT, they observed a heterogeneous distribution of ICG-labeled PEGylated liposomes in tumor-bearing mice and a marked variation on the ability to localize within the tumor of different vascularization—HT29 human adenocarcinoma cells (restricted microvascular) and 4T1 murine breast cancer cells (leaky vasculature). In the HT29 tumor model, liposomal ICG (75 μM) did not appear in the tumor core but remained peripheral to the tumor vasculature. It is well documented that significant physiological and anatomical differences exist among tumor types (Hanahan and Weinberg, 2011). MSOT allowed illustration of such differences, opening new paths toward personalization of treatment regimes. However, it is also foreseen that labeled liposomes can be employed to assess the efficacy of payload delivery over longer periods of time, in a theranostics scenario. In this role, liposomal ICG (75 μM) can be used to optimize therapies and better assess therapeutic efficacy.

A high purity, clinical grade hCTM01 antibody has shown great potential for the delivery of anticancer drug with reduced systemic toxicity due to its cellular internalization capacity, as was the case of clinically tested hCTM01-calicheamicin immunoconjugate (Gillespie et al., 2000). Lozano et al. (2015) incorporated ICG as photo-absorbent/fluorescent probe and DOX as anticancer therapeutics in PEGylated liposomes. To improve tumor-specific targeting and distribution of DOX, the liposomal surface was conjugated with the humanized monoclonal antibody hCTM01. The antibody was first conjugated to DSPE-PEG2000-Malemide micelles followed by postinsertion into the preformed PEGylated liposome-ICG. The hydrodynamic diameter increased from 110 to 130 nm for both systems after postinsertion but still within the suitable range for intravenous administration and evaluation with MSOT imaging.

To evaluate the tumor distribution of the targeted and nontargeted PEGylated liposome-ICG in HT-29 human colon adenocarcinoma (slow growth) and 4T1 murine breast tumor (fast growth), in vivo MSOT imaging was used. Preferential accumulation was observed for the 4T1 tumor model, known for its faster growth compare to HT-29, regardless the presence of the antibody. However, the targeted liposomes accumulated faster at early time points. The nontargeted PEGylated liposomes accumulated in the center of the tumor at later time points while a more diffuse tissue distribution in the periphery of the tumor and decrease of visible signal in the vasculature was observed for targeted liposomes. These observations indicated the potential of liposome-ICG in combination with MSOT live imaging to provide a noninvasive and better understanding of liposomal cancer treatment efficacy. They encapsulated DOX into both targeted and nontargeted PEGylated liposome-ICG to combine both therapy and diagnosis. Intratumoral distribution of targeted DOX-loaded PEGylated liposome-ICG was then evaluated in 4T1 murine breast tumor-bearing mice in vivo using MSOT live imaging. 4T1 tumor model was selected for this study as it showed higher tumor accumulation of liposome-ICG compared to HT-29 tumor model. MSOT images showed similar tumoral distribution of DOX-loaded targeted PEGylated liposome-ICG into 4T1 tumor model compared to those without DOX. As early as 5 min after intravenous injection, ICG signal was detected in the vasculature of the tumor. However, after prolonged time points (4 and 24 h), a reduction in ICG signal in the vasculature was noticed with higher localized homogenous tumor accumulation. ICG-labeled liposomes with and without antibody coupled with MSOT imaging allowed both immediate and long-term detection of liposomes in the tumor. This system in combination with MSOT imaging allowed the noninvasive real-time monitoring of the vesicle accumulation in tumor. Therefore, a monoclonal antibody-targeted PEGylated liposome-ICG system containing DOX could be a potential theranostic anticancer drug-delivery system. In addition, the biocompatibility of both targeted and nontargeted PEGylated liposome–ICG systems and their long-term optical stability could facilitate their clinical translation.

Pang and coworkers (2016) reported a dual-modal imaging-guided active targeted thermal sensitive liposomes (TSLs) based on ICG and anti-angiogenesis agent rapamycin (RAPA). The system satisfactorily released the drug from the liposomes under hyperthermia conditions induced by NIR fluorescence. The RAPA-loaded and folate-conjugated ICG TSLs [folic acid (FA)-ICG/RAPA-TSLs] plus NIR laser exhibited efficient drug accumulation and cytotoxicity in tumor cells and epithelial cells. After 24 h intravenous injection of liposomes, the margins of tumor and normal tissue were accurately identified via in vivo NIR fluorescence and photoacoustic dual-modal imaging. In addition, the liposomes combined with NIR irradiation inhibited tumor growth to a great extent in tumor-bearing nude mice and showed much lower side effects to normal organs. Thus, the liposomal systems could be a therapeutic tool for diagnostics as well as the treatment of tumors. The structure of FA-ICG/RAPA-TSLs system is shown in Scheme 7.1a. The antitumor mechanism of the FA-ICG/RAPA-TSLs is shown in Scheme 7.1b. It was shown that selective targeting to the tumor site and uptake into tumor cells and endothelial cells could be achieved through EPR and active targeting effects. FA-ICG/RAPA-TSLs could serve as theranostic probe to track the tumor region, which was essential to focus the laser beam on the tumor areas and adequately convert NIR light energy to hyperthermia, thus reaching satisfactory anti-tumor performance. The hyperthermia produced by activated ICG under NIR irradiation could trigger the release of RAPA from the TSLs. Finally, the released RAPA would block the regulation effect of *mTOR (*mammalian target of rapamycin), thus impeding the proliferation of cancer cells and endothelial cells.

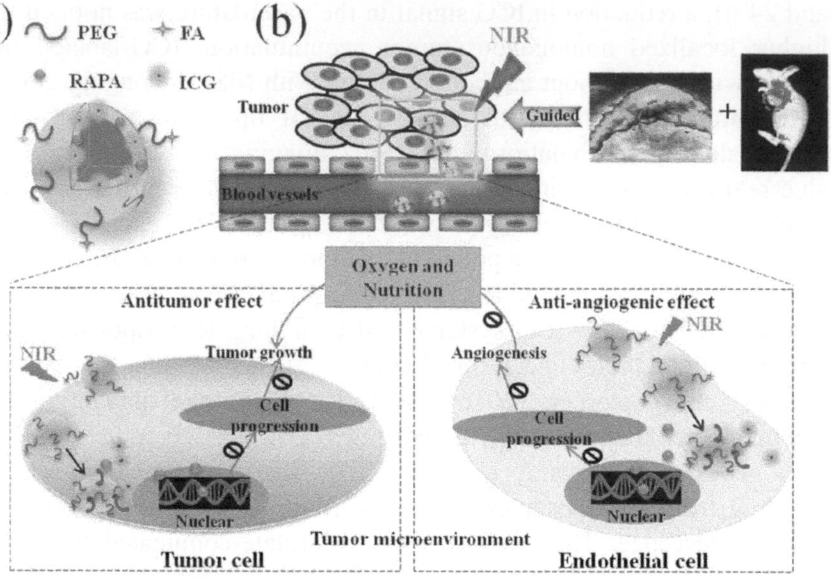


**SCHEME 7.1 (See color insert.)** Illustration of the theranostic TSLs: (a) structure of FA-ICG/RAPA-TSLs and (b) antitumor mechanism of FA-ICG/RAPA-TSLs.

[Reprinted with permission from Pang, X.; Wang, J.; Tan, X.; Guo, F.; Lei, M.; Ma, M.; Yu, M.; Tan, F.; Li, N. Dual-Modal Imaging-Guided Theranostic Nanocarriers Based on Indocyanine Green and mTOR Inhibitor Rapamycin. *ACS Appl. Mater. Interfaces* **2016,** *8*, 13819–13829. Copyright (2016) American Chemical Society.]

Cerebral malaria (CM) is a major cause of death due to *Plasmodium falciparum* infection. Misdiagnosis of CM often leads to treatment delay and mortality. The ICG-liposomes are expected to be uptake by vascular and perivascular activated phagocytes, prominent in CM (Hearn et al., 2000). In a study, Portnoy et al. (2016) showed that ICG-liposomes accumulated in brains of mice, depicting CM, and provided a convenient way of CM diagnosis. The drug, artemisone, which is effective against experimental CM (Waknine-Grinberg et al., 2010), diminished the fluorescence of ICG-liposomes in the brains of infected mice, highlighting the potential of liposomal ICG for therapeutic monitoring. ICG's whole body and cerebral emission intensity was greater for liposomal than free ICG in mice. The liposomal ICG cleared slowly compared to the free ICG. Four hours after injection, the excised were imaged. Image analysis demonstrated higher emission of liposomal-ICG compared to free ICG in the blood, brain, liver, spleen, and muscle. The results clearly suggested the advantage of ICG-liposomes for in vivo optical imaging. They evaluated organ emission intensity of ICG in *Plasmodium berghei* (PbA)-infected mice as a function of the day postinfection. ICG-liposomes were injected to naïve and infected mice and sacrificed 4 h later. The blood intensity was persistent throughout the experiment; however, the infected/naïve mice brain emission intensity on days 6 and 7 was significantly higher compared to the previous days. Increased emission intensity was also observed in the liver and kidneys toward the latest stage of CM. Based on these findings, mice that developed CM were compared for relative emission intensity with malaria-infected mice that did not develop CM. In contrast to the profound brain emission intensity in mice with CM, in infected mice that did not develop CM, the emission intensity was similar to that of naïve controls. Liver and kidney emission was significantly increased in mice demonstrating CM or anemic malaria compared to naïve mice.

ICG-liposome fluorescence was evaluated as a marker for monitoring therapy effectiveness. The artemisone-treated mice did not develop typical neurological symptoms and no parasites were detected in their blood smears. Imaging with ICG-liposomes revealed high emission intensity in the brains of infected mice compared to the drug-treated mice. Therefore, liposomal ICG provided an effective indicator for successful treatment of CM, as demonstrated by the higher brain emission intensity in vehicle treated, compared to artemisone-treated mice. Histological analyses suggest that the accumulation of liposomal ICG in the cerebral vasculature is due to extensive uptake mediated by activated phagocytes. To evaluate the possibility of uptake of liposomes by peripheral phagocytes, the murine monocyte cells loaded with ICG-liposomes were injected to infected mice on day 6, postinfection. The accumulation of the labeled liposomes in the brains of the infected mice supported the hypothesis that liposome uptake and transport to the brain was mediated by monocytes. The monocytes migrating to the brain vasculature in infected mice participated in CM induction (Mac-Daniel and Menard, 2015). The non-PEGylated liposomes accumulation in brains of infected mice was more pronounced.

The iRGD, a cyclic nanopeptide with amino acid sequence of CRGD-KGPDC, has shown efficient tumor-targeting and tumor-penetrating capabilities in the chemically conjugated formulation or the coadministrated one with therapeutic agents (Sugahara et al., 2010; Wang et al., 2014). To increase tumor accumulation and subsequently achieve NIR fluorescence molecular imaging-guided PTT and PDT therapy against breast tumor, Yan et al. (2016) reported internalized RGD (iRGD)-modified ICG liposomes (iRGD-ICG-LPs). DPPC, cholesterol, 1,2-distearoyl-*sn*-glycero-3-glycero-3-phosphoethanolamine-*N*-[(polyethylene glycol)-2000] (DSPE-PEG2000), and DSPE-PEG2000-iRGD were used in the lipid composition. DSPE-PEG2000-iRGD was synthesized through a reaction between iRGD and DSPE-PEG2000-maleimide. Because iRGD can specifically bind to the integrin alfa(v)beta(3) receptor which is over-expressed on both tumor neovasculature and some tumor cells (Desgrosellier and Cheresh, 2010), the system could show great promise in achieving molecular imaging for early diagnosis.

The average diameter of iRGD-ICG-LPs and ICG-LPs were 115.91 and 117.43 nm, respectively, with a narrow size distribution and thus were found suitable for targeted drug delivery.

To investigate the uptake behaviors of the ICG formations, 4T1 cells and human umbilical vein endothelial cells (HUVECs) were incubated with ICG-LPs and iRGD-ICG-LPs. The fluorescence signal from the cells was stronger for iRGD–ICG-LPs than ICG-LPs. The ICG fluorescence intensity for iRGD–ICG-LPs was about 1.86-fold and 1.69-fold higher in HUVECs and murine breast cancer 4T1 cells than that for ICG-LPs, respectively. Incubation with iRGD-ICGLPs exhibited a stronger fluorescence signal in HUVECs in comparison with 4T1 cells.

Given the fact that the integrin receptors are overexpressed in HUVECs and 4T1 cells, it is reasonable to obtain higher cellular uptake efficiency of iRGD-ICG-LPs than ICG-LPs. Because the integrin receptors are highly expressed in HUVECs than in 4T1 cells, the cellular uptake efficiency was higher in HUVECs. The viability of 4T1 cells significantly decreased with the increase of concentrations of iRGD-ICG-LPs. The cell viability decreased from 60% to only 7.1% when the amount of ICG in iRGD–ICG-LPs was increased from 20 to 80 μg/ml.

The tumor cells treated with iRGD-ICG-LPs plus laser caused significant increased level of singlet oxygen, with 1.91-fold, 1.75-fold, and 2.1-fold higher than that of the tumor cells treated with only iRGD-ICG-LPs, only laser, and phosphate-buffered saline (PBS) control, respectively. This cytotoxic effect was attributed to abundant generation of singlet oxygen in the cells receiving iRGD-ICG-LPs and laser irradiation.

The molecular imaging of iRGD-ICG-LPs was assessed in 4T1 tumor-bearing mice. The fluorescence signal intensity of the tumor gradually increased in the iRGD-ICG-LP group, reaching maximum signal intensity after 24 h. Although tumor fluorescence was also observed in the mice receiving ICG-LPs, significantly much weaker fluorescence signals were observed in the tumors compared with these receiving iRGD-ICG-LPs at the same time. The mice receiving free ICG revealed a gradual decrease of fluorescence signals in the tumors and no evident fluorescence signals in the tumors after 24 h injection. After being imaged, the mice were sacrificed and tumor tissue as well as other main organs was imaged to examine tumor accumulation. The images revealed an obvious tumor accumulation of the iRGD-ICG-LPs, but not ICG-LPs and the free ICG.

The temperature increase plays an important role for the photothermal effect since the hyperthermia-induced cell death has different mechanisms. It has been reported that the temperature between 42°C and 45°C could upregulate the expression of apoptosis genes, activate caspases, suppress TNF-alpha resistance, and damage mitochondrial (Yan and Qiu, 2015), resulting in the sublethal and reversible cell response. On contrary, the temperature above 50°C would cause irreversible cell injury, such as cell membrane collapse, protein denaturation, and halt in enzyme activity and DNA polymerase function, producing cell coagulative necrosis (Chu and Dupuy, 2014). Therefore, the temperature in tumor regions was monitored with the infrared thermal imaging camera during laser irradiation. The tumor temperature rapidly reached to 51.7°C and remained relatively stable under laser exposure for the mice receiving iRGD-ICG-LPs. In contrast, the tumor temperature of the mice receiving ICG-LPs peaked at a temperature of 47.6°C, followed by gradual decrease.

The mice treated with PBS or free ICG, the tumor temperature was not significantly changed when using the same laser irradiations. The photodynamic effects were also confirmed in vivo. The tumor treated with iRGD-ICG-LPs, ICG-LPs, and free ICG plus laser caused significant increased level of ROS, with 3.82-, 1.86-, and 1.07-fold higher than that of the tumor treated with PBS plus laser, respectively.

The in vivo phototherapy effect was further evaluated on the 4T1 subcutaneous tumor model. The tumor-bearing mice were intravenously injected with PBS, free ICG, ICG-LPs, or iRGD-ICG-LPs, followed by laser irradiation, respectively. The tumor growth of mice receiving iRGD-ICG-LPs was significantly inhibited, achieving almost complete tumor regression. Histological analysis of the organs (heart, liver, spleen, lung, and kidney) from mice treated with iRGD-ICG-LPs plus laser did not reveal any appreciable abnormality. The mice tolerated iRGD-ICGLPs well without any obvious acute toxicity. The targeting therapy mediated by iRGD provided almost equivalent antitumor efficacy at a 12.5-fold lower drug dose than that by monoclonal antibody (Zheng et al., 2012).

Therefore, the iRGD-ICG-LP could better be utilized as multifunctional probe with integrated molecular imaging and phototherapy capability. The accumulation of iRGD-ICG-LP in tumors could achieve tumor diagnosis and location through sensitive molecular imaging, and then the visualized tumor could be ablated selectively and efficiently while sparing the surrounding normal tissues.

Recently, the antitumor effect of a naturally occurring coumestan, wedelolactone, has been reported (Benes et al., 2011, 2012; Lim et al., 2012). ICG could convert NIR light into heat (Zheng et al., 2013), and thus, the hyperthermia produced could preferentially accelerate the death of cancer cells but not normal cells as a consequence of different degrees of temperature sensitivity (Tang and Aj, 2009). The hyperthermia from ICG-liposomes leads to phase transition of lipid bilayer membrane from the gel phase to the liquid crystalline phase (Lin et al., 2014) and promote the release of drug from the liposomes. Therefore, the ICG-loaded liposomes would provide a new platform for cancer therapy to achieve a combinative effect of chemotherapy and photo-thermotherapy.

Zhang et al. (2016) successfully loaded wedelolactone into the lipid bilayers of soybean phosphatidylcholine-cholesterol liposomes and ICG in the hydrophilic cavity by the ethanol injection method. In this method, the lipid phase was gradually dropped into the aqueous phase at the hydration temperature of 38°C. PEG 2000 was used in the aqueous phase for PEGylation. The PEGylated liposomal wedelolactone was uniform and round in shape with an average diameter of approximately 145.8 nm. The average diameter of ICG-liposomal wedelolactone was around 146.1 nm. The strong negative charges (−25.7 mV) on liposome surfaces generated electrostatic repulsive forces and indicated stability, nonaggregation behaviors of the carriers. The liposomes lost their structures after NIR treatment and did not recover their initial structures when the NIR irradiation was halted.

The extent of drug release from ICG-liposomes without laser irradiation was 23.32% over 1 h and 52.07% over 8 h; however, the system accelerated drug release up to 56.86% by laser irradiation over 1 h, and the release amount reached to 96.74% over 8 h in PBS (pH 7.4). There was no significant difference between the amount of drug release from PEGylated liposomes with and without laser. Similarly, the release of ICG from liposomal wedelolactone was also enhanced with laser irradiation.

Upon 808 nm NIR irradiation at 1-h interval (2.0 W/cm$^2$ for 5 min), ICG-liposomal wedelolactone converted the absorptive optical energy into heat and the elevated temperature promoted the release of wedelolactone from liposomes rapidly.

Upon laser irradiation, ICG-liposomal wedelolactone inhibited the growth of HepG2 cells up to 72.02%, at a rate higher than that observed without laser (60.84%). Therefore, combined effect of hyperthermia and wedelolactone could inhibit tumor cells growth effectively.

Moreover, the antitumor effect of ICG-liposomal wedelolactone with laser was significantly increased compared to that of free wedelolactone or PEGylated liposomal wedelolactone. On the other hand, HepG2 cells, when treated with wedelolactone-free ICG liposomes, demonstrated low inhibition rate (~5%) and this was far less than that of ICG-liposomal wedelolactone. However, the inhibition rate for tumor cells treated with ICG-L liposomes was obviously increased after laser irradiation (ICG-L + laser group), and it was also higher than that in the group of laser alone.

The early apoptotic rate in ICG-WL-L + laser group was 33.74%, which was about 10-fold compared with that of cyclophosphamide (PC) group (3.20%) and 15-fold compared with those of laser alone group (2.08%) and WL group (2.47%). The temperature of ICG-liposomal wedelolactone group was gradually improved after irradiation, and over 4 min, the temperature reached to 48°C. The mice treated with ICG liposomal wedelolactone under NIR irradiation showed the smallest tumor size and the lowest weight at the day 12. The antitumor effect of ICG-liposomal wedelolactone with laser was significant in vivo, and the inhibition rate of U14 xenograft tumor reached up to 81%. They reasoned that a temperature of 42°C was sufficient to kill cancer cells, but not normal cells (Tang and Aj, 2009). The local hyperthermia could induce phase transition of tumor cell membrane and increase membrane permeability and fluidity, therefore enhancing cellular uptake of liposomes (Zheng et al., 2011; An et al., 2013). The ICG-liposomal wedelolactone combining with NIR irradiation showed an early apoptotic rate on HepG2 cells and an excellent antitumor efficacy to U14 tumors in vivo.

In another attempt, lipid nanoparticles were produced following drop wise addition of poly(D,L-lactide-*co*-glycolide)/resveratrol (RSV) solution into ethanol-aqueous solutions of lecithin/DSPE-PEG2000-FA (4:1) and lecithin/DSPE-PEG2000-ICG (4:1) under gentle stirring (Xin et al., 2016). The ICG and FA-conjugated and RSV-encapsulated lipid nanoparticles (PNPs) were evaluated on U87 glioma for theranostic application. They used nanoparticles as NIR probe to monitor cell internalization, in vivo biodistribution, and tumor accumulation of RSV. Further, the targeted anticancer efficacy of decorated nanoparticles was evaluated in vitro and in vivo. The targeted nanoparticles at the same RSV concentration exhibited considerable glioma cell death rate. The fluorescence intensity in the tumor treated with targeted nanoparticles was approximately 2.8-fold and 12.6-fold higher than that of free ICG and nontargeted particles, respectively, and remained very high for up to 48 h.

It was confirmed that the main metabolic conversion of the compounds followed a hepatic pathway because strong fluorescence signals were detected in liver tissues. Highly selective accumulation behavior in tumor tissue was likely due to the EPR effect of solid tumors and targeting ligand attachment (Lin et al., 2015). The anticancer efficacy of targeted nanoparticles was tested in U87 tumor-bearing mice. Intravenous injection exhibited remarkable tumor growth suppression, and no tumor relapse was observed after about 1 month of treatment. Further, the targeted nanoparticles exhibited a powerful in vivo anticancer efficacy, biocompatibility, and nontoxicity in mice. The system represented a versatile nanoplatform useful as a potential tumor theranostic agent that simultaneously enables precise diagnosis for clinical surgery and selective chemotherapeutic applications.

## 7.3 CONCLUSION

Taking into consideration the reports on the field of theranostic applications of liposomal carriers, a number of benefits have been revealed for ICG-liposomal carriers over free ICG including increased fluorescent signal and improved stability in solution/biological fluids. Moreover, the liposomes enable the conjugation of ligands or antibodies for target-specific therapy and diagnosis. Most of the literatures related to theranostic applications of liposomes are focused on cancer diagnosis and treatment. The incorporation of NIR probe ICG into lipid vesicular delivery systems exhibited better performance in cancer detection, photothermal, and photodynamic effects. Apart from their composition, ICG-loaded nanocarriers enhance accumulation of the dye in cancer, due to the EPR effect and active targeting, allowing selective diagnosis or treatment. Some liposomal systems have demonstrated encouraging results, and therefore, ICG-liposomes could be a novel tool for image-guided and anticancer therapy. Indeed, the progress in this field has focused especially on academic research. Only limited information is available on biodistribution, in vivo interactions, and toxicity, and henceforth, their regulatory approval is clearly far away. Further studies are warranted to extend their use for human clinical applications.

## KEYWORDS

- **liposomes**
- **indocyanine green**
- **theranostic**
- **NIR irradiation**
- **cancer targeting**
- **drug delivery**

## REFERENCES

Abels, C. Targeting of the Vascular System of Solid Tumours by Photodynamic Therapy (PDT). *Photochem. Photobiol. Sci.* **2004,** *3*, 765–771.

Ahmed, N.; Fessi, H.; Elaissari, A. Theranostic Applications of Nanoparticles in Cancer. *Drug Discov. Today* **2012,** *17*, 928–934.

Alander, J. T.; Kaartinen, I.; Laakso, A.; Pätilä, T.; Spillmann, T.; Tuchin, V. V.; Venermo, M.; Välisuo, P. A Review of Indocyanine Green Fluorescent Imaging in Surgery. *Int. J. Biomed. Imaging* **2012,** *2012*, 940585.

Al-Jamal, W. T.; Al-Jamal, K. T.; Tian, B.; Cakebread, A.; Halket, J. M.; Kostarelos, K. Tumor Targeting of Functionalized Quantum Dot-Liposome Hybrids by Intravenous Administration. *Mol. Pharm.* **2009,** *6*, 520–530.

Allen, C.; Dos Santos, N.; Gallagher, R.; Chiu, G.; Shu, Y.; Li, W., Johnstone, S.; Janoff, A.; Mayer, L.; Webb, M. Controlling the Physical Behavior and Biological Performance of Liposome Formulations Through Use of Surface Grafted Poly(Ethylene Glycol). *Biosci. Rep.* **2002,** *22*, 225–250.

An, X.; Zhan, F.; Zhu, Y. Smart Photothermal-Triggered Bilayer Phase Transition in AuNPs–Liposomes to Release Drug. *Langmuir* **2013,** *29*, 1061–1068.

Balgavý, P.; Dubničková, M.; Kučerka, N.; Kiselev, M. A.; Yaradaikin, S. P.; Uhríková, D. Bilayer Thickness and Lipid Interface Area in Unilamellar Extruded 1,2-Diacylphosphatidylcholine Liposomes: A Small Angle Neutron Scattering Study. *Biochim. Biophys. Acta* **2001,** *1512*, 40–52.

Banerjee, J.; Hanson, A. J.; Gadam, B.; Elegbede, A. I.; Tobwala, S.; Ganguly, B.; Wagh, A. V.; Muhonen, W. W.; Law, B.; Shabb, J. B. Release of Liposomal Contents by Cell-Secreted Matrix Metalloproteinase-9. *Bioconjug. Chem.* **2009,** *20*, 1332–1339.

Benes, P.; Knopfova, L.; Trcka, F.; Nemajerova, A.; Pinheiro, D.; Soucek, K.; Fojta, M.; Smarda, J. Inhibition of Topoisomerase IIα: Novel Function of Wedelolactone. *Cancer Lett.* **2011,** *303*, 29–38.

Benes, P.; Petra, A.; Lucia, K.; Alena, S.; Jan, S. Redox State Alters Anti-cancer Effects of Wedelolactone. *Environ. Mol. Mutagen.* **2012,** *53*, 515–524.

Beziere, N.; Lozano, N.; Nunes, A.; Salichs, J.; Queiros, D.; Kostarelos, K.; Ntziachristos, V. Dynamic Imaging of PEGylated Indocyanine Green (ICG) Liposomes within the Tumor Microenvironment Using Multi-spectral Optoacoustic Tomography (MSOT). *Biomaterials* **2015,** *37*, 415–424.

Charron, D. M.; Chen, J.; Zheng, G. Theranostic Lipid Nanoparticles for Cancer Medicine. *Cancer Treat. Res.* **2015,** *166*, 103–127.

Chen, Q.; Tong, S.; Dewhirst, M. W.; Yuan, F. Targeting Tumor Microvessels Using Doxorubicin Encapsulated in a Novel Thermosensitive Liposome. *Mol. Cancer Ther.* **2004,** *3*, 1311–1317.

Chen, S.; Yu, G.; Zhang, B.; Wang, Y.; Zhang, N.; Chen, Y. Human Serum Albumin (HSA) Coated Liposomal Indocyanine Green for *In Vivo* Tumor Imaging. *RSC Adv.* **2016,** *6*, 15220–15225.

Choi, K. Y.; Liu, G.; Lee, S.; Chen, X. Theranostic Platforms for Simultaneous Cancer Imaging and Therapy: Current Approaches and Future Perspectives. *Nanoscale* **2012,** *4*, 330–342.

Chonn, A.; Cullis, P. R. Recent Advances in Liposome Technologies and Their Applications for Systemic Gene Delivery. *Adv. Drug Deliv. Rev.* **1998,** *30*, 73–83.

Chonn, A.; Semple, S. C.; Cullis, P. R. Association of Blood Proteins with Large Unilamellar Liposomes In Vivo. Relation to Circulation Lifetimes. *J. Biol. Chem.* **1992,** *267*, 18759–18765.

Chono, S.; Tanino, T.; Seki, T.; Morimoto, K. Uptake Characteristics of Liposomes by Rat Alveolar Macrophages: Influence of Particle Size and Surface Mannose Modification. *J. Pharm. Pharmacol.* **2010,** *59*, 75–80.

Chu, C.; Dijkstra, J.; Lai, M.; Hong, K.; Szoka, F. C. Efficiency of Cytoplasmic Delivery by pH-Sensitive Liposomes to Cells in Culture. *Pharm. Res.* **1990,** *7*, 824–834.

Chu, K. F.; Dupuy, D. E. Thermal Ablation of Tumours: Biological Mechanisms and Advances in Therapy. *Nat. Rev. Cancer* **2014,** *14*, 199–208.

Costa, R. A.; Farah, M. E.; Freymüller, E.; Morales, P. H.; Smith, R.; Cardillo, J. A. Choriocapillaris Photodynamic Therapy Using Indocyanine Green. *Am. J. Ophthalmol.* **2001,** *132*, 557–565.

Delori, F. C.; Webb, R. H.; Sliney, D. H. Maximum Permissible Exposures for Ocular Safety (ANSI 2000), with Emphasis on Ophthalmic Devices. *J. Opt. Soc. Am. A: Opt. Image Sci. Vis.* **2007,** *24*, 1250–1265.

Desgrosellier, J. S.; Cheresh, D. A. Integrins in Cancer: Biological Implications and Therapeutic Opportunities. *Nat. Rev. Cancer* **2010,** *10*, 9–22.
Desmettre, T.; Devoisselle, J. M.; Mordon, S. Fluorescence Properties and Metabolic Features of Indocyanine Green (ICG) as Related to Angiography. *Surv. Ophthalmol.* **2000,** *45*, 15–27.
Dzurinko, V. L.; Gurwood, A. S.; Price, J. R. Intravenous and Indocyanine Green Angiography. *Optometry* **2004,** *75*, 743–755.
Engel, E.; Schraml, R.; Maisch, T.; Kobuch, K.; König, B.; Szeimies, R. M.; Hillenkamp, J.; Bäumler, W.; Vasold, R. Light Induced Decomposition of Indocyanine Green. *Invest. Ophthalmol. Vis. Sci.* **2008,** *49*, 1777–1783.
Erdogan, S.; Torchilin, V. P. Gadolinium-Loaded Polychelating Polymer-Containing Tumor-Targeted Liposomes. *Methods Mol. Biol.* **2010,** *605*, 321–334.
Fomina, N.; Sankaranarayanan, J.; Almutairi, A. Photochemical Mechanisms of Light-Triggered Release from Nanocarriers. *Adv. Drug Deliv. Rev.* **2012,** *64*, 1005–1020.
Gillespie, A. M.; Broadhead, T. J.; Chan, S. Y.; Owen, J.; Farnsworth, A. P.; Sopwith, M.; Coleman, R. E. Phase I Open Study of the Effects of Ascending Doses of the Cytotoxic Immunoconjugate CMB-401 (hCTMO1-Calicheamicin) in Patients with Epithelial Ovarian Cancer. *Ann. Oncol.* **2000,** *11*, 735–741.
Hanahan, D.; Weinberg, R. A. Hallmarks of Cancer: The Next Generation. *Cell* **2011,** *144*, 646–674.
Harashima, H.; Sakata, K.; Funato, K.; Kiwada, H. Enhanced Hepatic Uptake of Liposomes Through Complement Activation Depending on the Size of Liposomes. *Pharm. Res.* **1994,** *11*, 402–406.
Hearn, J.; Rayment, N.; Landon, D. N.; Katz, D. R.; de Souza, J. B. Immunopathology of Cerebral Malaria: Morphological Evidence of Parasite Sequestration in Murine Brain Microvasculature. *Infect. Immun.* **2000,** *68*, 5364–5376.
Hoshino, I.; Maruyama, T.; Fujito, H.; Tamura, Y.; Suganami, A.; Hayashi, H.; Toyota, T.; Akutsu, Y.; Murakami, K.; Isozaki, Y.; Akanuma, N.; Takeshita, N.; Toyozumi, T.; Komatsu, A.; Matsubara, H. Detection of Peritoneal Dissemination with Near-Infrared Fluorescence Laparoscopic Imaging Using a Liposomal Formulation of a Synthesized Indocyanine Green Liposomal Derivative. *Anticancer Res.* **2015,** *35*, 1353–1359.
Hua, J.; Gross, N.; Schulze, B.; Michaelis, U.; Bohnenkamp, H.; Guenzi, E.; Hansen, L. L.; Martin, G.; Agostini, H. T. In Vivo Imaging of Choroidal Angiogenesis Using Fluorescence-labeled Cationic Liposomes. *Mol. Vis.* **2012,** *18*, 1045–1054.
Immordino, M. L.; Dosio, F.; Cattel, L. Stealth Liposomes: Review of the Basic Science, Rationale, and Clinical Applications, Existing and Potential. *Int. J. Nanomed.* **2006,** *1*, 297–315.
Jeong, H.-S.; Lee, C.-M.; Cheong, S.-J.; Kim, E.-M.; Hwang, H.; Na, K. S.; Lim, S. T.; Sohn, M.-H.; Jeong, H.-J. The Effect of Mannosylation of Liposome-Encapsulated Indocyanine Green on Imaging of Sentinel Lymph Node. *J. Liposome Res.* **2013,** *23*, 291–297.
Jing, L.; Shi, J.; Fan, D.; Li, Y.; Liu, R.; Dai, Z.; Wang, F.; Tian, J. $^{177}$Lu-Labeled Cerasomes Encapsulating Indocyanine Green for Cancer Theranostics. *ACS Appl. Mater. Interfaces* **2015,** *7*, 22095–22105.
Juzenas, P.; Juzeniene, A.; Kaalhus, O.; Iani, V.; Moan, J. Noninvasive Fluorescence Excitation Spectroscopy During Application of 5-Aminolevulinic Acid In Vivo. *Photochem. Photobiol. Sci.* **2002,** *1*, 745–748.

Kirby, C.; Clarke, J.; Gregoriadis, G. Effect of the Cholesterol Content of Small Unilamellar Liposomes on their Stability In Vivo and In Vitro. *Biochem. J.* **1980,** *186*, 591–598.
Klohs, J.; Wunder, A.; Licha, K. Near-Infrared Fluorescent Probes for Imaging Vascular Pathophysiology. *Basic Res. Cardiol.* **2008,** *103*, 144–151.
Kochubey, V. I.; Kulyabina, T. V.; Tuchin, V. V.; Altshuler, G. B. Spectral Characteristics of Indocyanine Green upon Its Interaction with Biological Tissues. *Opt. Spectrosc.* **2005,** *99*, 560–566.
Kono, K.; Takashima, M.; Yuba, E.; Harada, A.; Hiramatsu, Y.; Kitagawa, H.; Otani, T.; Maruyama, K.; Aoshima, S. Multifunctional Liposomes Having Target Specificity, Temperature-Triggered Release, and Near-Infrared Fluorescence Imaging for Tumor-Specific Chemotherapy. *J. Control. Release* **2015,** *216*, 69–77.
Koo, O. M.; Rubinstein, I.; Onyuksel, H. Role of Nanotechnology in Targeted Drug Delivery and Imaging: A Concise Review. *Nanomedicine NBM* **2005,** *1*, 193–212.
Kraft, J. C.; Ho, R. J. Interactions of Indocyanine Green and Lipid in Enhancing Near Infrared Fluorescence Properties: The Basis for Near-Infrared Imaging In Vivo. *Biochemistry* **2014,** *53*, 1275–1283.
Lajunen, T.; Kontturi, L.-S.; Viitala, L.; Manna, M.; Cramariuc, O.; Róg, T.; Bunker, A.; Laaksonen, T.; Viitala, T.; Murtomäki, L.; Urtti, A. Indocyanine Green-Loaded Liposomes for Light-Triggered Drug Release. *Mol. Pharm.* **2016,** *13*, 2095–2107.
Lee, R. J.; Huang, L. Folate-Targeted, Anionic Liposome-Entrapped Polylysine-Condensed DNA for Tumor Cell-Specific Gene Transfer. *J. Biol. Chem.* **1996,** *271*, 8481–8487.
Lehtinen, J.; Magarkar, A.; Stepniewski, M.; Hakola, S.; Bergman, M.; Róg, T.; Yliperttula, M.; Urtti, A.; Bunker, A. Analysis of Cause of Failure of New Targeting Peptide in PEGylated Liposome: Molecular Modeling as Rational Design Tool for Nanomedicine. *Eur. J. Pharm. Sci.* **2012,** *46*, 121–130.
Licha, K.; Olbrich, C. Optical Imaging in Drug Discovery and Diagnostic Applications. *Adv. Drug Deliv. Rev.* **2005,** *57*, 1087–1108.
Lim, S.; Jang, H. J.; Park, E. H.; Kim, J. K.; Kim, J. M.; Kim, E. K.; Yea, K.; Kim, Y. H.; Lee-Kwon, W.; Ryu, S. H.; Suh, P. G. Wedelolactone Inhibits Adipogenesis Through the ERK Pathway in Human Adipose Tissue-Derived Mesenchymal Stem Cells. *J. Cell. Biochem.* **2012,** *113*, 3436–3445.
Lin, C.; Kuo, F. W.; Chavanich, S.; Viyakarn, V. Membrane Lipid Phase Transition Behavior of Oocytes from Three Gorgonian Corals in Relation to Chilling Injury. *PLoS ONE* **2014,** *9*, e92812.
Lin, G.; Wang, X.; Yin, F.; Yong, K. T. Passive Tumor Targeting and Imaging by Using Mercaptosuccinic Acid-Coated Near-Infrared Quantum Dots. *Int. J. Nanomed.* **2015,** *10*, 335–345.
Lozano, N.; Al-Ahmady, Z. S.; Beziere, N. S.; Ntziachristos, V.; Kostarelos, K. Monoclonal Antibody-Targeted PEGylated Liposome-ICG Encapsulating Doxorubicin as a Potential Theranostic Agent. *Int. J. Pharm.* **2015,** *482*, 2–10.
Ma, Y.; Tong, S.; Bao, G.; Gao, C.; Dai, Z. Indocyanine Green Loaded SPIO Nanoparticles with Phospholipid-PEG Coating for Dual-Modal Imaging and Photothermal Therapy. *Biomaterials* **2013,** *34*, 7706–7714.
Mac-Daniel, L.; Menard, R. Plasmodium and Mononuclear Phagocytes. *Microb. Pathog.* **2015,** *78*, 43–51.

Maruyama, K.; Takizawa, T.; Takahashi, N.; Tagawa, T.; Nagaike, K.; Iwatsuru, M. Targeting Efficiency of PEG-Immunoliposome-Conjugated Antibodies at PEG Terminals. *Adv. Drug Deliv. Rev.* **1997,** *24*, 235–242.

Maruyama, T.; Akutsu, Y.; Suganami, A.; Tamura, Y.; Fujito, H.; Ouchi, T.; Akanuma, N.; Isozaki, Y.; Takeshita, N.; Hoshino, I.; Uesato, M.; Toyota, T.; Hayashi, H.; Matsubara, H. Treatment of Near-Infrared Photodynamic Therapy Using a Liposomally Formulated Indocyanine Green Derivative for Squamous Cell Carcinoma. *PLoS ONE* **2015,** *10*, e0122849.

Medina-Kauwe, L.; Xie, J.; Hamm-Alvarez, J. Intracellular Trafficking of Nonviral Vectors. *Gene Ther.* **2005,** *12*, 1734–1751.

Miao, W.; Shim, G.; Kim, G.; Lee, S.; Lee, H. J.; Kim, Y. B.; Byun, Y.; Oh, Y.-K. Image-Guided Synergistic Photothermal Therapy Using Photoresponsive Imaging Agent-Loaded Graphene-Based Nanosheets. *J. Control. Release* **2015,** *211*, 28–36.

Mochizuki-Oda, N.; Kataoka, Y.; Cui, Y.; Yamada, H.; Heya, M.; Awazu, K. Effects of Near-Infra-Red Laser Irradiation on Adenosine Triphosphate and Adenosine Diphosphate Contents of Rat Brain Tissue. *Neurosci. Lett.* **2002,** *323*, 207–210.

Moghimi, S. M.; Szebeni, J. Stealth Liposomes and Long Circulating Nanoparticles: Critical Issues in Pharmacokinetics, Opsonization and Protein-Binding Properties. *Prog. Lipid Res.* **2003,** *42*, 463–478.

Mordon, S.; Desmettre, T.; Devoisselle, J. M.; Soulie, S. Fluorescence Measurement of 805 nm Laser-Induced Release of 5,6-CF from DSPC Liposomes for Real-Time Monitoring of Temperature: An In Vivo Study in Rat Liver Using Indocyanine Green Potentiation. *Lasers Surg. Med.* **1996,** *18*, 265–270.

Murata, M.; Nakano, K.; Tahara, K.; Tozuka, Y.; Takeuchi, H. Pulmonary Delivery of Elcatonin Using Surface-Modified Liposomes to Improve Systemic Absorption: Polyvinyl Alcohol with a Hydrophobic Anchor and Chitosan Oligosaccharide as Effective Surface Modifiers. *Eur. J. Pharm. Biopharm.* **2012,** *80*, 340–346.

Murata, M.; Tahara, K.; Takeuchi, H. Real-time In Vivo Imaging of Surface-Modified Liposomes to Evaluate Their Behavior After Pulmonary Administration. *Eur. J. Pharm. Biopharm.* **2014,** *86*, 115–119.

Muthu, M. S.; Feng, S. S. Theranostic Liposomes for Cancer Diagnosis and Treatment: Current Development and Pre-Clinical Success. *Expert Opin. Drug Deliv.* **2013,** *10*, 151–155.

Muthu, M. S.; Kulkarni, S. A.; Xiong, J.; Feng, S. S. Vitamin E TPGS Coated Liposomes Enhanced Cellular Uptake and Cytotoxicity of Docetaxel in Brain Cancer Cells. *Int. J. Pharm.* **2011,** *42*, 332–340.

Nagasaki, T.; Shinkai, S. The Concept of Molecular Machinery Is Useful for Design of Stimuli-Responsive Gene Delivery Systems in the Mammalian Cell. *J. Incl. Phenom. Macrocycl. Chem.* **2007,** *58*, 205–219.

Nakano, K.; Tozuka, Y.; Takeuchi, H. Effect of Surface Properties of Liposomes Coated with a Modified Polyvinyl Alcohol (PVA-R) on the Interaction with Macrophage Cells. *Int. J. Pharm.* **2008,** *354*, 174–179.

Noble, G. T.; Stefanick, J. F.; Ashley, J. D.; Kiziltepe, T.; Bilgicer, B. Ligand-Targeted Liposome Design: Challenges and Fundamental Considerations. *Trends Biotechnol.* **2014,** *32*, 32–45.

Ott, P. Hepatic Elimination of Indocyanine Green with Special Reference to Distribution Kinetics and the Influence of Plasma Protein Binding. *Pharmacol. Toxicol.* **1998,** *83*, 1–48.

Oussoren, C.; Zuidema, J.; Crommelin, D.; Storm, G. Lymphatic Uptake and Biodistribution of Liposomes After Subcutaneous Injection: II. Influence of Liposomal Size, Lipid Composition and Lipid Dose. *Biochim. Biophys. Acta* **1997,** *1328*, 261–272.

Pang, X.; Wang, J.; Tan, X.; Guo, F.; Lei, M.; Ma, M.; Yu, M.; Tan, F.; Li, N. Dual-Modal Imaging-Guided Theranostic Nanocarriers Based on Indocyanine Green and mTOR Inhibitor Rapamycin. *ACS Appl. Mater. Interfaces* **2016,** *8*, 13819–13829.

Park, Y. S. Tumor-Directed Targeting of Liposomes. *Biosci. Rep.* **2002,** *22*, 267–281.

Pauli, J.; Brehm, R.; Spieles, M.; Kaiser, W. A.; Hilger, I.; Resch-Genger, U. Novel Fluorophores as Building Blocks for Optical Probes for In Vivo near Infrared Fluorescence (NIRF) Imaging. *J. Fluoresc.* **2010,** *20*, 681–693.

Peyman, G. A.; Khoobehi, B.; Shaibani, S.; Shamsnia, S.; Ribeiro, I. A Fluorescent Vesicle System for the Measurement of Blood Velocity in the Choroidal Vessels. *Ophthalmic Surg. Lasers* **1996,** *27*, 459–466.

Ponce, A. M.; Vujaskovic, Z.; Yuan, F.; Needham, D.; Dewhirst, M. W. Hyperthermia Mediated Liposomal Drug Delivery. *Int. J. Hyperthermia* **2006,** *22*, 205–213.

Portnoy, E.; Lecht, S.; Lazarovici, P.; Danino, D.; Magdassi, S. Cetuximab-Labeled Liposomes Containing Near-Infrared Probe for In Vivo Imaging. *Nanomedicine NBM* **2011,** *7*, 480–488.

Portnoy, E.; Vakruk, N.; Bishara, A.; Shmuel, M.; Magdassi, S.; Golenser, J.; Eyal, S. Indocyanine Green Liposomes for Diagnosis and Therapeutic Monitoring of Cerebral Malaria. *Theranostics* **2016,** *6*, 167–176.

Proulx, S. T.; Luciani, P.; Derzsi, S.; Rinderknecht, M.; Mumprecht, V.; Leroux, J.-C.; Detmar, M. Quantitative Imaging of Lymphatic Function with Liposomal Indocyanine Green. *Cancer Res.* **2010,** *70*, 7053–7062.

Sawa, M.; Awazu, K.; Takahashi, T.; Sakaguchi, H.; Horiike, H.; Ohji, M.; Tano, Y. Application of Femtosecond Ultrashort Pulse Laser to Photodynamic Therapy Mediated by Indocyanine Green. *Br. J. Ophthalmol.* **2004,** *88*, 826–831.

Saxena, V.; Sadoqi, M.; Shao, J. Degradation Kinetics of Indocyanine Green in Aqueous Solution. *J. Pharm. Sci.* **2003,** *92*, 2090–2097.

Saxena, V.; Sadoqi, M.; Shao, J. Polymeric Nanoparticulate Delivery System for Indocyanine Green: Biodistribution in Healthy Mice. *Int. J. Pharm.* **2006,** *308*, 200–204.

Senior, J.; Gregoriadis, G. Is half-life of Circulating Liposomes Determined by Changes in their Permeability? *FEBS Lett.* **1982,** *145*, 109–114.

Shemesh, C. S.; Hardy, C. W.; David, S. Y.; Fernandez, B.; Zhang, H. Indocyanine Green Loaded Liposome Nanocarriers for Photodynamic Therapy Using Human Triple Negative Breast Cancer Cells. *Photodiagn. Photodyn. Ther.* **2014,** *11*, 193–203.

Shemesh, C. S.; Moshkelani, D.; Zhang, H. Thermosensitive Liposome Formulated Indocyanine Green for Near-Infrared Triggered Photodynamic Therapy: In Vivo Evaluation for Triple-Negative Breast Cancer. *Pharm. Res.* **2015,** *32*, 1604–1614.

Sheng, Z., Hu, D., Xue, M., He, M., Gong, P., Cai, L. Indocyanine Green Nanoparticles for Theranostic Applications. *Nano-Micro Lett.* **2013,** *5*, 145–150.

Shimada, S.; Ohtsubo, S.; Ogasawara, K.; Kusano, M. Macro- and Microscopic Findings of ICG Fluorescence in Liver Tumors. *World J. Surg. Oncol.* **2015,** *13*, 198.

Shum, P.; Kim, J.; Thompson, D. H. Phototriggering of Liposomal Drug Delivery Systems. *Adv. Drug Deliv. Rev.* **2001,** *53*, 273–284.

Skřivanová, K.; Škorpíková, J.; Švihálek, J.; Mornstein, V.; Janisch, R. Photochemical Properties of a Potential Photosensitiser Indocyanine Green In Vitro. *J. Photochem. Photobiol. B* **2006,** *85*, 150–154.

Still, J. M.; Law, E. J.; Klavuhn, K. G.; Island, T. C.; Holtz, J. Z. Diagnosis of Burn Depth Using Laser-Induced Indocyanine Green Fluorescence: A Preliminary Clinical Trial. *Burns* **2001,** *27*, 364–371.

Straubinger, R. M.; Hong, K.; Friend, D. S.; Papahadjopoulos, D. Endocytosis of Liposomes and Intracellular Fate of Encapsulated Molecules: Encounter with a Low pH Compartment After Internalization in Coated Vesicles. *Cell* **1983,** *32*, 1069–1079.

Strijkers, G. J.; Kluza, E.; Van Tilborg, G. A.; van der Schaft, D. W.; Griffioen, A. W.; Mulder, W. J.; Nicolay, K. Paramagnetic and Fluorescent Liposomes for Target-Specific Imaging and Therapy of Tumor Angiogenesis. *Angiogenesis* **2010,** *13*, 161–173.

Sugahara, K. N.; Teesalu, T.; Karmali, P. P.; Kotamraju, V. R.; Agemy, L.; Greenwald, D. R.; Ruoslahti, E. Coadministration of a Tumor-Penetrating Peptide Enhances the Efficacy of Cancer Drugs. *Science* **2010,** *328*, 1031–1035.

Suganami, A.; Toyota, T.; Okazaki, S.; Saito, K.; Miyamoto, K.; Akutsu, Y.; Kawahira, H.; Aoki, A.; Muraki, Y.; Madono, T.; Hayashi, H.; Matsubara, H.; Omatsu, T.; Shirasawa, H.; Tamura, Y. Preparation and Characterization of Phospholipid-Conjugated Indocyanine Green as a Near-Infrared Probe. *Bioorg. Med. Chem. Lett.* **2012,** *22*, 7481–7485.

Sumer, B.; Gao, J. Theranostic Nanomedicine for Cancer. *Nanomedicine* **2008,** *3*, 137–140.

Summerton, D. J.; Kitrey, N. D.; Lumen, N.; Serafetinidis, E.; Djakovic, N. European Association of Urology. EAU Guidelines on Iatrogenic Trauma. *Eur. Urol.* **2012,** *62*, 628–639.

Szoka, Jr., F.; Papahadjopoulos, D. Comparative Properties and Methods of Preparation of Lipid Vesicles (Liposomes). *Annu. Rev. Biophys. Bioeng.* **1980,** *9*, 467–508.

Tagami, T.; Foltz, W. D.; Ernsting, M. J.; Lee, C. M.; Tannock, I. F.; May, J. P.; Li, S. MRI Monitoring of Intratumoral Drug Delivery and Prediction of the Therapeutic Effect with a Multifunctional Thermosensitive Liposome. *Biomaterials* **2011,** *32*, 6570–6578.

Takeuchi, H.; Kojima, H.; Yamamoto, H.; Kawashima, Y. Polymer Coating of Liposomes with a Modified Polyvinyl Alcohol and Their Systemic Circulation and RES Uptake in Rats. *J. Control. Release* **2000,** *68*, 195–205.

Takeuchi, H.; Matsui, Y.; Sugihara, H.; Yamamoto, H.; Kawashima, Y. Effectiveness of Submicronsized, Chitosan-Coated Liposomes in Oral Administration of Peptide Drugs. *Int. J. Pharm.* **2005a,** *303*, 160–170.

Takeuchi, H.; Thongborisute, J.; Matsui, Y.; Sugihara, H.; Yamamoto, H.; Kawashima, Y. Novel Mucoadhesion Tests for Polymers and Polymer-Coated Particles to Design Optimal Mucoadhesive Drug Delivery Systems. *Adv. Drug Deliv. Rev.* **2005b,** *57*, 1583–1594.

Tang, Y.; Aj, M. Combined Effects of Laser-ICG Photothermotherapy and Doxorubicin Chemotherapy on Ovarian Cancer Cells. *J. Photochem. Photobiol. B* **2009,** *97*, 138–144.

Torchilin, V. P. How Do Polymers Prolong Circulation Times of Liposomes. *J. Liposome Res.* **1996,** *9*, 99–116.

Torchilin, V. P. Recent Advances with Liposomes as Pharmaceutical Carriers. *Nat. Rev. Drug Discov.* **2005,** *4*, 145–160.

Toyota, T.; Fujito, H.; Suganami, A.; Ouchi, T.; Ooishi, A.; Aoki, A.; Onoue, K.; Muraki, Y.; Madono, T.; Fujinami, M.; Tamura, Y.; Hayashi, H. Near-Infrared-Fluorescence Imaging of Lymph Nodes by Using Liposomally Formulated Indocyanine Green Derivatives. *Bioorg. Med. Chem.* **2014,** *22*, 721–727.

Turner, D. C.; Moshkelani, D.; Shemesh, C. S.; Luc, D.; Zhang, H. Near-Infrared Image Guided Delivery and Controlled Release Using Optimized Thermosensitive Liposomes. *Pharm. Res.* **2012,** *29*, 2092–2103.

Ulrich, A. S. Biophysical Aspects of Using Liposomes as Delivery Vehicles. *Biosci. Rep.* **2002,** *22*, 129–150.

Waknine-Grinberg, J. H.; Hunt, N.; Bentura-Marciano, A.; McQuillan, J. A.; Chan, H. W.; Chan, W. C.; Barenholz, Y.; Haynes, R. K.; Golenser, J. Artemisone Effective Against Murine Cerebral Malaria. *Malar. J.* **2010,** *9*, 227–241.

Wang, K.; Zhang, X. F.; Liu, Y.; Liu, C.; Jiang, B. H.; Jiang, Y. Y. Tumor Penetrability and Antiangiogenesis Using iRGD-Mediated Delivery of Doxorubicin–Polymer Conjugates. *Biomaterials* **2014,** *35*, 8735–8747.

Warner, S. Diagnostics Plus Therapy = Theranostics. *Scientist* **2004,** *18*, 38–39.

Weissleder, R. A Clearer Vision for In Vivo Imaging. *Nat. Biotechnol.* **2001,** *19*, 316–317.

Weissleder, R.; Ntziachristos, V. Shedding Light onto Live Molecular Targets. *Nat. Med.* **2003,** *9*, 123–128.

Xie, J.; Lee, S.; Chen, X. Nanoparticle-Based Theranostic Agents. *Adv. Drug Deliv. Rev.* **2010,** *62*, 1064–1079.

Xin, Y.; Liu, T.; Yang, C. Development of PLGA-Lipid Nanoparticles with Covalently Conjugated Indocyanine Green as a Versatile Nanoplatform for Tumor-Targeted Imaging and Drug Delivery. *Int. J. Nanomed.* **2016,** *11*, 5807–5821.

Yan, F.; Wu, H.; Liu, H.; Deng, Z.; Liu, H.; Duan, W.; Liu, X.; Zheng, H. Molecular Imaging-Guided Photothermal/Photodynamic Therapy Against Tumor by iRGD-Modified Indocyanine Green Nanoparticles. *J. Control. Release* **2016,** *224*, 217–228.

Yan, L.; Qiu, L. Indocyanine Green Targeted Micelles with Improved Stability for Near Infrared Image-Guided Photothermal Tumor Therapy. *Nanomedicine* **2015,** *10*, 361–373.

Zanganeh, S.; Xu, Y.; Hamby, C. V.; Backer, M. V.; Backer, J. M.; Zhu, Q. Enhanced Fluorescence Diffuse Optical Tomography with Indocyanine Green-Encapsulating Liposomes Targeted to Receptors for Vascular Endothelial Growth Factor in Tumor Vasculature. *J. Biomed. Opt.* **2013,** *18*, 126014.

Zhang, X.; Li, N.; Liu, Y.; Ji, B.; Wang, Q.; Wang, M.; Dai, K.; Gao, D. On-Demand Drug Release of ICG-Liposomal Wedelolactone Combined Photothermal Therapy for Tumor. *Nanomedicine: NBM* **2016,** *12*, 2019–2029.

Zhao, P.; Zheng, M.; Luo, Z.; Gong, P.; Gao, G.; Sheng, Z.; Zheng, C.; Ma, Y.; Cai, L. NIR-Driven Smart Theranostic Nanomedicine for On-Demand Drug Release and Synergistic Antitumour Therapy. *Sci. Rep.* **2015,** *5*, 14258.

Zheng, M.; Yue, C.; Ma, Y.; Gong, P.; Zhao, P.; Zheng, C.; Sheng, Z.; Zhang, P.; Wang, Z.; Cai, L. Single-Step Assembly of DOX/ICG Loaded Lipid-Polymer Nanoparticles for Highly Effective Chemo-photothermal Combination Therapy. *ACS Nano* **2013,** *7*, 2056–2067.

Zheng, X.; Xing, D.; Zhou, F.; Wu, B.; Chen, W. R. Indocyanine Green-Containing Nanostructure as Near Infrared Dual-Functional Targeting Probes for Optical Imaging and Photothermal Therapy. *Mol. Pharm.* **2011,** *8*, 447–456.

Zheng, X.; Zhou, F.; Wu, B.; Chen, W. R.; Xing, D. Enhanced Tumor Treatment Using Biofunctional Indocyanine Green-Containing Nanostructure by Intratumoral or Intravenous Injection. *Mol. Pharm.* **2012,** *9*, 514–522.

## CHAPTER 8

# AQUASOMES: A NANOCARRIER SYSTEM

BHUSHAN RAJENDRA RANE[1*], ASHISH S. JAIN[1], NAYAN A. GUJARATHI[2], and RAJ K. KESERVANI[3]

[1]*Shri. D. D. Vispute, College of Pharmacy & Research Center, New Panvel, Mumbai University, Mumbai, India*

[2]*Sandip Foundation, College of Pharmacy, Nasik, India*

[3]*Faculty of Pharmaceutics, Sagar Institute of Research and Technology-Pharmacy, Bhopal 462041, India*

[*]*Corresponding author. E-mail: rane7dec@gmail.com*

## CONTENTS

## ABSTRACT

Aquasomes are spherical, nanosized particles in the range between 5 and 925 nm, self-assembling biopharmaceutical carrier system, and contain nanosized crystalline core of calcium phosphate or ceramic diamond. Inert core is coated with a glassy oligomeric compound containing polyhydroxyl groups. Aquasomes can be used during delivery of drug and antigens. Aquasomes protect and preserve fragile biological molecules, so known as "bodies of water"; also, it possess conformational integrity and surface exposure. As a result, it is a successful carrier system for delivery of various biochemically active molecules like hormones, protein, peptide, genes, and antigens to specific sites. Targeting of bioactive molecules at specific sites is achieved due to its high degree of surface exposure. Calcium phosphate is the core of interest, due to its natural presence in the body. The brushite is used as core material for the preparation aquasomes to further formulate into implants as they are unstable and convert to hydroxyapatite upon prolong storage. Aquasomes have different applications such as red blood cell substitute, targeted system in gene therapy via intracellular route, and in delivery of viral antigen through vaccination. Aquasome acts as novel carrier for enzymes (DNAse), pigment, and dyes due to its sensitivity and enzyme activity toward molecular conformation.


## 8.1 INTRODUCTION

In the recent era, the research is oriented toward the fields of biotechnology and genetic research by using the various enzymes, proteins, and peptides as a key carrier and therapeutic agents. It is a challenge to administer biologically active moiety in active state; thus, it is difficult for all research and formulation development-based biotechnological and pharmaceutical industries to make stable system. During formulation development, they faced various problems which are associated with suitable route of drug administration, formulations physical and chemical instability, bioavailability of drug, side effects, and serious toxic effects of these biotechnologically derived entities shown possible restrictions on their successful formulation. All the problems associated with the formulation can be overcome by combining nanotechnology and biotechnology approach by developing novel drug-delivery system in the form of aquasomes (Kossovsky et al., 1991, 1994a,b).

Various vesicular approaches were developed with an aim to deliver the drug at targeted site with desired therapeutic response. During development, the various novel vesicular approaches possess some problems which can be solved by developing newer system. Aquasomes is one of the vesicular approaches to deliver the drug at desired site with increased entrapment of drug due to presence of increased binding sites on the surface and stability. Carbohydrate-stabilized nanoparticulate ceramic system called an aquasome was first developed by Nir Kossovsky. Aquasomes are three-layered nanoparticulate carrier system which self-assembled, composed of a solid-phase nanocrystalline core coated with carbohydrate moiety to form oligomeric film to which biochemically active molecules are adsorbed on the surface with or without modification. Aquasomes are also known as "bodies of water"; their water-like properties guard and protect biological molecules of those which are fragile in nature (Kossovsky et al., 1993, 1994a,b).

Due to its protection property, it maintains conformational integrity as well as high degree of surface exposure is used in targeting of bioactive molecules like antigens, hormones, protein, peptide, and genes to specific sites. It consist of a ceramic core which is made of diamond or ceramic, brushite whose surface is noncovalently modified with oligomeric film of carbohydrates to obtain a sugar ball, which is then placed into drug solution for adsorption of a therapeutically active agent. This core provides structural stability to a largely immutable solid (Dunitz, 1994; Kossovsky et al., 1994a,b).

Aquasomes put forward a smart mode of drug delivery for active drug molecules which belonging to the group of proteins and peptides, since they are able to overcome some inherent problems associated with these molecules. The pharmacologically active molecules are often incorporated by copolymerization, diffusion, or adsorption to carbohydrate surface of preformed nanoparticles. These three-layered structures are self-assembled by noncovalent bonds. Principle of "self-assembly of macromolecule" is governed by three physiochemical processes (Jain, 2008; Kossovsky et al., 1990, 1991, 1994a,b), that is, (1) interaction between charged groups, (2) hydrogen bonding and dehydration effect, and, (3) structural stability.

Self-assembly, broadly defined as the spontaneous fabrication of multi-component molecular structures, is the elegant mechanism through which the most complex biological molecules achieve their ultimate form. As an approach to macromolecular synthesis, self-assembly is appealing because biomimetic processes imply more biochemically functional products. This report reviews the principles of self-assembly, the challenges of maintaining both the conformational integrity and biochemical activity of immobilized surface pairs, and the convergence of these principles into a single functional composition (Bauman and Gauldie, 1994).

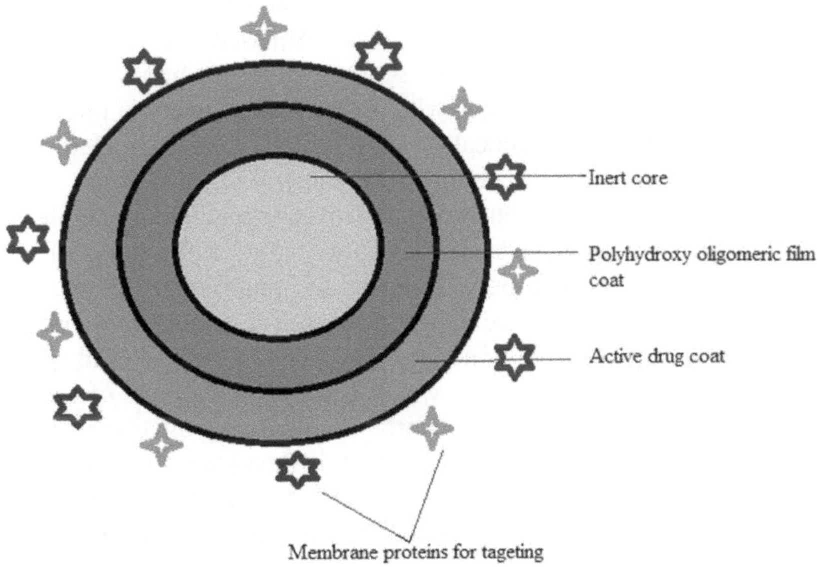


**FIGURE 8.1** Aquasome basic core-based structure.

Aquasome basic core-based structure is as per Figure 8.1. Three types of core materials are mainly used for producing aquasomes: tin oxide, nanocrystalline carbon ceramics, and brushite. Various vesicular approaches in novel drug delivery during development of pharmaceutical systems are listed in Table 8.1.

### *8.1.1 ADVANTAGES*

1. Aquasomes act as reservoirs to release the bioactive molecules either in a continuous or a pulsatile manner, avoiding a multiple-injection therapy.

**TABLE 8.1** Novel Approaches Used in the Various Vesicular Systems.

| Sr. no. | Carrier | Description | Use |
|---|---|---|---|
| 1 | Aquasomes | Three-layered self-assembly compositions with ceramic nanocrystalline particulate core loaded with glassy layer of polyhydroxy compounds | Molecular shielding, specific targeting |
| 2 | Archeosomes | Vesicles composed of glycerolipids of archea with potent adjuvant activity | Potent adjuvant activity |
| 3 | Cryptosomes | Lipid vesicles with a surface coat composed of PC and of suitable polyoxyethylene derivative | Ligand-mediated drug targeting |
| 4 | Discomes | Niosomes solubilized with nonionic surfactant solution | – |
| 5 | Emulsomes | Nanosized lipid particles consisting of microscopic lipid assembly with a polar core | Parenteral delivery of poorly water soluble drugs |
| 6 | Enzymosomes | Liposomes designed to provide a mini bioenvironment in which enzymes are covalently immobilized or coupled to the surface of liposomes | Targeted delivery to tumor cells |
| 7 | Erythrosomes | Human erythrocyte cytoskeletons used as a support to which lipid bilayer is coated | Effective targeting of macromolecular drugs |
| 8 | Ethosomes | Lipid-based soft, malleable vesicles containing a permeation enhancer and composed of phospholipids, ethanol, and water | Targeted delivery to deep skin layers |
| 9 | Genosomes | Artificial macromolecular complexes for functional gene transfer. Cationic lipids are most suitable because they possess high biodegradability and stability in the blood stream | Cell-specific gene transfer |
| 10 | Novasomes | Consist of glyceryl dialurate, cholesterol and polyoxyethylene 10-stearyl ether at a weight percent ratio of 57:15:28, respectively | Drug delivery to pilosebaceous compartment |

**TABLE 8.1** *(Continued)*

| Sr. no. | Carrier | Description | Use |
|---|---|---|---|
| 11 | Photosomes | Photolyase encapsulated in liposomes that release the contents by photo triggered changes in membrane permeability characteristics | Photodynamic therapy |
| 12 | Proteosomes | High-molecular-weight multisubunit enzyme complexes with catalytic activity that is specifically due to assembly pattern of enzymes | Better catalytic activity turnover than nonassociated enzymes may serve as adjuvant as well as protein carrier |
| 13 | Transferosomes (elastic liposomes) | Modified lipid-based soft, malleable carriers tailored for enhanced systemic delivery of drugs | Noninvasive delivery of drugs into or across the deeper skin layers and/or the systemic circulation |
| 14 | Vesosomes | Nested-bilayer compartments with "interdigitated" bilayer phase formed by adding ethanol to a variety of saturated phospholipids | Multiple compartments of the vesosomes give better protection to the interior contents in serum |
| 15 | Virosomes | Liposomes spiked with virus glycoprotein, incorporated into the liposome bilayers based on retrovirus-derived lipids | Immunological adjuvants |

2. This vesicular system in which presence of inorganic core is coated with polyhydroxy oligomeric film responsible for their hydrophilic behavior provides favorable environment for proteins, thereby avoiding their denaturalization.
3. Aquasome increases the release rate of therapeutically and pharmaceutically active agents and protects the drug from phagocytosis and degradation.
4. Multilayered aquasomes conjugated with bioactive molecules such as antibodies, peptides, and nucleic acid which are known as biological labels can be used for various imaging tests.
5. Enzyme activity and sensitivity toward molecular conformation made aquasome a novel carrier for enzymes such as DNAse and pigment/dyes.
6. Aquasome-based vaccines offer many advantages as a vaccine-delivery system. Both cellular and humoral immune responses can be elicited to antigens adsorbed on to the surface of aquasomes.

### *8.1.2 PROPERTIES OF AQUASOMES (KOSSOVSKY et al., 1995a,b)*

1. Aquasomes with water-like properties provide a platform for preserving the conformational integrity of bioactive substances.
2. These systems deliver their contents through a combination of specific targeting, molecular shielding, and slow sustained release.
3. Calcium phosphate is biodegradable in nature and its degradation can be achieved by monocytes and osteoclasts.
4. These carriers also protect the drug/antigen/protein from harsh pH conditions and enzymatic degradation, thus requiring lower doses.
5. The structure stability of aquasomes and their size avoid its clearance by reticuloendothelial system or degradation by other environmental challenges.
6. Mechanism of aquasomes is controlled by their surface chemistry and delivers their contents through the combination of specific targeting, molecular shielding, slow, and sustained release process.

### 8.1.3 MECHANISM

1. Aquasomes have ability to protect biochemical entity which was responsible for certain therapeutic activity while other vesicular systems (like prodrugs and liposomes) as a carrier were unable to protect the active moiety. Sometimes, these are prone to destructive interactions between drug and carrier, in such case aquasomes being a choice for better carrier; as in aquasomes, the carbohydrate coating prevents destructive denaturing interaction between drug and solid carriers (Green and Angel, 1989).
2. Aquasomes have the ability to maintain molecular confirmation and optimum pharmacological activity for drug molecule. Normally, drug molecules possess a unique three-dimensional conformation, a freedom of internal molecular rearrangement induced by molecular interactions, and a freedom of bulk movement but proteins undergo irreversible denaturation when desiccated, even unstable in aqueous state. In the aqueous state pH, temperature, solvents, and salts cause denaturation; hence, bioactive faces many biophysical constrain. In such case, aquasomes with natural stabilizers like various polyhydroxy sugars act as dehydroprotectant that maintains water-like state, thereby preserves molecules in dry solid state (Cherian et al., 2000).

### 8.1.4 RATIONALE BEHIND DEVELOPMENT OF AQUASOMES

First, aquasomes protect bioactives. Many other carriers like prodrugs and liposomes utilized but these are prone to destructive interactions between drug and carrier. The drugs are often inevitable and these always bring limitation to drug-delivery system. In such case, aquasomes proof to be worthy carrier, which are composed of solid carriers whose film has been treated with a film of carbohydrate to prevent destructive denaturing interaction between drug and solid carriers (Cherian et al., 2000). Second, aquasomes maintain molecular confirmation and optimum pharmacological activity. Normally, active molecules possess following qualities, that is, a unique three-dimensional conformation, a freedom of internal molecular rearrangement induced by molecular interactions, and a freedom of bulk movement. This is to be maintained for optimal pharmacological activity. Dehydration, degradation, and decomposition can change these spatial qualities. Many of the biological molecules like proteins undergo irreversible denaturation and become nonfunctional when desiccated; at the same time, they are not resistant to denaturation for a long time in aqueous state. In the aqueous state pH, temperature, solvents, salts, etc. can cause denaturation. So, the challenge is to maintain water-like circumstance; otherwise, it may lead to dehydration and conformational changes, which in turn lead to degradation and alteration of chemical composition. The intrinsic biophysical constraints, dehydration, and conformational changes caused by the drug-delivery system can lead to adverse or allergic reaction with suboptimal pharmacological activity. By incorporating such biological molecules on aquasomes with natural stabilizers, one can preserve the molecular conformation since these natural sugar acts as dehydro protectant. Sugars and polyols stabilize protein against heat denaturation and stabilization is due to the effect of sugars and polyols on hydrophobic interactions. The extent of stabilization by different sugars and polyols is explained by different influences on structure of water. The hydroxyl group on carbohydrate interacts with polar and charged groups of biological molecules in a manner similar to water molecules alone and preserves the aqueous structure of biological molecules like protein on dehydration. Since these disaccharides are rich in hydroxyl groups and help to replace water around the polar residues in proteins, thus maintaining their integrity in the absence of water, the free mobility associated with rich hydroxyl component creates a unique hydrogen-binding substrate that produces glassy aqueous state (Dunitz, 1994). There are many systemic biophysical and intrinsic biophysical constraints, which tend to destabilize the drug.

### *8.1.4.1 SYSTEMIC BIOPHYSICAL CONSTRAINTS*

There are physical and chemical degradative agents, which cause compositional changes and loss of spatial activity by breaking chemical bonds in the drug candidate. Such agents include UV radiation, heat, ozone, peroxide, and other free radicals. Likewise, mammalian body also contains certain agents, namely, inflammatory, peroxides, free radicals, and degradative enzymes related to serine proteases. Other than these physical and chemical degradative agents, those agents that promote dehydration also cause molecular inactivation. Since water is critical structural component of most biochemically reactive molecules, its loss leads to change in energies and results in altered molecular conformation and impaired spatial qualities. Exposure and surface immobilization often promotes dehydration. Degradative agents present in mammals can destroy rapidly complex and expensive polypeptide biopharmaceuticals, while denaturation during dehydration can impair polypeptides on long-term storage (Lyklema and Norde, 1991).

#### *8.1.4.2 INTRINSIC BIOPHYSICAL CONSTRAINTS*

The intrinsic biophysical constraint is normally posed by drug-delivery system. When drug candidates are immobilized to nanoparticulate substrate, it can cause surface-induced dehydration and, in turn, molecular conformation. The altered molecular conformation can produce adverse or allergic reaction with suboptimal pharmacological activity. In short, biochemically active molecules lose their functional properties in either case, means in a "dry" or "wet" state. At the same time, a water environment is vital for molecular activity. Therefore, the challenge is to store and transport promising and useful biomolecules in the dry state without causing them to lose too much of their potential activity. In such case, aquasomes with natural stabilizers like various polyhydroxy sugars act as dehydro protectant, maintain water-like state, and thereby help to preserves the molecular conformation of bioactive molecules in dry solid state. Fungal spores producing ergot alkaloids were stabilized by sucrose-rich solution. Desiccation-induced molecular denaturation is reported to be prevented by certain disaccharides (Crowe et al., 1988).

## 8.2 COMPOSITION OF AQUASOMES

### *8.2.1 CORE MATERIAL*

Ceramic and polymers are most widely used core materials. Polymers such as albumin, gelatin, or acrylate are used. Ceramics such as diamond particles, brushite (calcium phosphate), and tin oxide are used. The ceramics are structurally the most regular materials known for core; being crystalline, high degree of order ensures that bulk properties of ceramic will be preserved because any surface modification will have only limited effect on nature of atoms below surface layer. The surface will exhibit high level of surface energy that will favor the binding of polyhydroxy oligomer surface film. Within a very less time, the freshly prepared particles possess good property of adsorbing molecules.

### *8.2.2 COATING MATERIAL*

Coating materials commonly used are cellobiose, pyridoxal 5 phosphate, sucrose, trehalose, chitosan, citrate, etc.; presence of carbohydrate film prevents soft drug from changing shape and being damaged when surface-bound. Carbohydrate plays important role and acts as natural stabilizer; its stabilization efficiency has been reported. Preformed carbon ceramic nanoparticle and self-assembled calcium phosphate dihydrate particles (colloidal precipitation) are taken to which glassy carbohydrates are then allowed to adsorb as a nanometer thick surface coating in order to formulate molecular carrier. Second step is followed by coating of carbohydrate epitaxially over nanocrystalline ceramic core.

### *8.2.3 BIOACTIVE*

Third, bioactive molecules are adsorbed onto surface which possess property of interacting with film via noncovalent and ionic interactions.

Among three layers of aquasomes, carbohydrate fulfills the objective of aquasomes. The hydroxyl groups on oligomer interact with polar and charged groups of proteins, in a same way as with water, thus preserve the aqueous structure of proteins on dehydration. These disaccharides rich in hydroxyl group help to replace the water around polar residues in protein, maintaining integrity in absence of water. The free bound mobility associated with a rich hydroxyl component creates unique hydrogen binding substrate that produces a glassy aqueous state.

In the formulations of aquasomes, the role of various core materials, coating materials, and bioactive moieties is mentioned in Table 8.2.

**TABLE 8.2** Material of Composition and Role.

| Sr. no. | Material of composition | Examples | Role |
|---|---|---|---|
| 1 | Core material | Albumin, gelatin, or acrylates are polymers diamond particles, brushite, calcium phosphate, and tin oxide | Inert matrix have high degree of order |
| 2 | Coating material | Cellobiose, pyridoxal-5-phosphate, chitosan, citrate, sucrose, and trehalose | Oligomeric films have number of binding sites for attachment and act as natural stabilizer |
| 3 | Bioactive material | Antigen, antibody, protein, enzyme, nucleic acid, peptides, various therapeutic agents | Bioactive molecules |

## 8.3 FORMULATION OF AQUASOMES

### *8.3.1 PRINCIPLES OF SELF-ASSEMBLY (VYAS et al., 2008)*

Self-assembly implies that the constituent parts of some final product assume spontaneously prescribed structural orientations in two- or three-dimensional space. The self-assembly of macromolecules in the aqueous environment, either for the purpose of creating smart nanostructured materials or in the course of naturally occurring biochemistry, is governed basically by three physicochemical processes; the interactions of charged groups, dehydration effects, and structural stability.

#### *8.3.1.1 INTERACTIONS BETWEEN CHARGED GROUPS*

The interaction of charged group facilitates long-range approach of self-assembly; subunits charge group also plays a role in stabilizing tertiary structures of folded proteins. The intrinsic chemical groups or adsorbed ions from the biological milieu lend to most biological and synthetic surfaces, a charge polarity. Most biochemically relevant molecules, in fact, are amphoteric. The interactions of charged groups, such as amino-, carboxyl-, sulfate-, and phosphate groups, facilitate the long-range approach of self-assembling subunits. The long-range interaction of constituent subunits, beginning at an intermolecular distance of around 15 nm, is the necessary first phase of self-assembly. With hydrophobic structures, long-range forces may extend up to 25 nm. Charged groups also play a role in stabilizing tertiary structures of folded proteins.

#### *8.3.1.2 HYDROGEN BONDING AND DEHYDRATION EFFECTS*

Hydrogen bond helps in base pair matching and stabilization of secondary protein structure such as alpha helices and beta sheets. Molecules forming hydrogen bonds are hydrophilic and this confers a significant degree of organization to surrounding water molecules. In case of hydrophobic molecules, which are incapable of forming hydrogen bond, their tendency to repel water helps to organize the moiety within surrounding environment. The organized water decreases level of entropy and is thermodynamically unfavorable, the molecules dehydrate, and get self-assembled.

#### *8.3.1.3 STRUCTURAL STABILITY*

Structural stability of protein in biological environment determined by interaction between charged group and hydrogen bonds largely external to molecule and by van der Waals forces largely internal to molecule experienced by hydrophobic molecules, responsible for hardness and softness of molecule and maintenance of internal secondary structures, provides sufficient softness and allows maintenance of conformation during self-assembly. Self-assembly leads to altered biological activity; van der Waals needs to be buffered. In aquasomes, sugars help in molecular plasticization. van der Waals forces, most often experienced by the relatively hydrophobic molecular regions that are shielded from water, play a subtle but critical role in maintaining molecular conformation during self-assembly. van der Waals forces largely internal to the molecule also play a small but measurable role in the interaction of polypeptides with carbohydrates and related polyhydroxyl oligomers. When molecules change their shape substantially following an interaction, the energy minima assumed upon conformational denaturation tend to preclude reversal.

## 8.3.2 METHOD OF PREPARATION OF AQUASOMES (JAIN, 2008; JAIN AND UMAMAHESHWARI, 2011)

The general procedure consists of an inorganic core formation, which will be coated with lactose forming the polyhydroxylated core that finally will be loaded by model drug.

By using the principle of self-assembly, the aquasomes are prepared in three steps, that is, preparation of core, coating of core, and immobilization of drug molecule as shown in Figure 8.2.

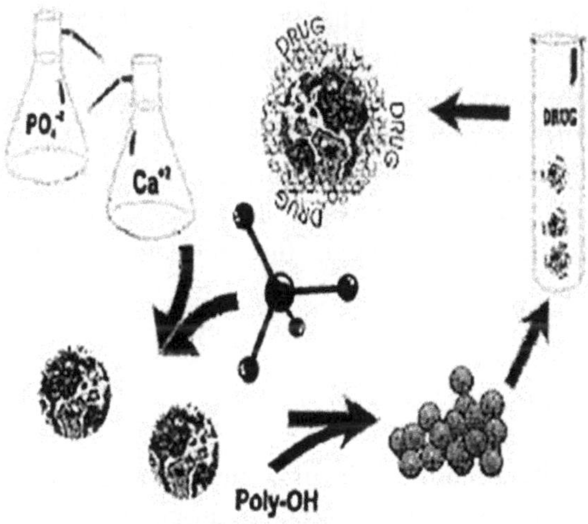


**FIGURE 8.2** Synthesis of aquasomes.

The flow chart for the preparation of aquasome is as per Figure 8.3.

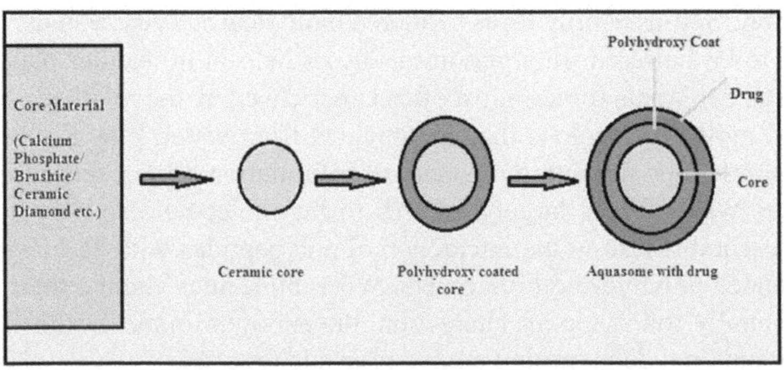


**FIGURE 8.3** Diagrammatic flow chart for preparation of aquasomes.

### 8.3.2.1 PREPARATION OF THE CORE

The first step of aquasome preparation is the fabrication of the ceramic core. The process of ceramic core preparation depends on the selection of the materials for core. These ceramic cores can be fabricated by colloidal precipitation and sonication, inverted magnetron sputtering, plasma condensation, and other processes. For the core, ceramic materials were widely used because ceramics are structurally the most regular materials known. Being crystalline, the high degree of order in ceramics ensures that any surface modification will have only a limited effect on the nature of the atoms below the surface layer and thus the bulk properties of the ceramic will be preserved. The high degree of order also ensures that the surfaces will exhibit high level of surface energy that will favor the binding of polyhydroxy oligomeric surface film. Two ceramic cores that are most often used are diamond and calcium phosphate.

#### *8.3.2.1.1 Preparation of Ceramic Core Using Coprecipitation*

In this method, diammonium hydrogen phosphate solution is added drop wise to calcium nitrate solution with continuous solution. The temperature of the solution is maintained at 75°C in a flask bearing a charge funnel, a thermometer, and a reflux condenser fitted with a carbon dioxide trap. The synthesis can be described by the following equation: 75°C.

$$(NH_4)_2HPO_4 + 3\ Ca(NO_3)_2 \parallel Ca_3(PO_4)_2 + 6\ NH_4NO_3 + H_3PO_4 \quad (8.1)$$

During the synthesis, the pH of calcium nitrate has to be maintained between 8 and 10 using concentrated aqueous ammonia solution. The mixture is then magnetically stirred by maintaining the temperature and pH conditions as detailed above. The precipitates are then filtered, washed, and finally dried overnight. The powder was then sintered by heating to 800–900°C in an electric furnace.

#### *8.3.2.1.2 Preparation of Ceramic Core Using Sonication*

The synthesis can be described by the following equation: 4°C, 2 h.

$$3\ Na_2HPO_4 + 3\ CaCl_2 \parallel Ca_3(PO)_4 + 6\ NaCl + H_3PO_4 \quad (8.2)$$

Based on the above reaction stoichiometry, equivalent moles of the reagents were used. The solutions of disodium hydrogen phosphate and calcium chloride are mixed and sonicated using an ultrasonic bath. The ceramic core can be separated by centrifugation. After the decantation of supernatant, the core is washed, resuspended in distilled water, and filtered. The core material retained on the filter medium is collected, dried, and then percentage yield is calculated.

#### *8.3.2.1.3 Poly(amidoamine)*

Dendrimers with carboxylate terminals, that is, half-generation dendrimers, can be used to study crystallization of calcium carbonate in aqueous solution.

### *8.3.2.2 CARBOHYDRATE COATINGS*

The second step involves coating by carbohydrate on the surface of ceramic cores. There are number of processes to enable the carbohydrate (polyhydroxy oligomers) coating to adsorb epitaxially on to the surface of the nanocrystalline ceramic cores. The processes generally entail the addition of polyhydroxy oligomer to dispersion of meticulously cleaned ceramics in ultra-pure water, sonication, and then lyophilization to promote the largely irreversible adsorption of carbohydrate on to the ceramic surfaces. Excess and readily desorbing carbohydrate is removed by stir cell ultrafiltration. The commonly used coating materials are cellobiose, citrate, pyridoxal-5-phosphate, sucrose, and trehalose.

### *8.3.2.3 IMMOBILIZATION OF DRUGS*

The surface-modified nanocrystalline cores provide the solid phase for the subsequent nondenaturing self-assembly for broad range of biochemically active molecules. The drug can be loaded by partial adsorption (Mizushima et al., 2006). The stepwise procedure for the preparation of aquasome is as per Figure 8.4.

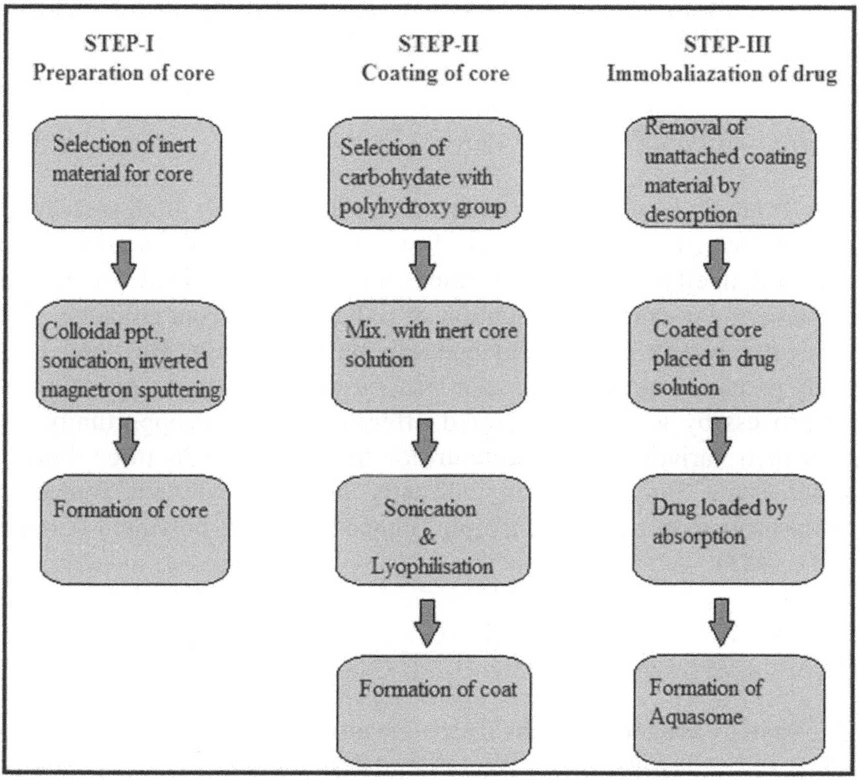


**FIGURE 8.4** Steps in the preparation of aquasomes.

### *8.3.3 STRATEGIES USED IN CHEMICAL SYNTHESIS OF NANOSTRUCTURE (VYAS et al., 2006)*

Aquasomes are self-assembled three-layered nanostructures. Therefore, the strategies involved in chemical synthesis of nanostructure need elaboration. The strategies normally used in the chemical synthesis of nanostructures are discussed below.

#### *8.3.3.1 SEQUENTIAL COVALENT SYNTHESIS*

This can be used to generate arrays of covalently linked atoms generated with well-defined composition, connectivity, and shape, that is, vitamin

B12. It can generate the structures that are far from the thermodynamic minimum for that collection of atoms (Frankel et al., 1989).

#### *8.3.3.2 COVALENT POLYMERIZATION*

This strategy is used for preparing molecules with high-molecular weight. Here, a relatively simple low weight substance is allowed to react with itself to produce molecule comprising many covalently linked monomers, for example, formation of polyethylene from ethylene. The molecular weight of polyethylene can be high (>106 Da), and it is easily prepared, but the molecular structure is simple and repetitive and the process by which it is formed offers only limited opportunity for controlled variation in the structure or for control of its three dimensional shape. Polymerization indirectly provides synthetic routes to stable nanostructures, for example, phase-separated polymers (Crowe et al., 1983).

#### *8.3.3.3 SELF-ORGANIZING SYNTHESIS*

This strategy abandons the covalent bond as required connection between atoms and relies instead on weaker and less directional bonds such as ionic, hydrogen, and van der Waals interactions to organize atoms, ions, or molecules into structures. The different type of structures prepared by this strategy includes molecular crystals, ligand crystals, colloids, micelles, emulsions, phase separated polymers, and self-assembled monolayer. Self-organization is the peculiar feature of these methods. The molecules or ions adjust their own position to reach thermodynamic minimum. By self-organization, true nanostructures can be prepared (Haberland et al., 1992).

#### *8.3.3.4 MOLECULAR SELF-ASSEMBLY*

It is the spontaneous assembly of molecules into structured, stable, and noncovalently joined aggregates. Molecular self-assembly combines features of each preceding strategies to make large structurally well-defined assemblies of atoms; formation of well-defined molecules of intermediate structural complexity through sequential covalent synthesis. Formation of large, stable, and structurally defined aggregates of these molecules results through ionic, hydrogen, and van der Waals interactions or other noncovalent links. The key to this type of synthesis is to understand and overcome intrinsically unfavorable entropy together in a single aggregate. For final assembly to be stable and to have well-defined shape, the noncovalent connection between molecules must be stable. The strength of the individual van der Waals interactions and hydrogen bonds is weak (0.1–5 kcal/mol) relative to typical covalent bonds (40–100 kcal/mol) and comparable to thermal energies. Thus, to achieve acceptable stability, molecules in self-assembled aggregates must be joined by many of these weak noncovalent interactions or by multiple hydrogen bonds or both (Jain, 2001).

### *8.3.4 FATE OF AQUASOME*

The drug-delivery vehicle aquasome is colloidal range biodegradable nanoparticles, so that they will be more concentrated in liver and muscles. Since the drug is adsorbed on to the surface of the system without further surface modification, they may not find any difficulty in receptor recognition on the active site so that the pharmacological or biological activity can be achieved immediately. In normal system, the calcium phosphate is a biodegradable ceramic. Biodegradation of ceramic in vivo is achieved essentially by monocytes and multicellular cells called osteoclasts because they intervene first at the biomaterial implantation site during inflammatory reaction. Two types of phagocytosis were reported when cells come in contact with biomaterial; calcium phosphate crystals were taken up alone and then dissolved in the cytoplasm after disappearance of the phagosome membrane or dissolution after formation of heterophagosomes. Phagocytosis of calcium phosphate coincided with autophagy and the accumulation of residual bodies in the cell (Luo et al., 2004).

## 8.4 CHARACTERIZATION

Aquasomes are characterized chiefly for their structural and morphological properties, particle-size distribution, and drug-loading capacity.

## 8.4.1 CHARACTERIZATION OF CERAMIC CORE

### 8.4.1.1 SIZE DISTRIBUTION

For morphological characterization and size distribution analysis, scanning electron microscopy and transmission electron microscopy are generally used. Core, coated core, as well as drug-loaded aquasomes are analyzed by these techniques. Mean particle size and zeta potential of the particles can also be determined by using photocorrelation spectroscopy (Rawat et al., 2008; Patil et al., 2004).

### 8.4.1.2 STRUCTURAL ANALYSIS

FT-IR spectroscopy can be used for structural analysis. Using the potassium bromide sample disk method, the core as well as the coated core can be analyzed by recording their IR spectra in the wave number range 4000–400 $cm^{-1}$; the characteristic peaks observed are then matched with reference peaks. Identification of sugar and drug loaded over the ceramic core can also be confirmed by FT-IR analysis of the sample (Kossovsky et al., 2009).

### 8.4.1.3 CRYSTALLINITY

The prepared ceramic core can be analyzed for its crystalline or amorphous behavior using X-ray diffraction. In this technique, the X-ray diffraction pattern of the sample is compared with the standard diffractogram, based on which the interpretations are made (Rawat et al., 2008; Patil et al., 2004; Kossovsky et al., 2009).

## 8.4.2 CHARACTERIZATION OF COATED CORE

### 8.4.2.1 CARBOHYDRATE COATING

Coating of sugar over the ceramic core can be confirmed by concanavalin A-induced aggregation method (determines the amount of sugar coated over core) or by anthrone method (determines the residual sugar unbound or residual sugar remaining after coating). Furthermore, the adsorption of sugar over the core can also be confirmed by measurement of zeta potential.

#### *8.4.2.2 GLASS-TRANSITION TEMPERATURE*

Differential scanning calorimetry (DSC) can be used to analyze the effect of carbohydrate on the drug loaded to aquasomes. DSC studies have been extensively used to study glass transition temperature of carbohydrates and proteins. The transition from glass to rubber state can be measured using a DSC analyzer as a change in temperature upon melting of glass.

### *8.4.3 CHARACTERIZATION OF DRUG-LOADED AQUASOMES*

#### *8.4.3.1 DRUG PAYLOAD*

The drug loading can be determined by incubating the basic aquasome formulation (i.e., without drug) in a known concentration of the drug solution for 24 h at 4°C. The supernatant is then separated by high-speed centrifugation for 1 h at low temperature in a refrigerated centrifuge. The drug remaining in the supernatant liquid after loading can be estimated by any suitable method of analysis.

#### *8.4.3.2 IN VITRO DRUG RELEASE STUDIES (GOYAL et al., 2009; HE et al., 2000)*

The in vitro release kinetics of the loaded drug is determined to study the release pattern of drug from the aquasomes by incubating a known quantity of drug-loaded aquasomes in a buffer of suitable pH at 37°C with continuous stirring. Samples are withdrawn periodically and centrifuged at high speed for certain lengths of time. Equal volumes of medium must be replaced after each withdrawal. The supernatants are then analyzed for the amount of drug released by any suitable method.

#### *8.4.3.3 IN-PROCESS STABILITY STUDIES*

Sodium dodecyl sulfate polyacryl amide gel electrophoresis can be performed to determine the stability and integrity of protein during the formulation of the aquasomes.

#### *8.4.3.4 ROLE OF DISACCHARIDES*

Among three layers of aquasomes, carbohydrate fulfills the objective of aquasomes. The hydroxyl groups on oligomer interact with polar and charged groups of proteins, in a same way as with water, and thus preserve the aqueous structure of proteins on dehydration. These disaccharides rich in hydroxyl group help to replace the water around polar residues in protein, maintaining integrity in absence of water. The free bound mobility associated with a rich hydroxyl component creates unique hydrogen-binding substrate that produces a glassy aqueous state.

### ***8.4.4 OTHER EVALUATION PARAMETERS***

#### *8.4.4.1 IN VITRO DRUG RELEASE STUDIES*

The in vitro release kinetics of the loaded drug is determined to study the release pattern of drug from the aquasomes by incubating a known quantity of drug-loaded aquasomes in a buffer of suitable pH at 37°C with continuous stirring. Samples are withdrawn periodically and centrifuged at high speed for certain lengths of time. Equal volumes of medium must be replaced after each withdrawal. The supernatants are then analyzed for the amount of drug released by any suitable method.

#### *8.4.4.2 DRUG-LOADING EFFICIENCY*

This test is done to ensure the amount of drug which is bound on the surface of aquasomes. Spectrophotometric analysis of hydrophobic drugs like indomethacin and piroxicam is done by using 0.1-N methanolic hydrochloric acid solutions.

#### *8.4.4.3 THE HB-LOADING CAPACITY (KHOPADE et al., 2002)*

It is estimated by the difference between the control sample (HbA solution) and the free hemoglobin contained in all fractions without nanoparticles. The spectrophotometric measurements of hemoglobin are done according to Drabkin's method.

#### *8.4.4.4 THE ANTIGEN-LOADING EFFICIENCY FOR THE AQUASOMES*

The formulation's loading efficiency can be determined as reported in literature. Accurately weighted antigen-loaded aquasome formulations were suspended in Triton X-100 and incubated in a wrist shaker for 1 h. Then, samples are centrifuged and absorbance is determined using micro-BCA methods with set a blank of unloaded aquasomes formulation. Antigen loading is expressed as per unit weight of aquasomes particles (g of antigen/mg of sample; Vyas et al., 2006).

#### *8.4.4.5 EFFECT OF CELLOBIOSE AND TREHALOSE ON ANTIGEN*

DSC analysis of aquasome formulations is done by DSC analyzer having a sample cell (containing formulation) and a reference cell (filled with buffer only).

## 8.5 APPLICATIONS OF AQUASOMES (JULIANO, 2005; PRAUSNITZ, 2004; PAUL AND SHARMA, 2001)

### *8.5.1 AQUASOMES AS RED BLOOD CELL SUBSTITUTES AND OXYGEN CARRIER*

Hemoglobin immobilized on oligomer surface because release of oxygen by hemoglobin is conformationally sensitive. By this toxicity is reduced, hemoglobin concentration of 80% achieved and reported to deliver blood in nonlinear manner like natural blood cells. The hemoglobin-adsorbed aquasomes can carry the oxygen satisfactorily, and it also establishes the superiority of hemoglobin aquasome formulation over the other methods acting as artificial blood substitute. The self-assembling surface-modified nanocrystalline ceramic core capable of nondenaturing attachment can be used for various applications like delivery of bioactive molecules as well as viruses (Bauman and Gauldie, 1994; Batz et al., 1974; Horbett and Brash, 1987; Jain and Umamaheshwari, 2011).

Khopade et al. prepared hydroxyapatite core by using carboxylic acid-terminated half-generation poly(amidoamine) dendrimers as templates or crystal modifiers. These cores were further coated with trehalose followed by adsorption of hemoglobin. Studies carried out in rats showed that aquasomes possess good potential for use as an oxygen carrier. The hemoglobin loading to various sugar-coated particles has shown better results and such formulations was able to retain the hemoglobin over a period of 30 days (Israelachvili, 1985; Khopade et al., 2002; Vyas and Khar, 2004).

### *8.5.2 AQUASOMES USED IN IMMUNOTHERAPY*

As vaccines for delivery of viral antigen, that is, Epstein–Barr and immune deficiency virus to evoke correct antibody, objective of vaccine therapy must be triggered by conformationally specific target molecules (Pandey et al., 2011).

The adjuvants generally used to enhance the immunity to antigens have a tendency either to alter the conformation of the antigen through surface adsorption or to shield the functional groups. So, Kossovsky et al. (1996) demonstrated the efficacy of a new organically modified ceramic antigen-delivery vehicle. Diamond, being a material with high surface energy, was the first choice for adsorption and adhesion of cellobiose. It provided a colloidal surface capable of hydrogen bonding to the proteinaceous antigen. The disaccharide, being a dehydro-protectant, helps to minimize the surface-induced denaturation of adsorbed antigens (muscle adhesive protein, MAP). For MAP, conventional adjuvants had proven only marginally successful in evoking an immune response. However, with the help of these aquasomes, a strong and specific immune response could be elicited by enhancing the availability and in vivo activity of antigen.

Vyas et al. (2006) prepared aquasomes by self-assembling of hydroxyl apatite using the coprecipitation method. The core was coated with cellobiose and trehalose, and finally bovine serum albumin was adsorbed as model antigen on to the coated core. When the immunological activity of the prepared formulation was compared to plain bovine serum albumin, the former was found to exhibit a better response. In view of these results, aquasomes were proposed to have superior surface immutability, in that they protect the conformation of protein structure and present it in such a way to immune cells that it triggers a better immunological response.

The use of ceramic core-based nanodecoy systems was proposed by Vyas et al. as an adjuvant and delivery vehicle for hepatitis B vaccine for effective immunization. Self-assembling hydroxyapatite core was coated with cellobiose, and finally hepatitis B surface antigen was adsorbed over the coated core. The nanodecoy systems were also found to be able to elicit a combined Th1 and Th2 immune response.

Vyas et al. demonstrated the immunoadjuvant properties of hydroxyapatite by administering it with malarial merozoite surface protein-119 (MSP-119). Prepared nanoceramic formulations also showed slower in vitro antigen release and slower biodegradability behavior, which may lead to a prolonged exposure to antigen presenting cells and lymphocytes. Furthermore, addition of mannose in nanoceramic formulation may additionally lead to increased stability and immunological reactions. Immunization with MSP-119 in nanoceramic-based adjuvant systems induced a vigorous IgG response, with higher IgG2a than IgG1 titers. In addition, a considerable amount of interferon $\gamma$ (IFN$\gamma$) and interleukin 2 was observed in spleen cells of mice immunized with nanoceramic-based vaccines. In contrast, mice immunized with MSP-119 alone or with alum did not show a significant cytotoxic response. The prepared hydroxyapatite nanoparticles exhibit physicochemical properties that point toward their potential as a suitable immunoadjuvant for used as antigen carriers for immunopotentiation.

The uses of drug-delivery systems in allergen specific immunotherapy appear to be a promising approach due to their ability to act as adjuvants, transport the allergens to competent cells and tissues, and reduce the number of administrations. This suggests that aquasomes could have possible implications in the future of peptide-based vaccines against allergic disorders.

### 8.5.3 AQUASOMES USED IN IMMUNOPOTENTIATION

Goyal et al. (2008) prepared aquasomes that develop immune responses to recombinant or synthetic epitopes which is of considerable importance in vaccine research for immunopotentiation.

### 8.5.4 AQUASOMES IN ANTITHROMBIC ACTIVITY

Chauvierre et al. (2004) formulated nanoparticles based on heparin–poly(isobutyl cyano acrylate) copolymers to carry hemoglobin. His work constitutes the demonstration of hemoglobin loaded on nanoparticle surface, rather than being encapsulated.

### 8.5.5 AQUASOMES IN GENE DELIVERY

Aquasomes have been used for successful targeted intracellular gene therapy, a five-layered composition composed of ceramic core, polyoxy oligomeric film, therapeutic gene segment, additional carbohydrate film, and a targeting layer of conformationally conserved viral membrane (Arakawa and Timasheff, 1982).

### 8.5.6 AQUASOMES FOR PHARMACEUTICALS DELIVERY

Insulin is developed because drug activity is conformationally specific. Bioactivity is preserved and activity is increased to 60% as compared to i.v. administration and toxicity is not reported.

Cherian et al. (2000) prepared aquasomes using a calcium phosphate ceramic core for the parenteral delivery of insulin. The utility of nanocarriers for effective delivery of insulin was also proved by Paul and Sharma. They prepared porous hydroxyapatite nanoparticles entrapped in alginate matrix containing insulin for oral administration. The optimum controlled release of insulin was also achieved in his study.

### 8.5.7 AQUASOMES FOR ENZYMES DELIVERY

Aquasomes are also used for delivery of enzymes like DNAse and pigments/dyes because enzymes activity fluctuates with molecular conformation and cosmetic properties of pigments are sensitive to molecular conformation (Vyas and Khar, 2002).

Kommineni et al. carried out a technological innovation for the delivery aquasomes via the peroral route. Piroxicam-loaded aquasomes with their nanometric dimensions, low drug dose, and water-like properties were prepared by using two techniques, namely, coprecipitation by refluxing and coprecipitation by sonication.

Rawat et al. proposed the use of a nanosized ceramic core-based system for oral administration of the acid-labile enzyme serratiopeptidase. Aquasomes were found to be protecting the structural integrity of enzymes so as to obtain a better therapeutic effect.

### *8.5.8 MISCELLANEOUS*

Mizushima and coworkers prepared sphericalporous hydroxyapatite particles by spray-drying. These particles were tried as a carrier for the delivery of drugs such as IFNα, testosterone enanthate, and cyclosporin A. IFNα was adsorbed well to spherical hydroxyapatite particles. Addition of HAS and zinc (for reinforcement) to IFNα-adsorbed hydroxyapatite particles caused marked prolongation of release in vivo.

Oviedo et al. (2007) prepared aquasomes loaded with indomethacin through the formation of an inorganic core of calcium phosphate covered with a lactose film and further adsorption of indomethacin as a low-solubility drug. Bioactive molecules incorporated into aquasome systems and their use are mentioned in Table 8.3.

**TABLE 8.3** Bioactive Molecules Incorporated into Aquasomes and Their Use.

| Sr. no. | Bioactive molecules | Use |
|---|---|---|
| 1 | Insulin | To increase the bioavailability of insulin |
| 2 | Indomethacin | To increase the solubility as well as to increase the release of the drug |
| 3 | Testosterone enanthate | Sustain release of drug |
| 4 | Oxygen | To increase the oxygen carrying capacity |
| 5 | Cyclosporine A | Sustain release of drug |
| 6 | IFN-α | Prolongation of release in vivo |
| 7 | Acid labile enzyme—serratiopeptidase | To better therapeutic activity and better structural integrity of the enzyme |
| 8 | Merozoite surface protein-119 (MSP-199), hepatitis B | To increase the immunity |
| 9 | Dithranol | Treatment of psoriasis |
| 10 | Delivery of gene | Protection and maintenance of structural integrity of the gene segment and prevention of the risk of irrelevant gene integration |

## 8.6 RECENT DEVELOPMENT

1. Aquasomes act as a novel carrier for the delivery drug as they possess properties like protection and preservation of fragile biological molecules, conformational integrity, and surface exposure which made it a successful carrier system for bioactive molecules like peptide, protein, hormones, antigens, and genes to specific sites (Frankel et al., 1989).
2. *Nanocarrier technology*: The challenges of maintaining both the conformational integrity and biochemical activity of immobilized surface pairs, and the convergence of these principles into a single functional composition achieved by aquasomes. Aquasomes is widely used for the preparation of implants for drug delivery as well as for delivery vaccines, viral antigen (Epstein–Barr and immune deficiency virus) to evoke correct antibody and as targeted system for intracellular gene therapy.
3. Aquasomes is a promising carrier for peptides and protein delivery. Enzyme activity and sensitivity toward molecular conformation made aquasome a novel carrier for enzymes like DNAse and pigment/dyes.
4. As novel carrier system, aquasomes are three-layered self-assembling composition with ceramic carbon nanocrystalline particle core coated with glassy ceroboise or degradable calcium phosphate monomer crystalline particle coated with glassy pyridoxal-5-phosphate. Subsequently drug/enzymes covalently bonded to outer coating. Aquasome delivers their content through specific targeting, molecular shielding, and slow sustained release process. Aquasome technology represents a platform system for conformational integrality and biochemical stability of bioactives (Johnson et al., 1985).
5. Aquasomes are self-assembled nanotechnology-based colloidal range nanoparticles that provide high degree of surface exposure and help to maintain molecular confirmation and optimum pharmacological activity. Various patented formulations based on aquasome, their trade names, and approval year are mentioned in Table 8.4.

**TABLE 8.4** Trade Names of Patented Formulations and Approval Year.

| Trade name | Recombinant product | Approval year |
|---|---|---|
| Helixate FS | Sucrose formulation | 2000 US |
| Refacto | B-domain-detected clotting factor VIII | 2000 US |
| Ovidrel or Ovitrelle | Human chorionic gonadotropin α | 2000 US |
| Novoseven | Clotting factor VII a | 1999 US |
| Thyogen | Thyotropin α | 1998 US |
| Enbrel | TNF-α receptor–IgG fusion protein | 1998 US |
| Herceptin | Anti-HER 2 humanized mAb | 1998 US |
| Remicade | Anti-TNF-α chimeric mAb | 1998 US |
| Simulect | Anti-IL2 receptor-α chimeric mAb | 1998 US |
| Gonal-f | Follicle stimulating hormone (follitropin α) | 1997 US<br>1995 EU |
| Rituxan (US)/Mabthera (EU) | Anti-CD20 chimeric mAb | 1997 US |
| Beneflx | Clotting factor IX | 1997 US |
| Avonox | IFN-β-1a | 1996 US |
| Cerezyme | B-glucocerebrosidase | 1994 US |
| Kogenate FS | Clotting factor VIII | 1993 US |
| Kogenate, Helixate | Clotting factor VIII | 1993 US |
| Recombinate | Clotting factor VIII | 1992 US |
| Epogen/Procrit | Erythropoietin (epoetin α) | 1989/1990 US |
| Activase | Tissue plasmingen activator | 1987 US |

## 8.7 CONCLUSION

Aquasomes classified under colloidal vesicular approaches to deliver the drug at desired site with increased entrapment of drug due to presence of increased binding sites on the surface and stability. Aquasomes put forward a smart mode of drug delivery for active drug molecules which belonging to the group of proteins and peptides, since they are able to overcome some inherent problems associated with these molecules. Aquasomes can be used in the delivery of wide variety of drug of different nature and category due to availability of carbohydrate coat which provides different binding sites for attachment of drug at surface. Bioactive molecules, enzymes, proteins, peptides, or many more chemical entities can be targeted via this approach.

## KEYWORDS


- **aquasomes**
- **carbon ceramics (diamonds)**
- **oligomeric film**
- **brushite**
- **self-assembling carrier system**
- **targeted drug delivery**


## REFERENCES


Arakawa, T.; Timasheff, S. N. Stabilization of Protein Structure by Sugars. *Biochemistry* **1982,** *21* (0), 6536–6544.

Batz, H. G.; Ringsford, H.; Ritter, H. Pharmacologically Active Polymers. *Macromol. Chem.* **1974,** *175* (8), 2229–2239.

Bauman, H.; Gauldie, J. The Acute Phase Response. *Immunol. Today* **1994,** *15*, 74–78.

Chauvierre, C.; Marden, M. C.; Vauthier, C.; Labarre, D.; Couvreur, P. Leclerc, L. Heparin Coated Poly(Alkylcyanoacrylate) Nanoparticles Coupled to Hemoglobin: A New Oxygen Carrier. *Biomaterials* **2004,** *25*, 3081–3086.

Cherian, A.; Rana, A. C.; Jain, S. K. Self-Assembled Carbohydrate Stabilized Ceramic Nanoparticles for the Parenteral Drug Delivery of Insulin. *Drug Dev. Ind. Pharm.* **2000,** *26*, 459–463.

Crowe, J. H.; Crowe, L. M.; Carpenter, J. F.; Rudolph, A. S.; Wistrom, C. A.; Spargo, B. J.; Acnhordoguy, T. J. Interaction of Sugars with Membrane. *Biochem. Biophys. Sin. Acta* **1988,** *1947*, 367–384.

Dunitz, J. D. The Entropic Cost of Bound Water in Crystals and Biomolecules. *Science* **1994,** *264* (5159), 264–670.

Frankel, D. A.; Lamparski, H.; Liman, U.; O'Brien, D. F. Photoinduced Destabilization of Bilayer Vesicles. *J. Am. Chem. Soc.* **1989,** *111* (26), 9262–9263.

Goyal, A.; Khatri, K.; Mishra, N.; Mehta, A.; Vaidya, B.; Tiwari, S. Development of Self-Assembled Nanoceramic Carrier Construct(s) for Vaccine Delivery. *J. Biomater. Appl.* **2009,** *24*, 65–84.

Green, J. L.; Angel, C. A. Phase Relations and Vitrification in Sacchride Solutions and Trehalose Anomaly. *J. Phys. Chem.* **1989,** *93*, 2880–2882.

Haberland, M. E.; Fless, G. M.; Scannu, A. M.; Fogelman, A. M. Malondialdehyde Modification of Lipoprotein Produces avid Uptake by Human Monocytes Macrophages. *J. Boil. Chem.* **1992,** *267*, 4143–4159.

He, Q.; Mitchell, A.; Johnson, S.; Wagner-Bartak, C.; Morcol, T.; Bell, S. Calcium Phosphate Nanoparticle Adjuvant. *Clin. Diagn. Lab. Immunol.* **2000,** *7*, 899–903.

Horbett, T. A.; Brash, J. L. Proteins at Interface: Current Issues and Future Prospects. In *Proteins at Interfaces Physiochemical and Biological Studies*; Brash, J. L., Horbett, T. A., Eds.; *ACS Symposium Series*, ACS: Washington, 1987; vol 343, pp 1–33.

Israelachvili, J. N. *Intermolecular and Surface Force*; Academic Press: New York, 1985.

Jain, N. K.; Umamaheshwari, R. B. Control and Novel Drug Delivery Systems. In *Pharmaceutical Product Development*, 2nd ed.; Jain, N. K., Ed.; CBS Publishers & Distributors: New Delhi, 2011; pp 521–564.

Jain, N. K. *Advances in Controlled and Novel Drug Delivery System*, 1st ed. CBC Publisher & Distributors: New Delhi, 2001; pp 317–328.

Johnson, L. N.; Cheetham, J.; Mclaunglin, P. J.; Acharya, K. R.; Barford, D.; Philips, D. C. Protein Oligosacchride Interactions, Lysozyme Phosphorylase Amylase. *Curr. Top.* **1985,** *139*, 81–86.

Juliano, R. L. Microparticulate Drug Carriers: Liposomes, Microspheres and Cells. In *Controlled Drug Delivery*, 2nd ed.; Robinson, J. R., Lee, V. H. L.; Marcel Dekker: New York, NY, 2005; pp 555–580.

Khopade, A. J.; Khopade, S.; Jain, N. K. Development of Haemoglobin Aquasomes from Spherical Hydroxyapatite Cores Precipitated in the Presence of Poly(Amidoamine) Dendrimer. *Drug Dev. Ind. Pharm.* **2002,** *241*, 145–154.

Kossovsky, N.; Bunshah, R. F.; Gelmm, A; Sponsler, E. D.; Dmarjee, D. M.; Suh; T. G.; Pralash, S.; Doel, H. J.; Deshpandey, C. A Non-denaturing Solid Phase Pharmaceutical Carrier Comprised of Surface Modified Nanocrystalline Materials. *J. Appl. Biomater.* **1990,** *1*, 289–294.

Kossovsky, N.; Gelman, A.; Hnatyszyn, H. J.; Rajguru, S.; Garrell, L. R.; Torbati, S. Surface-Modified Diamond Nanoparticles as Antigen Delivery Vehicles. *Bioconj. Chem.* **1995a,** *6*, 507–510.

Kossovsky, N.; Gelman, A.; Rajguru, S.; Nguyan, R.; Sponsler, E.; Hnatyszyn, C. K. Control of Molecular Polymorphism by a Structured Carbohydrate/Ceramic Delivery Vehicle-Aquasomes. *J. Control. Release* **1996,** *39*, 383–388.

Kossovsky, N.; Gelman, A.; Sponsler, E.; Rajguro, S.; Tones, M; Mena, E.; Ly, K.; Festeljian, A. Preservation of Surface-Dependent Properties of Viral Antigens Following Immobilization on Particulate Ceramic Delivery Vehicles. *Biomed. Mater. Res.* **1995b,** *29* (5), 561–573.

Kossovsky, N.; Gelman, A.; Sponsler, E. D.; Millett, D. Nano-crystalline Epstein-Bar Vims decoys. *Appl. Biomater.* **1991,** *2*, 251–259.

Kossovsky, N.; Gelman, A; Sponsler, E. E.; Hnatyszyn, A. J.; Rajguro, S.; Torres, M.; Pham, M.; Crowder, J.; Zemanovich, J.; Chung, A.; Shah, R. Surface Modified Nanocrystalline Ceramic for Drug Delivery Applications. *Biomaterials* **1994a,** *15*, 1201–1207.

Kossovsky, N.; Gelman, A.; Sponsler, E. E. Cross Linking Encapsulated Haemoglobin Solid Phase Supports: Lipid Enveloped Haemoglobin Adsorbed to Surface Modified Ceramic Particles Exhibit Physiological Oxygen Lability *Artif. Cells, Blood Subst. Biotechnol.* **1994b,** *223*, 479–485.

Kossovsky, N.; Millett, D.; Gelman, L. A.; Sponsler, E. D.; Huatyszyn, H. J. Self-Assembling Nanostructures. *Biotechnology* **1993,** *11*, 1534.

Luo, D.; Han, E.; Belcheva, N.; Saltzman, W. M. A Self-Assembled, Modular Delivery System Mediated by Silica Nanoparticles. *J. Control. Release* **2004,** *95*, 333–341.

Lyklema, J.; Norde, W. Why Proteins Prefer Interfaces. *J. Biomater. Sci. Polym. Ed.* **1991,** *2*, 183–202.

Mizushima, Y.; Ikoma, T.; Tanaka, J.; Hoshi, K.; Ishihara, T.; Ogawa, Y. Injectable Porous Hydroxyapatite Microparticles as a New Carrier for Protein and Lipophilic Drugs. *J. Control. Release* **2006,** *110*, 260–265.

Oviedo, R. I.; Lopez, S. A. D.; Gasga, R. J.; Barreda, C. T. Q. Elaboration and Structural Analysis of Aquasomes Loaded with Indomethecin. *Eur. J. Pharm. Sci.* **2007,** *32*, 223–230.

Pandey, R. S.; Sahu, S.; Sudheesh, M.; Madan, J.; Manoj, K.; Dixit, V. Carbohydrate Modified Ultrafine Ceramic Nanoparticles for Allergen Immunotherapy. *Int. J. Immunopharmacol.* **2011,** *11*, 925–931.

Patil, S.; Pancholli, S.; Agrawal, S.; Agrawal, G. Surface-Modified Mesoporous Ceramics as Delivery Vehicle for Haemoglobin. *Drug Deliv.* **2004,** *11*, 193–199.

Paul, W.; Sharma, C. P. Porous Hydroxyapatite Nanoparticles for Intestinal Delivery of Insulin. *Trends Biomater. Artif. Organs* **2001,** *14*, 37–38.

Prausnitz, M. R. Microneedles for Transdermal Drug Delivery. *Adv. Drug Deliv. Rev.* **2004,** *56*, 581–587.

Rawat, M.; Singh, D.; Saraf, S.; Saraf, S. Development and In Vitro Evaluation of Alginate Gel-Encapsulated, Chitosan-Coated Ceramic Nanocores for Oral Delivery of Enzyme. *Drug Dev. Ind. Pharm.* **2008,** *34*, 181–188.

Vyas, S. P.; Khar, R. K. *Targeted & Controlled Drug Delivery*; CBC Publisher & Distributors: New Delhi, 2004; pp 28–30.

Vyas, S. P.; Goyal, A. K.; Khatri, K.; Mishra, N.; Mehta, A.; Vaidya, B. Aquasomes—A Nanoparticulate Approach for the Delivery of Antigen. *Drug Dev. Ind. Pharm.* **2008,** *34*, 1297–1305.

Vyas, S. P.; Goyal, A. K.; Rawat, A.; Mahor, S.; Gupta, P. N.; Khatri, K. Nanodecoy System: A Novel Approach to Design Hepatitis B Vaccine for Immunopotentiation. *Int. J. Pharm.* **2006,** *309*, 227–233.

Vyas, S. P.; Khar, R. K. Introduction to Parenteral Drug Delivery. In *Targeted and Controlled Drug Delivery*; Vyas, S. P., Khar, R. K., Eds.; CBS Publishers & Distributors: New Delhi, 2002; pp 3–37.

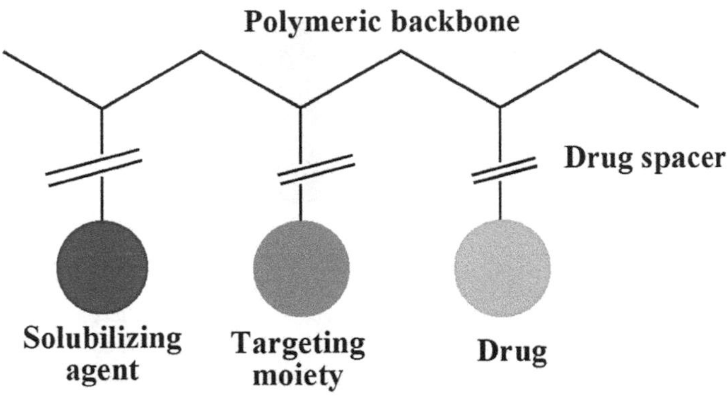


**FIGURE 1.5** A schematic diagram of polymer–drug conjugates.

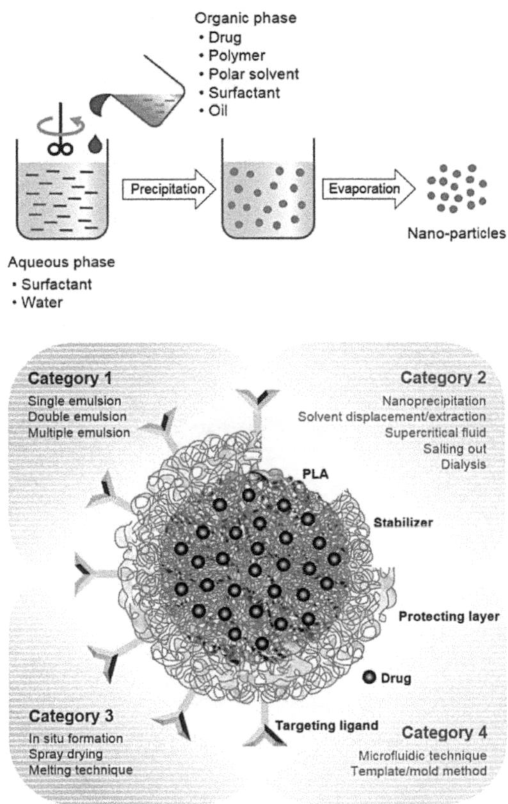


**FIGURE 2.1** Representative micro- and nanoparticle structure with the four categories of preparation techniques and a schematic description of the nanoprecipitation method. (Reprinted with permission from Lee et al. *Adv. Drug Deliv. Rev.* **2016,** in press. © 2016, Elsevier.)

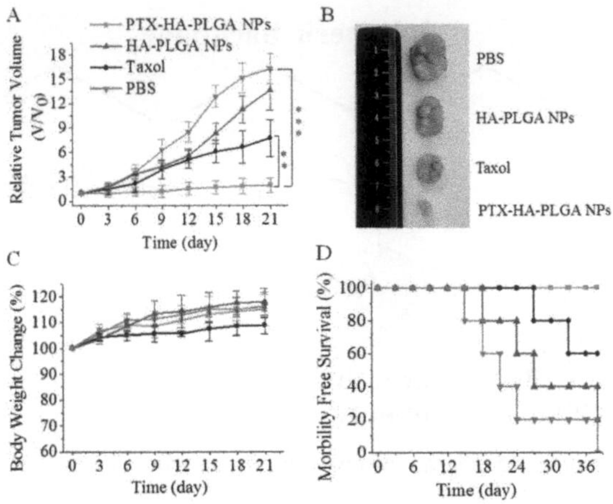


**FIGURE 2.3** In vivo antitumor activity of PTX-HA-PLGA NPs in MCF-7 tumor-bearing mice. Tumor volume changes of mice and photographs of typical tumor blocks isolated on day 21. (Reprinted with permission from Wu et al. *Biomacromolecules* **2016,** *17*, 2367. © 2016, American Chemical Society.)

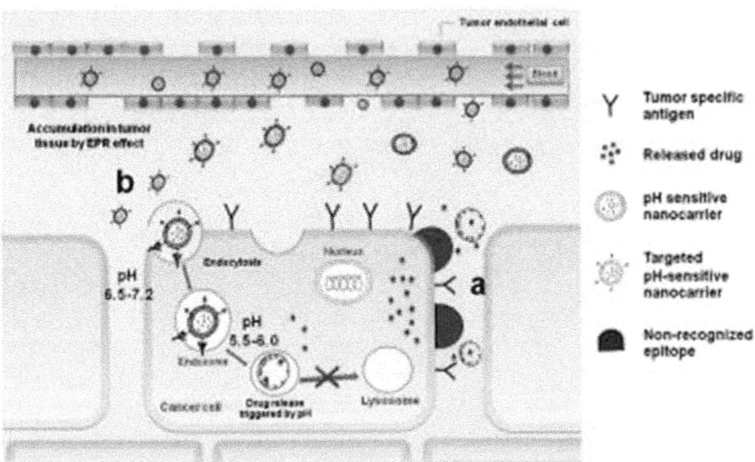


**FIGURE 2.7** Schematic representation of the proposed mechanism of pH-responsive polymeric micelles. (Reprinted with permission from Liu et al. *Asian J. Pharm. Sci.* **2013,** *8*, 159.)

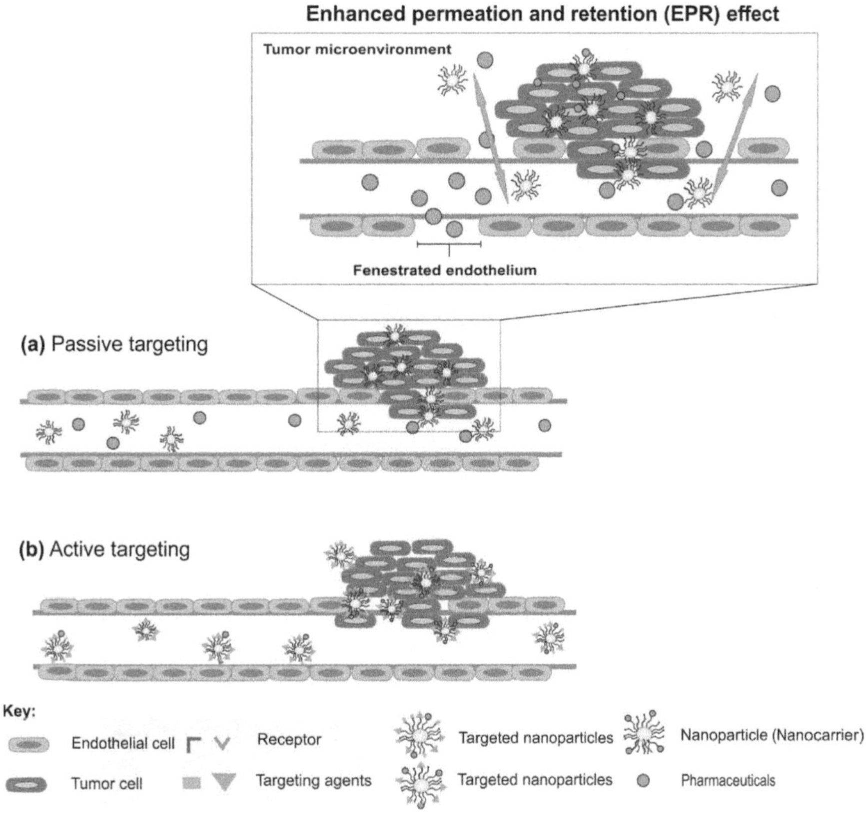


**FIGURE 4.1** Schematic illustration of active and passive-targeting mechanisms. PEGylated target-specific ligand-armed gold nanoparticles (GNPs) loaded with drugs. The tumor microvasculature in the tumor microenvironment is fenestrated with gaps and pores (120–1200 nm), in which GNPs can be accumulated and deliver the loaded therapeutic agents.

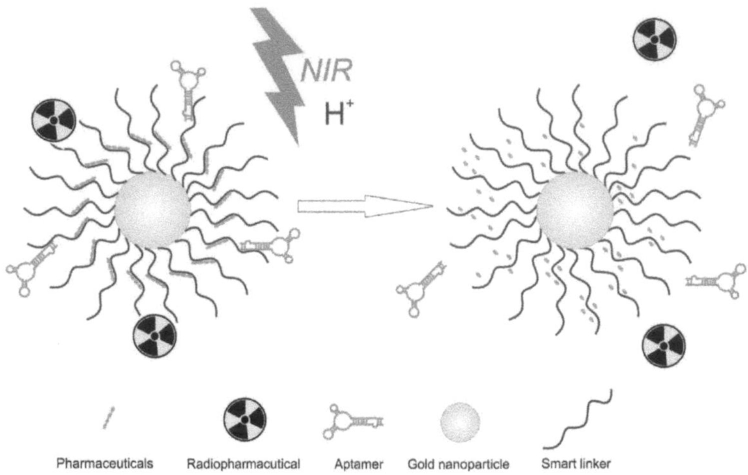


**FIGURE 4.4** Schematic illustration of NIR-light- and pH dual-stimuli-triggered GNPs.

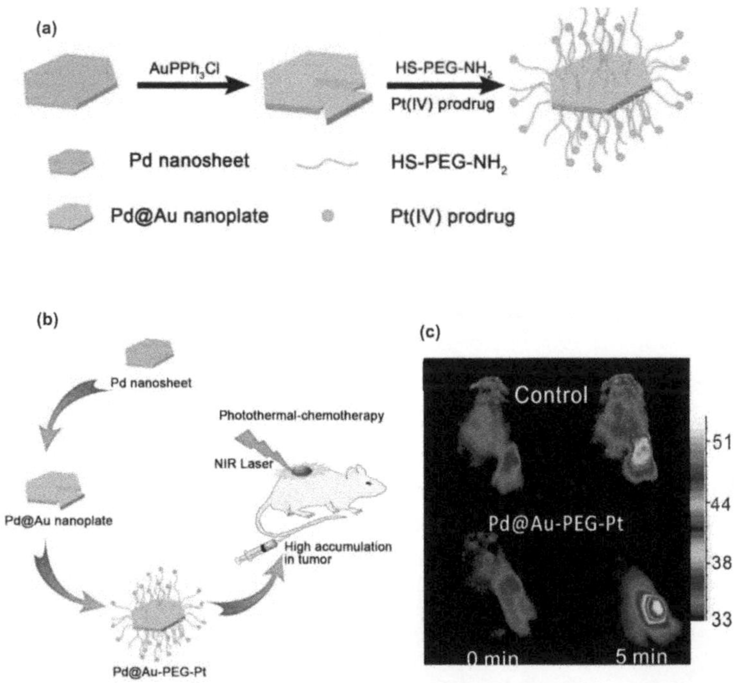


**FIGURE 4.5** Gold nanoplates as theranostics. (a) Process for preparation of Pd@Au–PEG–Pt nanoplate. (b) Prodrug conjugated Pd@Au nanoparticle for photothermal therapy and chemotherapy. (c) NIR-thermal images of tumor-bearing mice after Pd@Au–PEG–Pt injection at different times. (Data adapted with permission from Shi et al. © 2016, Royal Society of Chemistry.)

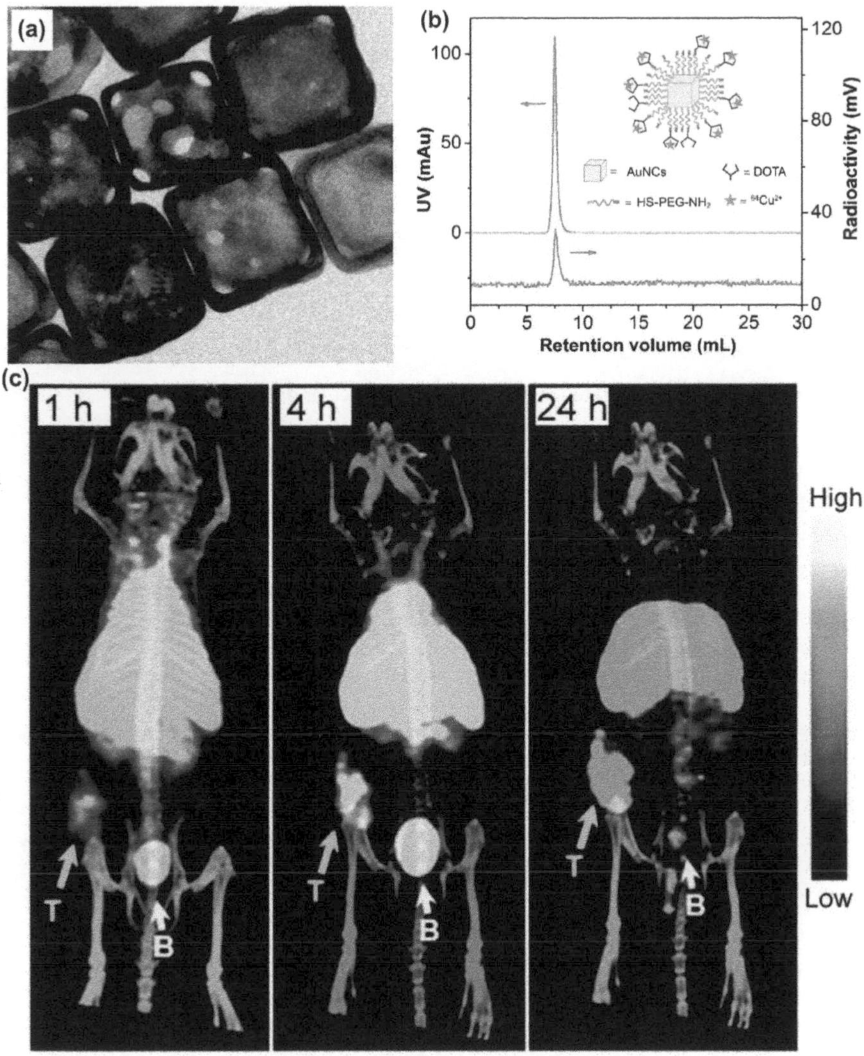


**FIGURE 4.8** A $^{64}$Cu-DOTA-PEG-Au nanocage. (a) Transmission electron microscopy micrograph of $^{64}$Cu-DOTA-PEG-Au nanocage. (b) Radio-HPLC analysis of the $^{64}$Cu-DOTA-PEG-GNP. (c) PET/CT images of the $^{64}$Cu-DOTA-PEG-GNP in a mouse bearing an EMT-6 tumor at different times. *B*, bladder; *T*, tumor. (Data adapted with permission from Wang et al., 2012, © 2012, American Chemical Society.)

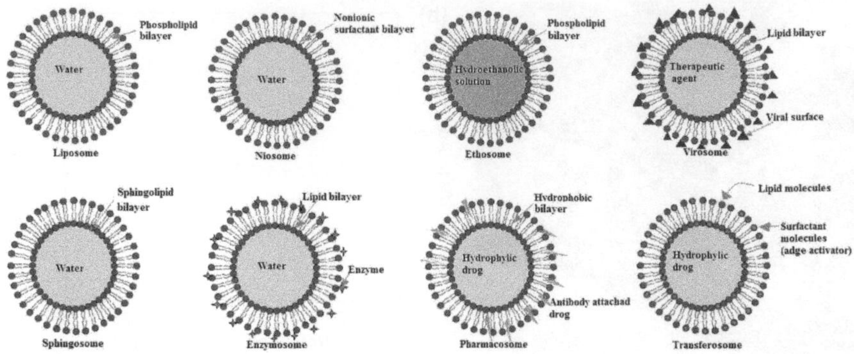


**FIGURE 5.2** Structure differences between some vesicular drug carriers.

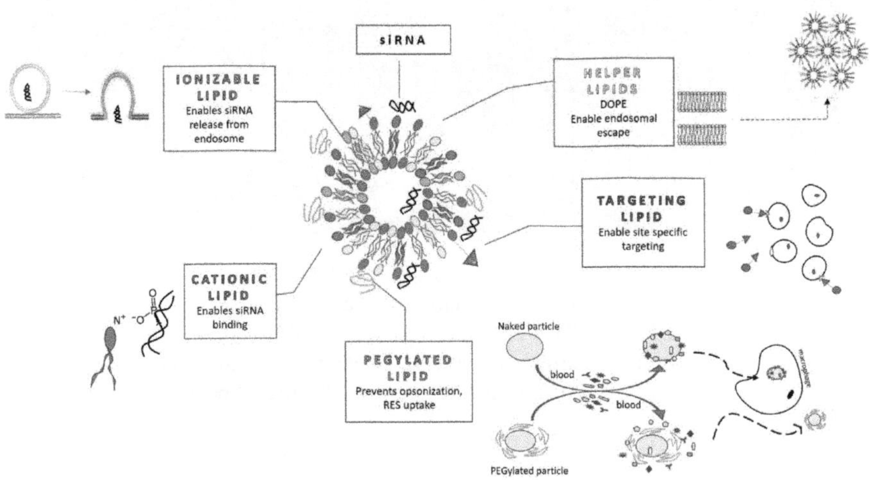


**FIGURE 6.2** Role of excipient in liposomal siRNA delivery vehicle.

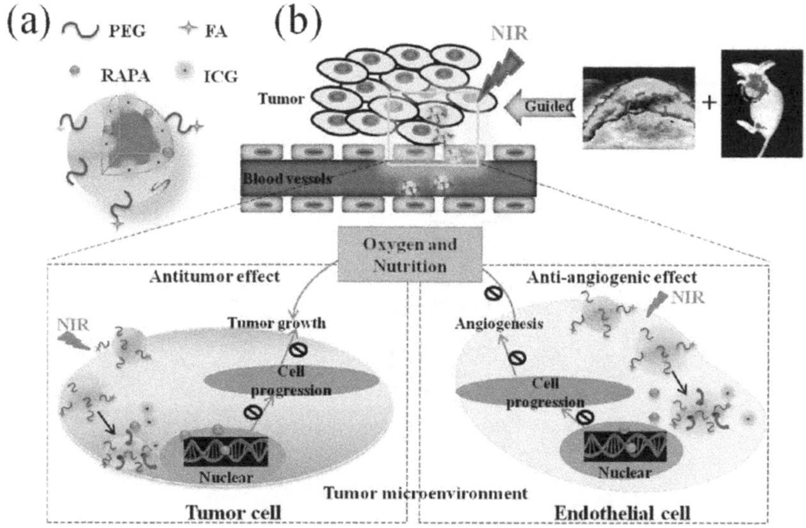


**SCHEME 7.1** Illustration of the theranostic TSLs: (a) structure of FA-ICG/RAPA-TSLs and (b) antitumor mechanism of FA-ICG/RAPA-TSLs.

[Reprinted with permission from Pang, X.; Wang, J.; Tan, X.; Guo, F.; Lei, M.; Ma, M.; Yu, M.; Tan, F.; Li, N. Dual-Modal Imaging-Guided Theranostic Nanocarriers Based on Indocyanine Green and mTOR Inhibitor Rapamycin. *ACS Appl. Mater. Interfaces* **2016,** *8*, 13819–13829. Copyright (2016) American Chemical Society.]

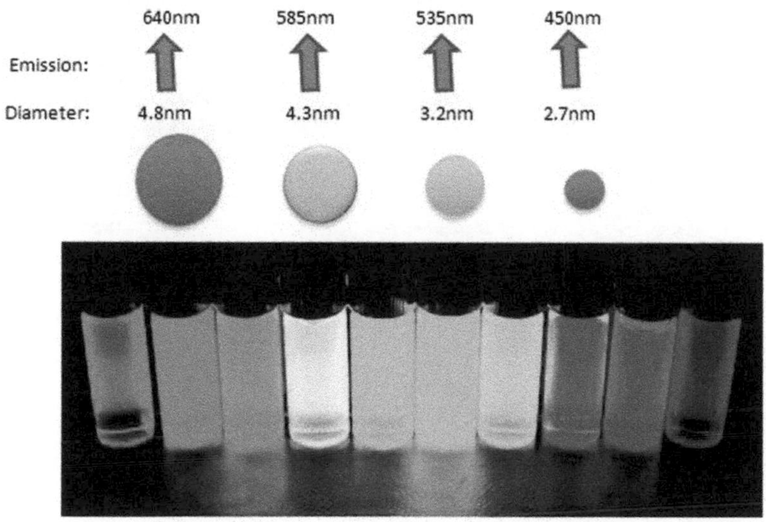


**FIGURE 9.1** Fluorescence emitted by different size quantum dots.

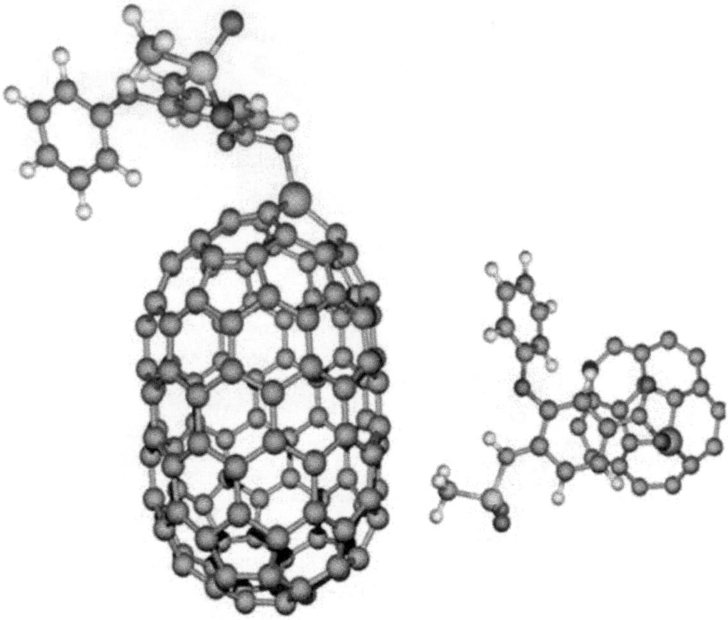

**FIGURE 12.8** Relaxed atomic structure of nimesulide on the cap region of CNTs. (Reprinted with permission from Zanella, 2007.)

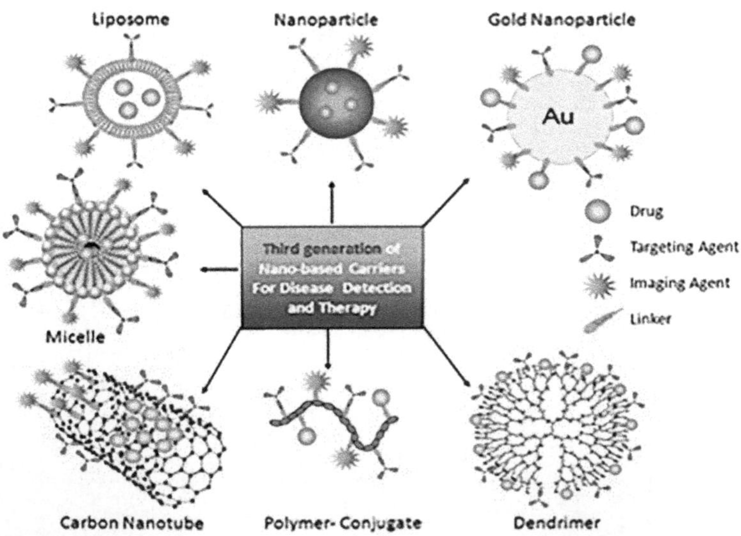


**FIGURE 15.1** Few nanocarriers used in drug delivery. (Adapted from http://slideplayer.com/slide/8751147/.)

# PART III
# Nanocarriers Derived from Carbon

## CHAPTER 9

# QUANTUM DOTS FOR DRUG DELIVERY

KOMAL SHARMA, ARUSHI VERMA, AYUSHI GUPTA, and NIDHI MISHRA*

*Indian Institute of Information Technology, Devghat, Jhalwa, Allahabad 211012, India*

*Corresponding author. E-mail: nidhimishra@iiita.ac.in*

## CONTENTS

## ABSTRACT

Due to their unique optical and surface properties, quantum dots (QDs) are explored for their potentials in biomedical applications. They have excellent photostability, high luminous intensity of fluorescence, which makes them desirable, and economic labels when compared to traditional organic dyes used in bioimaging and fluorescent bioassays. Conventional medicinal therapy suffers heavily due to nonspecificity in delivering drug molecule; this puts patients to discomfort due to acute drug side effects. The aftermath of chemotherapy in cancer treatment is one such example. Results of cadmium derived QD studies in biomedical applications have shown great potentials of cadmium quantum dots in cancer treatment and diagnosis. However, thorough investigation data of their in-vivo toxicity and pharmacokinetic activity must be procured and analyzed for their successful application in drug delivery. This chapter discusses properties, synthesis methods, and potential toxicity of QDs along with their applications in biomedical field in light of their promising potential as drug delivery carriers.

## 9.1 INTRODUCTION

Semiconductor quantum dots (QDs) are referred to as nanosized molecules that are made up of inorganic semiconducting material, for example, CdSe, Si, etc. They are approximately of 1 nm and contain less than 100 atoms which implies they are larger in size than other traditional molecular clusters. QDs are photostable and have a longer half-life than any other organic dyes and thus, it can be used effectively as dyes. This property has a great advantage in maintaining their brightness and makes it usable in targeted drug delivery, medical imaging, and monitoring. The QDs are limited in use due to narrow excitation wavelength of light and the overlapping of emitted fluorescence spectrum (Fig. 9.1).

QDs are hydrophobic initially after synthesis and might form a precipitate when exposed to aqueous environment which limits their biological applications. Despite of this shortcoming, the QDs are still used to locate tumor as well as track any metastatic growth in the body with the help of several imaging techniques. This helps in cancer detection, targeted drug delivery, and image guided surgery. QDs can also function as photosensitizers as they possess energy levels in the range of 1–5 eV and have the ability to absorb high energy photons from X-rays or gamma rays. This, in turn, helps in focusing and improving the radiation therapy techniques (Juzenas et al., 2008). QDs thereby possess the ability to enhance and improve the medical diagnostics and therapeutics. The only drawback of semiconductor-based QDs is their high toxicity, which has to be strictly kept in mind by the biologist or pharmacologist before exploiting it (Ghaderi et al., 2011).

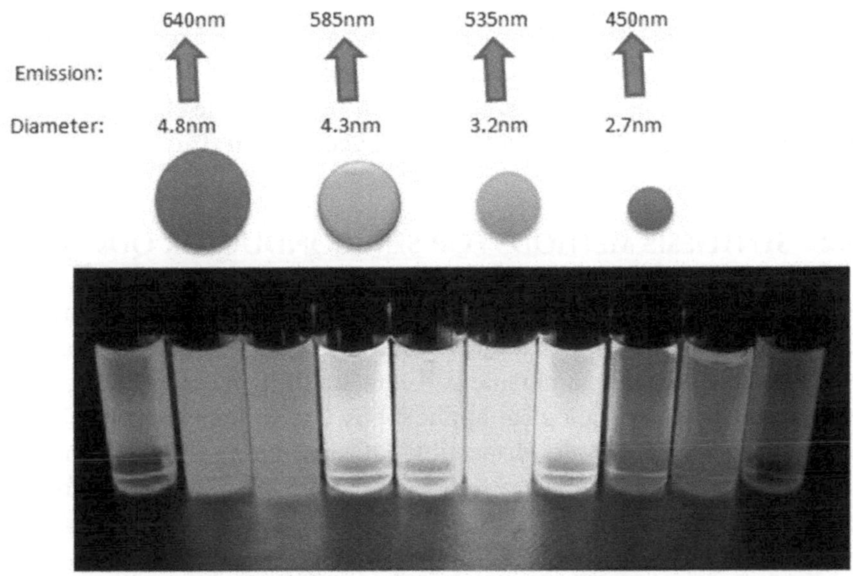


**FIGURE 9.1 (See color insert.)** Fluorescence emitted by different size quantum dots.

Carbon quantum dots (C-dots) on the other hand have high solubility, high resistance to photobleaching, and superior chemical inertness along with required facile modifications, as compared to traditional semiconductor QDs. The biocompatibility and low toxicity of C-dots make it even more superior and suitable for use in biological domain. C-dots have gained immense popularity these days due to their environment friendly nature, low cost, and low toxicity. Surface functionalization and passivation enables us to control their physical and chemical properties. C-dots find extensive use in the fields of bioimaging, nanomedicine, biosensing, and drug delivery, electrocatalysis, and photocatalysis (Lim et al., 2015; Wang and Hu, 2014).

Nanotechnology is expanding its domain in field of drug delivery systems by using these QDs as delivery agents in targeted delivery of drugs. Conventional drug delivery systems used gold nanoparticles (AuNPs) but their toxicity leads to a limited application as biological agents. Thiol group requirement for loading the drug further limited the use of AuNPs. Also, quenching of fluorophores in in-vivo systems causes difficulty in tracking the administered AuNPs. Thus, C-dots can be used as a good alternative to solve all the above problems and serve the same purpose with lesser complications. The ease of different functionalization available in C-dots enables us to understand and utilize many possibilities for conjugation with drug molecules in combination with other targeting agents. This flexibility in the system thereby expands the choices for drug delivery, making QDs a new and efficient vector for it (Lim et al., 2015).

## 9.2 SYNTHESIS METHODS FOR SEMICONDUCTOR QDs

Several methods are available for synthesis of QDs including organic and aqueous protocols (for highly luminescent nanocrystals formation). Nanocrystals prepared in organic solvents are of high quality and cannot be directly used for biological applications due to their hydrophobicity. Strongly fluorescent, water dispersible QDs are prepared by exchanging the ligand with hydrophilic moiety. Nonpolar organic solvents are considered to be the best for QD synthesis. Functionalization of QDs is done with secondary coating or capping material in order to make them water soluble. These materials include PEG, mercaptosuccinic acid, mercaptopropionic acid, etc. and are known to maintain the QDs in their monovalent form. Targeted drug delivery system includes these types of coating as their conjugation with targeting molecules such as antibodies, receptor ligands, and peptides facilitate the QD to stay in the targeted tissue or organ (Hu and Zhu, 2015).

### *9.2.1 HOT INJECTION SYNTHESIS*

Hot injection approach of QDs synthesis is being used these days for high-quality production. The process implies fast injection of a precursor into a hot solution comprising another precursor. A drawback associated with this approach is that it requires an instant homogeneous reaction which is hard to attain in large-scale reactions and thus, there are also difficulties in reproducing the process. Thereby this approach is avoided for large-scale synthesis (Hu and Zhu, 2015).

### *9.2.2 NON-INJECTION ORGANIC SYNTHESIS*

The major part of QD synthesis is the way to initiate the reaction. For example, a monodispersed QD requires formation of a uniform nanocrystals nucleus in a very short span of time. Due to the known limitations of fast-injection approach an alternative method has been developed for QD synthesis, that is, the noninjection approach. It uses two different precursors which are present in the system simultaneously before the start of the reaction at a certain temperature. A clear separation between growth and nucleation is required just as in injection approach for synthesis of monodispersed QDs. The elevated temperature maintained by a heating system with a certain temperature growth rate enables the growth of colloidal nanocrystals. This process enables the precursor to nucleate partially and also obtain synchronous nucleation and growth.

The basic process of QD formation requires both nucleation and growth process. The onset of nucleation is superseded by super saturation of the monomers. Consumption of additional monomers in the system is done after nuclear growth has taken place. The concentration of different monomers present determines the shape and size of the QD being formed. Effective nanocrystal growth is due to the contribution of both thermodynamic and kinetic properties of the system. For instance, slow growth limit under thermodynamic control leads to formation of nanocrystals with low aspect ratio and vice versa (Hu and Zhu, 2015).

### *9.2.3 SOLVOTHERMAL SYNTHESIS (ORGANIC MEDIUM) OF SEMICONDUCTOR NANOCRYSTAL QDs*

Solvothermal method is a widely used synthetic approach for nanoparticle synthesis. Surpassing the conventional methods of QD synthesis due to their inferior quality, this method provides an optimum temperature and pressure for nanocrystal growth. The methodology used is almost the same as normal batch synthesis. The process is initiated at a lower temperature by mixing of two different precursors. The temperature is then slowly increased as per the desired growth rate. Thus, the QDs are synthesized in a sealed autoclave providing unique conditions for desired results (Hu and Zhu, 2015).

### *9.2.4 AQUEOUS AND HYDROTHERMAL SYNTHESIS*

Aqueously synthesized QDs are widely used in biological industry as they possess excellent compatibility with water. Aqueous synthesis is preferred over organic-based synthesis as it is less expensive, less toxic, and also environment friendly. The only flaw associated with this type of synthesis is that the QDs obtained have low quantum yield and large and random size distribution. Further quality improvement may be done by surface modification, size selective precipitation, and selective photochemical etching. However, this would lead to a more complex and expensive approach of QD synthesis, which is undesirable in large-scale production.

The process involved in this type of synthesis includes mixing the two precursors comprising anions and cations at a considerably high temperature to react with each other. Nucleation and growth stages occur as in the conventional processes. Water is used instead of organic solvents in hydrothermal synthesis as the reaction medium. Thus occurs the formation of QDs. However, QDs thus formed are hard to possess desired properties for biological applications, so the production through this method is limited. Despite of these limitations, the lower cost, lower toxicity, and the environment friendly processing cannot be ignored altogether (Hu and Zhu, 2015).

## 9.3 SYNTHESIS METHODS FOR C-DOTS

The high toxicity of the semiconductor QDs even in relatively low levels is due to usage of heavy metals, for their production is a major limitation in biological domain. This leads to replacement of semiconductor-based QDs by C-dots. These C-dots are known to possess similar fluorescence properties with considerably low toxicity, biocompatibility, chemically inert behavior, and low-cost production. Some of the methods used for synthesis of C-dots are chemical ablation, electrochemical carbonization, laser ablation, solvothermal treatment, microwave irradiation, etc. In chemical ablation, small organic materials are carbonized by strong oxidizing acids to carbonaceous materials. These are later cut into thin sheets by using controlled oxidation. Bulk carbon materials are used as precursors to prepare C-dots via electrochemical soaking in electrochemical carbonization method. C-dots via laser ablation were first prepared by Sun and coworkers by using a carbon target in presence of water vapor and argon gas (as carrier) at a certain pressure and temperature. In solvothermal process, a hydrothermal reactor is used, which is sealed in with a solution of organic precursors, at high temperature. A rapid and a cost-effective process used for C-dot synthesis is microwave irradiation. It is a highly preferred method of synthesis for researchers (Wang and Hu, 2014).

Similarly, various other methods are also available for synthesis of both semiconductor and C-dots. The most common and extensively used methods have been briefly described in this chapter. There are also certain on-going researches on possible methods for large-scale production of QDs and are expected to give fruitful results.

## 9.4 PROPERTIES

QDs are fluorescent semiconductor crystals of zero-dimensional electron system varying from 1 to 10 nm in size, often called as "artificial atoms" due to their extremely less space occupancy. They are known for their exquisite quantum mechanical behavior. Their interesting electrical, optical, and catalytic properties stem from their shape, size, and electronic bandgap energy. QDs have electron-filled valence band which is separated from empty conduction band by bandgap. When a semiconducting material is illuminated with incident light above certain threshold value, the electrons absorb the incident light, become excited, and jump to higher energy level leaving positively charged holes in their places. These oppositely charged entities, holes and electrons, are attracted to one another by Coulomb's force and are collectively called as excitons. The distance between electrons in conduction band and holes in valence band is called Bohr's radius (rb), which is crucial for the optical properties of QDs as quantum confinement of spatial motion of electrons and holes takes place in QDs with radius lower than or comparable to their exciton Bohr radius. This results in increase in the electron–hole transition energy which contributes to blue shift and luminescence in QDs (Shukla, 2013). The energy levels of valence and conduction band in QDs are highly quantized and are directly related to the size of QDs (Samir et al., 2012). These semiconducting materials are named QDs after a characteristic phenomenon they exhibit, known as quantum confinement, which is responsible for their unique optical properties (Jamieson et al., 2007; Michalet et al., 2005; Azzazy et al., 2007). The excited electron stays in the higher energy level for nanoseconds after which it reverts to its hole in valence band emitting the absorbed light which is responsible for fluorescence of the QDs. However, surface imperfections in QDs hinder the movement of electrons back to their valence band temporarily and result in intermittent fluorescence called blinking which affects the quantum yield of QDs (Jamieson et al., 2007; Michalet et al., 2005).

Widely accepted types of QDs are ZnSe, CdSe, CdTe, and ZnS, metals such as In, Ga, and even natural materials such as curcumin, sucrose, etc. can also be used for the synthesis of QDs (Green and O'Brien, 1998; Ryvolova et al., 2011). QDs usually have a layer outside them which protects them from oxidation, strengthen their quantum yield, and also provides them photostability (Alivisatos et al., 2005; Jaiswal and Simon, 2004). Bare crystalline QDs are highly reactive due to large surface to volume ratio, consequently making them susceptible to surface imperfection and prone to losing their functional integrity upon taking part in undesired reactions. Capping of QDs with semiconductor, such as ZnS, having higher bandgap than that of core is done to overcome these problems of bare core QDs (Manna et al., 2002). The optical properties of QDs depend upon their shape and size as the small-sized QDs are excited at higher energy due to their larger bandgap compared to those of larger size QDs. Consequently, they emit light of higher energy compared to large size QDs while returning to valence band. This phenomenon is responsible for fluorescence of QDs. Due to their excellent optical properties, they have applications in optoelectronic and biomedical field. Furthermore, they can be used in multiplexed bioassays as they have wide-absorption spectra which enable QDs of varied colors to be excited by incident light of a single wavelength (Chan et al., 2002; Han et al., 2001). Also, they are known to have emission spectra in ultraviolet (UV) and infrared region (Jamieson et al., 2007); multiple colors and intensities of emitted light can be used to encode biomolecules such as proteins, genes, and other biomarkers (Han et al., 2001; Gao and Nie, 2003). QDs are desirable fluorophores than most of the conventional dyes in use because of their high photostability and comparatively long postexcitation lifetime (Arora et al., 2015; Wu et al., 2003; Chan and Nie, 1998). Organic fluorophores bleach after a few minutes of exposure to light, as a result of which their brightness decays shortly, whereas QDs have high level of photo bleaching threshold and hence have excellent photostability, which helps them retain their fluorescence even after hours of being exposed to light (Han et al., 2001; Alivisatos, 1996).

QDs are new-generation nanomaterials having wide applications in biomedical field. They have superior electrical and optical and transport properties suitable for use in diagnostics (as sensing and imaging probe), drug delivery, and targeted therapy. However for using them in therapeutics, their biocompatibility, cytotoxicity, and ability to conjugate with therapeutic agents must be evaluated. Surface passivation is done to take care of these issues. Water solubility and dispersibility are desirable properties in a QD for drug delivery purpose. QDs such as CdTe are dispersible in water, whereas some QDs such as CdSe, which are synthesized in organic solvent, are either sparingly soluble or insoluble in water. Such QDs are made water soluble by passivating their surface either electrostatically or following ligand exchange method. Chemicals such as mercaptopropionic acid, cystamine, or charged surfactants having capability to intercalate between hydrophobic ligands are used to make QDs water dispersible electrostatically (Chan and Nie, 1998). In ligand exchange method, ligands are exchanged with polysilanes or thiols groups (Bruchez et al., 1998). They can be made dispersible also by using bifunctional cross-linker molecules which connects from end to the QD surface and its other end is free to react with biomolecule; encapsulation of QDs in phospholipid micelles or coating QDs with polysaccharide layer is another method to improve their solubility (Nitin et al., 2004; Zhou et al., 2007). Mecracptocarbonic acid and 2-aminoethanethiol are two of the bifunctional cross-linker molecules used to enhance solubility of QDs (Wuister et al., 2003). Also, the QDs are layered with polymeric sequential shells around their core following layer-by-layer technique to enhance their stability and water dispersibility. In this method, the core of QD is surrounded by alternating charged layers which optimize the zeta potential while maintaining its absorbance characteristics (Jaffar et al., 2004).

## 9.5 CHARACTERIZATION

After QDs synthesis, they must be characterized for their size and photoluminescence capabilities. Usually UV–visible light and spectrophotometer is used for characterization of their photoluminescence. They exhibit light in visible region upon UV irradiation. Also, photoluminescence spectrometry can be used to study the absorption and emission characteristics of

QDs. All of these properties depend upon the size of QDs which is responsible for the purity of photoluminescence and spectral position.

Since QDs are nanosized materials, electron microscopy techniques are employed for their size characterization. Scanning electron microscope (SEM), transmission electron microscope (TEM), and dynamic light scattering (DLS) studies are done for their characterization. A complementary technique to conventional characterization of water-soluble QDs is field-flow fractionation technique (Rameshwar et al., 2006). For monitoring the epitaxial growth of QDs, AFM, TEM, and scanning tunneling microscope (STM) are used. STM can not only provide the information about the size of QDs but also is capable of providing their morphological data (Drbohalvova et al., 2009).

## 9.6 TOXICITY

Toxicity of QDs is an important parameter that must be evaluated for their in-vivo use in pharmaceuticals. QDs are particles with size in nanoscale, due to which they have tremendous surface energy, therefore less stability. To ensure their stability, they are functionalized with ligands such as carboxylic acid, amine, polyether, organophosphate, etc., which helps them from forming QD aggregates; these QD aggregates can range from small to large size. Functional integrity of QD changes when in aggregation in-vivo. Also, due to their small size they are capable to reach organelles which cannot be accessed otherwise with molecules of relatively larger size. This is a potential feature which renders QDs cytotoxicity as post cellular uptake of QDs, they are transported to the perinuclear space from cell periphery packaged in small vesicles (Salata, 2004).

For CdSe QDs, it has been shown that on exposure to air or upon UV irradiation QDs undergo oxidation reaction which results in the release of free cadmium ions. These free ions are toxic to liver cells and studies show they cause primary cell death in liver (Derfus et al., 2004). Studies also show that upon UV or air exposure they produce oxygen free radicals in the reaction mixture which are often called as reactive oxygen species (ROS). These ROS causes serious damage to biomolecules such as proteins, DNA, lipid, etc. It is most likely that free metal ions release from the core of QD and ROS produce work in unison to cause QD cytotoxicity (Lovrić et al., 2005). Most researched and popular QDs are CdSe QDs; toxicity data is available for CdSe QDs in both cell culture and animal models. Cell culture studies for QD toxicity showed that they are toxic to be used as they are made up of cadmium which is a carcinogen and they have a tendency to get accumulated inside the cell in organelles such as lysosomes and endosomes and thus get exposed to oxidative environment (Godt et al., 2006; Liu et al., 2015). However, animal studies for CdSe QDs show that they are less toxic in-vivo. This could be because of continuous movement of QDs inside the body of animals compared to cells in culture medium which is continuously exposed to fixed amount of QD concentration and their toxic by-products produced due to UV exposure and hostile reaction occurring due to physiological environment of the culture medium (Sayes et al., 2007).

C-dots are made from organic molecules. They are used in biosensing and bioimaging application and are known to have good stability. However, to be useful in in-vivo applications including imaging and drug delivery, their biocompatibility must be ensured. Biocompatibility has been tested for both bare and surface-passivated C-dots and it has been found that biocompatibility is still a crucial issue for surface-functionalized C-dots (Wang and Hu, 2014). Polyethylene glycol (PEG), poly (propionylethyleneimine-*co*-ethyleneimine) (PPEI-EI), polyethylenimine (PEI), branched poly (ethylenimine) (BPEI), and poly(acrylic acid)] (PAA) are used for passivation of C-dots to study their cytotoxicity. It has been found that PEG- and PPEI-EI-passivated C-dots are suitable for in vivo imaging and biosensing. However, higher cytotoxic molecules, such as BPEI and PAA, can still be used to functionalize C-dots used in-vivo at lower concentrations for shorter incubation time (Wang et al., 2011).

## 9.7 MODE OF ACTION OF QDs

### *9.7.1 QDs AS TAGS FOR OTHER DRUG CARRIERS*

Many polymers, for instance PEI and poly(lactic-*co*-glycolic acid) have been known to act as major drug carriers. Apart from these, a few inorganic materials too have been incorporated for the same. But these types of drug carriers have drawbacks which tend to constrain the innate signals in case of real-time imaging. A solution to this complication was taken care of by using organic fluorophores. However, tracking the signal for a long-duration and real-time imaging was unattainable. This was due to the property of photobleaching present in almost all organic dyes and fluorescent proteins.

The quick fix to all the obstacles stated above are QDs due to their exceptional spectral attributes. They have been used to tag drug carriers (both organic and inorganic) as well as viruses and bacteria (Chen et al., 2005; Tan et al., 2007; Jia et al., 2007; Akin et al., 2006). Several activities have been described in case of delivery related with siRNA and oligodeoxynucleotide (Qi and Gao, 2008).

Small interfering RNA (siRNA) or silencing RNA is a 25 base pair long double-stranded RNA molecule. It is generally involved in disrupting the mRNA for a particular gene that has a complementary sequence (Agarwal et al., 2003). Therefore the unwanted protein is not formed as faulty transcription or no transcription occurs. This property of siRNA has been exploited to get rid of undesirable and disease-causing protein to be formed.

The siRNA delivery to deeper tissues is not feasible and they require a vector through which it can be delivered. And to this, QDs have come to our aid. For example, delivery of siRNA is achieved through cotransfection along with lipofectamine and QDs (Chen et al., 2005). In a similar manner, highly homogeneous QDs were synthesized with chitosan nanobeads. This was done in order to achieve traceable siRNA delivery (Tan et al., 2007). Recent studies such as the carbon nanotubes which are coated with PEI, helpful in targeted RNA and DNA delivery, have break open a new area (Fig. 9.2; Jia et al., 2007).

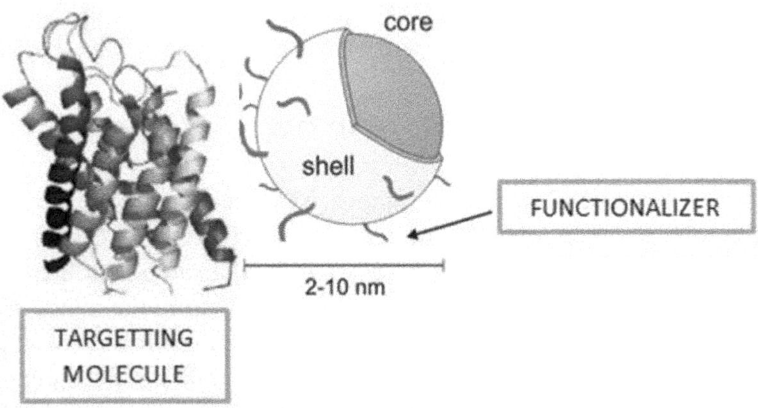


**FIGURE 9.2** Quantum dots used as drug delivery agents.

### 9.7.2 QDs AS CARRIERS WITH INTEGRATED FUNCTIONALITIES

QDs vary in size from 2 to 10 nm. As they then undergo polymer encapsulation, the size of these particles escalates to almost 10–20 nm in diameter. The result of size modulation of QDs is that particles with size less than 5 nm are filtered out through renal system and particles having larger size are absorbed ahead of their reaching the site of interest by the reticuloendothelial system. Thus, it is inferred that larger, solid mass of QDs cannot penetrate the tissues and reach the target site. Due to the above-stated difficulties, studies have been made to determine the perfect size of QDs so as to effectively deliver it to the target site. The large surface to volume ratio helps the QDs to have integrated numerous objectives in a relatively small size (Qi and Gao, 2008). This is why, QDs act as scaffold in which any hydrophobic molecule can be encapsulated. This insertion occurs between the amphiphilic layer of the polymer and the inorganic core. The molecule is held within by means of either covalent or noncovalent bonds toward the hydrophilic side of the polymer surface.

### 9.7.3 FLUORESCENT QDs

QDs and their optical properties have been defined by the particle size, surface chemistry, and the material that is being used. This is due to the presence of dangling bonds as the number of dangling bonds represents the nonradioactive neutralization (Dabbousi, 1997). Mostly materials such as CdSe and CdTe have been used for biological utilization. Although QDs of InP and InGaP are available in the market, their number of dangling bonds and other properties have to be amended according to the use (Resch-Genger et al., 2008). QDs act as nanocrystals with fluorophore properties. The fluorescence property of QDs is due to the excitation of electrons from the valence band and thus moving to the conduction band of the semiconductor material used. As a result of this, a hole is formed in the valence band. The electron–hole pair is then called an exciton. When the electron returns back and combines with the hole, ultimately releasing a photon. The color of the photon is modulated by changing the energy gap between the valence and the conduction band which is decided by the size of the nanocrystal. The generic notion being, smaller structures emit shorter wavelength. In general, materials such as CdSe/ZnS are being used. They mostly are applicable as secondary antibody conjugates. As QDs have a polymer coating, to become water soluble, they are subjected to zinc sulfide passivation. This facilitates the bioconjugation of molecules such as secondary antibodies, IgG, peptides, and others to the target molecules (Deerinck, 2008).

### 9.7.4 ENCAPSULATED QDs

The properties of QDs are defined by the surface chemistry and the specifications associated with it. Therefore to make the QDs' use feasible in the area of biological imaging, the surface properties are being modulated. It is achieved by changing the surface ligands and includes various others that are desirable, such as water solubility, functional molecule for targeted delivery, etc. There are mostly two ways in which this is done—by surface encapsulation and by surfactant exchange (Green, 2010; Medintz et al., 2005). During surfactant exchange, there occurs an exchange of the ligands present at the surface—generally hydrophobic to hydrophilic. The functional molecule that has to be attached is adhered to this new ligand. Thus, this enables the water solubility of the molecules and as well as desirable targeting (Taniguchi et al., 2016).

### 9.7.5 BIMODAL MOLECULAR IMAGING

In today's medical industry, imaging procedures such as computed tomography (CT), positron emission tomography (PET), ultrasound, and magnetic resonance imaging (MRI) are the most efficient in detection of critical diseases. But in most of the abovementioned diagnostic devices, the sensitivity decreases with increase in resolution. To overcome this, nowadays, people have started using these biomedical imaging modalities in several combinations. This enables us to use the best feature of one paired with that of the other. However, the problem still remains.

The characteristics of both resolution and sensitivity are encompassed in QDs. They are being used as fluorescent entities for various biomedical utilizations (Yu and Schanze, 2013). The usability of QDs in imaging devices is due to the high photostability against photobleaching and exceptionally high quantum yield (Kairdolf et al., 2013; Shao et al., 2011). Ions such as Mn and Gd, which are paramagnetic in nature, and nanoparticles of iron oxide are used to improve the quality of MR. Several compounds have been derived via ion doping, encapsulation, and bond conjugation (Liu et al., 2011). These changes in the design of the imaging modalities help in constructing a robust and stable device (Yang et al., 2015).

## 9.8 CONCLUSION

To conclude this chapter, we could say that QDs are being applied for the development in a diverse area of research. These new developments have helped us to achieve various objectives set before us in area of diagnosis and imaging. This has led us to acknowledge the various objectives of real-time imaging and the visualization of the vital activities that are taking place in the living organisms. Several new methods have emerged for labeling of clinical sample and therefore helping in diagnosing of the disease. QDs have also helped in the enhancement of drugs and several other drug delivery methods to bring out the best in the incorporation of different drug delivery methods. On the whole, we could say that QDs have helped us maximize the scope of several clinical applications and in the near future, it will provide the boost that is needed for breakthrough in several problems associated with the clinical research.

## KEYWORDS

- **bioimaging**
- **carbon quantum dots**
- **semiconductor quantum dots**
- **synthesis protocols for quantum dots**
- **targeted drug delivery**
- **toxicity**

## REFERENCES

Agarwal, N.; Dasaradhi, P. V. N.; Mohmmed, A.; Malhotra, P.; Bhatnagar, R. K.; Mukherjee, S. K. RNA Interference: Biology, Mechanism, and Applications. *Microbiol. Mol. Biol. Rev.* **2003,** 67(4), 657–685.

Akin, D.; Sturgis, J.; Ragheb, K. Bacteria-mediated Delivery of Nanoparticles and Cargo into Cells. *Nat. Nanotechnol.* **2006,** *2*, 441–449.

Alivisatos, A. P. Semiconductor Clusters, Nanocrystals, and Quantum Dots. *Science* **1996,** *271*, 933–937.

Alivisatos, A. P.; Gu, W.; Larabell, C. Quantum Dots as Cellular Probes. *Annu. Rev. Biomed. Eng.* **2005,** *7*, 55–76.

Arora, S.; Latwal, M.; Singhal, S.; Agarwal, S.; Reddy, K. M.; Kumar, D. Quantum Dots: A Potential Candidate as a Biomedical Material. *J. Chem. Pharm. Res.* **2015,** *7*(4), 810–814.

Azzazy, H. M.; Mansour, M. N.; Kazmierczak, S. C. From Diagnostics to Therapy: Prospects of Quantum Dots. *Clin. Biochem.* **2007,** *40*(13–14), 917–927.

Bruchez, M. J.; Moronne, M.; Gin, P.; Weiss, S.; Alivisatos, S. P. Semiconductor Nanocrystals as Fluorescent Biological Labels. *Science* **1998,** *281*(5385), 2013–2016.

Chan, W. C.; Nie, S. Quantum Dot Bioconjugates for Ultrasensitive Nonisotopic Detection. *Science* **1998,** *281*(5385), 2016–2018.

Chan, W. C.; Maxwell, D. J.; Gao, X.; Bailey, R. E.; Han, M.; Nie, S. Luminescent Quantum Dots for Multiplexed Biological Detection and Imaging. *Curr. Opin. Biotechnol.* **2002,** *13*(1), 40–46.

Chen, A. A.; Derfus, A. M.; Khetani, S. R.; Bhatia, S. N. Quantum Dots to Monitor RNAi Delivery and Improve Gene Silencing. *Nucleic Acids Res.* **2005,** *33*(22), 190.

Dabbousi, B. O. (CdSe) ZnS Core–Shell Quantum Dots: Synthesis and Characterization of a Size Series of Highly Luminescent Nanocrystallites. *J. Phys. Chem. B* **1997,** *101*, 9463–9475.

Deerinck, T. J. The Application of Fluorescent Quantum Dots to Confocal, Multiphoton, and Electron Microscopic Imaging. *Toxicol. Pathol.* **2008,** *36*(1), 112–116.

Derfus, A. M.; Chan, W. C. W.; Bhatia, S. N. Probing the Cytotoxicity of Semiconductor Quantum Dots. *Nano Lett.* **2004,** *4*(1), 11–18.

Drbohalvova, J.; Adam, V.; Kizek, R.; Hubalek, J. Quantum Dots—Characterization, Preparation and Usage in Biological Systems. *Int. J. Mol. Sci.* **2009,** *10*(2), 656–673.

Gao, X.; Nie, S. Doping Mesoporous Materials with Multicolor Quantum Dots. *J. Phys. Chem. B* **2003,** *107*(42), 11575–11578.

Ghaderi, S.; Ramesh, B.; Seifalian, A. M. Fluorescence Nanoparticles "Quantum Dots" as Drug Delivery System and Their Toxicity: A Review. *J. Drug Target.* **2011,** *19*(7), 475–486.

Godt, J.; Scheidig, F.; Gross-Siestrup, C.; Esche, V.; Brandenburg, P.; Reich, A.; Groneberg, D. A. The Toxicity of Cadmium and Resulting Hazards for Human Health. *J. Occup. Med. Toxicol.* **2006,** *1*, 22.

Green, M. The Nature of Quantum Dot Capping Ligands. *J. Mater. Chem.* **2010,** *20*, 5797–5809.

Green, M.; O'Brien, P. A Novel Metalorganic Route for the Direct and Rapid Synthesis of Monodispersed Quantum Dots of Indium Phosphide. *Chem. Commun.* **1998,** *22*, 2459–2460.

Han, M.; Gao, X.; Su, J. Z.; Nie, S. Quantum-dot-tagged Microbeads for Multiplexed Optical Coding of Biomolecules. *Nat. Niotechnol.* **2001,** *19*(7), 631–635.

Hu, M. Z.; Zhu, T. Semiconductor Nanocrystal Quantum Dot Synthesis Approaches Towards Large-scale Production for Energy Applications. *Nanoscale. Res. Lett.* **2015,** *10*, 469.

Jaffar, S.; Nam, K. T.; Khademhosseini, A.; Xing, J.; Langer, R. S.; Belcher, A. M. Layer-by-layer Surface Modifications and Patterned Electrostatic Deposition of Quantum Dots. *Nano Lett.* **2004,** *4*(8), 1421–1425.

Jaiswal, J. K.; Simon, S. M. Potentials and Pitfalls of Fluorescent Quantum Dots for Biological Imaging. *Trends Cell. Biol.* **2004,** *14*(9), 497–504.

Jamieson, T.; Bakhshi, R.; Petrova, D.; Pocock, R.; Imani, M.; Seifalian, A. M. Biological Applications of Quantum Dots. *Biomaterials* **2007,** *28*(31), 4717–4732.

Jia, N.; Lian, Q.; Shen, H.; Wang, C.; Li, X.; Yang, Z. Intracellular Delivery of Quantum Dots Tagged Antisense Oligodeoxynucleotides by Functionalized Multiwalled Carbon Nanotubes. *Nano Lett.* **2007,** *7*(10), 2976.

Juzenas, P.; Chen, W.; Sun, Y. P.; Coelho, M. A.; Generalov, N.; Christensen, I. L. Quantum Dots and Nanoparticles for Photodynamic and Radiation Therapies of Cancer. *Adv. Drug Deliv. Rev.* **2008,** *60*(15), 1600–1614.

Kairdolf, B. A.; Smith, A. M.; Stokes, T. H.; Wang, M. D.; Young, A. N.; Nie, S. M. Semiconductor Quantum Dots for Bioimaging and Biodiagnostic Applications. *Annu. Rev. Anal. Chem.* **2013,** *6*, 143–162.

Lim, S. Y.; Shen, W.; Gao, Z. Carbon Quantum Dots and Their Applications. *Chem. Soc. Rev.* **2015,** *44*, 362–381.

Liu, Y. L.; Ai, K. L.; Yuan, Q. H.; Lu, L. H. Fluorescence-enhanced Gadolinium-doped Zinc Oxide Quantum Dots for Magnetic Resonance and Fluorescence Imaging. *Biomaterials* **2011,** *32*, 1185–1192.

Liu, Q.; Li, H.; Xia, Q.; Liu, Y.; Xiao, K. Role of Surface Charge in Determining the Biological Effects of CdSe/ZnS Quantum Dots. *Int. J. Nanomed.* **2015,** *10*, 7073–7088.

Lovrić, J.; Cho, S. J.; Winnik, F. M.; Maysinger, D. Unmodified Cadmium Telluride Quantum Dots Induce Reactive Oxygen Species Formation Leading to Multiple Organelle Damage and Cell Death. *Chem. Biol.* **2005,** *12*(11), 1227–1234.

Manna, L.; Scher, E. C.; Li, S. C.; Alivisatos, A. P. Epitaxial Growth and Photochemical Annealing of Graded CdS/ZnS Shells on Colloidal CdSe Nanorods. *J. Am. Chem. Soc.* **2002,** *124*(24), 7136–7145.

Medintz, I. L.; Tetsuo Uyeda, H.; Goldman, E. R.; Mattoussi, H. Quantum Dot Bioconjugates for Imaging, Labeling and Sensing. *Nat. Mater.* **2005,** *4*, 435–446.

Michalet, X.; Pinaud, F. F.; Bentolila, L. A.; Tsay, J. M.; Doose, S.; Li, J. J.; Sundaresan, G.; Wu, A. M.; Gambhir, S. S.; Weiss, S. Quantum Dots for Live Cells, In Vivo Imaging, and Diagnostics. *Science* **2005,** *307*(5709), 538–544.

Nitin, N.; LaConte, L. E. W.; Zurkiya, O.; Hu, X.; Bao, G. Functionalization and Peptide-based Delivery of Magnetic Nanoparticles as an Intracellular MRI Contrast Agent. *J. Biol. Inorg. Chem.* **2004,** *9*(6), 706–712.

Qi, L.; Gao, X. Emerging Application of Quantum Dots for Drug Delivery and Therapy. *Expert Opin. Drug Deliv.* **2008,** *5*(3), 263–267.

Rameshwar, T.; Samal, S.; Lee, S.; Kim, S.; Cho, J.; Kim, I. S. Determination of the Size of Water-soluble Nanoparticles and Quantum Dots by Field-flow Fractionation. *J. Nanosci. Nanotechnol.* **2006,** *6*(8), 2461–2467.

Resch-Genger, U.; Grabolle, M.; Cavaliere-Jaricot, S.; Nitschke, R.; Nann, T. Quantum Dots Versus Organic Dyes as Fluorescent Labels. *Nat. Methods* **2008,** *5*, 763–775.

Ryvolova, M.; Chomoucka, J.; Janu, L.; Drbohlavova, J.; Adam, V.; Hubalek, J.; Kizek, R. Biotin-modified Glutathione as a Functionalized Coating for Bioconjugation of CdTe-based Quantum Dots. *Electrophoresis* **2011,** *32*(13), 1619–1622.
Salata, O. V. Applications of Nanoparticles in Biology and Medicine. *J. Nanobiotechnol.* **2004,** *2*, 3.
Samir, T. M.; Mansour, M. M. H.; Kazmierczak, S. C.; Azzazy, H. M. E. Quantum Dots: Heralding a Brighter Future for Clinical Diagnostics. *Nanomedicine* **2012,** *7*(11), 1755–1769.
Sayes, C. M.; Reed, K. L.; Warheit, D. B. Assessing Toxicity of Fine and Nanoparticles: Comparing In Vitro Measurements to In Vivo Pulmonary Toxicity Profiles. *Toxicol. Sci.* **2007,** *97*(1), 163–180.
Shao, L. J.; Gao, Y. F.; Yan, F. Semiconductor Quantum Dots for Biomedical Applications. *Sensors* **2011,** *11*, 11736–11751.
Shukla, S. K. Recent Development in Biomedical Applications of Quantum Dots. *Adv. Mater. Rev.* **2013,** *1*(1), 2–12.
Tan, W. B.; Jiang, S.; Zhang, Y. Quantum-dot Based Nanoparticles for Targeted Silencing of HER2/neu Gene Via RNA Interference. *Biomaterials* **2007,** *28*(8), 1565.
Taniguchi, S.; Sandiford, L.; Cooper, M.; Rosca, E. V.; Khanbeigi, R. A.; Fairclough, S. M.; Thanou, M.; Dailey, L. A.; Wohlleben, W.; Vacano, B. V.; de Rosales, R. T. M.; Dobson, P. J.; Owen, D. M.; Green, M. Hydrophobin-encapsulated Quantum Dots. *ACS Appl. Mater. Interfaces* **2016,** *8*(7), 4887–4893.
Wang, Y.; Hu, A. Carbon Quantum Dots: Synthesis, Properties and Applications. *J. Mater. Chem.* **2014,** *2*, 6921–6939.
Wang, Y.; Anilkumar, P.; Cao, L.; Liu, J. H.; Luo, P. G.; Tackett, K. N. 2nd.; Sahu, S.; Wang, P.; Wang, X.; Sun, Y. P. Carbon Dots of Different Composition and Surface Functionalization: Cytotoxicity Issues Relevant to Fluorescence Cell Imaging. *Exp. Biol. Med. (Maywood)* **2011,** *236*(11), 1231–1238.
Wu, X.; Liu, H.; Liu, J.; Haley, K. N.; Treadway, J. A.; Larson, J. P.; Ge, N.; Peale, F.; Bruchez, M. P. Immunofluorescent Labeling of Cancer Marker Her2 and Other Cellular Targets with Semiconductor Quantum Dots. *Nat. Biotechnol.* **2003,** *21*(1), 41–46.
Wuister, S. F.; Swart, I.; Driel, F. V.; Hickey, S. G.; Donega, C. D. M. Highly Luminescent Water-soluble CdTe Quantum Dots. *Nano Lett.* **2003,** *3*(4), 503–507.
Yang, W.; Guo, W.; Gong, X.; Zhang, B.; Wang, S.; Chen, N.; Yang, W.; Tu, Y.; Fang, X.; Chang, J. Facile Synthesis of Gd–Cu–In–S/ZnS Bimodal Quantum Dots with Optimized Properties for Tumour Targeted Fluorescence/MR In Vivo Imaging. *ACS Appl. Mater. Interfaces* **2015,** *7*(33), 18759–18768.
Yu, K.; Schanze, K. S. Preface: Forum on Biomedical Applications of Colloidal Photoluminescent Quantum Dots. *ACS Appl. Mater. Interfaces* **2013,** *5*(8), 2785–2785.
Zhou, M.; Nakatani, E.; Gronenberg, L. S.; Tokimoto, T.; Wirth, M. J.; Hruby, V. J.; Roberts, A.; Lynch, R. M.; Ghosh, I. Peptide-labeled Quantum Dots for Imaging GPCRs in Whole Cells and as Single Molecule. *Bioconjug. Chem.* **2007,** *18*(2), 323–332.
Zunger, A. Semiconductor Quantum Dots. *MRS Bull.* **1998,** *23*(2), 15–17.

**CHAPTER 10**

# GRAPHENE AND GRAPHENE-BASED MATERIALS: SYNTHESIS, CHARACTERIZATION, TOXICITY, AND BIOMEDICAL APPLICATIONS

GAZALI, SANDEEP KAUR, ARJU DHAWAN, and INDERBIR SINGH*

*Department of Pharmaceutics, Chitkara College of Pharmacy, Chitkara University, Rajpura 140401, Patiala, Punjab, India*

**Corresponding author. E-mail: inderbirsingh@gmail.com*

## CONTENTS

## ABSTRACT

Graphene is the newly discovered nanomaterial which has a wide use in drug delivery and newer drug formulations. It is a single layer of $sp^2$ bonded carbon atoms and has various properties that make graphene a unique and a widely accepted substance in drug delivery. Various researches have been made in this area to develop uses and applications of graphene in drug delivery. One of the major approaches made is the formulation of graphene as nanocarriers in the form of graphene quantum dots (GQDs). Apart from these various derivatives of graphene, such as monolayer and bilayer graphene, graphene oxide, a graphene nanomaterial, has also been synthesized which have found to contribute efficiently in the application of drug and gene delivery. In this chapter, we have discussed various methods of synthesis of all these approaches including their characterization techniques and various properties that they exhibit, which makes them a potential candidate in drug and gene delivery. Nanocarriers of graphene and various forms of graphene are also known to cause certain toxicities to various organs which are discussed and marked in this chapter which potentially cover all the major toxicities that are caused by forms such as GQDs and graphene oxide. A plethora of applications of graphene and its derivatives in gene and drug delivery have been exclusively discussed including their uses in biomaging and biosensing. To put an end, future perspectives of graphene and graphene-generated forms have also been discussed.

## 10.1 INTRODUCTION

Graphene is the first truly found nanomaterial ever discovered and it came to stardom since being first outlayed by Andre K. Geim and Nobel Laureate Konstantin Novoselov of University of Manchester in 2004. Graphene is a single layer of carbon atoms, which is arranged in a flat honey-combed lattice. Graphene is a single layer of $sp^2$ bonded carbon atoms; it has high ballistic transport, current density, high thermal conductivity, chemical inertness, and super hydrophobicity. Graphene is one of the prodigy materials in 21st century. Introduction of graphene into the drug delivery forms one of its biomedical applications. Graphene along with its type graphene oxide (GO) has been widely and extensively used for drug delivery purposes. The above-expressed properties of graphene and its forms along with its good biocompatibility endow their promising use in the advancement of drug delivery systems and the delivery of broad range of therapeutics. Various researches have been made for the application of graphene into the drug delivery systems. One such approach is the use of graphene and GO as nanocarriers in drug delivery as an application. Additionally, there are new drug delivery concepts based on controlling various mechanisms which include targeting and stimulation with pH, thermal-, photo-, and magnetic induction, and chemical interactions.

In the past few years, it has been observed that graphene is being used as a substitution of silicon or as the new and next genre of organic solar cells and also as the material in fast-charging batteries. It is regarded as one of the strongest materials that have been unearthed yet (Bianco, 2016).

Graphene was primarily discussed in the year 1946 by Philip Russel Wallace, a Canadian physicist. He studied graphite, the stuff in the pencil lead and ended up making the three-dimensional honeycomb structural graphene. It is a 2D grid of carbon atoms exhibited in a hexagonal structure. In 2004, Geim and Novoselov while playing with graphite cubes and scotch tape found out that graphite separated into thinner and thinner flakes and concluded up with graphite flakes with only one atom thick (materials do not get any thinner than one atom thick, therefore, this stuff was close to two dimension as physically as possible).

Novoselov and colleagues made revolution in scientific community. Physicists wanted to know how the laws of physics transformed when the electrons were forced to survive in two dimensions instead of three. It is observed that graphene has a distinctive band structure that forces electrons in the materials to travel in the same speed and never lets them to terminate. This is the phenomenon which makes them behave a lot like light. As graphene is so thin therefore, it is known as transparent electrode and graphite is an enormously good conductor and responds quickly to applied voltage (Novoselov et al., 2012).

Applications of graphene in drug delivery and biomedicine were first studied by the Hongjie Dai group at Stanford University, the United States in the year 2008. Branched structure of polyethylene glycol was used to fabricate GO making PEGylated nanographene which has a higher stability in physiological solutions. Graphene surface was used for the loading of aromatic cancer drugs for intracellular drug delivery.

The in vivo study of graphene was recently reported in which PEGylated nanographene was docketed by near infrared (NIR) fluorescence dye for the purpose of in vivo imaging. Therefore, the recent knowledge and discovery of graphene has set a bench mark in the use and application of graphene in the drug discovery.

Great interest in graphene has been observed in various fields including drug and gene delivery. It has various properties such as ultrahigh surface area, easy surface functionalization, and also a single-layered graphene is extensively used for drug and gene delivery. Due to their intrinsic high near-infrared absorbance, graphene and several of its derivatives have been examined to be great candidates for cancer photothermal, chemo-dynamic, and photodynamic therapies. The proper surface functionalization and controlled sizes of nanomaterials are the two factors which are crucial for drug and gene delivery. A lot of successful deliveries of anticancer drugs and genes have been achieved by the use of graphene-based nanomaterials as carriers, but more information on its surface functionalization, size distribution, and also therapeutic outcomes need to be discovered.

## 10.2 SYNTHESIS OF GRAPHENE

Graphene can be synthesized by various methods. There are mainly two methods for the synthesis of graphene. These are basically two approaches that have been laid down, these are:

1. Top-down approach
2. Bottom-up approach

### *10.2.1 TOP-DOWN METHOD*

The top-down method usually involves the synthesis of graphene from graphite. This method takes place via the oxidative exfoliation of graphite. Also thermal expansion of graphite oxide is carried out which is useful for unzipping the carbon nanotubes (CNTs) which on treatment with acid reactions, liquid ammonia, plasma treatment, and catalytic approaches results in elongated graphene strips. It usually involves the electrochemical oxidation that controls the accuracy and degree of synthesis of single-layered graphene, an electrochemical method which reduces GO under the application of constant potential. GO can also be reduced by radiation induction by using sunlight, UV radiations, and laser. These were the exfoliation methods of graphene. Apart from these graphene, can also be prepared by the exfoliation–reinteraction–expansion of graphite, wherein it is treated with oleum. Photoexfoliation is a method in which graphene monolayers are produced which are free from defects and contamination. Various methods are discussed below to carry out the reduction of GO in order to produce graphene by top-down method (Magan, 2015).

#### *10.2.1.1 ELECTROCHEMICAL OXIDATION*

GO was primarily exfoliated in nanopure distilled water. For instance, 20 mg of GO was combined with 20 mL water and then it was ultrasonicated for 1 h. The homogenous yellow-brown GO solution was resulted. This solution can remain stable for longer periods as the surface of GO is negatively charged. This solution was then applied to the prepolished electrodes made of glassy carbon or gold disk and then, these were permitted to dry in air. To these, 10 μL of 0.05 wt% Nafion was then casted, and permitted to dry. The electrolytic reduction was negotiated in cyclic volammetry in 0.1 M sodium sulfate solution which was put in a standard electrode having three cells which contain mercury/mercury sulfate and platinum foil as the reference and counter electrode, respectively (Hummers and Offeman, 1958).

#### *10.2.1.2 RADIATION INDUCTION*

A stable colloidal suspension is formed when GO is mixed with water. This aqueous suspension was then put through to ultrasonic analysis and single-layered GO was formed. The solution of GO was then bared to sunlight for a few hours. In order to treat the solution under UV radiation, it was irradiated with a Philips low-pressure mercury lamp (254 nm, 25 W, 90 $\mu W/cm^2$). A KrF excimer laser (248 nm, 5 Hz) was subjected to illuminate the aqueous solution of GO which is taken in a vial. Then a metallic slit of aluminum was removed during laser illumination of solution because it usually tends to give a rectangular beam. Thus, the laser is made uniform over that area where there is presence of GO (Jiao et al., 2009; Kosynkin et al., 2009; Kumar et al., 2012).

### 10.2.2 BOTTOM-UP APPROACH

Bottom-up approach is not a conventional method for the synthesis of graphene and other useful materials used for the production of graphene-based substances. Chemical vapor deposition (CVD) is a method used under bottom-up approach and it necessitates a transfer process which involves the use of solvents, water solutions, and polymers which includes the introduction of defects such as vacancies or grain boundaries and graphene multilayers. Another method is the atomic layer deposition which is used to instantly synthesize layers of oxide on the top of graphene. This primarily requires chemical modification of graphene to avoid the problems associated with wetting. In this process, again, impurities are introduced or cause cleavages in the C–C bond which leads to notable electron mobility degradation. There are more novel strategies that serve as the alternative for the production of graphene which are based on the deposition of Hf and Si and succeeding oxidation have been adopted which exists in combination with expensive transition metal supports say, for instance, platinum and rubidium (Omiciuolo et al, 2014).

Graphene has various properties that are useful in many areas of this era. As graphene has a unique $sp^2$ configuration and various characteristics, it is possible to use graphene in almost every new program. Graphene has found great space in the application of drug delivery due to its fundamental properties which makes it a good carrier for the delivery of various drugs in various physical forms. The properties of graphene that give us such an advantage are discussed here along with some of the properties that enable graphene to be used in almost any research area. Graphene has many distinctive properties such as it has high carrier mobility at room temperature itself, a good optical transparency along with high Young's modulus, and eminent thermal conductivity ranging between 3000 and 5000 W/m K. These properties make graphene easily integrated into polymers and make composites that have enhanced properties than the original or pure polymers. Graphene has a tendency to exist in several physical forms and the most useful form of graphene used for drug delivery is the nanoparticles form. The advantage provided by graphene when existing in nanoform is primarily that it has a large surface to volume ratio which results in enhanced penetration sites. The most important property is the good biocompatibility of graphene. Graphene in its forms such as GO, graphene nanoparticles (GFNs), graphene quantum dots (GQDs), etc. has a very good biocompatibility with the human body. A distinctive property of graphene is the immobilization of a large number of substances, for example, drugs, metals, biomolecules, cells, and fluorescent probes. This property is being used nowadays to develop newer approaches where graphene can be a useful candidate. Therefore, graphene is emerging as a new and preferable candidate for drug delivery and other applications as it has a wide variety of properties which enable it to enter any research area. As it has a unique structure and properties, graphene is attracting the attention of both practical application and fundamental research communities. The emerging graphene-based products have promising features for a variety of applications including drug delivery.

## 10.3 FORMS OF GRAPHENE

Graphene is available in various forms which provide thousands of applications in today's world. One or the other form of graphene is widely used in today's industrial era. Some of the forms of graphene are discussed below.

### *10.3.1 MONOLAYER SHEETS*

In the year 2013, Polish scientists observed a unit that was capable of forming continuous monolayer sheets as graphene. The process on which this presentation was based was the growth of metallic liquid matrix. The product of this process was high strength metallurgical graphene.

The main comparison of this process over the others is that it allows manufacturing defect-free graphene structures. The formation of monolayer sheets has found a great advantage in the application and use of graphene to enhance the electrical properties of a substance or a polymer. Therefore, this form of graphene has more usefulness toward conducting applications as compared to drug delivery.

### *10.3.2 BILAYER GRAPHENE*

Bilayer graphene is advance and is used in the electromechanical areas where it displays the anomalous quantum Hall effect that is caused by transverse conductivity formed by the magnetic field. It has optoelectronic and nanoelectronic applications. It is either found in twisted configurations or in graphitic Bernal-stacked configurations.

CVD is the most widely used way or technique of synthesizing bilayer graphene and also for synthesizing various other forms of graphene. In this method, graphene is revealed to one or more volatile precursors such as methane which gets reacted to the surface of the substrate (in this case, graphene) and a bilayer graphene is synthesized under suitable conditions of temperature and pressure. Similar to the monolayer form of graphene, bilayer graphene also find its most of the uses in the approach of graphene toward electrical conductivity rather than in drug delivery (Barlas et al., 2010; Min et al., 2007).

### *10.3.3 GRAPHENE SUPERLATTICES*

The graphene structure can be spontaneously synthesized or altered in the form of a crystal lattice and can prove to be of huge significance. Thus, proving that with the change in the conditions, graphene has the capability of changing its shape. Therefore, this knowledge is applied to examine the changes in the shape of graphene when it is exposed to toxins and further finding out new ways or techniques of detection of the diseases. It is examined that particles exhibiting magnetic behavior are widely used for drug delivery operations but they cannot always be used as these can be toxic under various physiological conditions; thus, making them a suitable material for drug delivery as they do not contain any magnetic property. Furthermore, it is estimated that graphene can be used for a new kind of drug delivery systems.

#### *10.3.3.1 SYNTHESIS*

CVD method was used to grow twisted bilayer graphene on copper foils in a furnace of quartz at an ambient pressure. To assist the growth of bilayer graphene, larger flow rate of methane was used. The two layers of graphene have relative rotation angles from nearly 0° to 3°. Transmission electron micoscope (TEM) and scanning tunneling microscope (STM) were the methods used to study the confirmation of lattice rotation by which electron diffraction and lattice patterns can be observed.

Such bilayer graphene structures as compared to the structures observed before show that the misorientation of the two hexagons reveals their actual lattice rotation.

#### *10.3.3.2 PROPERTIES*

The graphene superlattices are the forms of graphene that have a great use in the upcoming novel applications in drug delivery. As the graphene superlattices have optical properties, they are widely being used in many biological applications. These optical properties are useful in monitoring the delivery of the drug into the body. This property is very useful in the cases when the drug delivery in chemotherapy for brain tumors and for the delivery of drugs in central nervous system is to be monitored (Wang et al., 2013).

### *10.3.4 GRAPHENE QUANTUM DOTS*

Quantum dots are a new revolution in drug delivery. The discovery of quantum dots has led to the evolution of drug delivery through nanoparticles. Quantum dots are small sized and allow a versatile surface area for the delivery and absorption of the drug; their small size allows their incorporation within virtually any nanodrug delivery system and has minimal effect on overall characteristics. Graphene-mediated quantum dots or quantum dots have various biomedical applications that focus on sensing, molecular imaging, and other optical properties. Size of the quantum dots vary between 2 and 10 nm, which when encapsulated with polymer generally increases up to 5–20 nm in diameter. The quantum dots that have the size of about 5 nm can be cleared easily by renal filtration, whereas the bigger particles before reaching the desired sites are in due course bound up by the reticuloendothelial system; also, larger particles cannot easily penetrate into the solid tissues.

Quantum dots are a form of traceable drug delivery as they are photostable fluorescent reporters they are labeled over the conventional drug carrier. The drug carriers are usually made up of polymers such as polyethylenimine (PEI) and poly(lactic-co-glycolic) acid and others are based on inorganic materials. The sole limitation of this type is the abridgement of congenital signal by perennial and actual-time imaging of drug transport.

GQDs are the ones that are made of both graphene and carbon dots. The quantum confinement and edge effects of carbon dots are combined with the structure of graphene to obtain GQDs. GQDs have the characteristics of both graphene and carbon dots. Mainly, the GQDs are used in the therapy and drug delivery of cancer. These are used in targeted quantum dots, magnetic imaging, and drug delivery are studied for fluorescent imaging. Graphene dots are usually synthesized by microwave-assisted hydrothermal method, soft template method, the ultrasonic exfoliation method, the chemical synthesis, and electron beam lithography method.

#### 10.3.4.1 SYNTHESIS

##### 10.3.4.1.1 Microwave-Assisted Hydrothermal Method

In the microwave-assisted hydrothermal method, GQDs are prepared by the hydrothermal reaction at 190°C for 10 h. In this, if GO sheets are used, they are kept for 9 min under 800 W and hydrothermal reaction was carried out at 190° for 6 h, and then the XRD patterns were observed.

Hydrothermal method forms its basis due to the ability of aqueous solutions to dilute at high pressures (10–80 MPa, or even up to 300 MPa) and at high temperatures (500°) for the substances that are insoluble under normal conditions: some oxides, silicates, and sulfides. The parameters that need to be controlled as they determine the properties of the products and process kinetics are: duration of the temperature, pH of the medium, and pressure of the system. This synthesis is carried out in autoclaves as they can withstand high temperatures and pressures for a long time.

##### 10.3.4.1.2 Chemical Synthesis

The main methods that involve the chemical synthesis of graphene are the improved Hummer's method and solvothermal method or route.

In this method, 480 mg of GO is added in 48 mL of N–N dimethyl formamide (DMF) to produce 10 mg/mL GO/DMF suspension. The suspension made was then treated with ultrasonic dispersion (500 W) for 1 h and was then transferred into a 60 mL Teflon bottle which was held in a stainless steel autoclave; this was then heated in the metal furnace for 200°C for 8 h. The final product was obtained by vacuum filtration using

0.22 μm pore filter membrane. This product so formed is a suspension of GQD/DMF product. Now, this suspension was evaporated by rotation to remove DMF and obtain GQDs which are further dissolved in various solvents such as normal or saline water, pure water, and phosphate buffer saline in order to produce different suspensions.

#### *10.3.4.1.3 Electrochemical Exfoliation Approach*

GQDs of the size 3–5 nm are synthesized through electrochemical method. This method involves the treatment of graphene film with oxygen plasma and then breaking it up in order to increase hydrophillicity. Then the synthesized GQDs have enhanced stability in water dispersion and exhibit green luminescence.

Another approach is the graphite electrochemical exfoliation probing the reduction of nanoscale GQDs at room temperature by hydrazine which is in antithesis to the earlier observed procedures where high temperatures are used. It yields GQDs with yellow luminescence and they are of uniform sizes.

#### *10.3.4.1.4 Nanolithography*

Nanolithography requires high quality but it commits a lesser yield of the GQDs and requires an expensive instrumentation. Therefore, this technique is less widely used. In an approach of this sort, ultrahigh resolution electron beam lithography is used to slice the graphene into wanted sizes. A new work with the chemical deposition technique was used to create GQDs.

### *10.3.4.2 PROPERTIES*

Quantum dots of graphene have the structural similarities with the parent molecule but they have some unique optical and certain electrical properties that work as a function of their size. These nanostructures acquire a size even lesser than the Bohr's radius.

GQDs are recognized for their photoluminescence properties even in the drug delivery. They generally depict a wide absorbance in the ultraviolet region and a prominent peak of around 230 nm.

GQDs are the types or forms of the carbon that show a low range of toxicities. They have a great potential in the areas of bioimaging and disease diagnosis. It is to be referred that GQDs produced by nanotubes of carbon have higher toxicity than the ones synthesized from amino acids and GO. Surface modification and other surface properties are important criteria that are widely used. These are the materials that have a potential in future for biomedical applications (Zhang et al., 2016).

### *10.3.5 GRAPHENE OXIDE*

Graphene is synthesized from graphite and therefore, to gather the knowledge about GO, graphite oxide needs to be studied. Graphite oxide is a compound formed by variable ratios of hydrogen, carbon, and oxygen. It is produced by reacting graphite with oxidizers. The product obtained in bulk is maximally oxidized and is solid and yellow in color with C:O between 2.1 and 2.9 this causes a layered structure of graphite but it is large and has irregular spacing.

The bulk material is then diffused into basic solutions which leads to the formation of monomolecular sheets called GO. GO has set a revolution in drug delivery systems as ordinary graphene was hydrophobic and impermeable to all gases and liquids, that is, it is vacuum tight but with the discovery of GO, both liquid water and water vapor flow quickly through it.

#### *10.3.5.1 SYNTHESIS*

GO or graphite oxide is principally synthesized or prepared by three principal methods given by scientists Brodie, Hummers, and Staudenmeier. GO is composed of epoxide and hydroxyl group attached over graphene sheets. It is usually formed by the oxidative reaction of graphite which is done by one of the techniques given by the respective scientists.

##### *10.3.5.1.1 Brodie's Method*

The synthesis of GO was first given by Brodie in 1859. He treated graphitic powder with potassium chlorate in fuming concentrated nitric acid to yield GO.

### *10.3.5.1.2 Staudenmeier's Method*

Staudenmeier method was developed in 1899 by using higher excess of oxidizing agent and addition of concentrated sulfuric acid. This was an advantage as it allowed the process to be carried out continuously without the necessity of adding nitric acid. This method also had a great deal of disadvantage since this method was highly dangerous as it may cause explosions.

### *10.3.5.1.3 Hummer's Method*

This is safe, fast and most efficient method for the formation or synthesis of GO. It was given in 1958 by William S. Hummers and Richard E. Offeman who developed this technique as a replacement to the other ways for synthesizing GO. This method involves addition of graphite into a concentrated acid. They simplified the ingredients to graphite, concentrated sulfuric acid, potassium permanganate, and sodium nitrate. The temperatures used were not more than 98°C. Under this procedure 100 g graphite and 50 g sodium nitrite were added to sulfuric acid at a temperature of 66°C and it was then allowed to cool at 0°C. To it, 300 g of potassium permanganate was added and stirred. The solution is then made up to 32 L with the addition of water in increments. The final solution contains 0.5% of solids that needs to be removed from impurities and was dehydrated with phosphorus pentoxide and GO is obtained. The synthesis of GO quantum dots and GQDs by Hummer's and Staudenmeier's method has been depicted in Figure 10.1.

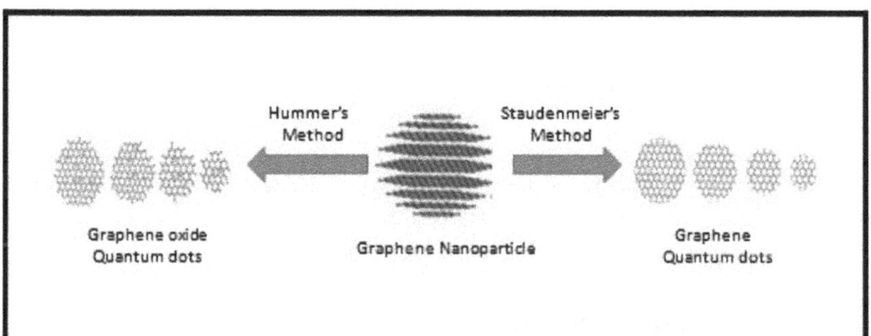


**FIGURE 10.1** Schematic representations of quantum dots by Hummer's and Staudenmeier's method.

##### 10.3.5.2 PROPERTIES

There are various properties or characterizations of GO or graphite oxide that has made it relevant in today's era. GO absorbs moisture that is proportional to humidity, that is, increase in humidity will increase the absorption of moisture in GO and thus, will lead to the swelling of GO. Also, the amount of water absorbed will depend upon the particular synthesis method and also on the temperature.

In the area of drug delivery, GO is extensively used as it has various fundamental properties that make it adaptable for various drug delivery systems. It has various unique properties such as large surface area, chemical, and mechanical stability, two-dimensional planar structure, good compatibility, and conductivity. Due to all these extensively rare properties, GO is used very potentially in the field of nanocarriers of drug delivery (Tran et al., 2015).

### 10.3.6 GRAPHENE NANOPARTICLES

Nanoparticles are the ones which are nanosized and colloidal particles which are composed of synthetic or organosynthetic polymers. The nanoparticles range is between 1 and 1000 nm. The methodology used to create nanoparticles is nanotechnology. In general, for the preparation of nanoparticles, the drug is dissolved or entrapped, or it can be encapsulated or attached to the matrix of the nanoparticles. The formation of nanoparticles allows the change of pharmacokinetic properties of the drug without altering the active compound. There are different types of nanoparticles such as dendrimer, liposomes, biological nanoparticles, CNTs, quantum dots, etc. (Goenka et al., 2013; Torrs, 2014).

#### 10.3.6.1 SYNTHESIS

There are various methods used for the synthesis of GFNs or graphene nanocomposites. Some of these are discussed underneath.

##### 10.3.6.1.1 Reduction Method

This method is used for the synthesis of metallic nanoparticles. The metals that are generally used are noble metals. This type of synthesis is a one-pot synthesis and is highly efficient and easy to perform. This method is performed by the reduction of chemical salts with the help of chemical agents such as ethylene glycol, sodium borohydrate, and sodium citrate. The main mechanism that comes into action is the nucleation of positively charged metal salts with the negatively charged functional groups that are present on the surface of GO. This results in the growth of nanoparticles on the surface of GO. The problem with this type of method is that the size of nanoparticles and their morphology are difficult to control but this can be facilitated by the microwave-assisted method (Gogotsi and Presser, 2014).

### *10.3.6.1.2 Ex Situ Method*

It is used for the synthesis of inorganic nanoparticles, again especially for noble metals. In this method, using other methods of synthesis of nanoparticles, new nanoparticles are generated whose size, shape, and density are already controlled according to the previously synthesized nanoparticles and therefore, hybrids are produced. This method includes the covalent and noncovalent interactions such as the van der Waal forces, $\pi$–$\pi$ stacking, hydrogen bonding, electrostatic interactions, etc.

#### *10.3.6.1.2.1 Covalent Interactions*

Various polymers used for covalent modifications are polyethylene glycol (PEG), dextran, hcitosan, polyvinylacrylonitrile, and so on. Covalent modifications can be achieved by nucleophilic substitution of the reactive sites such as the epoxy groups of GO which are further allowed to react with amino functionalities that have a lone pair. This method is generally used for the large-scale production of GFNs.

#### *10.3.6.1.2.2 Noncovalent Modifications*

Yang et al. in the year 2009 deposited ferric oxide particles over GO and produced a hybrid which exhibited paramagnetic properties and had high-loading capacity and targeted drug delivery. Then in the year 2012, Yang et al. developed a nanoparticle assembly by folic acid (FA)-modified beta-cyclodextrin and GO were noncovalently linked by admantane-grafted porphyrin with the help of $\pi$–$\pi$ stacking interactions. Also, Feng et al. did the functionalization of graphene by PEI polymers by noncovalent interactions, such as electrostatic interactions, yielding complexes of GO–PEI with strong positive charges. These were highly stable in physiological solutions and reduced the cytotoxicity effects to the cells.

#### *10.3.6.2 PROPERTIES*

There are various essential properties of nanoparticles which allow them to be of greater advantage in the application of drug delivery. These properties include a ratio of large surface area and volume which results in enhanced interaction sites. They have a tendency to release drug in a controlled manner resulting in the manufacture of controlled drug delivery systems. These have an advantage as they can be administered by almost all routes of drug delivery including oral, ocular, and nasal routes, etc. As they have a large surface area and small size, they have more efficient cellular uptake. Another approach is attaching specific ligands to the surface of the nanoparticles which can be used for administering drugs to specific target sites. Nanoparticles have an advantage of improving therapeutic index and stability and reducing toxic effects in the body. Various approaches for synthesis of different types of graphene along with their properties are depicted in Table 10.1.

## 10.4 CHARACTERIZATION

Characterization is also termed as material science which generally refers to the process by which a material's structure and properties are measured and scrutinized. It is the science and a fundamental process without which the study of a material engineered is incomplete. The characterization of graphene involves the study of physical and structural elucidation of graphene and its forms and gives a detailed knowledge regarding the use of these in various applications including drug delivery. Various methods (Table 10.2) used for the characterization of substances involve the X-ray diffraction (XRD) technique, scanning electron microscopy (SEM), optical microscope, transmission electron microscope, field ion microscope, STM, atomic force microscope (AFM), etc. Herein, a detailed account on the characterization of graphene and its synthesized forms is provided.

**TABLE 10.1** Approaches for the Synthesis and Properties of Different Types of Graphene.

| Type | Synthesis | Procedure for synthesis | Properties |
|---|---|---|---|
| Graphene | Top-down method | Oxidative exfoliation of graphite and thermal expansion of graphite oxide<br>*Electrochemical oxidation*<br>*Radiation induction* | High carrier mobility, high surface to volume ratio, good biocompatibility, optical properties, and good penetration power |
| | Bottom-up approach | Layers of oxide are developed over the surface of graphene and again impurities are added in graphene or cleavages are made<br>*Chemical vapor deposition*<br>*Atomic layer deposition* | |
| Monolayer graphene | Conventional method | Growth of graphene on liquid metallic matrix resulting in the formation of high strength metallurgical graphene | Useful for providing electrical properties to graphene |
| Bilayer graphene | Chemical vapor deposition | Graphene is revealed to one or more than one volatile precursors such as methane which gets reacted to graphene and produces bilayer graphene under various conditions of temperature and pressure | Possesses electrical properties and is used for the manufacture of other forms of graphene used in drug delivery |
| Graphene superlattices | Chemical vapor deposition | Bilayer graphene was grown on the surface of copper foils in a quartz furnace at an ambient pressure | Possesses optical properties used in cell imaging and for optical surveillance of drug delivery in chemotherapy |
| GQDs | Hydrothermal method | *Chemical synthesis*<br>*Electrochemical exfoliation technique*<br>*Nanolithography* | GQDs have both optical and electrical properties. Photoluminescence properties are also very helpful in drug delivery |

**TABLE 10.1** *(Continued)*

| Type | Synthesis | Procedure for synthesis | Properties |
|---|---|---|---|
| Graphene oxide | Brodie's method | Graphite powder treated with potassium chloride in the presence of fuming nitric acid to yield graphene oxide | Good chemical and mechanical properties with large surface area and 2D planar structure |
| | Staudenmeier's method | Excess of oxidizing agent is added to concentrated sulfuric acid to yield graphene oxide | |
| | Hummer's method | Graphite and sodium nitrate are added in concentrated sulfuric acid and potassium permanganate to yield graphene oxide | |
| Graphene nanoparticles | Reduction method | The interaction of positively charged metal ions with negatively charged graphene surface results in the production of nanoparticles | Large surface area to volume ratio resulting in enhanced interaction, controlled release, efficient uptake by cells. Has good stability and therapeutic index with less toxic effects |
| | Ex situ method | In covalent modification, nucleophilic substitution occurs over the reactive sites of graphene and bonding within the functionalities involves van der Waals forces while in noncovalent methods $\pi$–$\pi$ stacking is involved | |

GQDs, graphene quantum dots.

### 10.4.1 X-RAY DIFFRACTION

Graphene and its substituents have catched the eye of various scientists for their use in various applications. This is primarily due to various exceptional properties that graphene and its substituents tend to possess. For the synthesis of graphene and its forms, various methods have been developed such as the physical, chemical, and mechanical methods. After graphene and other forms have been synthesized using these methods, it is important for the scientists to characterize the synthesize products. There are many techniques that are used for the characterization as mentioned above but XRD is one of the most important methods. This process is carried out by taking the samples of graphene and its forms that have been synthesized by various different methods of synthesis. Therefore, using these samples XRD patterns were obtained with the use of a diffractometer having Cu K alpha radiations. Same program is used to obtain various patterns for all the samples which are fixed with step counter time and step interval method by the scanning line detector mode of the detector used (PIXcel detector).

Using the XRD technique various results were concluded. One of the most important parameters which defines the structure and shape of graphene and graphene-based substances is the interlayer distance. XRD pattern confirmed the crystalline nature of graphene and its forms. A detailed difference in the structure of graphene and graphene-based substituents is observed through this technique. For example; XRD patterns of GO showed strong and sharp peaks and also show the existence of oxygen rich groups on the either sides of the sheets and water molecules which are entrapped between the sheets. Also when GO is reduced, the peaks disappear indicating the absence or the removal of the oxygen-containing functional groups, whereas in graphene two adjacent layers having van der Waal forces of attraction becomes dominant in the peaks as compared to the functional groups present in graphene and also graphene patterns showed broad peaks.

### 10.4.2 SCANNING ELECTRON MICROSCOPY

SEM is a technique that is widely used for the characterization of the substances and is used for the imaging of the structure of the substances. The pharmaceutical application of SEM is the analysis of shape of the particles of the substance. The basic principle that is involved in the SEM technique is coating of the particles with gold foil this is done to provide earthing and to allow conductance through nonconducting particles. The SEM technique used for graphene and graphene-based particles provide the images of the shape and the structure of particles for a particular compound. The intensity of laze is varied and then the images of graphene and its forms are obtained. If there is rough surface over the graphene or graphene-based particles it indicates enhanced light absorbance and diffusion reflection. The SEM shows the difference in normal graphene and various different techniques used for the synthesis of graphene and its substituents. As the techniques vary, the SEM imaging of the substance varies.

### *10.4.3 OPTICAL MICROSCOPY*

Optical microscope is the technique in which the typical microscope uses visible light and a system of lenses to magnify the images of small samples. Optical microscope is one of the oldest forms of microscope which are used for the characterization of the substances. Graphene characterization has also been successfully done with the help of electron microscope. Optical microscope is basically used for the large area quality growth of graphene. It helps to characterize the growth of graphene and several forms of graphene formed by different methods of synthesis over a large area and thus, illustrations of the images are drawn in the form of results (Jia et al., 2012).

### *10.4.4 TRANSMISSION ELECTRON MICROSCOPY*

In transmission electron microscopy, a broad beam is passed through a thin sample and an image is produced due to the phenomenon of mass-thickness contrast or diffraction. This image is then eventually magnified by a set of electromagnetic lenses followed by the detection of the image over a charged coupled device (CCD) camera or a photographic film. The schematic diagram formed by the means of a transmission electron microscope shows the thickness and the symmetry of graphene and graphene-based particles and the arrangement of these particles. A simple experiment is also carried out in which the effect on the orientation of graphene and graphene-based nanoparticles is studied by causing variations in the electron beam intensity. Also the number of layers along with the thickness in the structure of graphene and graphene-based particles is studied.

### *10.4.5 FIELD ION MICROSCOPY*

Field ion microscope has the basic and general principle that an image is formed of the arrangement of atoms at the surface of a sharp metallic tip. This characterization does not have much approach in the area of drug delivery but its little detail is discussed in this chapter. In a field ion microscope, a sharp metallic tip is produced and is placed under an ultrahigh vacuum chamber and is backfilled with gases such as neon and helium, that is, inert gases which are used for imaging (Paul et al., 2012).

### *10.4.6 SCANNING TUNNELING MICROSCOPY*

STM is used for studying the surfaces of graphene and graphene-based products. The experimental studies conducted using the STM primarily show the interaction between the layers of graphene and its derivatives and also the underlying substrate. Using the STM technique, various layers of graphene and its forms are observed, whereas the peaks are observed through Raman spectroscopy. An optical image of the sample is observed after electron deposition on the surface of the sample. Most of the single and multilayer structure of the graphene and its substituents are visible through the technique of STM (Stolyarova et al., 2007).

### *10.4.7 ATOMIC FORCE MICROSCOPY*

AFM is one of the very high resolution providing microscopes and is more than 1000 times better than optical diffraction microscope. Information is gathered by touching the surface of a mechanical probe. It measures and determines each precise movement of the particle and gives highly precise and high-resolution scanning. It has three strong and major abilities which include force measurement, manipulation, and imaging. The basic principle of AFM is to determine the force between the probe and the sample. In order to form the image, there is a reaction of the probe to the forces imposed by the sample and therefore, a 3D shape of the sample surface is observed at a very high resolution. Thus, the highly resolved surfaces of graphene and graphene-based derivatives are studied using this method. A brief of characterization methods is provided in Table 10.2.

**TABLE 10.2** Various Characterization Techniques for Different Types of Graphene.

| Type of graphene | Characterization methods |
|---|---|
| Graphene | Raman spectroscopy, SEM, TEM, STEM |
| Monolayer graphene | TEM, CTLM, AFM |
| Bilayer graphene | Raman spectroscopy, STM, electrical methods |
| Graphene superlattices | ARPES, STS, Raman spectroscopy |
| Graphene quantum dots | AFM, SEM, TEM, FTIR |
| Graphene oxide | TEM, FTIR spectroscopy, SEM, elemental analysis |
| Graphene nanoparticles | SEM, TGA, XRD, TEM |

AFM, atomic force microscopy; ARPES, angle resolved photo electron spectroscopy; CTLM, circular transmission line model; FTIR, Fourier-transform infrared spectroscopy; SEM, scanning electron microscope; STEM, scanning electron transmission microscopy; STM, scanning tunneling microscope; STS, scanning tunneling spectroscopy; TEM, transmission electron microscope; TGA, thermogravimetric analysis; XRD, X-ray diffraction.

## 10.5 TOXICITY OF GRAPHENE AND GRAPHENE-BASED MATERIALS

Graphene has proved to be a novel and advantageous approach in drug delivery and pharmaceutical research. But, before a substance can be successfully used in drug delivery or in any other approach, it needs to be tested for safety purposes for the use in drug delivery. Graphene has great strength, approximately zero thickness, and fascinating optical and electrical properties. If graphene is used in drug delivery of nanoparticles the question of safety and toxicity of the substance is of prime importance. As it is the nature of graphene, namely, thin and light weight, it has tough and intractable particles; these properties are worrisome in terms of detrimental effects the particles can have on our health when particularly breathed in. As graphene is a novel approach, especially in the terms of nanodelivery, cautious release of novel materials has to be determined.

The interaction of graphene with the biological systems is in different ways. It is observed in the studies carried out that pristine GO (p-GO) was not found to enter into the cell or damage the membrane but its presence increased reactive oxygen species (ROS), DNA damage, and T-lymphocyte apoptosis (Nezakati et al., 2014).

Toxicity of graphene is basically due to the graphene-based materials and their interactions in biological systems such as cells and tissues. The interaction of graphene with the biological systems is the newest and fastest growing areas of carbon-based nanomaterial research. The major drawback of graphene and graphene-based materials is that there is a limited knowledge of their toxicity and biological safety profile. A wide range of testing is nowadays essential for materials made from graphene for both present and future assess of their biological safety profile, which is dependent on various physicochemical factors which are related to their surface chemistry, shape, relative concentration, charge, and size. There are many unresolved issues which need to be clarified before the use of graphene-based systems in health care applications.

Toxicology studies are therefore becoming advanced in large- and small-scale animals and human cell in vivo and in vitro cell lines. For example, lethal dose of graphite (LD50), CNTs, are reported as 2 g/kg, 2 mg/kg, and 1.2 g/kg, respectively, in animals. Usually the most common cytotoxicity assays are used to evaluate the toxicity of graphene and its materials. The examples of such assays are caspase-3,7 to measure cell adhesion, hemocompatbility, cell death, cytokine adhesion, morphology, hemolysis and lactate dehydrogenase assay which assess the membrane integrity (Horvath et al., 2013; Sasidharan et al., 2011).

GO is potentially used in the drug delivery of pharmaceuticals and is used in approximately every approach or method used for the delivery of pharmaceuticals such as nanodelivery of drug, etc. The morphology of graphene is one of the most important parameters for the indication of cell status. Through the experimentation conducted by Chang et al. (2011), the GO cells were found out to be in normal spindle shape and showed great adhesion. GO showed toxicity toward cell viability. Greater the concentrations of GO, the viability of cells is lost. This causes an indirect effect over the mortality of cells, as the viability of cells decreases the mortality of cells increase as it affects the activity over mitochondria. The use of

GFNs and their toxicology studies have been made. In the biological microenvironment, the nanoparticles and the biomolecules bind with each other to form a cornea. Graphene-based nanomaterials are toxic in both bacteria and fungi as well. Graphene effectively inhibits the growth of Gram-negative and -positive bacteria at minimal doses. It is stated that at the smallest size, GO show the greatest hemolytic activity, whereas the sheets show lowest toxicity. Various toxicity issues of graphene and its derivatives are given in Table 10.3. Horvath et al. studied the toxicity of graphene derivatives on the cells present at luminal surface of lungs. Graphene although possesses antimicrobial or antibacterial properties but causes a serious problem of lung cytotoxicity in mammals. The mammalian cell toxicity was first studied in pheochromocytoma-derived (PC12) cells and screened with the effect of graphene over single-walled CNT. The toxicity of graphene-based particles in the mammalian lungs was as such that it decreased the cell number/metabolic activity of epithelial cells of the lungs. The cell number/metabolic activity of graphene and its derivatives and macrophages was assayed by the MTT assay.

As mentioned earlier about p-GO causing toxicity in the DNA or other areas of the body including the damage in ROS and also causing T-lymphocyte apoptosis, the mechanism causing toxicity is of great interest. The toxic mechanism is that p-GO inhibits the ligand binding ability of the protein receptor by directly interacting with it which leads to ROS-dependent passive apoptosis. It was further observed that the interaction of GO and GO–COOH have good biocompatibility but at doses above 50 μg per mL induced cytotoxicity. The nanoplates of graphene are formed by the method of sonication of covalently PEGylated GO sheets. It was observed that the human stem cells were destroyed. The genotoxicity and cytotoxicity was observed for the first time as a side effect of large size, that is, it is size dependent. The result of cytotoxicity is that it causes degradation of stem cells (Roberts et al., 2016; Chang et al., 2011; Guo and Mei, 2014).

Despite graphene has provided, advanced approaches in drug delivery such as bioimaging, biosensing, its safety issues are still unclear. GQDs induce in vitro and in vivo toxicity. Although GQDs have ultrasmall size, it shows very low cytotoxicity which was illustrated by in vitro studies. The material accumulation in mice and fast clearance through kidney was observed through in vivo studies. It is therefore, observed that quantum dots of graphene have lesser toxicity as compared to GO (Chng and Pumera, 2013; Singh, 2014; Olteanu et al., 2015).

**TABLE 10.3** Toxicity Studies on Graphene and Graphene-based Materials.

| Toxicity experiment | Type of graphene | Outcome of study |
|---|---|---|
| In vivo and in vitro experiments in mice (Chong et al., 2014) | GQDs and GO | In vitro studies showed less toxicity of GQD than GO. In vivo experiments depicted no accumulation of GQDs in the main organ fast clearance through kidney. However, GO showed toxic behavior and accumulation in the main organs |
| To measure the efflux of the cytoplasmic materials of Gram-negative bacteria (*Escherichia coli*) and Gram-positive bacteria (*Staphylococcus aureus*) | Graphene nanosheets | Gram-negative bacteria were more resistant to damage of cell membranes than Gram-positive bacteria |
| Lung epithelial cells toxicity studies. Mitochondrial activity was studied by MTT (methylthiazolyldiphenyl-tetrazolium bromide) assay and WST-8 (water-soluble tertazolium salt) assay | GO | GO showed dose-dependent cytotoxic response after an exposure of 24 h |
| MTT assay to determine the cytotoxicity and cell viability of the nanomaterials (Chang et al., 2011) | Graphene-based nanomaterials | These studies are considered as an important indicator of toxicity for graphene-based materials |
| Mitochondrial toxicity and cell membrane integrity was measured in neuronal PC12 cells | Graphene-based nanomaterials | Lower toxicity at higher concentrations of graphene was reported |
| Cytotoxicity studies using immortalized cells (immune cells, stem cells, blood components) | GO | A dose-dependent reduction in fluorescent intensity was observed indicating loss of structural integrity due to strong interaction of phospholipid bilayer of plasma membrane with GO |
| Trypan blue assay using A549 cells incubated for 24 h with different concentrations of GO and rGO. The dead cells stained blue and the live cells remained unchanged. | GO and rGO | The mortality of cells was monitored. The cells treated with GO showed no mortality |

GO, graphene oxide; GQD, graphene quantum dot; rGO, reduced graphene oxide.

## 10.6 APPLICATIONS OF GRAPHENE AND ITS DERIVATIVES

Graphene is a new 2D material which is $sp^3$ hybridized and its various forms are available which can be used in drug delivery and other biomedical applications. Graphene possesses various extraordinary properties these can be mechanical, thermal, optical, and electronic. As there are growing technical advancements in the functionalization and synthesis approaches graphene and its derivatives, graphene has shown outstanding advancements in various streams such as composite materials, nanomaterials, and may other applicable areas.

Graphene and its derivatives are widely used in biomedical applications and this approach is a novel and a trending one. It was observed in the review article given by Dai et al. that GO forms a potential and efficient nanocarriers in drug delivery. Various areas wherein graphene has been used in the biomedical applications are the ones such as bioimaging, drug/gene delivery, biosensing, antibacterial material and is widely used as a biocompatible skeleton for cell culture purposes. These research areas are possible due to various rare properties which are exhibited by graphene; these are: electronic conductivity of about 200,000 $cm^2/V$ s, mechanical strength of approximately 1100 GPa, a high surface area ranging 2630 $m^2/g$, thermal conductivity of around 5000 W/m K approximately, intrinsic compatibility, low scalable production, and chemical and biological functionalization of GO.

Last 3 years have proved to be exciting related to the latest researches being made on the applications and uses of graphene. Various applications of graphene related to drug delivery primarily includes all the biomedical applications of graphene which are described underneath.

Other applications included under biomedical applications include biosensing and bioimaging also (Nirunnabi, 2015; Pan et al., 2012).

### *10.6.1 BIOMEDICAL APPLICATIONS*

#### *10.6.1.1 DRUG DELIVERY*

GO that is prepared by Hummer's method is the most suitable candidate as a nanocarrier in the drug and gene delivery. Various parameters that need to be focused for the use of graphene in drug delivery are usually the thickness of 1–2 nm thick which is approximately one to three layers and its size that ranges from a few to a hundred nanometers. Unique characteristics of graphene such as excellent biocompatibility, physiological stability and solubility, and using chemical conjugation and physisorption approaches for loading of drugs or genes make it a purposeful candidate for delivery purposes. Conjugation with various systems such as biomolecules and polymers is achieved due to various functional groups such as COOH and OH over GO. The presence of different functional groups and the conjugation with various systems facilitates multifunctionalities and modalities for various medical applications. Dai et al. for the first time did a candid exploration over nanoscale GO (NGO) as a new carrier for the efficient carrier for the delivery of a water-soluble aromatic anticancer drug into the cells. This was inspired by the idea of carbon nanotube for the purpose of drug delivery. NGO was conjugated with polyethyleneglycol (PEG) which has amine terminated six-membered functionality. Subsequent to loading, a water-insoluble anticancer drug SN38 was adsorbed onto the surface of NGO by $\pi$–$\pi$ stacking and noncovalent adsorption technique. It was inferred that the PEG-induced NGO further loaded with SN38 exhibited a high level of toxicity for HCT-116 cells; also, it was found out that it was 1000 times more potent than camptothecin-11.

A widely accepted criterion for the clinical practice in cancer therapy is the combined use of drugs which is helpful in reducing the resistance of cancer cells. There are various research works that have been done to ensure the practicability of GO as a candidate of nanocarrier for the purpose of targeted drug delivery and controlled drug delivery of mixed drugs. One of the methods is the one in which GO conjugated with folic acid and $SO_3H$ groups was loaded with two drugs in a controlled way via $\pi$–$\pi$ stacking. It was revealed that this combination of drugs with GO exhibited specific target delivery and a greater toxicity to MCF-7 cells and also to human breast cancer cells with folic acid receptors. It was noted that remarkably high toxicity content was observed with the conjugation of two drugs than the toxicity level of a single dose (Nirrunabi, 2015; Shao et al., 2009).

#### *10.6.1.2 GENE DELIVERY*

Gene delivery is a new approach in this era which is used to treat various disorders prevailing in the body. These disorders can be genetic disorders which include cancer, Parkinson's disease, and cystic fibrosis. In the process of gene therapy, it is important to have gene vector which has the main function of protection of DNA from the degradation through nucleases and also to facilitate uptake of cellular DNA with high efficacy. The major challenge that is faced in the development of gene therapy is the choice of efficient and safe gene vectors. Various recent researches over gene delivery have been made in the recent years, one of which uses PEI-modified GO described here. In this, GO was conjugated with positively charged PEI and DNA plasmid was conjugated onto the surface of the sheets of GO through electrostatic interactions which arises from the polymer (cationic). The transfection efficiency of GO–PEI–10K and GO–PEI–1.2K was compared with the unbound polymers such as PEI–10K and PEI–1.2K. It was concluded that the grafting of GO polymer with the PEI polymer significantly lowered the cytotoxicity of the polymer and improved the transfection efficiency of the polymer. Also, much higher luciferase expression and transfection DNA efficiency was found in HeLa cells with the complexes of GO with the polymer than that of unbound polymer. In addition to it, further researches were made which concluded that PEI did not show higher transfection efficiency when conjugated with other systems.

A most recent approach was made in Singapore using chitosan (CS)-functionalized GO and its application in gene delivery. It was recorded from the results that the drug payload and GO–CS loaded with camptothecin (CPT) showed better ability to kill cancer cells than pure CPT. Simultaneously, stable nanoparticles through condensation of the same complex were made which showed a good transfection delivery in Henrietta Lacks (HeLa) cells at a particular nitrogen/phosphate ratio. Further, simultaneous delivery and loading of drug and gene by this nanocarriers gave high therapeutic efficacy for both chemo and gene therapy (Feng et al., 2015).

#### *10.6.1.3 CANCER THERAPY*

With the extraordinary success of graphene and graphene-based materials in drug and gene delivery it is widely used in the cancer therapy. For clinical purposes, the in vivo behavior of graphene loaded with drugs is investigated. For this purpose, Liu and his colleagues performed an experiment over the in vivo tumor apprehension and photothermal therapy with GO which have been PEGylated using xenograft tumor mouse models. Higher tumor intake of PEGylated GO was observed. This was due to a well-defined target of GO which was caused by EPR effect. Probing further, a well-ordered destruction was achieved by the use of low power (IR) as GO shows strong absorbance under IR (near).

With the combination of photo- and chemothermal technologies a complex (NGO–PEG–DOX) for the antitumor affect and was checked both in vivo and in vitro experiments. In the observations made, it was shown that synergistic effect was caused due to the combination of both chemo- and photothermal therapy. This synergistic effect led to a better killing of cancer cells than chemo- and photothermal effects alone. Also, folic acid and sulfonic acid when conjugated with GO which when loaded with porphyrin photosensitizers was used for targeting photodynamic therapy (PDT). PDT has been reviewed as a novel technique used for treating cancers and other diseases as it causes low toxicity and is stable under physiological conditions. For this purpose, loading of GO was done with a photosensitizer (Chlorin e6) and was achieved due to $\pi$–$\pi$ stacking and hydrophobic interactions. These systems tend to increase the hoarding of the photosensitizer in the tumor cells resulting in the concentration-dependent photodynamic effect under radiation.

To sum up, facile modifications of GO and structural improvements provided opportunities for loading and delivery of a variety of substances with the help of graphene. This is widely used for the treatment of cancer and other diseases by loading and delivery of chemical drugs, photosensitizers, etc. With the advancements in drug delivery and technologies of drug delivery, graphene and its derivatives are likely to be proved of great importance in the near future.

#### *10.6.1.4 BIOSENSING*

Graphene and its derivatives have been used widely for applications such as biosensing and biomolecular detection such as detection of thrombin, amino acid, ATP, etc. There are various types of biosensors for the purpose of biosensing; most common of those being: (1) FRET biosensors–fluorescence resonance energy transfer, (2) FET type biosensors based on the electronic properties of graphene, (3) ultrasensitive biosensors for the detection of DNA and other such molecules, (4) used as a matrix for detection of molecules for the purpose of MALDI-TOF-MS, and (5) biosensors based on GO having electrochemical principle are used that focus on the properties of graphene such as capability of loading biomolecules, high surface area, and appreciable electrical conductivity.

#### *10.6.1.5 BIOIMAGING*

Graphene and its derivatives such as GO or GQDs have been used for the application of bioimaging. Cellular uptake of GO (PEGylated) loaded with drugs that exhibited inflorescence in the near IR region was first examined by Dai et al. Not only GO and GQDs are used for bioimaging but also reduced graphene oxide (rGO) has been used widely. This is explained by a research under which rGO grafted with gelatin and labeled with fluorescence dye was studied for cellular imaging and drug delivery.

GQDs are nothing but smaller GO ranging within the size of <10 nm or equal to it and are prepared using bottom-up approach and chemical oxidation of graphite. It is of significance that these quantum dots exhibit inflorescence and are widely used for bioimaging. Hydrothermal cutting of GO has been used by Pan et al. for the development of GQDs having blue florescence. But these quantum dots obtained were weakly florescent even much lesser than organic dyes exhibiting florescence. Therefore, this problem was further overcome by reacting the synthesized GQDs with hydrazine vapor that significantly improved the florescence. Also, it has been observed that GQDs when surface functionalized with alkylamine gave good florescence. In the recent researches made by Zhu et al. and others, it was potentially observed that GQDs exhibited excellent physiological solubility and biocompatibility, low cytotoxicity levels, and are good candidates for direct use in intracellular imaging without any further processing needed. Adding to it, they possess optical properties such as upconversion florescence behavior and pH dependency; also, the upconversional property is responsible for the excitation of graphene in the near IR region, which makes it suitable for both in vivo and in vitro biodetection, safe, and image efficient without any interference from the autoflorescence effect of cells, tissues, or organs in that particular region (Nirunnabi et al., 2015).

#### *10.6.1.6 GO-BASED ANTIBACTERIAL MATERIAL*

GO and rGO papers were formulated from their suspension by vacuum filtration technique by Fan et al. and these papers were found to exhibit strong antibacterial effect; also, the GO paper is available in low costs therefore, it can easily be used for clinical and environmental purposes.

Antibacterial effect of graphene is also exhibited in the form of nanosheets and nanowalls which are deposited on substrates of stainless steel and possess this action for both Gram-positive and -negative bacteria. It was inferred that reduced GO nanowalls was more toxic than GO nanowalls and the reduced one showed better antibacterial activity than the other because of the better transfer of charge between the bacteria and more sharpened edges during the contact interaction. Later, a research conducted showed the antibacterial effects of four types of graphene, namely, graphite, graphite oxide, GO, and rGO. It was of the order GO > rGO > graphite > graphite oxide. Their effect was both over membrane and oxidative stress. This work enabled the knowledge of interaction of GO with bacteria and for the future production of graphene-based antibacterial materials.

#### *10.6.1.7 GO-BASED SCAFFOLD FOR CELL CULTURE*

Behavior of NIH 3T3 fibroblasts was studied as a model of mammalian cells which was allowed to grow on a supported film of GO. The work done over the GO film suggested that no harmful effect on the mammalian cells taking in account adhesion and it exhibited good gene transfection efficiency. The result indicated that GO material can be used for the surface coating of the implant.

A new research made over graphene/chitosan film which was produced by casting method was investigated as a Scaffold material for tissue engineering. The results showed that the film formed of graphene did not hamper the proliferation of human mesenchymal stem cells (hMSCs). It also speeds the differentiation of stem cells into the bone cells with the use of growth factors and inducers such as the oestrogenic inducers; this growth was accomplished in a controlled manner. These were then used for proliferation and transplantation of stem cells and were specifically differentiated into bones, muscles, and cartilages for the purpose of bone regeneration.

Using the CVD method, graphene film was grown as a substrate of neurites and was used as key structure for neural functions of the mice in hippocampal culture model. The observations made showed that the average length of the neurite numbers significantly enhanced in a period of 2–7 days after the seeding of cells as compared to tissue culture polystyrene substrates (TCPS). Also, the GAP–43 expression was increased as compared to the TCPS.

The interesting work described above gave the applications of graphene as antibacterial material and Scaffold material in the culture of cells (Nirunnabi et al., 2015).

## 10.7 CONCLUSIONS AND FUTURE PERSPECTIVES

Graphene has proved to be of great significance in the field of medicine and in various applications of the pharmaceutical industry. Graphene and its derivatives are widely used in the applications and the materials synthesized from these are widely used in drug delivery. Graphene has made outstanding advances in medicinal applications which are exciting but there are various challenges which need to be overcome. The major challenge faced is thorough understanding of the interactions of graphene with various cells, tissues, and organs especially understanding the cellular uptake mechanism. Such problems and challenges lead to the development of delivery systems, bioimaging and sensing, and various other applications.

Although graphene and its derivatives have provided benefits in the medicinal applications, toxicity of graphene and derivatives such as GO which are widely used in drug delivery pose a major threat. The primary observations suggest that charges and flat shape are the factors that majorly affect the cytotoxicity and also affect in vivo biodistribution of graphene and its forms.

Moving ahead, it is observed that graphene has unique physicochemical properties and structural features which provides new directions for discovery of its role in the delivery systems. New delivery systems can only be made into existence by the combined role of chemistry, material sciences, biomedicine, and nanotechnology. For the best performance of graphene and its derivatives, changes in the size distribution, morphology, and structural defects is urgently needed.

Various researches done over graphene and its forms-based materials used in cell culture require special attention. Also, graphene and its derivatives are used in the differentiation, growth, and proliferation of stem cells and have a major role in regenerative medicine. Therefore, it is believed that graphene and its forms will make major advances in biomedical and drug delivery applications in the near future.

## KEYWORDS


- **graphene**
- **graphene-based materials**
- **graphene oxide**
- **graphene quantum dots**
- **biomedical applications**


## REFERENCES


Barlas, Y.; Cote, R.; Lambert, J.; MacDonald, A. H. Anamolus Exciton Condensation of Graphene Bilayers. *Phys. Rev. Lett.* **2010,** *101*, 1–20.

Bianco, A. Graphene: Safe or Toxic? The Two Faces of the Medal. *Angew. Chem.* **2016,** *52*, 4986–4997.

Chang, Y.; Yang, S. T.; Liu, J. H.; Dong, E.; Wang, Y.; Cao, A.; Liu, Y.; Wang, H. In Vitro Toxicity Evaluation of Graphene Oxide on A549 Cells. *Toxicol. Lett.* **2011,** *200*, 2209–2216.

Chng, E. L. K.; Pumera, M. The Toxicity of Graphene Oxides: Dependence on the Oxidative Methods Used. *Chem. Eur. J.* **2013,** *19*, 8009–8351.

Chong, Y.; Ma, Y.; Shen, H.; Tu, X.; Zhou, X.; Xu, J.; Dai, J.; Fan, S.; Zhng, Z. The In Vitro and In Vivo Toxicity of Graphene Quantum Dots. *Biomaterials* **2014,** *35*, 5041–5048.

Goenka, S.; Sant, V.; Sant, S. Graphene-based Nanomaterials for Drug Delivery and Tissue Engineering. *J. Control. Release* **2013,** *173*, 2775–2785.

Gogotsi, Y.; Presser, V. Field Effect Transistors. *Carbon Nanomater* **2014,** *6*, 26–46.

Guo, X.; Mei, N. Assessment of the Toxic Potential of Graphene Family Nanomaterials. *J. Food Drug Anal.* **2014,** *22*, 105–115.

Horvath, L.; Magrez, A.; Burghard, M.; Kern, K.; Schwaller, L. F. B. Evaluation of the Toxicity of Graphene Derivatives on the Cells of Lung Luminal Surface. *Carbon* **2013,** *64*, 45–60.

Hummers, W. S.; Offeman, R. E. Preparation of Graphitic Oxide. *J. Am. Chem. Soc.* **1958,** *80*, 1339–1339.

Jia, C.; Jiang, J.; Gan, L.; Guo, X. Direct Optical Characterization of Graphene Growth and Domains on Growth Substrates. *Sci. Rep.* **2012,** *2*, 4369–4381.

Jiao, L.; Zhang, L.; Wang, X.; Diankov, G.; Dai, H. Narrow Graphene Nanoribbons from Carbon Nanotubes. *Nature* **2009,** *458*, 877–880.

Kosynkin, D. V.; Higgimbotham, A. L.; Sinitskii, A.; Lomeda, J. R.; Dimiev, A.; Price, B. K.; Tour, K. M. Longitudinal Unzipping of Carbon Nanotubes to form Graphene Nano Ribbons. *Nature* **2009,** *9*, 1527–1533.

Kumar, P.; Subhramanyam, K. S.; Rao, C. N. R. Graphene Produced by Radiation Induced Reduction of Graphene Oxide. *Macromol. Chem. Phys.* **2012,** *213*, 1146–1163.

Min, H.; Sahu, B.; Banerjee, S. K; MacDonald, A. H. Ab Initio Theory of Gate Induced Gaps in Graphene Bilayer. *Phys. Rev. B* **2007,** *75*, 2–8.

Nezakati, T.; Cousins, B. G.; Seifalian, A. M.; Toxicology of Chemically Modified Graphene-based Materials for Medical Application. *Arch. Toxicol.* **2014,** *88*, 1987–2012.

Novoselov, K. S.; Fal'ko, V. L.; Colombo, L.; Gallert, P. R.; Schwab, M. G.; Kim, K. A Roadmap for Graphene. *Nature* **2012,** *490*, 192–200.

Nurunnabi, M.; Parvez, K.; Nafiujjaman, M.; Revuri, V.; Khan, H.; Feng, X.; Lee, Y. Bioapplication of Graphene Oxide Derivatives: Drug/Gene Delivery, Imaging, Polymeric Modification, Toxicology, Therapeutics and Challenges. *RSC Adv.* **2015,** *5*, 2–10.

Olteanu, D.; Flip, A.; Socaci, C.; Biris AR.; Filip, X.; Coros, M.; Rosu, M. C.; Pogacean, F.; Alb, C.; Baldea, I.; Bolfa, P.; Pruneanu, S. Cytotoxicity Assessment of Graphene-based Nanomaterials on Human Dental Follicle Stem Cells. *Colloids Surf. B* **2015,** *136*, 791–798.

Omiciuolo, L.; Hernandez, E. R.; Miniussi, E.; Orlando, F.; Lacovig, P. Bottom-up Approach for Low-cost Synthesis of Graphene. *Nat. Commun.* **2014,** *5*, 1–10.

Pan, Y.; Sahoo, N. G.; Li, L. The Application of Graphene Oxide in Drug Delivery. *Expert Opin. Drug Deliv.* **2012,** *9*, 1365–1376.

Paul, W.; Miyahara, Y.; Grutter, P. Implementation of Automatically Defined Field Ion Microscopy Tips in Scanning Probe Microscopy. *Nanotechnology* **2012,** *23*, 335702–335708.

Roberts, J. R.; Mercer, R. R.; Stefaniak, A. B.; Seehra, M. S.; Geddam, U. K.; Chaudhuri, I. S.; et al. Evaluation of Pulmonary and Systemic Toxicity Following Lung Exposure to Graphite Nanoplates: A Member of the Graphene-based Nanomaterial Family. *Part. Fibre Toxicol.* **2016,** *13*, 34–55.

Sasidharan, A.; Panchakarla, L. S.; Chandran, P; Menon, D; Nair, S; Rao, C. N. R; Koyakutty, M. Differential Nano-bio Interactions and Toxicity Effects of Pristine Versus Functionalized Graphene. *Small* 2011, *8*, 1251–1263.

Shao, Y.; Wang, J.; Englehard, M.; Wang, C.; Lin, Y. Facile and Controllable of Graphene and Its Applications. *J. Mater. Chem.* **2009,** *20*, 743–748.

Singh, Z. Toxicity of Graphene and Its Nanocomposites to Human Cell Lines: The Present Scenario. *Int. J. Biomed. Clin. Sci.* **2014,** *1*, 24–29.

Stolyarova, E.; Rim, K. T.; Ryu, S.; Maultzsch, J.; Kim, P.; Brus, L. E.; Heinz, T. F.; Hybersten, M. S.; Flynn, G. W. High Resolution Scanning Tunneling Microscopy Images of Mesosomic Graphene Sheets on an Insulating Surface. *Proc. Natl. Acad. Sci.* **2007,** *104*, 9209–9212.

Torres, L.; Roche, S.; Charlier, J. C. Introduction to Graphene-based Nanomaterials: From Electronic Structure to Quantum Transport. *Nat. Phys.* **2014,** *880*, 1367–2360.

Tran, T.; Nagyun, H.; Pham, T.; Choi, J.; Choi, H.; Yong, C.; Kim, J. Development of a Graphene Oxide for Dual-drug Chemophototherapy to Overcome Drug Resistance in Cancer. *ACS Appl. Mater. Interfaces* **2015,** *46*, 325–339.

Wang, Y.; Su, Z.; Wu, W.; Nie, S.; Xie, N.; Gong, H.; Guo, Y.; Lee, J. H.; Xing, S.; Lu, X.; McCarty, K.; Pei, S.; Hernandez, F.; Hadjiev, V. K.; Bao, J. Twisted Bilayer Graphene Superlattices. *ACS Nano* **2013,** *7*, 2587–2594.

Zhang, W.; Wang, C.; Li, Z.; Lu, Z.; Li, Y.; Yin, J. J.; Zhou, Y. T.; Gao, X.; Fang, Y.; Nie, G.; Zhao, Y. Unraveling Stress-induced Toxicity Properties of Graphene Oxide and the Underlying Mechanism. *Adv. Mat.* **2012,** *24*, 5277–5397.

## CHAPTER 11

# GRAPHENE FOR DRUG DELIVERY: FOCUS ON ANTIMICROBIAL ACTIVITY

DARIANE JORNADA CLERICI[1], MÁRCIA EBLING DE SOUZA[1], and ROBERTO CHRIST VIANNA SANTOS[2*]

[1]*Nanotechnology Laboratory, Centro Universitário Franciscano, Santa Maria, Rio Grande do Sul, Brazil*

[2]*Laboratory of Oral Microbiology Research, Department of Microbiology and Parasitology, Universidade Federal de Santa Maria, Santa Maria, Rio Grande do Sul, Brazil*

[*]*Corresponding author. E-mail: robertochrist@gmail.com*

## CONTENTS

## ABSTRACT

Despite advancements in drug discovery and in pharmaceutical biotechnology, infections caused by microorganisms are a major cause of human morbidity and mortality. Antimicrobials are the first and foremost option for the treatment of infections. However, abuse of such drugs led to resistance, making it extremely difficult to treat infections. Therefore, new intelligent solutions are necessary to overcome such concerns. Carbon-based nanoparticles are part of it, since they present high antimicrobial activity. Recently, graphene has been proposed as a new and effective antimicrobial material having severe cytotoxic effect on bacteria, fungi, and plant pathogens with less resistance. Besides, compared with carbon nanotubes, graphene has a tolerable effect in mammalian cells. In the biomedical field, graphene materials have applications in therapy, diagnosis, and drug release, properties that are unique. Also, graphene has been widely used as an effective carrier for nanoadministering drugs including antimicrobials due to its rich chemical surface, high aspect ratio, and ability to cross the plasma membrane. Graphene-based nanocomposites have emerged as promising antibacterial materials since they can overcome the limitations of the individual components. In general, the antimicrobial action of graphene involves physical and chemical effects but the antibacterial graphene mechanism has not been fully elucidated.


## 11.1 INTRODUCTION

Despite advancements in drug discovery and in pharmaceutical biotechnology, infections caused by microorganisms are a major cause of human morbidity and mortality. Antimicrobials are the first and foremost option for the treatment of infections. However, abuse of such drugs led to resistance, making it extremely difficult to treat infections. Almost all organisms have an exceptional intrinsic ability to ignore many therapeutic interventions due to their genetic mechanisms, among other factors. At the same time, major pharmaceutical companies are losing interest in the development of new antimicrobial drugs, transferring their capital investments to more profitable research and development fields, such as drugs to control blood pressure, diabetes, and cancer. Therefore, new intelligent solutions are necessary to overcome such concerns and should match the viability of industrial production processes with low cost and effectiveness (Kwiatkowski, 2000; Rizzello and Pompa, 2014).

Microorganisms resistant to multiple drugs are a major problem worldwide. In 2009, the American Infectious Disease Society highlighted a list of recent multiresistant pathogens, being six pathogens most often found in hospitals, having difficult treatment: two Gram-positive bacteria (G+) (*Enterococcus faecium* and *Staphylococcus aureus*) and four Gram-negative species (G−) (*Klebsiella pneumoniae, Acinetobacter baumannii, Pseudomonas aeruginosa,* and *Enterobacter* spp.) (Rice, 2008). The resistance of microorganisms is a major problem in therapy since the multiresistant microorganisms group cannot be treated with traditional antibiotics. These antimicrobial agents are resistant to most drugs (e.g., *Escherichia coli* can become resistant to carbapenems and cephalosporins) (Ross et al., 2015). Therefore, there is a strong need to find alternative ways and new therapeutic agents (Taitt et al., 2015).

Carbon-based nanoparticles are part of it, since they present high antimicrobial activity. Nanomaterials have been applied to improve physicochemical and therapeutic efficacy of drugs. Similarly, nanotechnology in pharmaceuticals and microbiology showed promising solutions to overcome the problem of antimicrobial resistance (Adibkia et al., 2010; Mohammadi et al., 2011). Recently, graphene has been proposed as a new and effective antimicrobial material having severe cytotoxic effect on bacteria, fungi, and plant pathogens with less resistance. Besides, compared with carbon nanotubes, graphene has a tolerable effect in mammalian cells (He et al., 2015; Li et al., 2013a). Graphene is renewable, easier to obtain, and more expensive than metals and metal oxides. In general, the antimicrobial action of graphene involves physical and chemical effects. Physical damage is mainly induced by direct contact of its sharp edges with bacterial membranes and destruction of lipid molecules (Chen et al., 2014). The chemical effect is the result of oxidative stress created by reactive oxygen species (ROS) or charge transfer. In addition, graphene has been used as a carrier to disperse and stabilize various nanomaterials, such as metals, metal oxides, polymers and their composites, with high antimicrobial effectiveness due to the synergistic effect. Besides, platforms of distribution of antibiotics based on graphene were also reported (Chen et al., 2013; Wang et al., 2013a; Maktedar et al., 2014).

## 11.2 GRAPHENE: GENERAL ASPECTS

Strictly speaking, the term graphene refers to a single layer of graphite. However, the term is used to refer up to 10 layers of graphite. Similar to carbon nanotubes, fullerene, graphite, and diamond; it is a carbon allotrope of elemental carbon atoms in $sp^2$ hybridized with partially filled p orbitals above and below the sheet plane. Graphene materials may be classified into two ways: either based on their structure or stacking arrangement. It was first described as a monolayer of crystalline graphite films by Andre Geim and Konstantin Novoselov (Novoselov et al., 2004). Graphene has good electrical conductivity, high surface area and strength, excellent elastic properties, good thermal conductivity, ease of functionalization, chemical resistance, and gas impermeability. Its many applications in the fields of sensors, energy, storage devices, fuel cells, and high strength materials are remarkable (Edwards and Coleman, 2013; Suk et al., 2013).

The synthesis of graphene and its derivatives can be classified into two categories: top-down and bottom-up. The first approach employs an exfoliation of a layer of graphene from graphite. The second approach involves the construction of graphene using carbon-based materials. The bottom-up approach is simple, but produces more defective materials than the top-down approach (Wei and Liu, 2010; Zhu et al., 2010a). Top-down approaches separate the stacked sheets by interrupting van der Waals forces that hold the sheets together. The damage to the leaves during the exfoliation process and the re-agglomeration of separate sheets are some of the disadvantages of this technique. Another disadvantage is that the precursor (graphite) is scarce. On the other hand, the bottom-up approach requires a very high temperature. The top-down approaches include micro-mechanical exfoliation, electrochemical exfoliation, chemical and electro-chemical reduction strategy, exfoliation of graphite oxide, solvent-based exfoliation, arc discharge, and decompression of carbon nanotubes. The bottom-up approaches include epitaxial growth, chemical vapor deposition method without substrate, and carbonization (Kim et al., 2012; Sutter et al., 2008; Zhu et al., 2010b).

In the biomedical field, graphene materials have applications in therapy, diagnosis, and drug release, properties that are unique (Feng and Liu, 2011). While there has been some progress in the diagnosis and management of pharmaceuticals, therapeutic applications of graphene remain in the early stages. This difference in the application of graphene in biological and nonbiological sectors is due to the toxicity of chemically reduced graphene oxide (GrO). The toxic potential of graphene and its derivatives in biological systems range from prokaryotes to eukaryotes. Then comes the need to identify ecological and simple systems to prepare graphene biocompatible materials (Akhavan and Ghaderi, 2010; Hu et al., 2010; Sanchez et al., 2012).

## 11.3 GRAPHENE IN DRUG DELIVERY

The drug delivery controlled systems are designed to improve the therapeutic efficacy, and the advantages are convenience, high utilization rate, and reduced toxicity. Graphene has been widely used as an effective carrier for nanoadministering drugs including antimicrobials due to its rich chemical surface, high aspect ratio, and ability to cross the plasma membrane (Dreyer et al., 2010). Ma et al. (2011) prepared hybrid polybasic anhydride modified with graphene oxide-poly(styrene-*co*-acrylic acid) (GO-PSA) nanocomposites and investigated its kinetics of release to the antibacterial drug levofloxacin. The GO–PSA composite showed a significant increase in the release time as compared with pure PSA. Besides, PSA modified with 2% GO showed a linear release with a nearly ideal behavior with a time delayed release up to 80 days and more than 95% of the drug release rate. Pandey et al. (2011) investigated the release behavior of gentamicin sulfate from graphene derived from methanol. The results indicated a mechanism dominated by diffusion, following the model of Korsmeyer–Peppas. Furthermore, the prepared nanohybrid can be used to treat various bacterial infections as an only medication, increasing patient compliance due to their prolonged action. Zhang et al. (2013) prepared double-hybrid films of Mg–Al hydroxide interspersed with anion GO–benzilpenicillin and hydroxides of GO–benzylpenicillin in double layers and investigated benzilpenicillin anions release behavior and its antibacterial activity. The release of benzylpenicillin anions was sustained for a long time and followed the first-order equation. Compared to the single GO film, prepared hybrid film showed an increase in antibacterial synergistic effect of GO and release of benzylpenicillin anions. Li et al. (2013a) prepared a GO–balofloxacine nanocomposite and investigated the loading and release behavior of this drug. The prepared nanocomposite showed longer release time due to the interaction of hydrogen bonds, which led to an excellent effect against *E. coli*. Ghadim et al. (2013) systematically investigated the behavior of tetracycline adsorption in GO at different pH, temperatures, and sorption time. Xiao et al. (2014) produced a carboxylated inclusion of graphene-β-cyclodextrin/chlorhexidine acetate (GO-COO-β-CD/CA) as a drug carrier based on graphene. The inclusion showed large blood compatibility as indicated by hemolysis testing and recalcification test. The inclusion also showed an excellent antibacterial activity and no cytotoxic effect. Wang et al. (2014a) produced hybrid materials of GO–PSA for controlled release of lomefloxacin. Soon the prepared hybrid system presented a perfect release of zero order. Lomefloxacin release was controlled by strong interactions of π–π stacking and hydrogen bonds with GO and its release rate can be controlled by GO content. Abdelhamid et al. (2014) first reported GO functionalized with gramicidin (GD) with antibacterial activities against various bacterial strains. They confirmed that GO could adsorb GD through physical interactions and showed higher antibacterial activity than GO and GD.

## 11.4 ANTIMICROBIAL ACTIVITY OF GRAPHENE

The antimicrobial mechanism of graphene and its derivatives still has not been effectively explained. One of the most interesting and simpler modes of action is the mechanical damage in the cell wall of bacteria by some forms of carbon. According to Park et al. (2010) material composed by GrO presented binding to *Bacillus cereus* bacterium and subsequent interference with bacterial growth, showing no cytotoxicity at different cell lines. In contrast, Hu et al. (2010) demonstrated that the GO and GrO can inhibit the growth of *E. coli* bacteria while having a slight cytotoxicity. In the case of GO nanosheets deposited in stainless steel, it has been found that cell damage from bacteria (*E. coli* or *S. aureus*) caused by direct contact of bacteria with the extremely sharp edges of the nanosheets was an effective mechanism in the bacterial inactivation. Furthermore, GO nanosheets reduced by hydrazine were more toxic to the bacteria than the nonreduced GO, Gram-positive (G+) bacteria were more sensitive to these materials, although in other studies, it was observed that the particles of GrO in suspension were more toxic to Gram-negative bacteria (G−). GO has been described as biocompatible, and promotes cell growth in GO film. *E. coli* cells caused a reduction in GL (60% deoxygenation) in the metabolic process, so that GL sheets could act as biocompatible sites for adsorption and bacterial proliferation. However, the bacteria remained sensitive to GrO, inhibiting the proliferation of *E. coli* (Akhavan and Ghaderi, 2010).

Early studies showed that the edges of the boards have a significant role in mechanical damage in bacterial cell walls, due to the fact that when graphene-based surface was smoothed, no antimicrobial effect was observed. Furthermore, it was shown that the interaction of GO with the bacteria depends not only on the physicochemical properties of GL (size, shape, degree of oxidation, etc.) but also on the cell wall of the microorganism. The characteristics of individual microorganisms (structure, presence or absence of structures such as cell wall, cell age of the bacteria, stress conditions in the environment, and metabolism) have an important influence on the sensitivity to carbon materials (Krishnamoorthy et al., 2012).

Other more complex mechanisms of action have also been demonstrated for graphene-based structures. Besides the already mentioned damage to the cell membrane integrity (mechanical damage), these materials might: involve bacteria, isolating them; generate harmful ROS; extract phospholipid molecules of the bacteria by the presence of lipophilic graphene; and reduce metabolic activity of the bacteria (Akhavan and Ghaderi, 2011; Liu et al., 2011a). In the case of GO, ROS production is mentioned as the main cause of bacterial killing (Gurunathan et al., 2012).

The extraction of phospholipid molecules proved as another antimicrobial mechanism of graphene, combining experimental and theoretical approaches. According to transmission electron microscopy (TEM) images, *E. coli* cells supported about three stages during the process of incubation with graphene. Cell membranes gradually lost their integrity, leading to loss of cytoplasm. The simulation of the subsequent molecular dynamics confirmed this finding by observing three distinguishable modes: balance, insertion, and extraction. It has been shown that during its entry, graphene sheets can insert and cut cell membranes of bacteria with their edges as blades (Tu et al., 2013).

The capture of microorganisms is another antimicrobial mechanism of graphene. The bacteria are trapped in graphene sheets when added to the bacterial suspension (Chen et al., 2014). The trapped bacteria are inactivated since they are isolated from their environment and cannot proliferate. The analysis of atomic force microscopy (AFM) showed that OG sheets can interact with bacteria and easily cover their surfaces. Furthermore, OG larger sheets showed a stronger antimicrobial activity due to its ability to completely cover the bacteria, block the active sites and decrease its viability. In comparison, GO smaller sheets showed low antimicrobial activity since they only joined the bacterial surface without isolating them completely (Akhavan and Ghaderi, 2011; Liu et al., 2012.).

The photothermal ablation is an effective strategy to eliminate trapped bacteria. Studies have shown that graphene has excellent intrinsic thermal conductivity as well as good absorption properties in the region near infrared, being highly suitable for photothermal therapy. Furthermore, the integration of a target substance can increase the efficiency of bacterial death, thereby treating the target region and minimizing cellular damage (Wei and Qu, 2012). Likewise, nanomaterials with superparamagnetic properties cause rapid aggregation of bacteria. In turn, the cytotoxic effect on other cells would be improved since radiation in the infrared region could be concentrated only in the bacteria (Wang et al., 2015; Wu et al., 2013).

Graphene-induced oxidative stress has also been applied to cause damage in bacterial cells through the generation of ROS, charge transfer, or direct oxidation of cellular components. The role of ROS in the antimicrobial activity of graphene can be confirmed by lipid peroxidation assay (Gurunathan et al., 2012; Krishnamoorthy et al., 2012). Gurunathan et al. (2012) compared the antimicrobial action of GO and GrO, causing oxidative stress induced by ROS generation in its antimicrobial activity. This mechanism was further confirmed by nuclear fragmentation test.

The charge transfer is considered another form of graphene-induced oxidative stress causing damage to the bacterial cells. Li et al. (2014) investigated the antibacterial activity of graphene films of large domain size made on various substrates. Their results showed that bacterial growth can only be inhibited by the graphene films on the Cu conductor and Ge semiconductor. The damage of the membrane was further confirmed by bacterial morphology. They proposed that electron transfer from microbial membrane to the graphene instead of causing damage mediated by ROS, contributes to its antimicrobial activity. This mechanism was further confirmed by subsequent reports demonstrating loss of antibacterial action of GO when its basal plane is masked (Hui et al., 2014; Salas et al., 2010). Another way to independently generate ROS by oxidative stress has been reported. Graphene-based materials-induced oxidative stress for the endogenous antioxidant glutathione, which mediate bacterial redox state. However, the production of ROS induced by superoxide anion was not detected (Mangadlao et al., 2015).

The physicochemical characteristics of graphene, such as size of its side, number of layers, surface charges, surface functional groups, and oxygen levels can have a significant impact on their antimicrobial properties. Tu et al. (2013) found that the antimicrobial activity of GO with side sheets of different sizes was dependent on the lateral size. After incubation for 2.5 h under the same concentration, a strong antimicrobial activity for larger GO sheets was observed. The large sheets of GO completely cover the bacteria by blocking their active sites and reducing its viability (Liu et al., 2012). In contrast, when the GO was coated on the substrate surface, smaller GO showed strong antimicrobial activity due to its higher density of defects, which is mediated by oxidative mechanisms (Perreault et al., 2015). Antibacterial activities of graphene also depend on time and concentration. The increased concentration of GO led to a continuous increase in its antibacterial activity (Liu et al., 2012). When the antimicrobial activity of different types of graphene was compared, the results indicated that the dispersed GO had the highest antimicrobial activity, followed by GrO, graphite, and graphite oxide. Their antimicrobial activities were largely dependent on the degree of dispersion (Wang et al., 2013b). The bacteria were kept individually in thin layers of GO mainly being incorporated into large aggregates of GrO. Kurantowicz et al. (2015) obtained similar results. They reported that different forms of graphene interact differently with pathogenic bacteria. The bacteria can adhere to the GO surface, resulting in the highest antimicrobial activity. As for graphene and GrO, the bacteria were prepared at the edges. Although GrO presented less antimicrobial effect than GO, its cytotoxicity was higher. The difference in cytotoxicity may be due to the variety of functional groups and surface charge of GO and GrO (Hu et al., 2010).

## 11.5 ANTIMICROBIAL ACTIVITY OF GRAPHENE-BASED NANOCOMPOSITES

Graphene-based nanocomposites have emerged as promising antibacterial materials since they can overcome the limitations of the individual components. For instance, the antibacterial nanomaterials bonded to a substrate are more stable and disperse in a better way (Ji et al., 2016).

### 11.5.1 GRAPHENE-BASED METALLIC NANOCOMPOSITES

Silver (Ag) has been used for centuries as an antibacterial agent. The released Ag ions can destroy bacterial cell membranes and enter cells to inactivate the enzymes and cause cell death (Feng et al., 2000). Silver nanoparticles (AgNPs) can cause direct damage to the bacterial cell membrane. Bacterial death may be due to bactericidal effects of the combined Ag ions released and AgNPs (Panacek et al., 2006). However, when pure AgNPs aggregate come into contact with bacteria, they lose their active surface area and show weak antibacterial activity. To overcome this problem, nanocomposite compounds of graphene and AgNPs were manufactured using various reducing agents, such as sodium borohydride ($NaBH_4$), hydrazine monohydrate, ascorbic acid, dopamine, glucose, starch, hydroquinone, gelatin, sodium citrate, microorganisms, and plant extracts (Chook et al., 2012; Haldorai et al., 2014; Ma et al., 2011). Shen et al. (2010) obtained nanocomposites of GrO–Ag via chemical reduction of silver ions in GO sheets applying mixed reducers. The resulting nanocomposite dispersion presented extensive surfaces and enhanced antibacterial activity, thereby indicating their potential as graphene-based antibacterial agents. Under GO support, the dispersion of AgNPs in aqueous solution can be well maintained. Its synergistic effect gave excellent antibacterial activity to GO–Ag nanocomposite. The decrease in surface charge of GO–Ag facilitated the contact of GO and AgNPs with the cells (Das et al., 2011; Shen et al., 2010). Another synergistic effect involving the adsorption properties of GrO sheets and bactericidal action of AgNPs was reported by Xu et al. (2011). The GrO–Ag nanocomposite showed satisfactory antibacterial activity, and no visible edema or erythema was observed in in vivo skin irritation testing. Kholmanov et al. (2012) have developed a composite film of graphene and silver nanowires with antibacterial properties. Ocsoy et al. (2013) applied a new technique for making various GO-based hybrid metallic nanostructures using the DNA as a guide. Among them, the double-stranded GO–DNA nanostructure (dsDNA)–Ag showed strong antibacterial effect on pathogenic bacterium *Xanthomonas perforans*. This technique increased the synergistic effect of GO and AgNPs increasing the adhesion of bacteria to the GO–Ag composites controlling the size, distribution, and aggregation of AgNPs. The antibacterial activity of the nanohybrid GO–Ag involved a species-specific mechanism of cell wall integrity disruption in *E. coli* and inhibition of cell division in *S. aureus*. Zhou et al. (2013) prepared a nanohybrid GO–Ag by in situ reduction of silver ions in GO using $NaBH_4$ as a reducing agent. The prepared nanohybrid showed a remarkable antibacterial effect on *E. coli*. Furthermore, it could be easily converted into fibers and films for extensive use as an antibacterial material (He et al., 2014).

Copper (Cu) is also an efficient and inexpensive antibacterial agent that kills bacteria rapidly in a metal surface. Typically, the redox properties of the copper nanoparticles (CuNPs) lead to cellular damage by oxidation of lipids and proteins (Wei et al., 2010). However, its antibacterial applications were limited due to their aggregation. Thus, it is important to increase the stability of CuNPs and control the release of $Cu^{2+}$. To overcome this limitation, Ouyang et al. (2013) prepared a hybrid GrO–Cu anchoring CuNPs in GrO surface modified with poly-L-lysine (PLL). This hybrid showed a high antibacterial activity in long-term and outstanding water solubility, suitable for microbial control.

In addition to these metals, gold (Au) and lanthanum were combined with GO or GrO to prepare antibacterial materials. Hussain et al. (2014) prepared GrO–Au using an environmental friendly route, which showed higher antimicrobial activity than GO sheets. Bacteria were trapped by GrO sheets and its permeability was stopped by AuNPs, leading to leakage of sugars and proteins, and eventually cell death. The authors attributed the antibacterial activity to oxidative stress both in antioxidant systems and the membrane. The photothermal properties of AuNPs under laser irradiation in the infrared region have also been explored to kill bacteria. A nanocomposite GO–lanthanum carboxylate was prepared and found to possess outstanding antibacterial and anticoagulant activities.

### *11.5.2 GRAPHENE-BASED METAL OXIDES NANOCOMPOSITES*

When the metal oxide semiconductor is illuminated by a luminous energy higher than their energy band gap, the electron–hole pairs will spread on its surface. Subsequently, the generated positive holes can react with water to produce hydroxyl radicals and negative electrons can combine with oxygen to form superoxide anion. The two types of ROS generated can effectively inactivate bacteria. A number of well-defined hybrid nanostructures consisting of semiconductor graphene and metal oxides such as $TiO_2$ and ZnO, have recently been reported with significantly improved photocatalytic properties (Liu et al., 2010; Shah et al., 2012.).

Hybrid composites of metal oxide and graphene semiconductors present several advantages. The graphene could significantly inhibit the recombination rate of the support load. Graphene can generate ROS as it absorbs UV/visible increasing visible light activity. Among all the photocatalyst semiconductor metal oxide, $TiO_2$ is the most promising candidate due to its excellent chemical stability and photocatalytic efficiency (Chen et al., 2010; Ding et al., 2011). Akhavan and Ghaderi (2009) investigated graphene increased efficiency in *E. coli* photoinactivation under sunlight using thin films of GrO–$TiO_2$. After lighting with UV for 4 h, GrO–$TiO_2$ showed the highest antibacterial activity which was 7.5 times stronger than that of a $TiO_2$ film. GrO platelets could accept electrons from the conduction band of $TiO_2$ excited by UV and effectively inhibit charge carrier recombination rate acting as electron heat sinks. Liu et al. (2011b) obtained GO–$TiO_2$ nanorods on a small scale through the assembly in two stages. GO–$TiO_2$ composites showed higher photocatalytic and antibacterial activities against *E. coli* than simple $TiO_2$ nanorods under simulated sunlight. These activities were attributed to the greater number of facets and anti-charge recombination. A porous graphene film was produced by Yin et al. (2013) using the technique of "water scatter." This film showed excellent antibacterial activity of broad spectrum and could be transferred to any substrate of interest. The functional nanoparticles, such as $TiO_2$, can be readily made by premix scaffolding conductors based on graphene. The honeycomb structures increased light absorption properties, thereby increasing the efficiency of light conversion. Kim et al. (2014) fabricated an independent film of graphene with antibacterial activity, incorporating titanate nanosheets and performing filtration directed to flow. This was the first report of highly flexible hybrid films composed of nanosheets metal oxide and graphene. The prepared hybrid films had greater hydrophilicity, mechanical strength, and chemical stability than GrO films. In addition, hybrid films showed a significantly greater sterilization efficiency than other nanostructured carbon-based films. The incorporated titanate nanosheets could control effectively membrane stress and surface roughness of the graphene film, thereby increasing significantly the antibacterial properties of the film. Gao et al. (2014) investigated the photocatalytic activity of $TiO_2$ in various nanostructures and found that a 3D sphere composed of nanosheets showed the highest activity. The coupling with GO sheets might further increase the activity of light absorption and disinfection against *E. coli.* Cao et al. (2013) further extended the range of light absorption to visible light. The GO–$TiO_2$ activated by an internal light (400–700 nm) showed an antibacterial effect higher than the isolated $TiO_2$ nanoparticles, indicating their potential for air disinfection.

ZnO is another photocatalyst semiconductor metal oxide used as an antibacterial agent. However, their photocatalytic activity is significantly limited by certain inherent deficiencies. For example, it can only be activated under UV irradiation due to bandgap, and its photocatalytic efficiency reduces with a high recombination rate of electrons and holes produced by light. To overcome these limitations, several GO–ZnO hybrids have been made with photocatalytic and antibacterial activities increased due to effective load separation (Ren et al., 2010). Kavitha et al. (2012) produced a hybrid graphene–ZnO and investigated its antibacterial activity under UV irradiation. Wang et al. (2014b) prepared composites of high quality GO–ZnO through an easy approach, which showed higher antibacterial ability and low cytotoxicity. The high antibacterial properties are attributed to the synergistic effect of GO and ZnO. OG facilitated the dispersion of the ZnO nanoparticles and placed them in close contact with bacteria. The contact has increased the concentration of zinc around the bacteria and the permeability of the bacterial membrane, leading to death of bacterial cells.

Moreover, $SnO_2$ and $Fe_3O_4$ were also combined with graphene antibacterial applications. The nanosheets of graphene–$SnO_2$ tend to adsorb to bacteria cells, blocking the uptake of nutrients by them and resulting in cell death. The nanosheets showed a cytotoxic effect 3.6 times higher than graphene because of the synergistic effect. GO–$Fe_3O_4$ with magnetic nanoparticles of iron oxide dispersed in GO nanosheets showed excellent antibacterial activity, and its convenient separation property facilitated the disinfection of water. Its effect on *E. coli* was confirmed by the release of ROS and the total protein degradation (Santhosh et al., 2014; Wu ct al., 2014). GO–$Fe_3O_4$ hybrid compounds were easily penetrated or adsorbed on the bacterial cells, causing leakage of intercellular contents and loss of cellular integrity. Its antibacterial property was proposed to increase the stress of the membrane and oxidative stress during the incubation period (Deng et al., 2014).

### *11.5.3 GRAPHENE-BASED POLYMER NANOCOMPOSITES*

The poor solubility and processability severely limit the application of antibacterial graphene because it tends to aggregate due to strong interplanar interactions. This can be solved by incorporating the graphene in a polymer matrix. A highly stable dispersion of graphene and polymer can be obtained using π-electron-rich polymers. Since infection normally occurs during biomaterial implantation procedures, the inhibitory effect of polymer–graphene hybrids on bacterial proliferation on its surface has been studied. Various polymeric matrixes such as poly-*N*-vinyl carbazole (PVK), chitosan (CS), and PLL have been used to manufacture antibacterial hybrids to reduce bacterial cell viability (Ko et al., 2013; Lim et al., 2012; Lu et al., 2012; Musico et al., 2014).

Among all the available polymer, PVK is a promising candidate due to its excellent electronic and mechanical performance, corrosion resistance, ease of production of thin films, and the use as polymeric dispersant. Santos et al. (2012) built the first monohybrid polymer of GO and PVK to use as an antibacterial coating. The nanohybrids more efficiently inactivated the bacteria when they were electrodeposited on the surface of indium tin oxide (ITO) than on unmodified surfaces. The incorporation of PVK in nanocomposites of GO–PVK led to a well-defined and homogeneous coating on various surfaces but without significant negative effect on the antibacterial activity. Compared to the GO pristine, GO–PVK containing 3% GO showed a strong antibacterial property and no significant cytotoxic effect on human cells. The antibacterial effects of GO–PVK were up to 57% (biofilms) and 30% (planktonic cells) larger than the GO alone. GO–PVK can encapsulate bacteria, thereby reducing their metabolism and eventually resulting in cell death (Carpio et al., 2012; Some et al., 2012.). Nanocomposites of graphene–PVK also have antibacterial and biocompatible properties, with many potential biomedical applications. Musico et al. (2014) commercially modified membrane filter surfaces available with PVK, graphene, graphene–PVK, and GO–PVK to impart antibacterial properties to treat water. Among them, the membrane filter modified with GO–PVK killed more effectively *Bacillus subtilis* and *E. coli*.

Chitosan (CS) is a linear hydrophilic polysaccharide biopolymer composed of glucosamine and *N*-acetylglucosamine, which can be applied as an antibacterial agent. The GrO–CS mounted noncovalently showed improved antibacterial properties against *E. coli* JM109 compared to GO (Rabea et al., 2013). Sreeprasad et al. (2011) built hybrid based on graphene, anchoring CS, gold modified by CS, and native lactoferrin on the surface of GO/GrO. The combined materials showed maximum antibacterial activity due to the synergistic effect. Some et al. (2012) have prepared novel compounds of graphene–PLL with dual functionality and biocompatibility of antibacterial activity through covalent and electrostatic interactions between graphene and lipopolysaccharide-binding protein derivatives.

### 11.5.4 GRAPHENE-BASED MULTICOMPONENT NANOCOMPOSITES

Compounds were manufactured for antibacterial multicomponent application in order to combine the advantages of graphene, nanoparticles, and polymers. The nanohybrid GO–Au@Ag bidimensional were manufactured by nonselective electrolytic deposition of Ag on preassembled GO–Au (Wang et al., 2013). The prepared hybrid showed enhanced antibacterial activity against *E. coli*, higher than the activity of Ag ions and nanoparticles of Au@Ag. This improvement can be attributed to the increased local concentration of AgNPs in testing of bacteria and bacterial multivalent interactions with a surface. Due to the smaller nanoparticles random aggregation and reduced sensitivity to $Cl^-$, these nanocomposites can be used as curatives. The graphene oxide–iron oxide nanoparticles—silver (GO–IONP–Ag) nanocomposite were also synthesized, with significantly increased antibacterial activity. In this nanocomposite, the photothermal ablation of graphene showed a prominent synergistic antibacterial effect. Furthermore, these nanocomposites are economical and environmental friendly, and can be easily recycled by magnetic separation (Tian et al., 2014). Gao et al. (2013) prepared hierarchical GO–ZnO–Ag plasmonic sulfonated compounds. The composites showed significantly higher antibacterial activity against *E. coli* than ZnO, ZnO–Ag, and GO–ZnO sulfonated under visible light irradiation. Its excellent antibacterial activity is attributed to the synergistic effect of the sulfonated GO sheets, AgNPs, and arrays of nanorods of ZnO, respectively. The GrO–Au–$TiO_2$ hybrid was manufactured by He et al. (2013) which showed antibacterial activity higher than GrO, $TiO_2$, and GO–$TiO_2$ irradiation by sunlight. Madhavan et al. (2013) designed a type of electrical wiring based on graphene–$TiO_2$–ZnO composite nanofibers, which presented a ninefold increase in conductance values compared with $TiO_2$–ZnO nanofiber. Antibacterial studies have indicated the potential use of these nanofibers in antibacterial curatives. To prepare graphene metallic nanocomposites, uniform dispersion of metal nanoparticles on the surfaces of graphene and the strong interaction of carbon and metal are necessary (Williams et al., 2012). However, the insolubility of graphene and the interaction between graphene and metal nanoparticles due to the inactive surfaces of graphene restricted its development. This can be overcome by introducing a polymer as a medium. AgNPs were anchored to a GrO substrate modified with polyethyleneimine (PEI) to prepare a nanocomposite PEI–GrO–Ag. This nanocomposite presented a long-term antibacterial activity, which is ideally used as a sprayable antibacterial solution due to its lower cytotoxicity and excellent water solubility (Cai et al., 2012). Tai et al. (2012) attached nanocrystals of Ag to nanosheets of graphene poly(acrylic acid) to prepare nanocomposites of graphene–*g*-poly(acrylic acid)–Ag. This strategy generated AgNPs in situ without requiring additional reducers or a complicated process. Hybrid materials of graphene–polidopamine–Ag were manufactured using polidopamine layer as a guide and as a reducing dopamine, which showed strong antibacterial broad-spectrum activities.

Even though many of GO–polymer composite nanofibers have been prepared, their superficial use was very low since GO sheets used were curly or were covered by fiber. This problem can be solved using simple sheets of expanded GO and maintaining the functional surface. Liu et al. (2014) manufactured continuous and uniform nanofibers of poly(vinyl alcohol) (PVA)–CS–GO combining irradiation electron beam and a technique of electrical wiring. The nanofibers prepared have had an excellent antibacterial activity against *E. coli*, indicating the potential for tissue engineering, drug delivery, and wound healing. In addition, CS–PVA nanofibers containing graphene were prepared for wound healing by Lu et al. (2012). CS–PVA–graphene fibers led to rapid and complete healing of wounds compared to CS–PVA fibers and controls. The authors argued that any free electron in graphene could inhibit the growth of prokaryotic cells without affecting eucaryoticones.

## 11.6 CONCLUSIONS

The antibacterial graphene mechanism has not been fully elucidated. The antibacterial activity of graphene and GO has also been discussed with various groups claiming that GO has little antibacterial activity. However, most current findings and advances support its antibacterial activity. They can be well dispersed to produce thin sheets involving bacteria easily, inactivating bacteria later by the stress induced by the membrane edges and oxidative stress induced by basal planes. The physicochemical characteristics, such as OG lateral size, number of layers, and oxygen content, can have a significant impact on its antibacterial property. The large sheets of GO easily covered the bacteria and blocked their active sites, while lower GOs showed higher defect density. The thickness of GO increases as the number of layers increases, thereby weakening the effect of the GO blade edges. GO with a higher oxygen content on the edges may form a nanostructure transmembrane to reduce the repulsion between the hydrophobic lipid tails and hydrophilic atoms of the edge. In addition, the carboxyl and hydroxyl groups, which are anchored on the surface of GO and tend to dissociate, may slightly reduce the pH of bacterial microenvironments. In addition, the impurities generated during the synthesis process of GO due to careless washing, such as sulfur and manganese, may disrupt bacterial microenvironment and inhibit their proliferation. These controversial results are attributed to differences in preparation methods and lateral sizes of the graphene surface treatments. Further studies are to understand the mechanisms and factors that influence the antibacterial activity of the materials based on graphene. The nanosheets based on graphene emerged as broad-spectrum antibacterial environmental friendly materials. Compared to conventional chemical antimicrobials, they are renewable, easier to obtain, and cheaper with little bacterial resistance. Their antibacterial activities include physical damage from direct contact of its sharp edges with bacterial membranes and destructive extraction of lipid molecules and chemical damage due to oxidative stress. Furthermore, graphene can be used as a support to disperse and stabilize various nanomaterials, such as metals, metal oxides, and polymers, leading to a high antibacterial activity due to a synergistic effect. Graphene can be used even as a nanocarrier for controlling the administration of antibiotics and improving therapeutic efficacy, having advantages such as convenience, high utilization rate, and reduced toxicity. The main limitation of nanosheets based on graphene is its tendency to agglomerate due to their high surface energy, which can reduce their antimicrobial activity, inevitably altering their surface properties and edge. The addition of the nutrient medium, bovine serum albumin, and tryptophan will significantly inhibit the antibacterial activity since the adsorbed substances may prevent the graphene to interact with bacteria. Besides, it is important to developed more functionalized graphene and hybrid graphene with desirable properties, for instance, increasing graphene dispersibility in biological mediums and adjusting to graphene toxicity.

## KEYWORDS


- **antimicrobial**
- **drug delivery**
- **graphene**
- **microbiology**
- **nanoscience**
- **nanotechnology**


## REFERENCES


Abdelhamid, H. N.; Khan, M. S.; Wu, H.-F. Graphene Oxide as a Nanocarrier for Ramicidin (GOGD) for High Antibacterial Performance. *RSC Adv.* **2014,** *4*, 50035–50046.

Adibkia, K.; Barzegar-Jalali, M.; Nokhodchi, A.; SiahiShadbad, M.; Omidi, Y.; Javadzadeh, Y. A Review on the Methods of Preparation of Pharmaceutical Nanoparticles. *Pharm. Sci.* **2010,** *15*, 303–314.

Akhavan, O.; Ghaderi, E. Photocatalytic Reduction of Graphene Oxide Nanosheets on $TiO_2$ Thin Film for Photoinactivation of Bacteria in Solar Light Irradiation. *J. Phys. Chem. C* **2009,** *113*, 20214–20220.

Akhavan, O.; Ghaderi, E. Toxicity of Graphene and Graphene Oxide Nanowalls Against Bacteria. *ACS Nano* **2010,** *4*, 5731–5736.

Akhavan, O.; Ghaderi, E. *Escherichia coli* Bacteria Reduce Graphene Oxide to Bactericidal Graphene in a Self-limiting Manner. *Carbon* **2012,** *50*, 1853–1860.

Cai, X.; Lin, M.; Tan, S.; Mai, W.; Zhang, Y.; Liang, Z.; Lin, Z.; Zhang, X. The use of Polyethyleneimine-modified Reduced Graphene Oxide as a Substrate for Silver Nanoparticles to Produce a Material with Lower Cytotoxicity and Long-term Antibacterial Activity. *Carbon* **2012,** *50*, 3407–3415.

Cao, B.; Cao, S.; Dong, P.; Gao, J.; Wang, J. High Antibacterial Activity of Ultrafine $TiO_2$/Graphene Sheets Nanocomposites Under Visible Light Irradiation. *Mater. Lett.* **2013,** *93*, 349–352.

Carpio, I. E. M.; Santos, C. M.; Wei, X.; Rodrigues, D. F. Toxicity of a Polymer–Graphene Oxide Composite Against Bacterial Planktonic Cells, Biofilms, and Mammalian Cells. *Nanoscale* **2012,** *4*, 4746–4756.

Chen, C.; Cai, W.; Long, M.; Zhou, B.; Wu, Y.; Wu, D.; Feng, Y. Synthesis of Visible-light Responsive Graphene Oxide/$TiO_2$ Composites with p/n Heterojunction. *ACS Nano* **2010,** *4*, 6425–6432.

Chen, J.; Wang, X.; Han, H. A New Function of Graphene Oxide Emerges: Inactivating Phytopathogenic Bacterium *Xanthomonas oryzae* pv. *oryzae*. *J. Nanopart. Res.* **2013,** *15*, 1–14.

Chen, J.; Peng, H.; Wang, X.; Shao, F.; Yuan, Z.; Han, H. Graphene Oxide Exhibits Broad-spectrum Antimicrobial Activity Against Bacterial Phytopathogens and Fungal Conidia by Intertwining and Membrane Perturbation. *Nanoscale* **2014,** *6*, 1879–1889.
Chook, S. W.; Chia, C. H.; Zakaria, S.; Ayob, M. K.; Chee, K. L.; Neoh, H. M.; Huang, N. M. Silver Nanoparticles–Graphene Oxide Nanocomposite for Antibacterial. *Adv. Mater. Res.* **2012,** *364*, 439–443.
Das, M. R.; Sarma, R. K.; Saikia, R.; Kale, V. S.; Shelke, M. V.; Sengupta, P. Synthesis of Silver Nanoparticles in an Aqueous Suspension of Graphene Oxide Sheets and Its Antimicrobial Activity. *Colloids Surf. B Biointerfaces* **2011,** *83*, 16–22.
Deng, C. H.; Gong, J. L.; Zeng, G. M.; Niu, C. G.; Niu, Q. Y.; Zhang, W.; Liu, H. Y. Inactivation Performance and Mechanism of *Escherichia coli* in Aqueous System Exposed to Iron Oxide Loaded Graphene Nanocomposites. *J. Hazard. Mater*. **2014,** *276*, 66–76.
Ding, Y.; Zhang, P.; Zhuo, Q.; Ren, H.; Yang, Z.; Jiang, Y. A Green Approach to the Synthesis of Reduced Graphene Oxide Nanosheets Under UV Irradiation. *Nanotechnology* **2011,** 22 (21), 215601.
Dreyer, D. R.; Park, S.; Bielawski, C. W.; Ruoff, R. S. The Chemistry of Graphene Oxide. *Chem. Soc. Rev*. **2010,** *39*, 228–240.
Edwards, R. S.; Coleman, K. S. Graphene Synthesis: Relationship to Applications. *Nanoscale* **2013,** *5*, 38–51.
Feng, Q. L.; Wu, J.; Chen, G. Q.; Cui, F. Z.; Kim, T. N.; Kim, J. O. A Mechanistic Study of the Antibacterial Effect of Silver Ions on *Escherichia coli* and *Staphylococcus aureus*. *J. Biomed. Mater. Res*. **2000,** *52*, 662–668.
Feng, L. Z.; Liu, Z. A. Graphene in Biomedicine: Opportunities and Challenges. *Nanomedicine* **2011,** *6*, 317–324.
Gao, P.; Ng, K.; Sun, D. D. Sulfonated Graphene Oxide–ZnO–Ag Photocatalyst for Fast Photodegradation and Disinfection Under Visible Light. *J. Hazard. Mater*. **2013,** *262*, 826–835.
Gao, P.; Li, A. R.; Sun, D. D.; Ng, W. J. Effects of Various $TiO_2$ Nanostructures and Grapheme Oxide on Photocatalytic Activity of $TiO_2$. *J. Hazard. Mater*. **2014,** *279*, 96–104.
Ghadim, E. E.; Manouchehri, F.; Soleimani, G.; Hosseini, H.; Kimiagar, S.; Nafisi, S. Adsorption Properties of Tetracycline onto Graphene Oxide: Equilibrium, Kinetic and Thermodynamic Studies. *PLoS One* **2013,** *8* (11), e79254.
Gurunathan, S.; Han, J. W.; Dayem, A. A.; Eppakayala, V.; Kim, J. H. Oxidative Stress-mediated Antibacterial Activity of Graphene Oxide and Reduced Graphene Oxide in *Pseudomonas aeruginosa. Int. J. Nanomed.* **2012,** *7*, 5901–14.
Haldorai, Y.; Kim, B.-K.; Jo, Y.-L.; Shim, J.-J. Ag@graphene Oxide Nanocomposite as an Efficient Visible-light Plasmonic Photocatalyst for the Degradation of Organic Pollutants: A Facile Green Synthetic Approach. *Mater. Chem. Phys*. **2014,** *143*, 1452–1461.
He, W.; Huang, H.; Yan, J.; Zhu, J. Photocatalytic and Antibacterial Properties of Au–$TiO_2$ Nanocomposite on Monolayer Graphene: From Experiment to Theory. *J. Appl. Phys*. **2013,** *114*, 204701.
He, T.; Liu, H.; Zhou, Y.; Yang, J.; Cheng, X.; Shi, H. Antibacterial Effect and Proteomic Analysis of Graphene-based Silver Nanoparticles on a Pathogenic Bacterium *Pseudomonas aeruginosa. Biometals* **2014,** *27*, 673–682.

He, J. L.; Zhu, X. D.; Qi, Z. N.; Wang, C.; Mao, X. J.; Zhu, C. L.; He, Z. Y.; Lo, M. Y.; Tang, Z. S. Killing Dental Pathogens Using Antibacterial Graphene Oxide. *ACS Appl. Mater. Interfaces* **2015,** *7*, 5605–5611.

Hu, W. B.; Peng, C.; Luo, W. J. Graphene-based Antibacterial Paper. *ACS Nano* **2010,** *4*, 4317–4323.

Hui, L.; Piao, J.; Auletta, J.; Hu, K.; Zhu, Y.; Meyer, T.; Liu, H.; Yang, L. Availability of the Basal Planes of Graphene Oxide Determines Whether It Is Antibacterial. *ACS Appl. Mater. Interfaces* **2014,** *6*, 13183–13190.

Hussain, N.; Gogoi, A.; Sarma, R. K.; Sharma, P.; Barras, A.; Boukherroub, R.; Saikia, R.; Sengupta, P.; Das, M. R. Reduced Graphene Oxide Nanosheets Decorated with Au Nanoparticles as an Effective Bactericide: Investigation of Biocompatibility and Leakage of Sugars and Proteins. *ChemPlusChem* **2014,** *79*, 1774–1784.

Ji, H.; Sun, H.; Qu, X. Antibacterial Applications of Graphene-based Nanomaterials: Recent Achievements and Challenges. *Adv. Drug Deliv. Rev.* **2016,** *105* (Pt B), 176–189.

Kavitha, T.; Gopalan, A. I.; Lee, K.-P.; Park, S.-Y. Glucose Sensing, Photocatalytic and Antibacterial Properties of Graphene–ZnO Nanoparticle Hybrids. *Carbon* **2012,** *50*, 2994–3000.

Kholmanov, I. N.; Stoller, M. D.; Edgeworth, J.; Lee, W. H.; Li, H.; Lee, J.; Barnhart, C.; Potts, J. R.; Piner, R.; Akinwande, D.; Barrick, J. E.; Ruoff, R. S. Nanostructured Hybrid Transparent Conductive Films with Antibacterial Properties. *ACS Nano* **2012,** *6*, 5157–5163.

Kim, Y. S.; Kumar, K.; Fisher, F. T.; Yang, E. H. Out-of-plane Growth of CNTs on Graphene for Supercapacitor Applications. *Nanotechnology* **2012,** *23* (1), 015301.

Kim, I. Y.; Park, S.; Kim, H.; Park, S.; Ruoff, R. S.; Hwang, S.-J. Strongly-coupled Free-standing Hybrid Films of Graphene and Layered Titanate Nanosheets: An Effective Way to Tailor the PHYSICOCHEMICAL and Antibacterial Properties of Graphene Film. *Adv. Funct. Mater.* **2014,** *24*, 2288–2294.

Krishnamoorthy, K.; Veerapandian, M.; Zhang, L. H.; Yun, K.; Jae, S. Antibacterial Efficiency of Graphene Nanosheets Against Pathogenic Bacteria via Lipid Peroxidation. *J. Phys. Chem. C* **2012,** *116*, 17280–17287.

Ko, T. Y.; Kim, S. Y.; Kim, H. G.; Moon, G.-S.; In, I. Antibacterial Activity of Chemically Reduced Graphene Oxide Assembly with Chitosan Through Noncovalent Interactions. *Chem. Lett.* **2013,** *42*, 66–67.

Kurantowicz, N.; Sawosz, E.; Jaworski, S.; Kutwin, M.; Strojny, B.; Wierzbicki, M.; Szeliga, J.; Hotowy, A.; Lipinska, L.; Kozinski, R.; Jagiello, J.; Chwalibog, A. Interaction of Graphene Family Materials with *Listeria monocytogenes* and *Salmonella enterica*. *Nanoscale Res. Lett.* **2015,** *10* (23).

Kwiatkowski, D. Science, Medicine, and the Future: Susceptibility to Infection. *BMJ* **2000,** *321*, 1061–1065.

Li, F.; Yang, C.; Liu, B.; Sun, X. Properties of a Graphene Oxide–Balofloxacin Composite and Its Effect on Bacteriostasis. *Anal. Lett.* **2013a,** *46*, 2279–2289.

Li, C.; Wang, X.; Chen, F.; Zhang, C.; Zhi, X.; Wang, K.; Cui, D. The Antifungal Activity of Graphene Oxide–Silver Nanocomposites. *Biomaterials* **2013b,** *34*, 3882–3890.

Li, J.; Wang, G.; Zhu, H.; Zhang, M.; Zheng, X.; Di, Z.; Liu, X.; Wang, X. Antibacterial Activity of Large-area Monolayer Graphene Film Manipulated by Charge Transfer. *Sci. Rep.* **2014,** *4* (4359), 1–8.

Lim, H. N.; Huang, N. M.; Loo, C. H. Facile Preparation of Graphene-based Chitosan Films: Enhanced Thermal, Mechanical and Antibacterial Properties. *J. Non-Cryst. Solids* **2012,** *358*.

Lin, D.; Qin, T.; Wang, Y.; Sun, X.; Chen, L. Graphene Oxide Wrapped SERS Tags: Multifunctional Platforms Toward Optical Labeling, Photothermal Ablation of Bacteria and the Monitoring of Killing Effect. *ACS Appl. Mater. Interfaces* **2014,** *6*, 1320–1329.

Liu, J. Bai, H.; Wang, Y.; Liu, Z.; Zhang, X.; Sun, D. D. Self-assembling $TiO_2$ Nanorods on Large Graphene Oxide Sheets at a Two-phase Interface and Their Anti-recombination in Photocatalytic Applications. *Adv. Funct. Mater.* **2010,** *20*, 4175–4181.

Liu, J.; Liu, L.; Bai, H.; Wang, Y.; Sun, D. D. Gram-scale Production of Graphene Oxide–$TiO_2$ Nanorod Composites: Towards High-activity Photocatalytic Materials. *Appl. Catal. B Environ.* **2011a,** *106*, 76–82.

Liu, S.; Zeng, T. H.; Hofmann, M.; Burcombe, E.; Wei, J.; Jiang, R.; Kong, J.; Chen, Y. Antibacterial Activity of Graphite, Graphite Oxide, Graphene Oxide, and Reduced Grapheme Oxide: Membrane and Oxidative Stress. *ACS Nano* **2011b,** *5*, 6971–6980.

Liu, S.; Hu, M.; Zeng, T. H.; Wu, R.; Jiang, R.; Wei, J.; Wang, L.; Kong, J.; Chen, Y. Lateral Dimension-dependent Antibacterial Activity of Graphene Oxide Sheets. *Langmuir* **2012,** *28*, 12364–12372.

Liu, Y.; Park, M.; Shin, H. K.; Pant, B.; Choi, J.; Park, Y. W.; Lee, J. Y.; Park, S.-J.; Kim, H.-Y. Facile Preparation and Characterization of Poly(Vinyl Alcohol)/Chitosan/Grapheme Oxide Biocomposite Nanofibers. *J. Ind. Eng. Chem.* **2014,** *20*, 4415–4420.

Lu, B.; Li, T.; Zhao, H.; Li, X.; Gao, C.; Zhang, S.; Xie, E. Graphene-based Composite Materials Beneficial to Wound Healing. *Nanoscale* **2012,** *4*, 2978–2982.

Ma, J.; Zhang, J.; Xiong, Z.; Yong, Y.; Zhao, X. S. Preparation, Characterization and Antibacterial Properties of Silver-modified Graphene Oxide. *J. Mater. Chem.* **2011,** *21*, 3350–3352.

Madhavan, A. A.; Mohandas, A.; Licciulli, A.; Sanosh, K. P.; Praveen, P.; Jayakumar, R.; Nair, S. V. A.; Nair, S.; Balakrishnan, A. Electrospun Continuous Nanofibers Based on a $TiO_2$–ZnO–Graphene Composite. *RSC Adv.* **2013,** *3*, 25312–25316.

Maktedar, S. S.; Mehetre, S. S.; Singh, M.; Kale, R. K. Ultrasound Irradiation: A Robust Approach for Direct Functionalization of Graphene Oxide with Thermal and Antimicrobial Aspects. *Ultrason. Sonochem.* **2014,** *2*, 11407–1416.

Mangadlao, J. D.; Santos, C. M.; Felipe, M. J. L.; de Leon, A. C. C.; Rodrigues, D. F.; Advincula, R. C. On the Antibacterial Mechanism of Graphene Oxide (GO) Langmuir-Blodgett Films. *Chem. Commun.* **2015,** *51*, 2886–2889.

Mohammadi, G.; Nokhodchi, A.; Barzegar-Jalali, M.; Lotfipour, F.; Adibkia, K.; Ehyaei, N. Physicochemical and Antibacterial Performance Characterization of Clarithromycin Nanoparticles as Colloidal Drug Delivery System. *Colloids Surf. B Biointerfaces* **2011,** *88*, 39–44.

Musico, Y. L. F.; Santos, C. M.; Dalida, M. L. P.; Rodrigues, D. F. Surface Modification of Membrane Filters Using Graphene and Graphene Oxide-based Nanomaterials for Bacterial Inactivation and Removal. *ACS Sustain. Chem. Eng.* **2014,** *2*, 1559–1565.

Novoselov, K. S.; Geim, A. K.; Morozov, S. V. Electric Field Effect in Atomically Thin Carbon Films. *Science* **2004,** *306*, 666–669.

Ocsoy, I.; Gulbakan, B.; Chen, T.; Zhu, G.; Chen, Z.; Sari, M. M.; Peng, L.; Xiong, X.; Fang, X.; Tan, W. DNA-guided Metal-nanoparticle Formation on Graphene Oxide Surface. *Adv. Mater*. **2013,** *25*, 2319–2325.

Ouyang, Y.; Cai, X.; Shi, Q.; Liu, L.; Wan, D.; Tan, S.; Ouyang, Y. Poly-l-lysine-Modified Reduced Graphene Oxide Stabilizes the Copper Nanoparticles with Higher Water-solubility and Long-term Additively Antibacterial Activity. *Colloids Surf. B Biointerfaces* **2013,** *107*, 107–114.

Panacek, A.; Kvítek, L.; Prucek, R.; Kolář, M.; Večeřová, R.; Pizúrová, N.; Sharma, V. K.; Nevěčná, T.; Zbořil, R. Silver Colloid Nanoparticles: Synthesis, Characterization, and Their Antibacterial Activity. *J. Phys. Chem. B* **2006,** *110* (33), 16248–16253.

Pandey, H.; Parashar, V.; Parashar, R.; Prakash, R.; Ramteke, P. W.; Pandey, A. C. Controlled Drug Release Characteristics and Enhanced Antibacterial Effect of Graphene Nanosheets Containing Gentamicin Sulfate. *Nanoscale* **2011,** *3*, 4104–4108.

Park, S.; Mohanty, N.; Suk, J. W.; Nagaraja, A.; An, J.; Piner, R. D.; Cai, W.; Dreyer, D. R.; Berry, V.; Ruoff, R. S. Biocompatible, Robust Free-standing Paper Composed of a TWEEN/Graphene Composite. *Adv. Mater*. **2010,** *22*, 1736–1740.

Perreault, F.; de Faria, A. F.; Nejati, S.; Elimelech, M. Antimicrobial Properties of Graphene Oxide Nanosheets: Why Size Matters. *ACS Nano*. **2015,** *9*, 7226–7236.

Rabea, E. I.; Badawy, M. E. T.; Stevens, C. V.; Smagghe, G.; Steurbaut, W. Chitosan as Antimicrobial Agent: Applications and Mode of Action. *Biomacromolecules* **2003,** *4*, 1457–1465.

Ren, C.; Yang, B.; Wu, M.; Xu, J.; Fu, Z.; lv, Y.; Guo, T.; Zhao, Y.; Zhu, C. Synthesis of Ag/ZnO Nanorods Array with Enhanced Photocatalytic Performance. *J. Hazard. Mater*. **2010,** *182*, 123–129.

Rice, L. B. Federal Funding for the Study of Antimicrobial Resistance in Nosocomial Pathogens: No ESKAPE. *J. Infect. Dis*. **2008,** *197*, 1079–81.

Rizzello, L.; Pompa, P. P. Nanosilver-based Antibacterial Drugs and Devices: Mechanisms, Methodological Drawbacks, and Guidelines. *Chem. Soc. Rev*. **2014,** *43*, 1501–1518.

Ross, A. S.; Baliga, C.; Verma, P.; Duchin, J.; Gluck, M. A Quarantine Process for the Resolution of Uodenoscope-associated Transmission of Multidrug Resistant *Escherichia coli*. *Gastrointest. Endosc*. **2015,** *82*, 477–483.

Salas, E. C.; Sun, Z. Z.; Luttge, A.; Tour, J. M. Reduction of Graphene Oxide via Bacterial Respiration. *ACS Nano* **2010,** *4*, 4852–4856.

Sanchez, V. C.; Jachak, A.; Hurt, R. H.; Kane, A. B. Biological Interactions of Graphene-family Nanomaterials: An Interdisciplinary Review. *Chem. Res. Toxicol*. **2012,** *25*, 15–34.

Santhosh, C.; Kollu, P.; Doshi, S.; Sharma, M.; Bahadur, D.; Vanchinathan, M. T.; Saravanan, P.; Kim, B.-S.; Grace, A. N. Adsorption, Photodegradation and Antibacterial Study of Graphene–$Fe_3O_4$ Nanocomposite for Multipurpose Water Purification Application. *RSC Adv*. **2014,** *4*, 28300–28308.

Santos, C. M.; Mangadlao, J.; Ahmed, F.; Leon, A.; Advincula, R. C.; Rodrigues, D. F. Graphene Nanocomposite for Biomedical Applications: Fabrication, Antimicrobial and Cytotoxic Investigations. *Nanotechnology* **2012,** *23* (39), 395101.

Shen, J.; Shi, M.; Li, N.; Yan, B.; Ma, H.; Hu, Y.; Ye, M. Facile Synthesis and Application of Ag-chemically Converted Graphene Nanocomposite. *Nano Res*. **2010,** *3*, 339–349.

Shah, M. S. A.; Park, A. R.; Zhang, K.; Park, J. H.; Yoo, P. J. Green Synthesis of Biphasic $TiO_2$–reduced Graphene Oxide Nanocomposites with Highly Enhanced Photocatalytic Activity. *ACS Appl. Mater. Interfaces* **2012,** *4*, 3893–3901.

Some, S.; Ho, S.-M.; Dua, P.; Hwang, E.; Shin, Y. H.; Yoo, H.; Kang, J.-S.; Lee, D.-K.; Lee, H. Dual Functions of Highly Potent Graphene Derivative—Poly-l-lysine Composites to Inhibit Bacteria and Support Human Cells. *ACS Nano* **2012,** *6*, 7151–7161.

Sreeprasad, T. S.; Maliyekkal, M. S.; Deepti, K.; Chaudhari, K.; Xavier, P. L.; Pradeep, T. Transparent, Luminescent, Antibacterial and Patternable Film Forming Composites of Graphene Oxide/Reduced Graphene Oxide. *ACS Appl. Mater. Interfaces* **2011,** *3*, 2643–2654.

Suk, J. W.; Lee, W. H.; Lee, J. Enhancement of the Electrical Properties of Graphene Grown by Chemical Vapor Deposition via Controlling the Effects of Polymer Residue. *Nano Lett.* **2013,** *13*, 1462–1467.

Sutter, P. W.; Flege, J. I.; Sutter, E. A. Epitaxial Graphene on Ruthenium. *Nat Mater*. **2008,** *7*, 406–411.

Tai, Z.; Ma, H.; Liu, B.; Yan, X.; Xue, Q. Facile Synthesis of Ag/GNS-g-PAA Nanohybrids for Antimicrobial Applications. *Colloids Surf. B Biointerfaces* **2012,** *89*, 147–151.

Taitt, C. R.; Leski, T. A.; Heang, V.; Ford, G. W.; Prouty, M. G.; Newell, S. W.; Vora, G. J. Antimicrobial Resistance Genotypes and Phenotypes from Multidrug-resistant Bacterial Wound Infection Isolates in Cambodia. *J. Blobal. Antimicrob. Res*. **2015,** *3*, 198–204.

Tian, T.; Shi, X.; Cheng, L.; Luo, Y.; Dong, Z.; Gong, H.; Xu, L.; Zhong, Z.; Peng, R.; Liu, Z. Graphene-based Nanocomposite as an Effective, Multifunctional, and Recyclable Antibacterial Agent. *ACS Appl. Mater. Interfaces* **2014,** *6*, 8542–8548.

Tu, Y.; Lv, M.; Xiu, P.; Huynh, T.; Zhang, M.; Castelli, M.; Liu, Z.; Huang, Q.; Fan, C.; Fang, H.; Zhou, R. Destructive Extraction of Phospholipids from *Escherichia coli* Membranes by Graphene Nanosheets, *Nat. Nanotechnol*. **2013,** *8*, 594–601.

Yin, S.; Goldovsky, Y.; Herzberg, M.; Liu, L.; Sun, H.; Zhang, Y.; Meng, F.; Cao, X.; Sun, D. D.; Chen, H.; Kushmaro, A.; Chen, X. Functional Free-standing Graphene Honeycomb Films. *Adv. Funct. Mater*. **2013,** *23*, 2972–2978.

Wang, X.; Zhou, N.; Yuan, J.; Wang, W.; Tang, Y.; Lu, C.; Zhang, J. Shen, J. Antibacterial and Anticoagulation Properties of Carboxylated Graphene Oxide–Lanthanum Complexes. *J. Mater. Chem*. **2012,** *22*, 1673–1678.

Wang, H.; Liu, J.; Wu, X.; Tong, Z.; Deng, Z. Tailor-made Au@Ag Core-shell Nanoparticle 2D Arrays on Protein-coated Graphene Oxide with Assembly Enhanced Antibacterial Activity. *Nanotechnology* **2013a,** *24*, 205102.

Wang, X.; Liu, X.; Han, H. Evaluation of Antibacterial Effects of Carbon Nanomaterials Against Copper-resistant *Ralstonia solanacearum*, *Colloids Surf. B: Biointerfaces*. **2013b,** *103*, 136–142.

Wang, Y. W.; Cao, A.; Jiang, Y.; Zhang, X.; Liu, J. H.; Liu, Y.; Wang, H. Superior Antibacterial Activity of Zinc Oxide/Graphene Oxide Composites Originating from High Zinc Concentration Localized Around Bacteria. *ACS Appl. Mater. Interfaces* **2014a,** *6*, 2791–2798.

Wang, T.; Zhang, Z.; Gao, J.; Yin, J.; Sun, R.; Bao, F.; Ma, R. Synthesis of Graphene Oxide Modified Poly(Sebacic Anhydride) Hybrid Materials for Controlled Release Applications. *Int. J. Polym. Mater. Polym. Biomater*. **2014b,** *63*, 726–732.

Wang, N.; Hu, B.; Chen, M.; Wang, J. Polyethylenimine-mediated Silver Nanoparticle-decorated Magnetic Graphene as a Promising Photothermal Antibacterial Agent. *Nanotechnology* **2015,** *26* (19), 195703.

Wei, Y.; Chen, S.; Kowalczyk, B.; Huda, S.; Gray, T. P.; Grzybowski, B. A. Synthesis of Stable, Low-dispersity Copper Nanoparticles and Nanorods and Their Antifungal and Catalytic Properties. *J. Phys. Chem. C* **2010,** *114*, 15612–15616.

Wei, D. C.; Liu, Y. Q. Controllable Synthesis of Graphene and Its Applications. *Adv Mater*. **2010,** *22*, 3225–3241.

Wei, W.; Qu, X. Extraordinary Physical Properties of Functionalized Graphene. *Small* **2012,** *8*, 2138–2151.

Williams, G.; Seger, B.; Kamat, P. V. $TiO_2$–Graphene Nanocomposites UV-assisted Photocatalytic Reduction of Graphene Oxide. *ACS Nano* **2012,** *2*, 1487–1491.

Wu, M.; Deokar, A. R.; Liao, J.; Shih, P.; Ling, Y. Graphene-based Photothermal Agent for Rapid and Effective Killing of Bacteria. *ACS Nano* **2013,** *7*, 1281–1290.

Wu, B.-S.; Abdelhamid, H. N.; Wu, H.-F. Synthesis and Antibacterial Activities of Graphene Decorated with Stannous Dioxide. *RSC Adv*. **2014,** *4*, 3722–3731.

Xiao, Y.; Fan, Y.; Wang, W.; Gu, H.; Zhou, N.; Shen, J. Novel GO-COO-beta-CD/CA Inclusion: Its Blood Compatibility, Antibacterial Property and Drug Delivery. *Drug Deliv*. **2014,** *21*, 362–369.

Xu, W.-P.; Zhang, L.-C.; Li, J.-P.; Lu, Y.; Li, H.-H.; Ma, Y.-N.; Wang, W.-D.; Yu, S.-H.; Facile Synthesis of Silver@Graphene Oxide Nanocomposites and Their Enhanced Antibacterial Properties. *J. Mater. Chem*. **2011,** *21*, 4593–4597.

Zhang, Z.; Zhang, J.; Zhang, B.; Tang, J. Mussel-inspired Functionalization of Graphene for Synthesizing Ag-polydopamine-graphene Nanosheets as Antibacterial Materials. *Nanoscale* **2013,** *5*, 118–123.

Zhou, Y.; Yang, J.; He, T.; Shi, H.; Cheng, X.; Lu, Y. Highly Stable and Dispersive Silver Nanoparticle–Graphene Composites by a Simple and Low-energy-consuming Approach and Their Antimicrobial Activity. *Small* **2013,** *9*, 3445–3454.

Zhu, Y. W.; Murali. S.; Cai, W. W. Graphene and Graphene Oxide: Synthesis, Properties, and Applications. *Adv. Mater*. **2010a,** *22*, 3906–3924.

Zhu, Y. W.; Murali, S.; Stoller, M. D.; Velamakanni, A.; Piner, R. D.; Ruoff, R. S. Microwave Assisted Exfoliation and Reduction of Graphite Oxide for Ultracapacitors. *Carbon* **2010b,** *48*, 2118–2122.

# CHAPTER 12

# CARBON NANOTUBES FOR DRUG DELIVERY

CEYDA TUBA SENGEL-TURK[1] and ONUR ALPTURK[2*]

[1]*Department of Pharmaceutical Technology, Faculty of Pharmacy, Ankara University, Turkey*

[2]*Department of Chemistry, Istanbul Technical University, Maslak 34469, Istanbul, Turkey*

[*]*Corresponding author. E-mail: onur.alpturk@itu.edu.tr*

## CONTENTS

## ABSTRACT

Carbon nanotubes offer an alternative and promising platform of drug delivery system due to their distinct and unified physicochemical properties. On the account of its great potential, these nanostructures have been employed as efficient delivery system of many therapeutic active compounds, ranging from antineoplastic agents, cardiovascular drugs, anti-infectives to anti-inflammatory molecules, and genes. In concert with their great promise, this chapter is devoted to the literature of chemistry of carbon nanotubes. Herein, we aim to discuss extensively many aspects of these materials including functionalization, drug loading capacity, and mechanisms. Subsequently, the literature on recent patents is surveyed to shed light on the current status of the field. Lastly, their potential applications as different therapeutic modalities in drug delivery, their toxicology profiles, and the developed strategies to overcome their cytotoxicities are also discussed with a detailed perspective.

## 12.1 TARGETED DRUG DELIVERY: WHERE DO WE STAND?

The word "therapeutic agents" or "drugs" fundamentally refer to the products (either natural or synthetic) designed to cure or prevent certain diseases or medical conditions. Their universal mode of action is to block a damaged or abnormal biological process upon which the physiological activities of the body are restored. Undoubtedly, this depiction sounds a little bit like an oversimplification, in the sense that how so-called therapeutics function in living systems is indeed fairly complex. "Therapeutic effect of a drug" begins with their administration, followed by their distribution within the body. However, the problems emerge only after these compounds are directed to their target tissue or organ to manifest their therapeutic effects. In doing this, either or any combination(s) of the following(s) could potentially happen: (1) drugs could travel to a normal tissue as well, besides the malfunctioning ones, (2) they might be excreted from body, and/or (3) they could be metabolized or even degraded even before making it to their destination. A simple solution to tackle all these problems at once is to administer these drugs to body at much elevated doses, with the hope that at least, a minor fraction of it would make it to their target in an intact form. However, this approach is rather unsought, as exposing body to a surplus of drug may conduce toward more side effects than what a tissue can tolerate (Torchilin, 1997).

Which way should the pharmaceutical companies track to resolve these problems? For one thing, they always have the option to formulate some novel therapeutic agents with superior properties (such as diminished toxicity, improved tissue distribution, and so forth), which translates into starting over from scratch. With all the progress in bioinformatics and high-throughput screening techniques, designing authentic compounds looks even more facile than ever. Granted, some fundamental problems that scientists experience turn the notion of "drug discovery" into an uneasy, pricy, and overwhelmingly time-taking procedure; still, the results from in vitro experiments or de novo design through in silico methods do not afford a congener success under in vivo conditions. Still, getting a grip on how these compounds interact with surrounding cells and tissues could be moderately puzzling. Surely, all these insights clarify why ca. 95% of the novel therapeutic agents suffer from inadequate biopharmaceutical properties (Koo et al., 2005).

Consequently, the focus of the scientific community has shifted on developing means to deliver drugs more efficiently; the underlying fundamentals of this notion dates back to Paul Ehrlich's vision to manufacture "tailored drugs" (Strebhardt and Ullrich, 2008). In modern medicine, his pioneering work has been a cornerstone of targeted drug delivery, which addresses the delivery of drugs by means of a carrier. What makes this line of delivery so alluring is that it considerably improves the therapeutic efficiency of drugs, as they tend to accumulate only in a certain region of body. Moreover, this methodology provides a control over the dose, and the release rate of drugs (Coelho et al., 2010). That being said, we should remark that a vast amount research has been devoted to targeted drug delivery in last few decades and numerous delivery systems (either of organic or inorganic origin) have emerged as candidates to fulfill the magic bullet of medicine. To date, targeted drug delivery, which is, in a way, a state-of-art trick for drugs, appears to be an asset to achieve optimum therapeutic efficiency with minimum amount of drugs.

In compliance with the concept of targeted drug delivery, this chapter is crafted to survey the literature of carbon nanotubes (CNTs), as a drug carrier. Firstly, we will discuss the basics of these materials to give the readers a feel for their structural aspects. Next is an overview on the functionalization of CNTs, which is designed to serve as initiative to "carbon nanotubes in medicine." The last concept we intend to cover is the toxicology profile of these materials, as it forms a link to "to which degree CNTs have been commercialized" and hence, "which direction the research on CNTs is headed in the future": the answers to these questions will embody our concluding remarks.

## 12.2 CHARACTERISTICS AND STRUCTURE

Carbon with $sp^2$-like hybridization is exclusive. This is because its valence permitted scientific community to engineer a large collection of molecular architectures over the years. What makes all these structures truly phenomenal is that they are indeed built from the same component and yet, they still can differ in shape, and dimensionality this greatly (Fig. 12.1). For instance, fullerenes structurally resemble a soccer ball, whereas graphene and CNTs are planner and cylindrical arrangement of carbon atoms, respectively (Terrones et al., 2010). No matter how unrelated their structures appear to be, these allotropes of carbon ultimately converge on numerous aspects, including thermal and electrical conductivity, respectable mechanical strength, and chemical inertness. Hence, these nanostructures are in demand by today's material science, and vital for nanoscience, electronics, and biomedical applications (De Volder et al., 2013).

Prior to reviewing why CNTs gained considerable attention in the context of targeted delivery, we first aim to discuss the fundamentals of CNTs. The morphology CNTs are cylinders formed by rolling one or multiple graphene layers. This description also hints on how CNTs are structurally classified: single-walled carbon nanotubes (SWCNs), and multiple-walled carbon tubes (MWCNs) (Fig. 12.2). SWCNs harbor one single cylinder of graphite sheet, with a diameter ranging from 0.4 to 3.0 nm (Iijima, 1991). By contrast, MWCNs are conventionally depicted as an array of tubes, which are coaxially aligned around a central hollow. The distance between layers seems invariable, and corresponds roughly to the distance of graphite-layer spacing. The diameters of MWCNs are dictated by the numbers of layers: the inner diameter can change from 0.4 nm to a few nanometers, while the outer diameter ranges from 2 to 100 nm (Ajayan and Ebbesen, 1997).

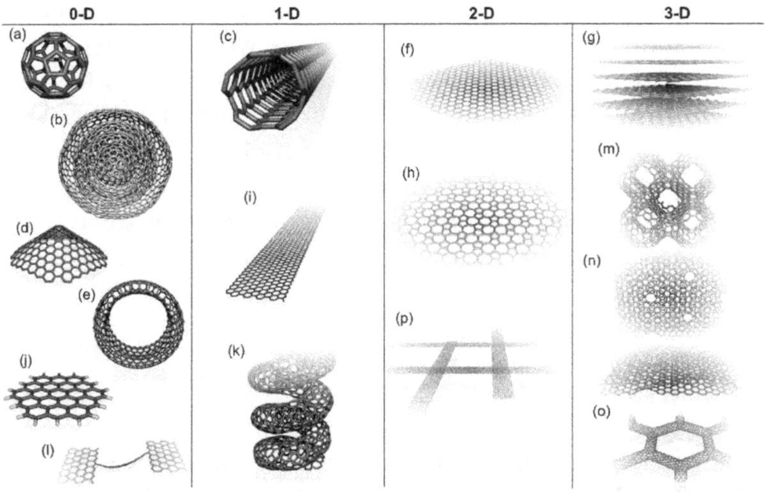


**FIGURE 12.1** Nanomaterials derived from sp$^2$-like hybridized carbon, with different dimensionalities: (a) $C_{60}$: buckminsterfullerene; (b) nested giant fullerenes or graphitic onions; (c) carbon nanotube; (d) nanocones or nanohorns; (e) nanotoroids; (f) graphene surface; (g) 3D graphite crystal; (h) Haeckelite surface; (i) graphene nanoribbons; (j) graphene clusters; (k) helicoidal carbon nanotube; (l) short carbon chains; (m) 3D Schwarzite crystals; (n) carbon nanofoams; (o) 3D nanotube networks, and (p) nanoribbons 2D networks. (Reprinted with permission from Terrones et al., 2010. © 2010, Elsevier.)

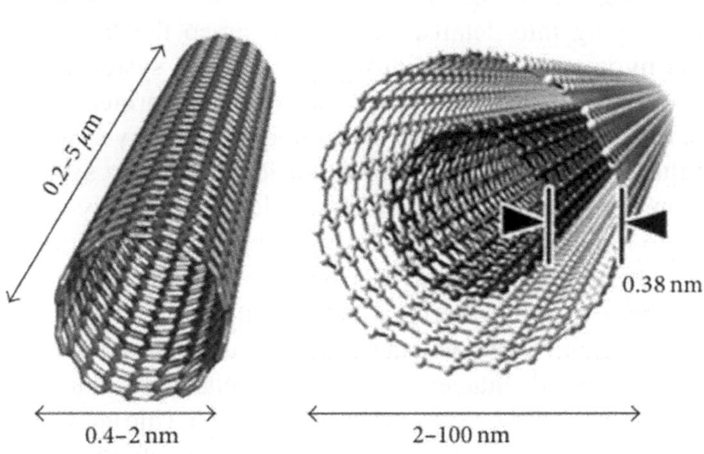


**FIGURE 12.2** The structures of SWCNs and MWCNs. (Reprinted from He, H.; Pham-Huy, L. A.; Dramou, P.; Xiao, D.; Zuo, P.; Pham-Huy, C. Carbon Nanotubes: Applications in Pharmacy and Medicine. *Biomed. Res. Int.* **2013,** *2013,* https://creativecommons.org/licenses/by/3.0/.)

As comparable as these nanostructures are in many ways, there exist some structural divergences among them as well. The first one is the length of the tube: relevant literature cites that MWCNs have generally a length in the range of micrometer, as that of SWCNTs varies from 1 μm to a few centimeters (Eatemadi et al., 2014). And the second one is the precision in diameter, such that SWNTs have well-defined diameters, whereas MWNTs are known to suffer from some structural defects, rendering them relatively less stable materials (Dresselhaus et al., 2004).

## 12.3 MERITS OF CNTs

From the pioneering works with silicone rubber in 1960s to nano era in the new millennium, drug delivery systems have evolved to a point where some made it through clinical studies (Hoffman, 2008). To take matters further, some of these systems (e.g., liposomes and nanoparticles) shine out and encouraged scientific community to launch the phase of commercialization (Pattni et al., 2015; Teekamp et al., 2015). Given how far drug delivery systems have come, one cannot help wondering whether CNTs are worth going back to square one or not.

Before diving into details, we need to grasp the fact that present effort and hype to turn CNTs into effective cargo systems stems from the need to remedy the deficiencies observed with former drug delivery systems. In reality, what is expected from drug delivery system goes far beyond than what Ehrlich described: an ideal delivery system should function in a way to release the therapeutic agents solely in a specific zone, while maintaining possibly minimum or no interaction with the immune system (Bae and Park, 2011). As unrealistic as these expectations may appear in short term, we are definitely on the right track by means of designing some new materials through derivatization. In that regard, CNTs are advantageous over other related structures, because they could imaginably be equipped with multiple (bio)molecules, each of which serves a different purpose: targeting agents to diminish side effects, drugs to provide therapeutic effects, stealth agents to avoid the immune system, and diagnostic agents to monitor a certain part of the body, if required. Their second profit concerns clinical chemistry because they have the capacity of retaining several copies of drugs

(Kushwaha et al., 2013). In theory, it looks like these merits might be the key to achieve appreciable therapeutic indexes with drug delivery systems and CNTs hold the potential to be as closest to the magic bullet as we will ever be.

For instance, Heisner and coworkers fabricated triply functionalized SWCNs conjugated to doxorubicin against colon cancer. Therein, the system consists of carcinoembryonic antigen (a monoclonal antibody) to recognize tumor makers and fluorescein dye to track nanotubes within cells. As the most central element in this design, doxorubicin is the anti-cancer agent, which accounts for the pharmaceutical action of the system. Following the application of this system to WiDr human colon cancer cells, fluorescence visualization via confocal microscopy revealed that SWCNs–doxorubicin conjugate was apparently taken up by cancerous cells and that upon the release of doxorubicin, SWCNs remained in the cytoplasmic region of the cancer cells, while active pharmaceutical ingredients (API) fully passed to the nucleus section where doxorubicin exerts its effects (Heister et al., 2009) (Fig. 12.3). This study alone proves that multiple functionalizations are truly beneficial to track the bioactivity of pharmaceutical compounds.

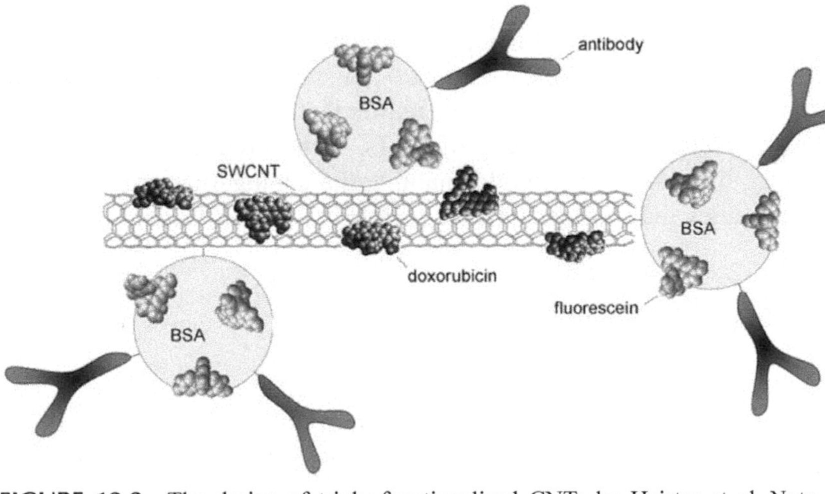


**FIGURE 12.3** The design of triply functionalized CNTs by Heister et al. Note that antibodies provide selectivity in drug delivery and doxorubicin is adsorbed on the graphitic surface. Also, the tubes can accommodate fluorescein dye for imaging purposes. (Reprinted with permission from Heister et al., 2009. © 2012, Elsevier.)

## 12.4 FUNCTIONALIZATION OF CNTs

Despite of all their merits, these nanostructures are not perfect either. Concerning their chemistry, the principal problem lies in that raw CNTs are insoluble virtually in any solvent or milieu. Because of strong van der Waals interactions among tubes, in conjunction with their hydrophobic character, CNTs do not disperse in solution but they rather form bundles or bundle aggregates. Undoubtedly, this complication could be resolved only if the medium can wet these hydrophobic surfaces or the intermolecular attractions between individual tubes are overcome by any means (Revathi et al., 2015). Overall, this condition constitutes a major impediment on two counts. On one hand, the lack of solubility or inability to disperse—especially, in water—vastly restricts their use in biological and biomedical applications, for an obvious reason. On the other hand, it precludes the whole notion of derivatization, which employs wet chemistry and relevant techniques (Battigelli et al., 2013). Hence, devising strategies to prevail these limitations is seemingly a prerequisite concerning the chemistry of CNTs and paves the way to the utilization of these organic materials as drug delivery systems.

Presently, efforts to disperse CNTs fall under four major categories: (1) surfactant-facilitated dispersion, (2) solvent dispersion, (3) the functionalization of CNTs sidewalls, and (4) biomolecular dispersion. Among those strategies, the approach of functionalization (either covalently or noncovalently) is well reputed by the reason that it may concurrently exercise effect on some other aspects of CNTs, such as reducing cytotoxicity, enhancing biocompability, while enabling the conjugation of some other (bio)molecules and pharmaceutical agents (Foldvari and Bagonluri, 2008).

## 12.5 COVALENT MODIFICATIONS

In general, covalent functionalization, which secures more stable derivatizations, is a powerful strategy to introduce numerous functional groups on CNTs (halogens, carbenes, and arynes, to name a few) (Zhang et al., 2011). From the standpoint of well-documented restraints to CNTs (vide supra), this strategy is eminently valuable on the grounds that it allows the derivatization of hydrophilic moieties, which remarkably facilitates the dispersion of CNTs. Presently, one approach is addition through 1,3-dipolar cycloaddition reactions, which was formerly developed to derivatize fullerenes (Maggini and Scorrano, 1993). Therein, the condensation of α-amino acids with an aldehyde affords azomethine ylides that are composed of a carbanion neighboring immonium ion. Once these ylides undergo cycloaddition reactions with dipolarophiles such as CNTs, they afford pyrrolidine intermediates, through which functional groups could be anchored on to CNTs (Fig. 12.3). To substantiate the significance of this chemistry, a prominent example is the derivatization of CNTs with triethylene glycol monomethyl ether chains, which render CNTs fully soluble in several solvents, including chloroform, acetone, ethanol, and more remarkably water. More importantly, the finding that the solubility of these functionalized materials is persistent and no precipitate was observed at least for 2 weeks is very promising, and should not be overlooked, giving the gravity of the problem (Georgakilas et al., 2002).

The other popular form of covalent modifications is the oxidation of CNTs, which inherently generates oxygen-containing functional groups (i.e., carbonyl, hydroxyl, and carboxylic acids being the predominant one) on both tips and defect sites (Fig. 12.4; Lin et al., 2003). Regarding their presumptive impact on dispersion, these polar groups are of twofold significance:

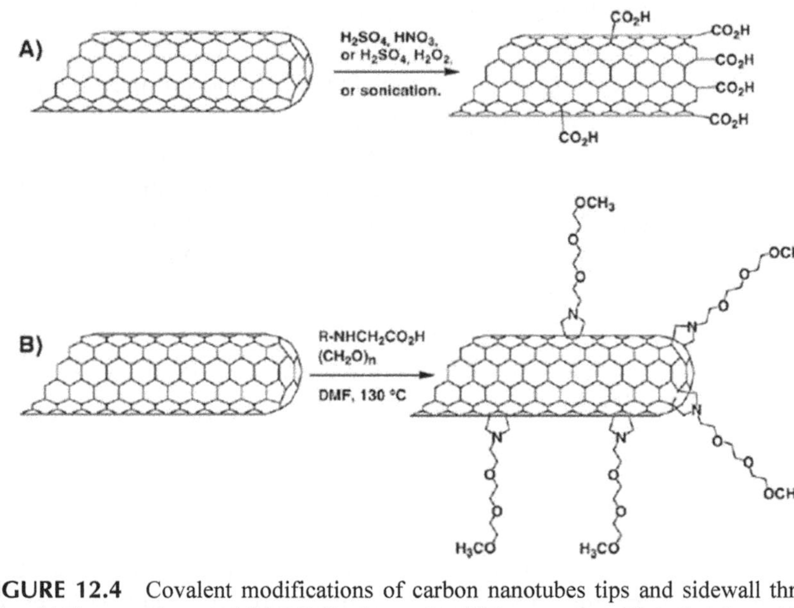


**FIGURE 12.4** Covalent modifications of carbon nanotubes tips and sidewall through (a) oxidation reactions and (b) 1,3-dipolar cycloaddition reaction. Note that the oxidation of CNTs also opens the tips. (Reprinted with permission from Klumpp et al., 2006, © Elsevier.)

apart from their characteristic hydrophilicity, carboxylic acids permit further incorporation of some other highly hydrophilic residues of choice. However, the poor reactivity of these functional groups doubtlessly necessitates their preactivation either by acyl chlorides (such as thionyl chloride or oxalyl chloride) or coupling agents such as carbodiimides and hydroxybenzotriazole, at first (Jain et al., 2003; Prato et al., 2008). For instance, Jain reported a cascade of chemical modifications whereby oxidized MWCNTs are galactosylated upon the activation of carboxylic acids with $SOCl_2$. As can be anticipated, the incorporation of this hydrophilic carbohydrate is reported to drastically improve the dispersion of the materials at different pH values ranging from 4 to 9, whereas as-synthesized materials exhibit a low dispersion profile under the same conditions (Jain et al., 2003).

Regardless, some major inadequacies or drawbacks considerably obstruct the yet-to-be-seen potential of these derivatization reactions, upon which its chemistry remained relatively less explored to date:

1. Oxidative chemistry requires some drastic reaction conditions, such as reflux in strong acids such as $HNO_3$ (Kyotani et al., 2001), $HNO_3/H_2SO_4$ (Liu et al., 1998), $KMnO_4/H_2SO_4$ (Hiura et al., 1995), or strong oxidizing agents such as $OsO_4$ (Hwang, 1995).
2. Chemoselectivity and regioselectivity in these reactions is hard to control, as a result of which the characterization of the final products remained a challenge (Chen et al., 2003).
3. Oxidation reactions notably damages the structural integrity of CNTs by causing structural defects over the tubular structure (Zhang et al., 2003), the shortening of tubes (Rinzler et al., 1998), the loss of small diameter CNTs (Yang et al., 2002), and even the loss of the entire material (Hu et al., 2003).
4. Upon altering the hybridization of carbon atoms, covalent modifications intrinsically disrupt the aromatic character of CNTs (Fischer, 2002).

## 12.6 NONCOVALENT MODIFICATIONS

Because both strategies of functionalization ultimately serve the same purpose, covalent and noncovalent functionalization are subject of comparison in terms of their benefits and drawbacks. In that frame, noncovalent functionalizations appear to be more favorable in that they are conventionally carried out under simple reactions conditions such as sonication and centrifuge, and obviate the necessity of harsh reactions conditions and strong reagents (Di Crescenzo et al., 2014). Besides, this mode of functionalization poses no threat to the aromaticity of CNTs and the structural integrity of both graphitic surface and tips (Zhao and Stoddart, 2009). However, all these merits do not necessarily imply that noncovalent functionalizations are trouble-free. For instance, noncovalent conjugates may dissociate from CNTs in biological fluids or may even undergo exchange with serum proteins. It is needless to say that this circumstance may raise an unquestionable concern of toxicity, as well as some other unsought outcomes (Battigelli et al., 2013).

At present, molecules prevalently utilized in noncovalent functionalizations are surfactants, polymers, and biological materials such as proteins, nucleic acids, and peptides. Their incorporation with CNTs is simply physical adsorption on to the outer walls of CNTs, which is driven by van der Walls forces, π–π, and CH–π, and similar interactions. Apart from generating steric clash (as always), these solubilizers function to procure dispersibility in aqueous medium through their hydrophilic parts, as their charge—if available—prohibits aggregate formation due to electrostatic repulsions (Kocharova et al., 2007). In light of extensive studies, it would appear that the molecules harboring aromatic groups tend to be more adequate dispersive agents in consequence of better π–π interactions. For instance, sodium dodecyl sulfate disperses CNTs at concentrations as high as just 0.1 mg/mL, as sodium dodecyl benzene sulfonate (SDBS) bearing aromatic residues can provide almost 10-fold more concentrated suspensions (Hu et al., 2009). In a similar way, biopolymers such as single-stranded nucleic acids and peptides containing aromatic acids turn out to be strong solubilizer for CNTs, exactly because of this reasoning (Zheng et al., 2003). We conclude this section by pointing out that these materials bind to CNTs in divergent geometries; while polymeric materials wrap CNTs to maximize van der Waals interactions, surfactants form micelle-like assemblies around graphitic surface whose absolute structure demands some clarification (Hu et al., 2009).

## 12.7 DRUG LOADING ONTO CNTs

Being one of the fundamental measures to be an expedient delivery system, drug loading is the process wherein active drugs are combined with the carriers to give the final form of the drug delivery system. CNTs are distinctly privileged from this aspect because their spherical shape and high surface area to volume ratio grant a tremendous potential to accommodate drugs (Heister et al., 2012). Besides, loading capacity can be improved through decorating hydrophilic or amphiphilic polymers on the surface, as a result of which some extra space is acquired (for instance, see: Liu et al., 2007). Another distinction of these nanomaterials is the fact that they are in position of carrying pharmaceutical agents through multiple ways, such as encapsulation inside the cavity (Arsawang et al., 2011), tethering on the surface upon functionalization, and adsorption on the wall or among the walls of CNTs (Chen et al., 2011). In overall, CNTs have been widely utilized as delivery system for many drugs because of these benefits (Table 12.1).

**TABLE 12.1** Drugs Encapsulated in CNTs.[a]

| Type of nanotubes | Drug | Method of immobilization |
|---|---|---|
| MWCNTs | Cisplatin | Encapsulation via capillary forces |
| f-CNTs | Amphotericin B | Conjugated to carbon nanotubes |
| SWCNTs | Gemcitabine | Encapsulation |
| MWNTs | Epirubicin hydrochloride | Adsorption |
| MWCNTs-poly(ethylene glycol-b-propylene sulfide) | Doxorubicin | Adsorption |
| f-CNTs | Sulfamethoxazole | Adsorption |
| SWNTs-PL-PEG-$NH_2$ | Pt(IV) prodrug-folic acid (FA) | Covalent amide linkages |
| SWNTs | Cisplatin-epidermal growth factor (EGF) | Attachment to carbon nanotubes via amide linkages |
| MWCNTs | Dexamethasone | Encapsulation |

[a]Adapted with permission from Wilczewska et al. © 2012 Elsevier.

Considering that the essential role of these systems is to ensure the structural integrity of drugs, encapsulation (or "endohedral functionalization") certainly diminishes the degradation of pharmaceutical agents and offers a platform to release drugs only under certain conditions

(Perry et al., 2011). This strategy is predominantly applicable to drugs with low surface tension with the reason that encapsulation is driven by capillary forces and hydrophobic forces. Currently, this mode of functionalization is very popular and further culminated with some smart strategies. For instance, Pastorin devised unprecedented "carbon nanobottles" wherein oxidized MWCNTs are capped with gold nanoparticles to halt the uncontrolled release of encapsulated cisplatin (Fig. 12.5) (Li et al., 2012).

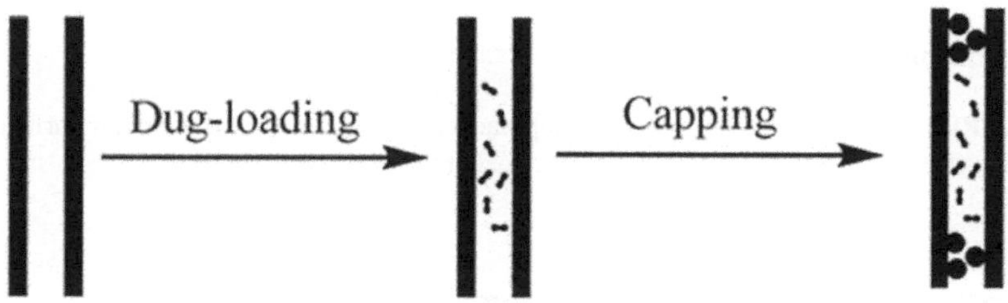


**FIGURE 12.5** The fabrication of carbon nanobottles. The ends of encapsulated carbon nanotubes are capped with gold nanoparticles. (Reprinted with permission from Liu, 2013. © 2013, American Chemical Society.)

In comparison to encapsulation, surface derivatization (i.e., exohedral functionalization) leaves the drugs explicitly exposed, which carves out a major handicap (Tsai et al., 2013). Notwithstanding this concern, the payoff is that this approach offers a rich repertoire of (bio)conjugation methods, ranging from tethering (i.e., covalent attachment of drugs) to noncovalent interactions (Fig. 12.6). Herein, we should remind the fact that so-called tethering requires an additional step of oxidation to create functional groups for conjugation and may not be as effortless as it appears on the paper. Besides, covalent bonds are likely to tune the molecular structure of drugs, and therefore their specificity and bioactivity could be altered (Tsai et al., 2013). Sure, it sounds displeasing but still the examples of tethering are well documented because covalent bonds provide an enduring link between drug and nanocarriers (for instance, see Lee et al., 2005). In contrast, noncovalent approaches are less toilless than tethering but they may eventually elicit the premature dissociation of therapeutic agents in biological fluids. To conclude, the choice of loading mechanism and location uncloaks a delicate balance between loading efficiency and therapeutic efficiency of pharmaceutical agents.

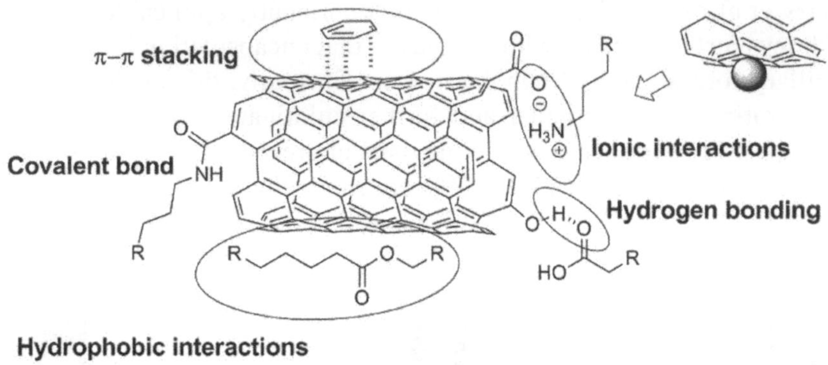


**FIGURE 12.6** An overview of all the approaches to attach a drug to the outer surface of CNTs. (Reprinted with permission from Boncel et al., 2013. © Elsevier.)

## 12.8 UPTAKE CAPACITY OF CNTs

A key factor in understanding of drug pharmacokinetics and pharmacodynamics is to comprehend how drugs are taken by projected cells, once they are applied to tissues. The big question that ought to be answered is whether a given drug is capable of passing through the cellular membrane or not, prior to all clinical studies and others. Under the condition where drugs are sufficiently water-soluble, diffusion through lipid bilayer is thought to be the principal mechanism of drug intake. But this mechanism is impaired under two instances: either drug is of exceedingly high molecular weight or it is not lipophilic enough to pass through bilayer structure (Dobson and Kell, 2008). In that case, the permeation of therapeutic agents could be salvaged or restored only through nanocarriers, which deeply facilitates the whole process in many ways.

In 2007, Lacerda and coworkers investigated the cell penetrating ability of CNTs in a general context. In light of their results, four parameters seem to impact the interactions of CNTs with cellular membranes: (1) their modes of manufacture and postprocess procedures, (2) their structural aspects (i.e., whether the materials are functionalized or not), (3) their surface properties (i.e., whether the materials are charged or not), and (4) the effect of functional groups on the surface (i.e., carboxylic acids or other oxygen containing groups). Besides, their work revealed an intriguing and equally unexpected finding that the uptake capacity of

CNTs seems independent of the type of cells (Lacerda et al., 2007). All these data laid the foundation on how to design drug delivery systems out of these organic nanomaterials and set the stage for detailed investigations which are yet to come. Later on, Rafa reported that MWCNs smaller than 1000 nm could be transported across cellular membranes via an energy-independent pathway. In the same study, the researchers also shown that submicron-sized MWCNs, which are commonly cited as "nanoneedles," enter the cells more readily than micron-sized MWCNs (Raffa et al., 2008). Another correlation between length and transport is that SWCNs with a size of up to 400 nm can be taken up through the cells via diffusion as a result of passive transport, while those with a size larger than 400 nm are internalized into the cell membranes through macrophages via endocytosis. Given that diffusion was the predominant mechanism of the internalization to that point, these conclusions are surely a game changer for the (bio)chemistry of CNTs, as well as the notion of drug delivery.

Sadly enough, the cellular uptake profile of these materials is far more perplexing that what we came to know. To make the scenario more complicated, nanotubes, which enter the cell through diffusion, tend to locate into the cytosolic compartment of the cell, whereas endocytotic vesicles are the main region for the CNTs entered into the cell after endocytosis (Antonelli et al., 2010). In other words, the mechanism of cellular uptake predetermines the intracellular location of materials upon entry. In an attempt to elucidate the rationale behind these key findings, the transport pathways of various MWCNs into HEK293 and the human embryonic kidney epithelial cells were elaborately researched. After all, it became perspicuous that regardless of their surface charge, single-type MWCNs penetrated into the cytoplasma via diffusion mechanism, while bundle-type MWCNs crossed the cellular membrane through endocytosis (Mu et al., 2009). With some parallel results from other researchers, we finally gathered some basic understanding on how these materials behave around and within cells (Chen and Schluesener, 2010).

Of course, it goes without saying that the acquisition of all these results would not be conceivable if we lacked proper techniques to evaluate the cellular entry of CNTs. In other words, the development of expeditious and accurate analytical techniques is in demand to fully explicate the biochemistry of CNTs and other nanosystems. In 2011, Draper and his coworkers developed an electrophoretic method to affirm and evaluate the cellular penetration of CNTs. In this multistep technique, cells that were exposed to nanocarriers are isolated first, lysed, and separated through gel electrophoresis. Upon processing gel images, this procedure enables one to quantify nanoparticles in a given region or cells (Draper et al, 2011). Likewise, Lin et al. developed a mass spectrometric technique to measure the CNTs uptake to mammalian cells via endocytosis. The lying principle of this "cell mass spectroscopy" is that the amount of particles (either in micro- or nanosize) taken up in cells reflects a proportional shift in mass/charge values of cells (Lin et al., 2011).

## 12.9 APPLICATIONS OF CNTs AS DRUG DELIVERY SYSTEM

In last two decades, CNTs drew attention in consequence of great interest in nanotechnology and the number of publications committed to these materials has skyrocketed upon realization of their unique physicochemical characteristics. In that frame, these materials have been utilized widely as nanovehicles to deliver various API (Bianco et al., 2005; Feazell et al., 2007; Liu et al., 2007a; Liu et al., 2008a; Kumar et al., 2014), peptides (Pantarotto et al., 2004), plasmid DNAs (Liu et al., 2005), proteins (Kam et al., 2005a), and siRNAs (Kam et al., 2005b; Liu et al., 2007b). We summarized the literature on the biomedical applications of CNTs in Table 12.2 to point out that CNTs (especially functionalized ones) are very versatile systems and compatible with many administration routes.

In keeping with the topic of this chapter, we next aim to go through the literature related of some APIs, and to discuss the fundamentals of how these therapeutic agents are transported through CNTs.

### *12.9.1 ANTINEOPLASTIC APIS*

After cardiovascular diseases, cancer ranks second among the causes of death in the world. It is now well established that the real challenge in curing this notorious disease is to deliver APIs specifically to cancerous region. Naturally, failure to discriminate healthy tissues induces various side effects such as cardiac or systemic toxicity and development of resistance to the APIs (Jain, 2012). Another challenge that we should pay regard to what is so-called multidrug resistance in chemotherapy. In order to survive antineoplastic agents, tumor cells express P-glycoprotein (P-gp) that pumps the therapeutic agents back outside tissue. Hence, antineoplastic

**TABLE 12.2** Literature Survey on the Use of CNTs as Drug Delivery Vehicles.

| Type of CNT | API | Administration route | Indication | References |
|---|---|---|---|---|
| SWCNs | Paclitaxel | Subcutaneous injection | Murine 4T1 breast cancer | Liu et al. (2008a) |
| SWCNs | Cisplatin | Intravenous injection | HNSCC tumors | Bhirde et al. (2009) |
| SWCNs | Doxorubicin | Intravenous injection | WiDr human colon cancer cell line | Heister et al. (2009) |
| SWCNs | Cisplatin | Intravenous injection | Prostate cancer cells (PC3 and DU145) | Tripisciano et al. (2009) |
| SWCNs | Doxorubicin | NR | HeLa cervical cancer cells | Zhang et al. (2009) |
| SWCNs | Pirarubicin | Intravasical injection | Human bladder cancer cell line BIU-87 xenograft rat bladder cancer | Chen (2012) |
| MWCNs | Oxaliplatin | Intravenous injection | HT-29 human colon cancer cell line | Wu et al. (2013) |
| MWCNs | Irinotecan | Oral delivery | HT-29 human colon cancer cell line | Zhou et al. (2014) |
| SWCNs | Nimesulide | Topical application | Inflammation | Zanella et al. (2007) |
| SWCNs | Dexamethasone phosphate | NR | Inflammation | Naficy et al. (2009) |
| SWCNs | Prednisolone | Local injection | Collagen-induced arthritis | Nakamura et al. (2011) |
| MWCNs | Indomethacin | NR | Inflammation | Madaenia et al. (2012) |
| SWCNs | Nifedipine | Oral delivery | Hypertension | Liu et al. (2009) |
| MWCNs | Clonidine | Topical application | Tachycardia and hypertension | Strasinger et al. (2009) |
| MWCNs | Captopril | Oral delivery | Hypertension | Ensafi et al. (2011) |
| MWCNs | Carvedilol-A | Oral delivery | Heart failure and hypertension | Li et al. (2011a) |
| MWCNs | Diltiazem hydrochloride | Transdermal delivery | Angina pectoris and hypertension | Bhunia et al. (2013) |

**TABLE 12.2** *(Continued)*

| Type of CNT | API | Administration route | Indication | References |
|---|---|---|---|---|
| MWCNs | Amphotericin B | NR | Chronic fungal infections | Wu et al. (2005) |
| SWCNs | Erythropoietin | Oral delivery | Anemia | Venkatesan et al. (2005) |
| SWCNs | Vitamin B3/C | NR | Vitamin deficiency | De Menezes et al. (2009) |
| SWCNs | siRNA | Intravenous injection | Human T cells and primary cells | Liu et al. (2007b) |
| SWCNs | Bioactive peptides | NR | Human 3T6 and murine 3T3 fibroblasts | Pantarotto et al. (2004) |
| MWCNs | DNA | NR | Gene delivery | Liu et al. (2005) |
| SWCNs | siRNA | NR | Gene delivery | Kam et al. (2005b) |

activity is obliterated even before the drug is given the chance of killing tumorous cells (Fabbro et al., 2012). With that in mind, novel technologies of drug delivery systems to cope with these concerns are of paramount significance to cure cancer once in for all (Rodzinski et al., 2016).

Up to present, our best chance to address this concern is what nanotechnology had to offer: nanoparticles-based drug delivery systems. As a material in nanosize, CNTs are utilized to carry many antineoplastic agents, including campthothecin, cisplatin, doxorubicin, methotrexate, and so on, and it appears at the moment that chemotherapy is the most investigated field of application for CNTs. Consequently, more and more systems are designed de novo each day to take over conventional chemotherapy, which suffer largely from detrimental side effects of plain anticancer agents (for detailed review, see Fabbro et al., 2012).

In this context, cisplatin sets the perfect example to rationalize how badly we need these delivery systems (Fig. 12.7). As one of the most popular medications, cisplatin impedes DNA replication and triggers cellular death by causing a cross-link among DNA strands. However, this platinum-based anticancer drug leads to many side effects, including nephrotoxicity, neurotoxicity, oxotoxicity, and so forth. Furthermore, the exchange of chloride ions with water in plasma effaces the pharmaceutical effect of this drug. Hence, confinement within the cavity of CNTs in the course of transport would be the perfect recipe to prevent deactivation, and to suppress aforementioned side effects of this drug. Inspired by this presumption, Tripisciano et al. developed cisplatin-embedded functionalized SWCNs to inhibit DU145 and PC3 prostate cancer cell lines. Their cellular uptake studies demonstrated that cisplatin was successfully penetrated from cellular membrane and selectively located into the cytoplasma of the prostate cancer cells (Tripisciano et al.; 2009). In concert with the principle of adsorption (vide supra), encapsulation is key to preserve clinically relevant form of cisplatin.

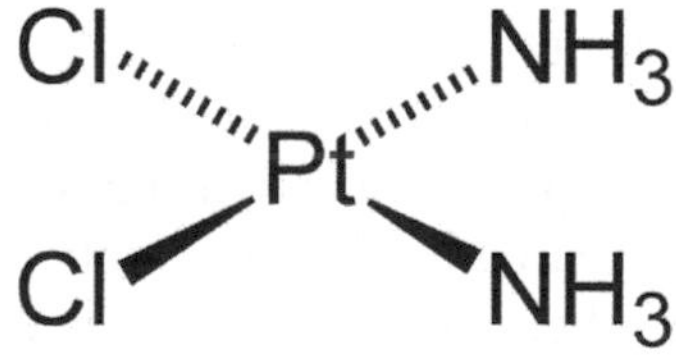


**FIGURE 12.7** The structure of cisplatin.

Doxorubicin is another anticancer agent that received significant attention. Early studies by Ali-Boucetta involved a supramolecular system wherein polyethylene glycol (PEG)-functionalized SWCNTs or copolymer-coated MWCNTs formed noncovalent complexes with doxorubicin. As with cisplatin, MWCNTs–doxorubicin construct turned out to be more active against cancer cells than doxorubicin alone (Ali-Boucetta et al., 2008). Without question, this result unambiguously signifies that CNTs-based delivery systems are applicable to a wide variety of antineoplastic drugs, ranging from inorganic to organic ones.

Before we finalize this section, we intend to discuss one last system, which was devised by Zhang (Zhang et al., 2009). In this article, the authors prepared SWCNTs, which were coated with two different polysaccharides (sodium alginate and chitosan) to enable the further conjugation of folic acid (i.e., targeting molecules) and doxorubicin. It was reported that the drug binds to the carrier at physiological pH and is released at pH values lower than 7.4, such as lysosomal or endosomal pH. Hence, tuning the surface potential of CNTs through functionalization with polysaccharides looks like a wise strategy to manipulate the loading efficiency and the release rate of drug. From what we have seen so far, CNTs show some promise for cancer treatment but for a better understanding on how these constructs behave within tumor cells, work is definitely needed to better the synthesis and characterization of functionalized CNTs (Fabbro et al., 2012).

### *12.9.2 ANTI-INFLAMMATORY APIS*

From a general point of view, the reasons why the transport of anti-inflammatory APIs through CNTs does not fall far from than what we have seen before with the others: to improve the release profile of molecules, to better cellular intake properties, and of course to reduce side effects (Sun et al., 2010). In a preliminary study, Zanella et al. studied the possible interactions of nimesulide onto both pristine and Si-doped capped SWCNs through first-principle calculations (Fig. 12.8). Nimesulide is a well-known anti-inflammatory agent as pain medication and with fewer reducing properties but the choice of this drug for density functional theory (DFT) calculations is a moot point. Their results revealed not only that CNTs-based materials could be a perfect delivery system for nimesulide harboring two aromatic residues but also pointed out that the physisorption event of the therapeutic agent was even more significant with Si-doped capped SWCNs due to its altered electronic properties (Zanella et al., 2007). So far, the doping of particular atoms have been mostly relevant to electronic or electronic-related applications but with this study, the clinical relevance of doped materials became known once again.

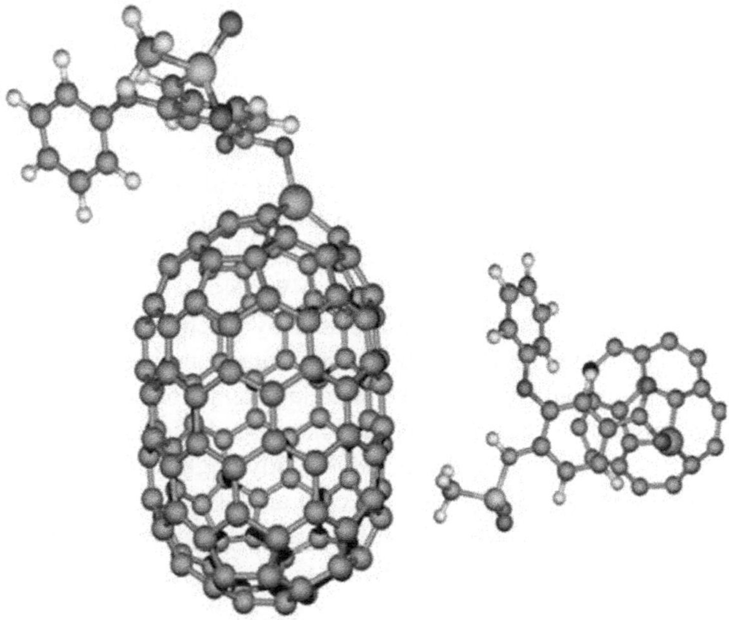

**FIGURE 12.8** **(See color insert.)** Relaxed atomic structure of nimesulide on the cap region of CNTs. (Reprinted with permission from Zanella, 2007. © Elsevier.)

In addition to the infamous problems such as toxicity and undesired side effects, the conventional administration fails to control the release rate and concentration of drugs at their targeted area. Thereupon, the plasma concentration of drugs becomes onerous to predict and may vary constantly. One approach to resolve the problem is drug delivery systems, which fits the purpose of extending the drug release over a longer period time and providing a constant concentration of the drug in plasma (Makhija and Vivia, 2013). In that regard, the design of these systems is predicated on the integration of a membrane around the drug-containing core to pump drugs out of the carrier through osmotic pump system. In order to demonstrate that CNTs are compatible with this notion, Madaeni et al. fabricated functionalized MWCNs, which were blended with cellulose acetate matrix. The authors stated that the presence of MWCNs on the tablet structure enhanced the hydrophilicity and altered the membrane porosity of the system, as a result of which the release of indomethacin is sustained (Madaenia et al., 2012).

Another intriguing idea regarding controlled released is "stimulated drug release," wherein a stimulant triggers the release of the drug from its carrier. Clearly, the nature of stimulant could range from a molecule (e.g., the release of insulin as a response to glucose), physiological conditions (such as pH or temperature) to even electrical signal. As electrical signal potentially opens the door to electronics and the integration of microsystems, the fabrication of intelligent and much sophisticated systems is within possibility (Murdan, 2003). Or maybe, we do not even need to go that far because nanosystems in our hands are more than efficacious to fabricate these systems. For instance, Naficy and coworkers developed chitosan–SWCNs hydrogel films, wherein the release of dexamethasone is stimulated by electrical signals. Therein, the authors based their design on the fact that functionalized SWCNs would contain indomethacin through electrostatic interactions. Once activated, the charge of CNTs is fully reversed and hence, indomethacin gets released because of electrostatic repulsion. This study highlighted that SWCNs generated an electrical field which affected and also controlled the release profiles of the loading API, and thus, bioavailability of the active molecule can be improved (Naficy et al., 2009).

### *12.9.3 CARDIOVASCULAR APIS*

Cardiovascular diseases are best described as a group of disorders related to heart and blood vessels. According to data from World Health Organization, these diseases are the primary cause of death in global scale. Thus far, the preferential treatment of cardiovascular disorders is the administration of drugs through conventional routes. However, the treatment of atherosclerosis and some other cardiovascular diseases including cardiomyopathy, rheumatic heart disease, and so is restrained by the failure to capably transport anticardiovascular medicaments across the endothelium (Mitragotri, 2013). For instance, rosiglitazone, which is a peroxisome proliferator-activated receptor agonist, serves to cure atheroma by diminishing macrophage infiltration into atherosclerotic lesions. However, the pharmaceutical effect of this drug is shadowed by its cytotoxicity to healthy cardiovascular tissues and some side effects ranging from heart failure to fluid retention (Gómez-Guerrero et al., 2012; Perampaladas et al., 2012). In consideration of these undesirable effects, more efficacious drug transport systems are vital to restore the therapeutic efficiency of anticardiovascular agents. Some other reasons of why we depend very much on novel transport systems in this context are: (1) to improve the solubility of APIs and thus, enhance the bioavailability of the active molecules, (2) to avoid the extreme API loss via urine, and lastly, and (3) to improve the physical stability of APIs (Liu et al., 2009; Ensafi et al., 2011; Li et al., 2011a).

Some emerging techniques to delivery anticardiovascular agents without any of these complications are macromolecular-aided, thiomer-mediated approaches (Bernkop-Schnürch et al., 2004), and silica particles (Galagudza et al., 2011; Laing, 2009). Among them, we concern ourselves with silica particles, as any success in drug delivery with this material would bring forward others such as CNTs. In this regard, the finding that silica-based nanomaterials are practicable in the delivery of annexin V expedites the use of nanomaterials in cardiovascular diseases (Galagudza et al., 2011). Naturally, CNTs did not go overlook in this context, given their distinct advantages. For instance, Liu and coworkers investigated possible interactions mechanism of nifedipine with SWCNs to explore: (1) whether nifedipine can be encapsulated within CNTs, (2) if so, does it a preference in the binding location, and (3) what drives the translocation of drug. As a calcium channel antagonist, this pharmaceutical agent is extensively utilized in the treatment of various cardiovascular diseases such as hypertension, angina pectoris, etc., and hence it is of broad clinical significance. Based on DFT calculations, the authors declared that nifedipine was adsorbed to internal cavity of the nanotubes structure spontaneously and that van der Waals forces, which is in competition with hydrophilic interactions, is the main driving force of encapsulation (Liu et al., 2009).

The study on anticardiovascular agents is not limited to theoretical studies, as one may assume. For instance, Li and his coworkers were intrigued by the drug-loading mechanism of carvedilol on CNTs (both pristine and oxidized ones, to be precise). As an vasodilator/adrenoceptor antagonist, carvedilol is commonly utilized for the treatment of hypertension but it is also known for some other biological functions such as neuroprotection and myocardial protection (Yue et al., 1993). Accordingly, the authors aimed to load these nanocarriers with carvedilol via three different methods (the fusion method, the incipient wetness impregnation method, and solvent method) and to study the impact of physical characteristics of CNTs, such as surface area and the pore size of the materials. In conclusion, the choice of loading method seems to directly impact the drug-loading capacity and of course the physical state of carvedilol. By contrast to pristine MWNTs, functionalized materials are more conformed to the encapsulation of this therapeutic agent, possibly because of better hydrophilicity (Li et al., 2011a). There are more to this story in the recent literature that illustrates the positive and improved effects of CNTs which prove their superiorities on the treatment of cardiovascular disorders (Strasinger et al., 2009; Bhunia et al., 2013). In summary, it would seem that the literature on this topic is not particularly rich and cardiovascular diseases did not benefit from nanotechnology sufficiently, despite of the promising results summarized within this chapter.

### *12.9.4 ANTI-INFECTIVE APIS*

In modern medicine, infectious diseases are still ranked among the leading causes of death. What is holding us back are the antibiotic resistance against bacteria and limitations/difficulties to prepare or devise some new antibiotics. To address these complications, the scientific community have put enormous time in understanding what renders a therapy rewarding. Within this scope, their first move was the synthesis of new antibiotics, to which bacteria are unfamiliar with. As noted before, this strategy may be overly difficult and time consuming in defiance of advanced computational techniques. Even if not, the chemical synthesis of materials could pose another challenge and may not be very straightforward, whereupon the notion of "new antibiotics" could simply not be that rewarding, after all.

We can then predicate our approach on the observation that a successful treatment of the infections is solely plausible with the presence of high therapeutic concentrations of drugs on the pathogenic cell region. Hence, the development of new conventional dosage forms for anti-infective APIs is another strategy but the limited intake of drugs into the intracellular compartments remained unresolved so far. More notably, most of the anti-infective APIs are inadequately taken by cells because of their poor solubility and cellular penetration ability. Besides, their intracellular retention is also restricted, and their subcellular distribution is unsatisfactory, all which hamper their intracellular activities (Imbuluzqueta et al., 2010; Lee, 2003; Pinto-Alphandary and Couvreur, 2000).

When viewed from this aspect, current nanotechnology-based drug delivery approaches, such as CNTs, are auspicious alternatives to concentrate drugs in pathogenic cell regions, and hence to cope with bacterial resistance (Banerjee et al., 2012; Gallo et al., 2007; Vukovic et al., 2010; Wu et al., 2005). An extra benefit of CNTs in this context is that they indirectly augment the solubility of therapeutic agents, which drug delivery systems are attached to. This aspect of delivery systems is especially handy for likes of amphotericin B; this antifungal agent is at disadvantage of low aqueous solubility hence, its parenteral use is largely prohibited because of the formation of aggregates in blood stream. For this reason, Wu et al. devised MWCN formulations of amphotericin B to tackle multiple complications at once: (1) to improve the water solubility, (2) to lower the aggregation in aqueous milieu, (3) to increase the cellular uptake, and (4) to tune the therapeutic effect of amphotericin B against different types of cells (Wu, 2005).

Without doubts, the most prominent motivation would be enhancing water solubility, inasmuch as the slightest improvement in solubility amounts to an overall rise in therapeutic efficiency, and the lowered toxicity of amphotericin B. As predicted, amphotericin B–MWCN conjugate displays a better antifungal activity against three pathogenic fungi species (*Cryptococcus neoformans* ATCC 90112, *Candida albicans*, and *Candida parapsilosis* ATCC 90118) than does the free form of API. Sure enough, API-functionalized MWCNs are taken-up rapidly through mammalian cells, and did not induce any apparent toxicity (Wu et al., 2005). In unrelated study, the incorporation of dapsone (an antibacterial agent) into MWCNTs was shown to shield the drug from being metabolized. Thus, the antimycobacterium activity of the API increased, whereas its systemic toxicity on liver reduced (Vukovic et al., 2010). Overall, the formulation of antibacterial agents in CNTs will likely revolve aforementioned drawbacks and antibacterial resistance in long run and spare us the trouble of designing new antibiotics.

### 12.9.5 GENES

Discovered in 1980, gene therapy is the therapeutic transport of nucleic acids into cells to prevent a disease and intends to correct genetic defects and alteration of the cellular genetic makeup. As always, the delivery of nucleic acids or its fragments requires the design, development of proper transportation system, but this time around, the need stems from the instability of exogenous nucleic acids. A second motivation is that failure to deliver DNA to correct cells comes to mean the transfection of some other cells. In this instance, we are blessed by some nonviral and viral gene systems and yet, we still fall short of the "perfect targeted delivery" of nucleic acids (Kostarelos and Miller, 2005). From this point on, it is understood that these complications could be resolved only throughout the intracellular transportation of genes via nano-sized vehicles such as solid or polymeric-based nanoparticles, liposomes, and CNTs. What promises success with these nanoarchitectures are their flexibility with regards to the particle diameter of nucleic acids, the suitability of scaling-up process, and the reduced immunogenicity compared to viral systems (Ojea-Jimenez et al., 2012).

As a generalization, cationic polyelectrolytes such as polylysine, protamine sulfate, and dendrimers makes good tools to formulate nonviral delivery systems for nucleic acids (Parker et al., 2002). Upon binding to DNA through noncovalent interactions, these systems enhance the cellular uptake of nucleic acids via endocytosis and assure the trafficking of their cargo to nucleus. In that regard, it would be safe to predict that polycationic CNTs constitute an elegant approach to pull off much-needed delivery of nucleic acids on account that they may function in a similar way to other polyelectrolytes. To attain this goal, Singh and coworkers developed various polycationic functionalized carbon nanotubes (f-CNTs) to deliver plasmid DNA to human cells. Their results show that cationic plasmid DNA is condensed around CNTs, and more remarkably, their delivery systems managed to induce upregulation of a marker gene, in comparison to normal cells. All these promising results certainly validate that we are on the right path concerning gene therapy (Singh et al., 2005).

## 12.10 RECENT PATENTS RELATED WITH CNTs AS A DRUG DELIVERY SYSTEM

In recent years, CNTs are one of the most investigated drug delivery vehicles among the submicron carriers and they give us another opportunity to cure diseases through pinpoint drug delivery. Encouraged by stunning progress, scientific community turned its attention on bettering the drug delivery profile of these materials, and reducing their plausible cytotoxicity. From this perspective, many patents related with CNTs have been issued on the subject of their synthesis, functionalization, purification methods, encapsulation strategies, and toxicological profiles. A general view of patents relevant to CNTs is presented in Table 12.3, which certainly highlights a remarkable consideration for CNTs and reveal exactly where we stand in this field.

## 12.11 TOXICOLOGY PROFILES OF CNTs

In modern era of medicine, CNTs procures an alternative platform whereupon numerous difficulties related to standard therapeutic methods are addressed at once (Prakash et al., 2011; Jain, 2012). Considering all the efforts which were exhaustively examined in this chapter, what is left to explore is the toxicological profile of these materials so that CNTs-based drug delivery systems can eventually hit the market. On this manner, some researchers firstly harped on the safety in the utilization of CNTs and insisted on the fact that CNTs cause no apparent toxicity whatsoever (Amenta et al., 2015; Chin et al., 2007; Schipper et al., 2008; Prakash et al., 2011). By contrast, it has been widely emphasized by subsequent articles that serious toxic effects were encountered, especially during the pulmonary and intravenous administration of these constructs (Jain et al., 2007; Mehra et al., 2008; Kayat et al., 2011; Bussy, 2013; He et al., 2013). Hence, the actual problem was inconsistency in the results, which also reprieved commercialization. In light of detailed analysis following an initial frustration, it came quite clear that research groups indeed studied different materials, rather than a standard one, because there was not a fully established procedure regarding the synthesis of CNTs yet. Of course, the use of different materials afforded different results, which should surprise no one. This picture also justifies the reason why we repeatedly underlined the fact that "a better understanding on the synthesis and characterization of these materials is a must to proceed any further" within this chapter.

In these circumstances, what needs to be done is simple. In long term, we must standardize the synthesis of these nanomaterials and put some serious efforts to improve the techniques of chemical characterization. In short term, we are obligated to evaluate each research independently. With this rationale in mind, diverse strategies were devised to reduce and/or annihilate the explicit toxicity and side effects of CNTs from this point onwards. Hence, covalent functionalization with PEG (Liu et al., 2008b; Yang et al., 2008), coating with immunoglobulins (Kagan et al., 2010), functionalization with diethylentriaminepentaacetic dianhydride (Lacerda

**TABLE 12.3** An Updated List of Patents Related with CNT.

| Patent number | Title | Date of publication | References |
|---|---|---|---|
| US20040166152A1 | Use of buckysome or carbon nanotube for drug delivery | August 26, 2004 | Hirsch et al. (2004) |
| EP1493714A1 | Encapsulation of biomolecules inside carbon nanotube | January 5, 2005 | Gao et al. (2005) |
| WO2006116065A1 | Method for molecular delivery into cells using nanotubes spearing | November 2, 2006 | Chiles (2006) |
| US7135158B2 | Method of purifying single wall carbon nanotubes | November 14, 2006 | Goto et al. (2006) |
| US20070280876A1 | Functionalization of carbon nanotubes in acidic media | December 6, 2007 | Tour et al. (2007) |
| US20080193490A1 | Use of carbon nanotube for drug delivery | August 14, 2008 | Hirsch et al. (2008) |
| US20080214494A1 | Method of drug delivery by carbon nanotube–chitosan nanocomplexes | September 4, 2008 | Mohapatra and Kumar (2008) |
| US20090170768A1 | Water-soluble carbon nanotube composition for drug delivery and medicinal applications | July 2, 2009 | Tour et al. (2009) |
| US20090269279A1 | Toxicology and cellular effect of manufactured nanomaterials | October 29, 2009 | Chen (2009) |
| US20090306427A1 | Chemical functionalization of carbon nanotubes | December 10, 2009 | Martinez-Rubi et al. (2009) |
| US20100021471A1 | Carbon nanotube-based drug delivery systems and methods of making same | January 28, 2010 | Chen et al. (2010) |
| US20100173376A1 | Functionalization of carbon nanotubes with metallic moieties | July 8, 2010 | Ostojic and Hersam (2010) |
| US20100184669A1 | Compositions and methods for cancer treatment using targeted carbon nanotubes | July 22, 2010 | Harrison and Resasco (2010) |
| US20100266694A1 | Chitosan/carbon nanotube composite scaffolds for drug delivery | October 21, 2010 | Jennings et al. (2010) |
| US20100324315A1 | Poly(citric acid) functionalized carbon nanotubes drug delivery system | December 23, 2010 | Atyabi et al. (2010) |
| US8029734B2 | Noncovalent sidewall functionalization of carbon nanotubes | October 4, 2011 | Dai and Chen (2011) |

**TABLE 12.3** *(Continued)*

| Patent number | Title | Date of publication | References |
|---|---|---|---|
| US20120058170A1 | Drug delivery by carbon nanotube arrays | March 8, 2012 | Gharib et al. (2012) |
| WO2012057511A2 | Method for preparing a highly dispersive carbon nanotube for reducing in vivo immunotoxicity | May 3, 2012 | Khang et al. (2012a) |
| WO2012060592A3 | Carbon nanotube polymer composite coating film which suppresses toxicity and inflammation and has improved biocompatibility and adjusted surface strength | August 16, 2012 | Khang et al. (2012b) |
| US20120220921A1 | Immunologically modified carbon nanotubes for cancer treatment | August 30, 2012 | Chen (2012) |
| US20130034610A1 | Hydrophobic nanotubes and nanoparticles as transporters for the delivery of drugs into cells | February 7, 2013 | Dai et al. (2013) |
| EP2594289A2 | Use of carbon nanotubes for preventing or treating brain diseases | May 22, 2013 | Jeong et al. (2013) |
| US8764681B2 | Sharp tip carbon nanotube microneedle devices and their fabrication | July 1, 2014 | Aria et al. (2014) |
| EP2797605A4 | Targeted self-assembly of functionalized carbon nanotubes on tumors | July 8, 2015 | Scheinberg et al. (2015) |
| US20150238742A1 | Drug delivery and substance transfer facilitated by nano-enhanced device having aligned carbon nanotubes protruding from device surface | August 27, 2015 | Gharib et al. (2015) |

et al., 2008a, 2008b), binding with blood proteins (Ge et al., 2011), surface-manipulation with genetic materials (Shvedova et al., 2012), coating with vitamin E (Wang et al., 2012), and conjugation with both PEG and RGD peptides (Liu et al., 2007c) are just a few of many strategies that brought success.

## 12.12 CONCLUDING REMARKS

In addition to their properties such as large aspect ratio and ability to pass through cellular membranes and so on (Table 12.4), the reason why these nanoparticles are marked out as distinctive is because they are compatible with the whole notion of multiple functionalization, which other delivery systems benefit very little from. In other words, CNTs could be decorated with various compounds each of which exhibits a different role: targeting agents to reduce side effects, drugs to provide therapeutic effects, stealth agents to avoid the immune system, and diagnostic agents to monitor

**TABLE 12.4** Advantages and Limitations of CNTs.[a]

| Pros | Cons |
|---|---|
| Unique mechanical properties offer in vivo stability | Nonbiodegradable |
| Extremely large aspect ratio, offers template for development of multimodal devices | Large available surface area for protein opsonization |
| Capacity to readily cross biological barriers; novel delivery systems | As-produced material insoluble in most solvents; need to surface treat preferably by covalent functionalization chemistries to confer aqueous solubility (i.e., biocompatibility) |
| Unique electrical and semiconducting properties; constitute advanced components for in vivo devices | Bundling; large structures with less than optimum biological behavior |
| Hollow, fibrous, and light structure with different flow dynamics properties; advantageous in vivo transport kinetics | Healthy tissue tolerance and accumulation; unknown parameters that require toxicological profiling of material |
| Mass production—low cost; attractive for drug development | Great variety of CNTs types; makes standardization and toxicological evaluation cumbersome |

[a]Reprinted with permission from Lacerda (2007). © Elsevier 2007.

the location of delivery system. In concert with the direction that drug delivery is headed, this notion is certainly of paramount significance as the drug delivery systems are ultimately sought to accomplish far more than spraying therapeutic agents to tissues.

As exceptional as they look to be, the (bio)chemistry of CNTs raise some serious concern, as well. The fundamental problem we face is their nonbiodegradable quality. Unlike liposomes, they are not disposable and hence the biological fate of drug delivery systems after releasing drugs in cells remains a little ambiguous. Also, there is no consensus on their toxic effects, which may vary depending on the study and the material utilized (vide supra). Hence, extensive studies on their toxicology profile and their synthesis is a must for us to fully comprehend how these materials behave in biological media. When viewed from this perspective, these disadvantages appear to outweigh all the previously mentioned profits and hold up the commercialization of CNTs-based systems. In short, what we accomplished so far with these nanostructures is a huge step forward, considering what is summarized in this chapter. However, there is still a long way to go before we make some tailored drugs delivery systems out of CNTs.

## KEYWORDS

- **carbon nanotubes**
- **functionalization**
- **drug delivery**
- **drug loading**
- **toxicity**

## REFERENCES

Ali-Boucetta, H.; Al-Jamal, K. T.; McCarthy, D.; Prato, M.; Bianco, A.; Kostarelos, K. Multiwalled Carbon Nanotube–Doxorubicin Supramolecular Complexes for Cancer Therapeutics. *Chem. Commun.* 2008, *28* (4), 459–461.

Ajayan, P.; Ebbesen, T. Nanometre-size Tubes of Carbon. *Rep. Prog. Phys.* **1997,** *60*, 1025–1062.

Amenta, V.; Aschberger, K. Carbon Nanotubes: Potential Medical Applications and Safety Concerns. *WIREs Nanomed. Nanobiotechnol.* **2015,** *7*, 371–386.

Antonelli, A.; Serafini, S.; Menotta, M.; Sfara, C.; Pierigé, F.; Giorgi, L.; Ambrosi, G.; Rossi, L.; Magnani, M. Improved Cellular Uptake of Functionalized Single-walled Carbon Nanotubes. *Nanotechnology* **2010,** *21*, 425101.

Aria, A. I.; Lyon, B.; Gharib, M. Sharp Tip Carbon Nanotube Microneedle Devices and Their Fabrication. US8764681B2, July 1, 2014.

Arsawang, U.; Saengsawang, O.; Rungrotmongkol, T.; Sornmee, P.; Wittayanarakul, K.; Remsungnen, T.; Hannongbua, S. How do Carbon Nanotubes Serve as Carriers for Gemcitabine Transport in a Drug Delivery System? *J. Mol. Graph. Model.* **2011,** *29*, 591–596.

Atyabi, F.; Ateli, M.; Sobhani, Z.; Dinarvand, R.; Ghahremani, M. H. Poly(Citric Acid) Functionalized Carbon Nanotube Drug Delivery System. US20100324315A1, December 23, 2010.

Bae, Y.; Park, K. Targeted Drug Delivery to Tumors: Myths, Reality and Possibility. *J. Control. Release* **2011,** *153* (3), 198–205.

Banerjee, I.; Douaisi, M. P.; Mondal, D.; Kane, R. S. Light-activated Nanotube–Porphyrin Conjugates as Effective Antiviral Agents. *Nanotechnology* **2012,** *23*, 105101.

Battigelli, A.; Ménard-Moyon, C.; Da Ros, T.; Prato, M.; Bianco, A. Endowing Carbon Nanotubes with Biological and Biomedical Properties by Chemical Modifications. *Adv. Drug Deliv. Rev.* **2013,** *65*, 1899–1920.

Bernkop-Schnürch, A.; Hoffer, M. H.; Kafedjiiski, K. Thiomers for Oral Delivery of Hydrophilic Macromolecular Drugs. *Expert Opin. Drug Deliv*. **2004,** *1* (1), 87–98.

Bhirde, A. A.; Patel, V.; Gavard, J.; Zhang, G.; Sousa, A. A.; Masedunskas, A.; Leapman, R. D.; Weigert, R.; Gutkind, J. S.; Rusling, J. F. Targeted Killing of Cancer Cells In Vivo and In Vitro with EGF-directed Carbon Nanotube-based Drug Delivery. *ACS Nano* **2009,** *3*, 307–316.

Bhunia, T.; Giri, A.; Nasim, T.; Chattopadhyay, D.; Bandyopadhyay, A. A Transdermal Diltiazem Hydrochloride Delivery Device Using Multi-walled Carbon Nanotube/Poly (Vinyl Alcohol) Composites. *Carbon* **2013,** *52*, 305–315.

Bianco, A.; Kostarelos, K.; Prato, M. Applications of Carbon Nanotubes in Drug Delivery. *Curr. Opin. Chem. Bio.* **2005,** *9*, 674–679.

Boncel, S.; Zając, P.; Koziol, K. Liberation of Drugs from Multi-wall Carbon Nanotube Carriers. *J. Control. Release* **2013,** *169*, 126–140.

Bussy, C. Carbon Nanotubes in Medicine and Biology—Safety and Toxicology. *Adv. Drug Deliv. Rev.* **2013,** *65*, 2061–2062.

Chen, F. Toxicology and Cellular Effect of Manufactured Nanomaterials. US20090269279A1, October 29, 2009.

Chen, W. R. Immunologically Modified Carbon Nanotubes for Cancer Treatment. US20120220921A1, August 30, 2012.

Chen, X.; Schluesener, H. J. Multi-walled Carbon Nanotubes Affect Drug Transport Across Cell Membrane in Rat Astrocytes. *Nanotechnology* **2010,** *21*, 105104.

Chen, Z.; Thiel, W.; Hirsch, A. Reactivity of the Convex and Concave Surfaces of Single-walled Carbon Nanotubes (SWCNTs) Towards Addition Reactions: Dependence on the Carbon-atom Pyramidalization. *ChemPhysChem* **2003,** *1*, 93–97.

Chen, J.; Wong, S. S.; Ojima, I. Carbon Nanotube-based Drug Delivery Systems and Methods of Making Same. US20100021471A1, January 28, 2010.

Chen, Z.; Pierre, D.; He, H.; Tan, S.; Pham-Huy, C.; Hong, H.; Huang, J. Adsorption Behavior of Epirubicin Hydrochloride on Carboxylated Carbon Nanotubes. *Int. J. Pharm.* **2011,** *28* (405), 153–161.

Chen, G.; He, Y.; Wu, X.; Zhang, Y.; Luo, C.; Jing, P. In Vitro and In Vivo Studies of Pirarubicin-loaded SWNT for the Treatment of Bladder Cancer. *Braz. J. Med. Biol. Res.* **2012,** *45*, 771–776.

Chiles, T. C. Method for Molecular Delivery into Cells Using Nanotubes Spearing. WO2006116065A1, November 2, 2006.

Chin, S. F.; Baughman, R. H.; Dalton, A. B.; Dieckmann, G. R.; Draper, R. K.; Mikoryak, C.; Musselman, I. H.; Poenitzsch, V. Z.; Xie, H.; Pantano, P. Amphiphilic Helical Peptide Enhances the Uptake of Single-walled Carbon Nanotubes by Living Cells. *Exp. Biol. Med. (Maywood)* **2007,** *232*, 1236–1244.

Coelho, J.; Ferreira, P.; Cordeiro, P.; Fonseca, A.; Góis, J.; Gil, M. Drug Delivery Systems: Advanced Technologies Potentially Applicable in Personalized Treatments. *EPMA J.* **2010,** *1*, 164–209.

Dobson, P. D.; Kell, D. B. Carrier-mediated Cellular Uptake of Pharmaceutical Drugs: An Exception or the Rule? *Nat. Rev. Drug Discov.* **2008,** *7* (3), 205–220.

Dai, H.; Chen, R. J. Noncovalent Sidewall Functionalization of Carbon Nanotubes. US8029734B2, October 4, 2011.

Dai, H.; Kam, N. W. S.; Wender, P. A.; Liu, Z. Hydrophobic Nanotubes and Nanoparticles as Transporters for the Delivery of Drugs into Cells. US20130034610A1, February 7, 2013.

De Menezes, V. M.; Fagan, S. B.; Zanella, I.; Mota, R. Carbon Nanotubes Interacting with Vitamins: First Principles Calculations. *Microelectron J.* **2009,** *40*, 877–879.

De Volder, M.; Tawfick, S.; Baughman, R.; Hart, A. Carbon Nanotubes: Present and Future Commercial Applications. *Science* **2013,** *339*, 535–539.

Di Crescenzo, A.; Ettorre, V.; Fontana, A. Non-covalent and Reversible Functionalization of Carbon Nanotubes. *Beilstein J. Nanotechnol.* **2014,** *5*, 1675–1690.

Draper, R. K.; Pantano, P.; Wang, R. H.; Mikoryak, C. Method for Measuring Carbon Nanotubes Taken-up by a Plurality of Living Cells. US20110203927A1, August 25, 2011.

Dresselhaus, M.; Dresselhaus, G.; Charlier, J.; Hernandez, E. Electronic, Thermal and Mechanical Properties of Carbon Nanotubes. *Philos. Trans. A Math. Phys. Eng. Sci.* **2004,** *362*, 2065–2098.

Eatemadi, A.; Daraee, H.; Karimkhanloo, H.; Kouhi, M.; Zarghami, N. Carbon Nanotubes: Properties, Synthesis, Purification, and Medical Applications. *Nanoscale Res. Lett.* **2014,** *9*, 393–404.

Ensafi, A. A.; Karimi-Maleh, H.; Mallakpour, S.; Rezaei, B. Highly Sensitive Voltammetric Sensor Based on Catechol-derivative-multiwall Carbon Nanotubes for the Catalytic Determination of Captopril in Patient Human Urine Samples. *Colloids Surf. B Biointerfaces* **2011,** *87*, 480–488.

Fabbro, C.; Ali-Boucetta, H.; Da Ros, T.; Kostarelos, K.; Bianco, A.; Prato, M. Targeting Carbon Nanotubes Against Cancer. *Chem. Commun.* **2012,** *48*, 3911–3926

Feazell, R. P.; Nakayama-Ratchford, N.; Dai, H.; Lippard, S. J. Soluble Single-walled Carbon Nanotubes as Longboat Delivery Systems for Platinum (IV) Anticancer Drug Design. *J. Am. Chem. Soc.* **2007,** *129*, 8438–8439.

Fischer, J. Chemical Doping of Single-wall Carbon Nanotubes. *Acc. Chem. Res.* **2002,** *35*, 1079–1086.

Foldvari, M.; Bagonluri, M. Carbon Nanotubes as Functional Excipients for Nanomedicines: II. Drug Delivery and Biocompatibility Issues. *Nanomedicine* **2008,** *4*, 183–200.

Galagudza, M. M.; Korolev, D. V.; Sonin, D. L.; Alexandrov, I. V.; Minasian, S. M.; Postnov, V. N.; Kirpicheva, E. B. Passive and Active Target Delivery of Drugs to Ischemic Myocardium. *Bull. Exp. Biol. Med.* **2011,** *152* (1), 105–107.

Gallo, M.; Favila, A.; Mitnik, D. G. DFT Studies of Functionalized Carbon Nanotubes and Fullerenes as Nanovectors for Drug Delivery of Antitubercular Compounds. *Chem. Phys. Lett.* **2007,** *447*, 105–109.

Gao, H.; Kong, Y.; Cui, D.; Ozkan, C. O. Encapsulation of Biomolecules Inside Carbon Nanotubes. EP1493714A1, January 5, 2005.

Ge, C.; Du, J.; Zhao, L.; Wang, L.; Liu, Y.; Li, D.; Yang, Y.; Zhou, R.; Zhao, Y.; Chai, Z.; Chen, C. Binding of Blood Proteins to Carbon Nanotubes Reduces Cytotoxicity. *PNAS* **2011,** *108*, 16963–16973.

Gharib, M.; Aria, A. M.; Beizai, M. Drug Delivery by Carbon Nanotube Arrays. US20120058170A1, March 8, 2012.

Gharib, M.; Aria, A. I.; Sansom, E. B. Drug Delivery and Substance Transfer Facilitated by Nano-enhanced Device Having Aligned Carbon Nanotubes Protruding from Device Surface. US20150238742A1, August 27, 2015.

Georgakilas, V.; Kordatos, K.; Prato, M.; Guldi, D. Organic Functionalization of Carbon Nanotubes. *J. Am. Chem. Soc.* **2002,** *124*, 760–761.

Gómez-Guerrero, C.; Mallavia, B.; Egido, J. Targeting Inflammation in Cardiovascular Diseases. Still a Neglected Field? *Cardiovasc. Ther*. **2012,** *30* (4), 189–197.

Goto, H.; Furuta, T.; Fujiwara, Y.; Ohashi, T. Method of Purifying Single Wall Carbon Nanotubes. US7135158B2, November 14, 2006.

Harrison, R. G.; Resasco, D. E. Compositions and Methods for Cancer Treatment Using Targeted Carbon Nanotubes. US20100184669A1, July 22, 2010.

He, H.; Pham-Huy, L. A.; Dramou, P.; Xiao, D.; Zuo, P.; Pham-Huy, C. Carbon Nanotubes: Applications in Pharmacy and Medicine. *Biomed. Res. Int.* **2013,** Vol. 2013, Article ID 578290, 12 pages.

Heister, E.; Neves, V.; Lamprecht, C.; Silva, S.; Coley, H.; McFadden, J. Drug Loading, Dispersion Stability, and Therapeutic Efficacy in Targeted Drug Delivery with Carbon Nanotubes. *Carbon* **2012,** *50*, 622–632.

Heister, E.; Neves, V.; Tîlmaciu, C.; Lipert, K.; Beltrán, V. S.; Coley, H. M.; Silva, S. R. P.; McFadden, J. Triple Functionalization of Single-walled Carbon Nanotubes with Doxorubicin, a Monoclonal Antibody, and a Fluorescent Marker for Targeted Cancer Therapy. *Carbon* **2009,** *47*, 2152–2160.

Hirsch, A.; Sagman, U.; Wilson, S. Use of Buckysome or Carbon Nanotube for Drug Delivery. US20040166152A1, August 26, 2004.

Hirsch, A.; Sagman, U.; Wilson, S. R.; Rosenblum, M. G.; Wilson, L. J. Use of Carbon Nanotube for Drug Delivery. US20080193490A1, August 14, 2008.

Hiura, H.; Ebbesen, T.; Tanigaki, K. Opening and Purification of Carbon Nanotubes in High Yields. *Adv. Mater.* **1995**, *7* (3), 275–276.

Hoffman, A. The Origins and Evolution of "Controlled" Drug Delivery Systems. *J. Control. Release* **2008,** *132*, 153–163.

Hu, H.; Zhao, B.; Itkis, M.; Haddon, R. Nitric Acid Purification of Single-walled Carbon Nanotubes. *J. Phys. Chem. B* **2003,** *107* (50), 13838–13842.

Hu, C.-Y.; Xu, Y.-J.; Duo, S.-W.; Zhang, R.-F.; Li, M.-S. Non-covalent Functionalization of Carbon Nanotubes with Surfactants and Polymers. *J. Chin. Chem. Soc.* **2009,** *56,* 234–239.

Hwang, K. Efficient Cleavage of Carbon Graphene Layers by Oxidants. *J. Chem. Soc. Chem. Commun.* **1995,** *2,* 173–174.

Iijima, S. Helical Microtubules of Graphitic Carbon. *Nature* **1991,** *354,* 56–58.

Imbuluzqueta, E.; Gamazo, C.; Ariza, J.; Blanco-Prieto, M. J. Drug Delivery Systems for Potential Treatment of Intracellular Bacterial Infections. *Front. Biosci.* **2010,** *15,* 397–417.

Jain, K. K. Advances in use of Functionalized Carbon Nanotubes for Drug Design and Discovery. *Expert Opin. Drug Discov.* **2012,** *7,* 1029–1037.

Jain, A.; Dubey, V.; Mehra, N.; Lodhi, N.; Nahar, M.; Mishra, D. K.; Jain, N. K. Carbohydrate-conjugated Multiwalled Carbon Nanotubes: Development and Characterization. *Nanomedicine* **2003,** *4,* 432–442.

Jain, A. K.; Mehra, N. K.; Lodhi, N.; Dubey, V.; Mishra, D. K.; Jain, P. K.; Jain, N. K. Carbon Nanotubes and Their Toxicity. *Nanotoxicology* **2007,** *1,* 167–197.

Jennings, J. A.; Haggard, W. O.; Bumgardner, J. D. Chitosan/Carbon Nanotube Composite Scaffolds for Drug Delivery. US20100266694A1, October 21, 2010.

Jeong, Y.; Lee, H. J.; Lee, D. Y. Use of Carbon Nanotubes for Preventing or Treating Brain Disease. EP2594289A2, May 22, 2013.

Kagan, V. E.; Konduru, N. V.; Feng, W.; Allen, B. L.; Conroy, J.; Volkov, Y.; Vlasova, I. I.; Belikova, N. A.; Yanamala, N.; Kapralov, A.; Tyurina, Y. Y.; Shi, J.; Kisin, E. R.; Murray, A. R.; Franks, J.; Stolz, D.; Gou, P.; Klein-Seetharaman, J.; Fadeel, B.; Star, A.; Shvedova, A. A. Carbon Nanotubes Degraded by Neutrophil Myeloperoxidase Induce Less Pulmonary Inflammation. *Nat. Nanotechnol.* **2010,** *5,* 354–359.

Kam, N. W. S.; Dai, H. Carbon Nanotubes as Intracellular Protein Transporters: Generality and Biological Functionality. *J. Am. Chem. Soc.* **2005a,** *127,* 6021–6026.

Kam, N. W. S.; Dai, H. Functionalization of Carbon Nanotubes via Cleavable Bonds for Efficient Intracellular Delivery of siRNA and Potent Gene Silencing. *J. Am. Chem. Soc.* **2005b,** *36,* 12492–12493.

Kayat, J.; Gajbhiye, V.; Tekade, R. K.; Jain, N. K. Pulmonary Toxicity of Carbon Nanotubes: A Systematic Report. *Nanomedicine* **2011,** *7,* 40–49.

Khang, D. W.; Nam, T. H.; Lee, S. Y.; Kim, S. H. Method for Preparing a Highly Dispersive Carbon Nanotube for Reducing *In Vivo* Immunotoxicity. WO2012057511A2, May 3, 2012a.

Khang, D. W.; Nam, T. H. Carbon Nanotube Polymer Composite Coating Film Which Suppresses Toxicity and Inflammation and Has Improved Biocompatibility and Adjusted Surface Strength. WO2012060592A3, August 16, 2012b.

Klumpp, C.; Kostarelos, K.; Prato, M.; Bianco, A. Functionalized Carbon Nanotubes as Emerging Nanovectors for the Delivery of Therapeutics. *Biochim. Biophys. Acta* **2006,** *1758,* 404–412.

Kocharova, N.; Aaritalo, T.; Leiro, J; Kankare, J.; Lukkari, J. Aqueous Dispersion, Surface Thiolation, and Direct Self-assembly of Carbon Nanotubes on Gold. *Langmuir* **2007,** *23,* 3363–3371.

Koo, O. M.; Rubinstein, I.; Onyuksel, H. Role of Nanotechnology in Targeted Drug Delivery and Imaging: A Concise Review. *Nanomedicine* **2005,** *1*, 193–212.

Kostarelos, K.; Miller, A. D. Synthetic, Self-assembly ABCD Nanoparticles: A Structural Paradigm for Viable Synthetic Non-viral Vectors. *Chem. Soc. Rev.* **2005,** *34*, 970–994.

Kumar, R.; Dhanawat, M.; Kumar, S.; Singh, B. N.; Pandit, J. K.; Sinha, V. R. Carbon Nanotubes: A Potential Concept for Drug Delivery Applications. *Recent Pat. Drug Deliv. Formul.* **2014,** *8*, 12–26.

Kushwaha, S.; Ghoshal, S.; Rai, A.; Singh, S. Carbon Nanotubes as a Novel Drug Delivery System for Anticancer Therapy: A Review. *Braz. J. Pharm. Sci.* **2013,** *49* (4), 629–643.

Kyotani, T.; Nakazaki, S.; Xu, W.-H.; Tomita, A. Chemical Modification of the Inner Walls of Carbon Nanotubes by $HNO_3$ Oxidation. *Carbon* **2001,** *39*, 771–785.

Lacerda, L.; Raffa, S.; Pratoc, M.; Bianco, A.; Kostarelos, K. Cell Penetrating CNTs for Delivery of Therapeutics. *Nano Today* **2007,** *2*, 38–43.

Lacerda, L.; Soundararajan, A.; Pastorin, G.; Al-Jamal, K. T.; Herrero, M. A.; Bao, A.; Emfietzoglou, D.; Mather, S.; Phillips, W. T.; Prato, M.; Bianco, A.; Goins, B.; Kostarelos, K. Dynamic Imaging of Functionalized Multi-walled Carbon Nanotube Systemic Circulation and Urinary Excretion. *Adv. Mater*. **2008a,** *20*, 225–230.

Lacerda, L.; Herrero, M. A.; Venner, K.; Bianco, A.; Prato, M.; Kostarelos, K. Carbon-nanotube Shape and Individualization Critical for Renal Excretion. *Small* **2008b,** *4*, 1130–1132.

Laing, S. T.; McPherson, D. Cardiovascular Therapeutic Uses of Targeted Ultrasound Contrast Agents. *Cardiovasc. Res*. **2009,** *83* (4), 626–635.

Lee, H. W. Interactions Between Antimicrobial Agents, Phagocytic Cells and Bacteria. *Curr. Med. Chem. Anti-Infect. Agents* **2003,** *2*, 73–82.

Lee, K.; Li, L.; Dai, L. Asymmetric End-functionalization of Multi-walled Carbon Nanotubes. *J. Am. Chem. Soc.* **2005,** *127*, 4122–4123.

Li, Y.; Wang, T.; Wang, J.; Jiang, T.; Cheng, G.; Wang, S. Functional and Unmodified MWNTs for Delivery of the Water-insoluble Drug Carvedilol—A Drug Loading Mechanism. *Appl. Surf. Sci.* **2011a,** *257*, 5663–5670.

Li, J.; Yap, S. Q.; Yoong, S. L.; Nayak, T. R.; Chandra, G. W.; Ang, W. H.; Panczyk, T.; Ramaprabhu, S.; Vashist, S. K.; Sheu, F.-S.; Tan, A.; Pastorin, G. Carbon Nanotube Bottles for Incorporation, Release and Enhanced Cytotoxic Effect of Cisplatin. *Carbon* **2012,** *50*, 1625–1634.

Lin, T.; Bajpai, V.; Ji, T.; Dai, L. Chemistry of Carbon Nanotubes. *Aust. J. Chem.* **2003,** *56*, 635–651.

Lin, H. C.; Lin, H. H.; Kao, C. Y.; Yu, L.; Peng, W. P.; Chen, C. H. Quantitative Measurement of Nano/Micro Particle Endocytosis with Cell Mass Spectrometry. US20110236882A1, September 29, 2011.

Liu, P. Modification Strategies for Carbon Nanotubes as a Drug Delivery System. *Ind. Eng. Chem. Res.* **2013,** *52*, 13517−13527.

Liu, Y.; Wu, D. C.; Zhang, W. D.; Jiang, X.; He, C. B.; Chung, T. S.; Goh, S. H.; Leong, K. W. Polyethylenimine-grafted Multiwalled Carbon Nanotubes for Secure Noncovalent Immobilization and Efficient Delivery of DNA. *Angew. Chem. Int. Ed. Engl.* **2005,** *44*, 4782.

Liu, Z.; Sun, X.; Nakayama-Ratchford, N.; Dai, H. Supramolecular Chemistry on Water-soluble Carbon Nanotubes for Drug Loading and Delivery. *ACS Nano* **2007a,** *1* (1), 50–59.

Liu, Z.; Winters, M.; Holodniy, M.; Dai, H. siRNA Delivery into Human T Cells and Primary Cells with Carbon-nanotube Transporters. *Angew. Chem. Int. Ed. Engl.* **2007b,** *46*, 2023–2027.

Liu, Z.; Cai, W; He, L.; Nakayama, N.; Chen, K.; Sun, X.; Chen, X.; Dai, H. In Vivo Biodistribution and Highly Efficient Tumor Targeting of Carbon Nanotubes in Mice. *Nat. Nanotechnol.* **2007c,** *2*, 47–52.

Liu, Z.; Chen, K.; Davis, C.; Sherlock, S.; Cao, Q.; Chen, X.; Dai, H. Drug Delivery with Carbon Nanotubes for In Vivo Cancer Treatment. *Cancer Res.* **2008a,** *68*, 6652–6660.

Liu, Z.; Davis, C.; Cai, W.; He, L.; Chen, X.; Dai, H. Circulation and Long-term Fate of Functionalized, Biocompatible Single-walled Carbon Nanotubes in Mice Probed by Raman Spectroscopy. *PNAS* **2008b,** *105*, 1410–1415.

Liu, H.; Bu, Y.; Mi, Y.; Wang, Y. Interaction Site Preference Between Carbon Nanotube and Nifedipine: A Combined Density Functional Theory and Classical Molecular Dynamics Study. *Comput. Theor. Chem.* **2009,** *90*, 163–168.

Madaenia, S. S.; Derakhshandeh, K.; Ahmadia, S.; Vatanpoura, V.; Zinadinia, S. Effect of Modified Multi-walled Carbon Nanotubes on Release Characteristics of Indomethacin from Symmetric Membrane Coated Tablets. *J. Memb. Sci.* **2012,** *389*, 110–116.

Maggini, M.; Scorrano, G. Addition of Azomethine Ylides to CM: Synthesis, Characterization, and Functionalization of Fullerene Pyrrolidines. *J. Am. Chem. Soc.* **1993,** *115*, 9798–9799.

Makhija, S. N.; Vavia, P. R. Controlled Porosity Osmotic Pump-based Controlled Release Systems of Pseudoephedrine. I. Cellulose Acetate as a Semipermeable Membrane. *J. Control. Release* **2003,** *89*, 5–18.

Martinez-Rubi, Y.; Guan, J.; Simard, B. Chemical Functionalization of Carbon Nanotubes. US20090306427A1, December 10, 2009.

Mehra, N. K.; Jain, A. K.; Lodhi, N.; Raj, R.; Dubey, V.; Mishra, D.; Nahar, M.; Jain, N. K. Challenges in the Use of Carbon Nanotubes for Biomedical Applications. *Crit. Rev. Ther. Drug Carrier Syst.* **2008,** *25*, 169–206.

Mitragotri, S. Devices for Overcoming Biological Barriers: the Use of Physical Forces to Disrupt the Barriers. *Adv. Drug Deliv. Rev*. **2013,** *65* (1), 100–103.

Mohapatra, S. S.; Kumar, A. Method of Drug Delivery by Carbon Nanotube–Chitosan Nanocomplexes. US20080214494A1, September 4, 2008.

Mu, Q.; Broughton, D. L.; Yan, B. Endosomal Leakage and Nuclear Translocation of Multiwalled Carbon Nanotubes: Developing a Model for Cell Uptake. *Nano Lett.* **2009,** *9*, 4370–4375.

Murdan, S. Electro-responsive Drug Delivery from Hydrogels. *J. Control. Rel.* **2003,** *92* (1–2), 1–17.

Naficy, S.; Razal, J. M.; Spinks, G. M.; Wallace, G. G. Modulated Release of Dexamethasone From Chitosan–Carbon Nanotube Films. *Sens. Actuators A Phys.* **2009,** *155*, 120–124.

Nakamura, M.; Tahara, Y.; Ikehara, Y.; Single-walled Carbon Nanohorns as Drug Carriers: Adsorption of Prednisolone and Anti-inflammatory Effects on Arthritis. *Nanotechnology* **2011,** *22*, 465102.

Ojea-Jimenez, I; Tort, O.; Lorenzo, J. Engineered Nonviral Nanocarriers for Intracellular Gene Delivery Applications. *Biomed. Mater.* **2012,** *7*, 054106.

Ostojic, G.; Hersam, M. C. Functionalization of Carbon Nanotubes with Metallic Moieties. US20100173376A1, July 8, 2010.

Pantarotto, D.; Briand, J. P.; Prato, M.; Bianco, A. Translocation of Bioactive Peptides Across Cell Membranes by Carbon Nanotubes. *Chem. Comm.* **2004,** *7*, 16–17.

Parker, A. L.; Oupicky, D.; Dash, P. R.; Seymour, L. W. Methodologies for Monitoring Nanoparticle Formation by Self-assembly of DNA with Poly (L-lysine). *Anal. Biochem.* **2002,** *302*, 75–80.

Pattni, B.; Chupin, V.; Torchilin, V. New Developments in Liposomal Drug Delivery. *Chem. Rev.* **2015,** *115*, 10938–10966.

Perampaladas, K.; Gori, T.; Parker, J. D. Rosiglitazone Causes Endothelial Dysfunction in Humans. *J. Cardiovasc. Pharmacol. Ther.* **2012,** *17* (3), 260–265.

Perry, J.; Martin, C.; Stewart, J. Drug-delivery Strategies by Using Template-synthesized Nanotubes. *Chemistry* **2011,** *17*, 6296–6302.

Pinto-Alphandary, A. A. H.; Couvreur, P. Targeted Delivery of Antibiotics Using Liposomes and Nanoparticles: Research and Applications. *Int. J. Antimicrob. Agents* **2000,** *13*, 155–168.

Prakash, S.; Malhotra, M.; Shao, W.; Tomaro-Duchesneau, C.; Abbasi, S. Polymeric Nanohybrids and Functionalized Carbon Nanotubes as Drug Delivery Carriers for Cancer Therapy. *Adv. Drug Deliv. Rev.* **2011,** *63*, 1340–1351.

Prato, M.; Kostarelos, K.; Bianco, A. Functionalized Carbon Nanotubes in Drug Design and Discovery. *Acc. Chem. Res.* **2008,** *41*, 60–68.

Raffa, V.; Ciofani, G.; Nitodas, S.; Karachalios, T.; D'Alessandro, D.; Masini, M.; Cuschieri, A. Can the Properties of Carbon Nanotubes Influence Their Internalization by Living Cells? *Carbon* **2008,** *46*, 1600–1610.

Revathi, S.; Vuyyuru, M.; Dhanaraju, M. Carbon Nanotube: A Flexible Approach for Nanomedicine and Drug Delivery. *Asian J. Pharm. Clin. Res.* **2015,** *8* (1), 25–31.

Rinzler, A.; Liu, J.; Dai, H.; Nikolaev, P.; Huffman, C.; Rodriguez-Macias, F.; Boul, P. J.; Lu, A. H.; Heymann, D.; Colbert, D. T.; Lee, R. S.; Fischer, J. E.; Rao, A. M.; Eklund, P. C.; Smalley, R. E. Large-scale Purification of Single-wall Carbon Nanotubes: Process, Product, and Characterization. *Appl. Phys. A* **1998,** *67*, 29–37.

Rodzinski, A.; Guduru, R.; Liang, P.; Hadjikhani, A.; Stewart, T.; Stimphil, E.; Runowicz, C.; Cote, R.; Altman, N.; Datar, R.; Khizroev, S. Targeted and Controlled Anticancer Drug Delivery and Release with Magnetoelectric Nanoparticles. *Sci. Rep.* **2016,** *6*, 20867.

Scheinberg, D. A.; Mcdevitt, M. R.; Villa, C. H.; Mulvey, J. J. Targeted Self-assembly of Functionalized Carbon Nanotubes on Tumors. EP2797605A4, July 8, 2015.

Schipper, M. L.; Nakayama-Ratchford, N.; Davis, C. R.; Kam, N. W. S.; Chu, P.; Liu, Z; Sun, X.; Dai, H.; Gambhir, S. S. A Pilot Toxicology Study of Single-walled Carbon Nanotubes in a Small Sample of Mice. *Nat. Nanotechnol.* **2008,** *3*, 216–221.

Singh, R.; Pantarotto, D.; McCarthy, D. Binding and Condensation of Plasmid DNA onto Functionalized Carbon Nanotubes: Towards the Construction of Nanotube-based Gene Delivery Vectors. *J. Am. Chem. Soc.* **2005,** *127*, 4388–4396.

Shvedova, A. A.; Kapralov, A. A.; Feng, W. H.; Kisin, E. R.; Murray, A. R.; Mercer, R. R.; Croix, C. M. S.; Lang, M. A.; Watkins, S. C.; Konduru, N. V.; Allen, B. L.; Conroy, J.; Kotchey, G. P.; Mohamed, B. M.; Meade, A. D.; Volkov, Y.; Star, A.; Fadeel, B.; Kagan, V. E. Impaired Clearance and Enhanced Pulmonary Inflammatory/Fibrotic Response to Carbon Nanotubes in Myeloperoxidase-deficient Mice. *PLoS One* **2012,** *7*, e30923.

Strasinger, C. L.; Scheff, N. N.; Wu, J.; Hinds, B. J.; Stinchcomb, A. L. Carbon Nanotube Membranes for Use in the Transdermal Treatment of Nicotine Addiction and Opioid Withdrawal Symptoms. *Subst. Abuse*. **2009,** *3*, 31–39.

Strebhardt, K.; Ullrich, A. Paul Ehrlich's Magic Bullet Concept: 100 Years of Progress. *Nat. Rev. Cancer* **2008,** *8* (6), 473–480.

Sun, D.; Zhuang, X.; Xiang, X.; Liu, Y.; Zhang, S.; Liu, C.; Barnes, S.; Grizzle, W.; Miller, D.; Zhang, H.-G. A Novel Nanoparticle Drug Delivery System: The Anti-inflammatory Activity of Curcumin Is Enhanced When Encapsulated in Exosomes. *Mol. Ther*. **2010,** *18* (9), 1606–1614.

Teekamp, N.; Duque, L; Frijlink, H.; Hinrichs, W; Olinga, P. Production Methods and Stabilization Strategies for Polymer-based Nanoparticles and Microparticles for Parenteral Delivery of Peptides and Proteins. *Expert Opin. Drug Deliv.* **2015,** *12*, 1311–1331.

Terrones, M.; Botello-Méndez, A. R.; Campos-Delgadoc, J.; López-Uríasd, F.; Vega-Cantúd, Y. I.; Rodríguez-Macíasd, F. J.; Elías, A. L.; Munoz-Sandovald, E.; Cano-Márquezd, A. G.; Charlier, J.-C.; Terrones, H. Graphene and Graphite Nanoribbons: Morphology, Properties, Synthesis, Defects and Applications. *Nano Today* **2010,** *5*, 351–372.

Torchilin, V. P. Passive and Active Drug Targeting: Drug Delivery to Tumors as an Example. In *Handbook of Experimental Pharmacology*; Schafer-Korting, E. P. Ed.; Springer: Heidelberg, Dordrecht, London, New York, 1997; pp 3–53.

Tour, J. M.; Hudson, J. L.; Dyke, C. R.; Stephenson, J. J. Functionalization of Carbonnanotubes in Acidic Media. US20070280876A1, December 6, 2007.

Tour, J. M.; Lucente-Schultz, R.; Leonard, A.; Kosynkin, D. V.; Price, B. K.; Hudson, J. L.; Conyers, J. L.; Moore, V. C.; Casscella, S. W.; Myers, J. N.; Milas, Z. L.; Mason, K. A.; Milas, L. Water-soluble Carbon Nanotube Composition for Drug Delivery and Medicinal Applications. US20090170768A1, July 2, 2009.

Tripisciano, C; Kraemer, K; Taylor, A; Borowiak-Palen, E. Single Wall Carbon Nanotubes-based Anticancer Drug Delivery System. *Chem. Phys. Lett.* **2009,** *478*, 200–205.

Tsai, H.-C.; Lin, J.-Y.; Maryani, F.; Huang, C.-C.; Imae, T. Drug-loading Capacity and Nuclear Targeting of Multiwalled Carbon Nanotubes Grafted with Anionic Amphiphilic Copolymers. *Int. J. Nanomed.* **2013,** *8*, 4427–4440.

Venkatesan, N.; Yoshimitsu, J.; Ito, Y.; Shibata, N.; Takada, K. Liquid Filled Nanoparticles as a Drug Delivery Tool for Protein Therapeutics. *Biomaterials* **2005,** *26*, 7154–7163.

Vukovic, G. D.; Tomic, S. Z.; Marinkovic, A. D.; Radmilovic, C.; Uskokovic, P. S.; Colic, M. The Response of Peritoneal Macrophages to Dapsone Covalently Attached on the Surface of Carbon Nanotubes. *Carbon* **2010,** *48*, 3066–3078.

Wang, J.; Sun, P.; Bao, Y.; Dou, B.; Song, D.; Li, Y. Vitamin E Renders Protection to PC12 Cells Against Oxidative Damage and Apoptosis Induced by Single-walled Carbon Nanotubes. *Toxic. In Vitro* **2012,** *26*, 32–41.

Wilczewska, A.; Niemirowicz, K.; Markiewicz, K.; Car, H. Nanoparticles as Drug Delivery Systems. *Pharmacol. Rep.* **2012,** *64*, 1020–1037.

Wu, W.; Wieckowski, S.; Pastorin, G.; Benincasa, M.; Klumpp, C.; Briand, J. P.; Gennaro, R.; Prato, M.; Bianco, A. Targeted Delivery of Amphotericin B to Cells by Using Functionalized Carbon Nanotubes. *Angew. Chem. Int. Engl.* **2005,** *44*, 6358–6362.

Wu, L.; Man, C.; Wang, H.; Lu, X.; Ma, Q.; Cai, Y.; Ma, W. PEGylated Multi-walled Carbon Nanotubes for Encapsulation and Sustained Release of Oxaliplatin. *Pharm. Res.* **2013,** *30*, 412–423.

Yang, Y.; Zou, H.; Wu, B.; Li, Q.; Zhang, J.; Liu, Z.; Guo, X.; Du, Z. Enrichment of Large-diameter Single-walled Carbon Nanotubes by Oxidative Acid Treatment. *J. Phys. Chem. B* **2002,** *106* (29), 7160–7162.

Yang, S. T.; Fernando, K. A. S.; Liu, J. H.; Wang, J.; Sun, H. F.; Liu, Y.; Chen, M.; Huang, Y.; Wang, X.; Wang, H.; Sun, Y. P. Covalently PEGylated Carbon Nanotubes with Stealth Character In Vivo. *Small* **2008,** *4*, 940–944.

Yue, T. L.; McKenna, P. J.; Gu, J. L.; Cheng, H. Y.; Ruffolo, Jr., R. R.; Feuerstein, G. Z.; Carvedilol, a New Antihypertensive Agent, Prevents Lipid Peroxidation and Oxidative Injury to Endothelial Cells. *Hypertension* **1993,** *22*, 922–928.

Zanella, I.; Fagan, S. B.; Mota, R.; Fazzio, A. Ab Initio Study of Pristine and Si-doped Capped Carbon Nanotubes Interacting with Nimesulide Molecules. *Chem. Phys. Lett.* **2007,** *439*, 348–353.

Zhang, X.; Meng. L; Lu, Q.; Fei, Z.; Dyson, P. J. Targeted Delivery and Controlled Release of Doxorubicin to Cancer Cells Using Modified Single Wall Carbon Nanotubes. *Biomaterials* **2009,** *30*, 6041–6047.

Zhao, Y.-L.; Stoddart, J. Noncovalent Functionalization of Single-walled Carbon Nanotubes. *Acc. Chem. Res.* **2009,** *42* (8), 1161–1171.

Zheng, M.; Jagota, A.; Semke, E. D.; Diner, B. A.; McLean, R. S.; Lustig, S. R.; Richardson, R. E.; Tassi, N. G. DNA-assisted Dispersion and Separation of Carbon Nanotubes. *Nat. Mater.* **2003,** *2*, 338–342.

Zhou, M.; Peng, Z.; Liao, S.; Li, P.; Li, S. Design of Micro Encapsulated Carbon Nanotube-based Microspheres and Its Application in Colon Targeted Drug Delivery. *Drug Deliv.* **2014,** *21*, 101–109.

# PART IV
# Additional Topics

## CHAPTER 13

# NANOEMULSION FOR DRUG DELIVERY

PREETI KHULBE*

*School of Pharmaceutical Sciences, Jaipur National University, Jaipur 302017, India*

**E-mail: khulbe.preeti@yahoo.in*

## CONTENTS

## ABSTRACT

Nanoemulsion can be defined as emulsion having droplet sizes with "nano" range. Both hydrophilic and lipophilic drugs can be used in nanoemulsion. It protects the drug from hydrolysis and oxidation due to encapsulation in oil droplet. It also provides taste masking. Different characterization parameters for nanoemulsion include transmission microscopy, nanoemulsion droplet size analysis, consistency determination, index of refraction, in vitro skin permeation studies, skin irritation check, in vivo efficacy study, thermodynamic stability studies, and surface characteristics. This chapter covers different applications, advantages, disadvantages, and methods of preparation of nanoemulsion. The future scopes of nanoemulsion are also discussed in this chapter.

## 13.1 INTRODUCTION

Nanoemulsions are the dispersions of nanoscale droplets fashioned by shear-induced rupturing. The current convention for nanoscale materials are the materials composed of structures having length scales within the range from 1 to 100 nm; below this range lies the angstrom unit scale, and higher than this range lies the small scale. Applying this convention to medium stable (metastable) emulsions, the defined nanoemulsions are the emulsions having droplet sizes with "nano" range. To be more precise regarding size, one should specify the droplet radius or diameter; generally, there are a lot of versatile definitions of a nanoemulsion during which the vast majority of droplets, both on a number- and volume-weighted basis, lie below 100 nm (Hansen, 1979). Since the diameter and radius solely differ by a factor of both, this can be a small distinction relative to the two decades of "nano" length scales. Though the extended definition of nanoemulsions to the 1 nm scale, it might be not possible to create a nanoemulsion smaller than the scale of a surface-active agent micelle, generally a few nanometers (El-Aasser and Sudol, 1997).

Nanoemulsions are the part of a broad category of polyphase colloidal dispersions. Though some lyotropic liquid crystalline phases, conjointly referred to as "micellar phases," "mesophases, and "microemulsions," could seem to be almost like nanoemulsions in composition and nanoscale structure, such phases are literally quite different. Lyotropic liquid crystals are equilibrium structures composed of liquids and surface-active agent, such as lamellar sheets, hexagonally packed columns, and worm-like micellar phases, that type spontaneously through thermodynamical self-assembly. In contrast, nanoemulsions do not form spontaneously; an external shear should be applied to rupture larger droplets into smaller ones. Compared to microemulsion phases, comparatively very little is known regarding creating and controlling nanoemulsions.

Despite their stability, nanoemulsions could be persisting over several months or years because of the presence of a stabilizing surface-active agent that inhibits the coalescence of the droplets (Walstra, 1996). Emulsions are the dispersions of one liquid phase in another unmixable liquid part that are created using mechanical shear. Because of differences in attractive interactions between the molecules of the two liquid phases, a surface tension, $\sigma$, exists between the two liquids everywhere they are in contact. This surface tension may be reduced considerably by adding amphiphilic active molecules, or "surfactants," that are extremely soluble in a minimum of one of the liquid phases.

Surfactants mostly adsorb around the interfaces, because the molecular structures is nonpolar organic compound tails that favor to be in nonpolar liquids, such as oils, and polar or charged head groups that favor to reside in polar liquids, such as water. Emulsions that have water as an eternal part and oil as a phase are referred to as "direct," "water-based," or "O/W" emulsions; for direct emulsions, the surface-active agent is usually soluble within the liquid part and provides a lot of stability of water films. In contrast, emulsions that have oil as an eternal part are referred to as "inverse," "oil-based," or "W/O" emulsions; for inverse emulsions, the surface-active agent is usually soluble within the oil phase and provides a lot of stability of oil films. For emulsions, the surface-active agent does not lower $\sigma$ to effectively zero; instead the surface tension, though somewhat reduced, remains at a large value, generally $\sigma$ 91 dyn/cm. At low $\varphi$, an isolated droplet is spherical and encompasses a radius, $a$. The curving interface exerts a pressure on the molecules within the droplet, and this pressure is termed the Laplace pressure, $L = 2\sigma/a$. For nonspherical droplets, $dl$ is proportional to the addition of the inverse of the two principal radii of curvature of the interface. Because of the inverse dependence of $dl$ on $a$, the molecules in smaller droplets expertise a higher pressure than those in larger droplets (Becher, 1965; Bibette et al., 1999).

"Microemulsions" represent a very much complex menagerie of fully different equilibrium systems. So-called spontaneous emulsification through the addition of a surface-active agent without shear are usually associated with the formation of equilibrium lyotropic liquid crystalline phases, during which the surface tension effectively vanishes and therefore the droplets are fashioned by thermodynamical molecular self-assembly from the "bottom up" (Miller, 2006). These lyotropic phases, generally named as "microemulsions," "mesophases," or "swollen micelles" are not emulsions within the classical sense. In most microemulsion phases, the surface-active agent is extremely soluble in each liquid phase, and therefore the two immiscible liquid phases themselves even have comparatively high mutual solubility. Microemulsions are true self-assembled thermodynamical phases which will have terribly interesting morphologies such as planar stacks of lamellae (lamellar phases), hexagonal-shaped tubes (hexagonal-type phases), and spherical-shaped droplets (spherical micellar phases).

Nanoemulsions are often characterized by specifying molecular constituents, quantities of those constituents, and therefore, the sizes of the droplet structures determine the formation of the emulsion. For the subsequent discussion, it tends to take into account direct nanoemulsions of oil droplets in water; the fundamental concepts also apply to inverse nanoemulsions. The relative molecular mass, MW, and molecular structure of the oil (e.g., alkane, silicone, or other) are generally chosen based on the application, yet MW should be large enough to inhibit Ostwald ripening. The surface-active agent type and concentration, $C$, within the liquid phase are chosen to produce good stability against coalescence. The droplet volume fraction, $\varphi$, describes the relative degree of concentration of the droplets. Finally, the droplet structure is usually reportable in terms of a size distribution of droplets, usually measured using dynamic light scattering (DLS) from a diluted sample that has been filtered to get rid of dust. After information collection, most DLS systems typically employ a regularization algorithm to suit decays within the intensity autocorrelation performed to get the size distribution, $p(\mathrm{a})$. Regularized fitting is ill-posed mathematically, thus it is a good plan to match unknown size distributions with control experiments using calibrated polymer spheres (Gabriel and Johnson, 1981).

Nanoemulsions have some interesting physical properties that distinguish them from normal microscale emulsions. For example, microscale emulsions usually exhibit strong multiple scattering of visible light, and, as a result, have a white look. Nanoemulsions have a much larger surface area to volume ratio than normal emulsions, thus phenomena associated with deformation of the droplets, such as the Laplace pressure and therefore the elastic modulus, are generally larger for nanoemulsions than normal emulsions. The amount of molecules of the dispersed phase in nanoemulsion droplets are a lot smaller than for normal emulsions and are generally several hundred to thousand, depending upon MW. Common inkjet technology uses picoliter (10–12 L) droplets having radii of many microns. In contrast, nanoemulsions that have a 10 nm radius contain of the order of a zeptoliter (10−21 L) of the dispersed phase. Thus, a single picoliter droplet might be ruptured into a billion nanoemulsion droplets (Cates and Webster, 2001).

Microemulsions are thermodynamic phases composed of self-assembled nanostructures, nanoemulsions are not equilibrium thermodynamic phases, but instead of dispersions of nanoscale droplets of one phase of liquid in another immiscible liquid. Here, the term nanoscale is used to refer to droplets that have radii when undeformed that are typically less than about 100 nm (Fig. 13.1).

| nanoemulsion | microemulsion |
|---|---|
| non-equilibrium dispersion of droplets | thermodynamic phase of nanostructures |
| formed by droplet rupturing typically | formed by self-assembly |
| extreme flow is typically required to form | forms spontaneously-no mixing is required |
| nanostructures are droplets of a dispersed phase coated with surfactant | nanostructures can be swollen spherical micelles, lamellae, columnar micelles,... |
| significant liquid-liquid interfacial tension | very low liquid-liquid interfacial tension |
| very low mutual solubility of immiscible liquid phases | significant mutual solubility of immiscible liquid phases |
| single surfactant stabilizes droplet interfaces against coalescence | a surfactant and usually a co-surfactant (e.g. an alcohol) reduce interfacial tension |
| little to no exchange of dispersed phase between droplets | rapid exchange of dispersed phase between micellar structures |

**FIGURE 13.1** Difference between microemulsion and nanoemulsion.

## 13.2 ADVANTAGE

1. Nanoemulsions are stable thermodynamically as well as kinetically; therefore flocculation occurs, large concentration of surfactants/cosurfactants is required for stabilization.
2. Aggregation, creaming, and coalescence do not occur.

3. It is nontoxic and nonirritant.
4. Various routes can be preferred for administration, such as oral, topical, parenteral, and transdermal, etc.
5. Both hydrophilic and lipophilic drugs can be used.
6. Droplet size is nano, so surface area is higher, thus increasing the rate of absorption and reduces variability, thus enhances bioavailability of drug.
7. Nanoemulsions are suitable for human and veterinary uses because they do not damage human or animal cell (Chang et al., 2006).
8. It protects the drug from hydrolysis and oxidation due to encapsulation in oil droplet. It also provides taste masking.
9. Nanoemulsion also enhances permeation of drug through skin.
10. The small size of the droplets allows them to deposit uniformly on substrates. Wetting as well as penetration may be also enhanced as a result of the low surface tension of the whole system and the low interfacial tension of the o/w droplets.
11. The very small droplet size causes a large reduction in the gravity force and the Brownian motion may be sufficient for overcoming gravity. This means that no creaming or sedimentation occurs on storage.
12. The small size of droplets also prevents the flocculation of the droplets. Weak flocculation is prevented and this enables the system to remain dispersed with no separation.
13. The small droplets also prevent their coalescence, since these droplets are elastic, surface fluctuations are prevented (Armstrong et al., 1977).
14. Nanoemulsions are suitable for efficient delivery of active ingredients via the skin. The large surface area of the emulsion system allows rapid penetration of actives.
15. The transparent nature of the system, their fluidity, as well as the nonpresence of any thickeners may give them a pleasant aesthetic character and skin feel.
16. Unlike microemulsions (which require a high surfactant concentration, usually at 20% and higher), nanoemulsions can be prepared using reasonable surfactant concentration. For a 20% o/w nanoemulsion, required surfactant concentration is 5–10% (Trotta et al., 1999).
17. Nanoemulsions can be applied for delivery of fragrance, which may be used in many cosmetic and care products. This could

also be applied in perfumes, which are desirable to be formulated alcohol free.

18. Nanoemulsions may also be applied as a substitute as liposomes and vesicles (which are less stable) and it is possible in some cases to build lamellar liquid crystalline phases around the nanoemulsion droplets (Adil et al., 2012).

## 13.3 DISADVANTAGE

1. The stability of nanoemulsion is always affected by temperature and its pH. The instability can also be caused because of Oswald ripening effect.
2. Formulation of nanoemulsion sometimes requires special application techniques, such as the use of high pressure homogenizers as well as ultrasonics. Such equipment (such as the microfluidizer) are available only in recent years and may be costly.
3. There is a perception in the drug and cosmetic industry that nanoemulsions are expensive to produce. Because in this, expensive equipment as well as the use of much high concentration of emulsifying agents may be required (Devarajan et al., 2011).
4. A major disadvantage includes the lack of understanding of the mechanism of production of very small, that is, submicron droplets and the role of surfactants and cosurfactants.
5. Lack of demonstration of the benefits is another important criterion that can be obtained from using nanoemulsions when compared with the classical macroemulsion systems.
6. Lack of knowledge of the interfacial chemistry that is involved in production of nanoemulsions (Alvarez-Figueroa et al., 2001).

## 13.4 FORMULATION

### *13.4.1 FORMULATION FACTORS THAT AFFECT THE STABILITY OF NANOEMULSIONS*

Although nanoemulsions enhance the physical and as chemical stability of medicine, stability of drug product is one of the issues related to the development of nanoemulsions. Stability studies have been done on nanoemulsions by storage at refrigerator and room temperatures over several months (Ahmad et al., 2007). The consistency, refractive index, and droplet size are determined throughout the time of storage. Insignificant changes in these parameters indicate formulation stability. Accelerated stability studies can even be performed on the nanoemulsions. During this instance, nanoemulsion formulation are kept at accelerated temperatures and samples are withdrawn at regular intervals and evaluated for drug content by stability indicating assay strategies. The number of drug degraded and remaining in nanoemulsion formulation is decided at each time interval (Shakeel et al., 2008). Stability of nanoemulsion formulation could also be increased by controlling factors such as type and concentration of surface-active agent and cosurfactant, type of oil phase, strategies used, process variables, and addition of additives. Overall, nanoemulsion formulation could also be considered as effective and safe, and have patient compliance for the delivery of pharmaceuticals.

Factors to be considered throughout preparation of nanoemulsion include the following:

a. The prime demand in nanoemulsion production is that an ultralow interfacial surface tension should be attained at the oil–water interface, thus surfactants should be carefully chosen.
b. Concentration of surface-active agent should be high enough to produce the amount of surface-active agent molecules required to stabilize the nanodroplets.
c. The interface should be versatile to promote the formation of nanoemulsion (Chang et al., 2006; Chakravarthi et al., 2013).

## 13.4.2 FORMULATION METHODS

### 13.4.2.1 HIGH-PRESSURE HOMOGENIZATION

This technique makes use of high-pressure homogenizer/piston homogenizer to formulate nanoemulsions of very low particle size (up to 1 nm). In an exceedingly high-pressure homogenizer, the dispersion of two liquids (oily phase and liquid phase) is achieved by forcing their mixture through a small inlet passage at terribly high pressure (500–5000 psi) that subjects the product to intense turbulence and hydraulic shear resulting in extremely fine particles of emulsion. Homogenizers of variable design are obtainable for research laboratory scale and industrial scale production of nanoemulsions. This method has great potency, the only disadvantage being high energy consumption and increase in temperature of emulsion throughout process (Constantinides et al., 1995; Khoshnevis et al., 1997).

#### *13.4.2.2 MICROFLUIDIZATION*

Microfluidization may be a patented admixture technology, that makes use of a tool referred to as microfluidizer. This device uses a high-pressure positive-displacement pump (500–20000 psi), that forces the product through the interaction chamber, that consists of small channels referred to as "microchannels." the product flows through the microchannels on to an impingement area resulting in very fine particles of submicron range. The two solutions (aqueous phase and oily phase) are combined along and processed in an inline homogenizer to yield a coarse emulsion. The coarse emulsion is added into a microfluidizer where it is further processed to get a stable nanoemulsion. The coarse emulsion is responded to the interaction chamber of the microfluidizer repeatedly till desired particle size is obtained. The majority emulsion is then filtered through a filter under nitrogen to get rid of large droplets leading to uniform nanoemulsions. Another methodology used for nanoemulsions preparation is that the phase-inversion temperature technique (Attwood, 1992).

## 13.5 CHARACTERIZATION OF NANOEMULSIONS

Different characterization parameters for nanoemulsion include transmission microscopy, nanoemulsion droplet size analysis, consistency determination, index of refraction, in vitro skin permeation studies, skin irritation check, in vivo efficacy study, thermodynamic stability studies, and surface characteristics. The surface charge of the nanoemulsion droplets includes a marked effect on the stability of the emulsion system and therefore the droplet in vivo disposition and clearance (Shen et al., 2006).

### 13.5.1 IN VITRO SKIN PERMEATION STUDIES

Franz diffusion cell is employed to get the drug release profile of the nanoemulsion formulation in the case of formulations for transdermal application. The extent or depth of skin penetration by the discharged content may be visualized by confocal scanning optical microscopy. In vitro drug release may be determined by dispersing an amount of the preparation within the donor compartment of a Franz cell having a membrane as barrier and observation of the appearance of the encapsulated drug within the reception medium, typically phosphate buffer saline (pH 7.4) and stirring on a magnetic stirrer at 100 revolutions per minute at 37°C ± 1°C. Samples (1 mL) of the dispersion are withdrawn from the medium and replaced with a similar quantity of the medium at definite intervals using 0.22–50 μm filter (e.g., Millipore, USA) and therefore the drug discharged then analyzed using UV–Visible spectroscopic analysis at wavelength of peak absorption of the drug. An alternative and fashionable technique of ex vivo release study is performed on diffusion cell. The skin is cut from the ear or abdomen and underlying animal tissue and fats are carefully removed. Appropriate size of skin is cut and placed on the diffusion cell that had earlier been filled with receptor solution. Samples of the vesicular preparation are then applied on the dorsal surface of the skin and thereafter the instrument is started. At intervals, up to 24 h, samples are withdrawn from the receptor medium and replaced with equal amounts of the medium and thereafter the withdrawn samples are analyzed for the drug permeated using high-performance liquid chromatography (HPLC) or ultraviolet radiation spectroscopy. Semipermeable membrane such as regenerated polysaccharide may be employed in place of skin for in vitro release studies. The flux $J$, of the drug across the skin or membrane is calculated from the formula:

$$J = D\, dc/dx \tag{13.1}$$

where $D$ is that the diffusion constant and may be a function of the size, form, and flexibility of the diffusing molecule as well as the membrane resistance, $c$ is the concentration of the diffusive species, $x$ is that the spatial coordinate. In vivo release study otherwise referred to as dermatopharmacokinetics, is carried out by applying or administering the preparation to whole live animal. Blood samples are then withdrawn at intervals, centrifuged, and thereafter the plasma is analyzed for the drug content using HPLC. Results obtained from in vitro and in vivo studies are extrapolated to reflect bioavailability of the drug formulation (Kuo et al., 2008).

### *13.5.2 THERMODYNAMIC STABILITY AND SURFACE CHARACTERISTICS*

Although the physical look of a nanoemulsion could match that of a microemulsion, each system could also be clear and of low consistency, there is a vital distinction between the two systems. A nanoemulsion is at the best, kinetically stable, whereas microemulsion is thermodynamically stable. Nanoemulsions owing to their tiny droplet size, possess higher stability against sedimentation or creaming than microemulsions. Both the systems are completely different since nanoemulsions are shaped by mechanical shear and microemulsion phases are fashioned by self-assembly (Agnieszka et al., 2009).

## 13.6 PROPERTIES

Nanoemulsions have distinctive properties such as tiny drop size, exceptional stability, clear look, and tunable rheology. These properties create nanoemulsions an attractive candidate for applications within the food, cosmetic, pharmaceutical industries, and in drug delivery applications. Furthermore, they will serve as the building blocks for designer advanced materials with distinctive properties (Mason and Wilking, 2007).

### *13.6.1 DROPLET SIZE AND STABILITY*

Nanoemulsions have droplets with diameter on the order of 100 nm. Because the droplet size is considerably smaller than the wavelength of visible radiation, nanoemulsions are usually clear in look. However, by controlling the droplet size of nanoemulsions, one will simply tune the looks of nanoemulsions to vary from clear to milky white. As mentioned in Section 13.1, nanoemulsions can even be tuned to possess strong stability with shelf life ranging from months to years. They have an extra advantage of being comparatively less sensitivity toward dilution, temperature, and pH changes than microemulsions (Bibette et al., 2011). These properties make nanoemulsions attractive in various industries, including in the food and cosmetic sectors.

### 13.6.2 TUNABLE RHEOLOGY

The rheologic properties of nanoemulsions may be tuned by dominancy of the dispersed particles volume fraction and droplet size and by the addition of salt and depletion agents. Nanoemulsions are significantly interesting from a rheologic point of view as they will exhibit considerably stronger physical property than macroemulsions because the physical property is roughly on the order of the Laplace pressure of an undeformed droplet. The standardization of the rheology of an emulsion usually dictates the consumer's perception of the product within the cosmetic industry. Mason and Wilking showed that one will tune the flow behavior of the nanoemulsions to be from a flowing fluid to a slowly relaxing fluid or a gel-like system by everchanging the amount of passes in a very high pressure homogenization. This irreversible flow-induced elastification happens because of excess droplet rupture resulting in electronic jamming of the nanoemulsion structure. One may also tune the rheologic response of nanoemulsions by adding salt or a depletion agent that results in gelling of droplets. Another way of calibrating rheologic behavior is to feature polymers which will physically associate among themselves or with nanoemulsion droplets. This reversible development is claimed to be due to changes in hydrophobic interactions between the polymer and therefore the droplets with temperature. At temperatures more than the gelling temperature ($T_g$), the chemical compound with two hydrophobic end groups bridges the nanoemulsion droplets, creating a percolated network. Once the temperature is brought down below $T_g$, the hydrophobic teams detach from the surface of nanoemulsion droplets and also the system returns to its initial clear and fluid-like state. The study showed that the relative worth of the interactions length (radius of gyration of the bridging polymer) to the droplet size plays a very important role. Strong gels can form once the length scale of the attractive interaction is comparable to the droplet size (Gao et al., 2015). This explains why macroemulsions form pastes with a low elastic modulus. The recent studies have highlighted that the nanoemulsions will undergo arrested spinodal decomposition and that the gelation is sensitive to the rate at that the engaging interactions (via temperature) is increased (Hsiao, 2015). Future studies ought to investigate the rheologic behavior of nanoemulsion-based gels in larger depth. Since the relative worth of interaction length to droplet size play a very important role, nanoemulsions with bi-modal or tri-modal distributions may be used to produce gels with rich rheology. Additional studies on rupture (yielding) of nanoemulsion-based gels might provide insights into the association and disassociation mechanism of the bridging gelator and, maybe, healing behavior of the system (Bose et al., 2014).

## 13.7 APPLICATIONS

### *13.7.1 ROLE OF NANOEMULSIONS IN COSMETICS*

Nanoemulsions have recently become progressively vital as potential vehicles for the controlled delivery of cosmetics and for the optimized dispersion of active ingredients especially skin layers. As a result of their lipophilic interior, nanoemulsions are a lot of suitable for the transport of lipophilic compounds than liposomes. Nanoemulsions are acceptable in cosmetics as a result of no inherent creaming, sedimentation, flocculation, or coalescence that are observed with macroemulsions. The incorporation of potentially irritating surfactants will usually be avoided by using high-energy equipment throughout manufacturing. Nanoemulsions have attracted extended attention in recent years for application in personal care product as potential vehicles for the controlled delivery of cosmetics and therefore the optimized dispersion of active ingredients in particular skin layers (Sintov, 2004).

### *13.7.2 ANTIMICROBIALAPPLICATIONSOFNANOEMULSIONS*

Antimicrobial nanoemulsions are oil-in-water droplets that vary from 200–600 nm. They are composed of oil and water and are stabilized by surfactants and alcohol. The nanoemulsion has a broad-spectrum activity against bacterium (e.g., *Escherichia coli* and enteric bacteria *Staphylococcus aureus*), enveloped viruses (e.g., HIV and Herpes simplex), fungi, and spores. The nanoemulsion particles are thermodynamically driven to fuse with lipid-containing organisms. This fusion is increased by the electrostatic attraction between the ion charge of the emulsion and therefore the anionic charge on the infectious agent. When enough nanoparticles fuse with the pathogens, they release a part of the energy trapped among the emulsion. Each of the active ingredient and therefore the free energy destabilize the pathogen lipid membrane, leading to cell lysis and death. In the case of spores, further germination enhancers are incorporated into the emulsion. Once the initiation of germination takes place, the germinating spores become susceptible to the antimicrobial action of the nanoemulsion. As a result, the nanoemulsion can do a level of topical antimicrobial activity that has only been previously achieved by systemic antibiotics (Tamilvanan, 2004).

### *13.7.3 NANOEMULSION AS MUCOSAL VACCINE*

Nanoemulsions are being employed to deliver either recombinant proteins or inactivated organisms to a tissue layer surface to supply an immunologic response. The primary applications, an influenza vaccine and an HIV vaccine, will proceed to clinical trials. The nanoemulsion causes proteins applied to the tissue layer surface to be adjuvant and it facilitates uptake by antigen-presenting cells. Further analysis is in progress to finish the proof of idea in animal trials for alternative vaccines as well as hepatitis B and anthrax (2016). Mice and guinea pigs are immunized intranasally by the application of recombinant HIV gp120 antigen mixed in nanoemulsion incontestable robust serum anti-gp120 immunoglobulin G, furthermore as bronchial, vaginal, and serum anti-gp120 immunoglobulin A in mice.

The serum of those animals demonstrated antibodies that cross-reacted with heterologous serotypes of gp120 and had significant neutralizing activity against two clade-B laboratory strains of HIV (HIVBaL and HIVSF162) and five primary HIV-1 isolates. This study suggests that nanoemulsion ought to be evaluated as a mucosal adjuvant for multivalent HIV vaccines (Bielinska et al., 2008a). Hepatitis B infection remains a vital world health concern despite the availability of safe and effective prophylactic vaccines. Limitations to those vaccines embrace requirement for refrigeration and three immunizations thereby limiting use within the developing world. A replacement nasal hepatitis B vaccine composed of recombinant hepatitis B surface antigen (HBsAg) during a novel nanoemulsion adjuvant (HBsAg-nanoemulsion, HBsAg-NE) might be effective with fewer administrations. Comprehensive preclinical pharmacological medicine analysis demonstrated that HBsAg-NE vaccine is safe and well tolerated in multiple animal models. The results recommend that needle-free nasal immunization with HBsAg-NE might be a secure and effective hepatitis B vaccine or as an alternate booster administration for the parenteral hepatitis B vaccines. This vaccine induces a Th1-associated cellular immunity and conjointly might provide therapeutic benefit to patients with chronic hepatitis B infection. World Health Organization lack cellular immune responses to control viral replication. Real-term stability (long term) of vaccine formulation at elevated temperatures suggests an immediate advantage within the field, since potential excursions from cold chain maintenance might be tolerated without a loss in therapeutic effectiveness (Bielinska et al., 2008b). A novel technique for vaccinating against a spread of infectious diseases using associate oil primarily based on emulsion is placed within the nose, instead of needles, has proved able to manufacture a strong immunologic response against smallpox and HIV in two new studies. Developing mucosal immunity may be important for protection against HIV. In this study, the nanoemulsion HIV vaccine showed that it was able to induce mucosal immunity, cellular immunity, and neutralizing antibody to various isolates of HIV virus. A protein employed by the team, gp120, is one of the key binding proteins under study in alternative HIV vaccine approaches (Arbor, 2008).

### *13.7.4 NANOEMULSION APPLICATION AS NONTOXIC DISINFECTANT CLEANER*

A breakthrough nontoxic disinfectant cleaner to be used in industrial markets that embrace healthcare, hospitality, travel, food process, and military applications has been developed by Enviro Systems, Inc. that kills tuberculosis and a wide spectrum of viruses, microorganisms, and fungi in 5–10 min with none of the hazards exhibited by different categories of disinfectants. The product needs no warning labels. It does not irritate eyes and might be absorbed through the skin, inhaled, or swallowed while showing not harmful effects. The disinfectant formulation is created from nanospheres of oil droplets of size 106 mm that are suspended in water requiring only minute amounts of the active ingredient, parachlorometaxylenol (PCMX). The nanospheres have surface charges that penetrate the charges present on surface on microorganisms' membranes—much like breaking through an electrical fence. Rather than "drowning" cells, the formulation permits PCMX to focus on and penetrate cell walls. As a result, PCMX is effective at concentration levels of 1–2 orders of magnitude below those of different disinfectants; hence, there are not any harmful effects on individuals, animals, or the atmosphere. Different microbial disinfectants require massive doses of their several active ingredients to surround pathogen cell walls, that cause them to disintegrate, essentially "drowning" them within the disinfectant solution. The formulation may be a broad-spectrum disinfectant cleaner which will be applied to any hard surface as well as equipment, counters, walls, fixtures, and floors. One product will currently take the place of many reducing product inventories and saving valuable storage space (ewire.com, 2016; infectioncontrol today.com, 2016).

### *13.7.5 NANOEMULSIONS IN CELL CULTURE TECHNOLOGY*

Cell cultures are used for in vitro assays or to produce biological compounds, such as antibodies or recombinant proteins. To optimize cell growth, the substance may be supplemented with variety of defined molecules or with serum. The benefits of using nanoemulsions in cell culture technology are much better uptake of oil-soluble supplements in cell cultures, improvement in growth and vitality of cultivated cells, and allowance of toxicity studies of oil-soluble drugs in cell cultures (mib-bio.com, 2016).

### *13.7.6 NANOEMULSION AS TARGETED DRUG DELIVERY*

The effects of the formulation and particle composition of gadolinium (Gd)-containing lipid nanoemulsion (Gd-nano-LE) on the biodistribution of Gd after intravenous (IV) injection into D1-179 melanoma-bearing hamsters were evaluated for its application in cancer neutron-capture therapy. Biodistribution information discovered that Brij 700 and HCO-60 prolonged the retention of Gd within the blood and increased its accumulation in tumors. Upon dermal application, the drug was predominantly localized in deeper skin layers, with least systemic escape. This has amounted to an absolute bioavailability of 70.62%. Inhibition of P-glycoprotein efflux by D-tocopheryl polyethyleneglycol 1000 succinate and labrasol would have contributed to the improved peroral bioavailability of paclitaxel (PCL). This investigation provides evidence on the localization of high-molecular-weight lipotropic drug, PCL, in dermis. Further, the nanoemulsion formulation has enhanced the peroral bioavailability significantly to more than 70%. The developed nanoemulsion formulation was safe and effective for both peroral and dermal delivery of PCL (Khandavilli et al., 2007). Camptothecin could be a topoisomerase-I inhibitor that acts against a broad spectrum of cancers. However, its clinical application is proscribed by its quality, instability, and toxicity. The aim of the current study was to develop acoustically active nanoemulsions for camptothecin encapsulation to bypass these delivery issues. The nanoemulsions were prepared using liquid perfluorocarbons and oil because the cores of the inner phase. These nanoemulsions were stabilized by phospholipids and/or Pluronic F68 (PF68) (Fang et al., 1998).

### *13.7.7 IMPROVED ORAL DELIVERY OF POORLY SOLUBLE DRUGS BY NANOEMULSION*

Nanoemulsions formulation was developed to enhance oral bioavailability of hydrophobic medication. Paclitaxel was selected as a model hydrophobic drug. The o/w nanoemulsions were created with pine nut oil as the internal oil part, water as the external part, and egg lecithin as the primary emulsifier. Stearylamine and deoxycholic acid were used to provide positive and negative charge to the emulsions, respectively. The formulated nanoemulsions had a particle size range of 100–120 nm and zeta potential starting from 34 to 245 mV. After oral administration of nanoemulsions, a considerably higher concentration of paclitaxel was determined within the circulation compare to regulate solution. The results of this study suggest that nanoemulsions are promising novel formulations which may promote the oral bioavailability of hydrophobic drugs (Cavello et al., 1987; Al-Zaagi et al., 2001).

### *13.7.8 NANOEMULSIONS IN OCULAR DRUG DELIVERY*

Ophthalmic drug delivery is one of the foremost interesting and difficult endeavors facing the pharmaceutical scientist. It is a typical knowledge that the application of eye drops as typical ophthalmic delivery systems leads to poor bioavailability and therapeutic response due to lachrymal secretion and nasolacrimal drain within the eye. Most of the drug is drained from the precorneal area in couple of minutes. As a result, frequent instillation of targeted solutions is required to attain the specified therapeutic effects. But, by the tear drain, the most a part of the administered drug is transported via the duct to the gastric intestinal tract where it gets absorbed, typically causing side effects. So as to extend the effectiveness of the drug, a dosage type ought to be chosen that will increase the contact time of the drug within the eye. This could then increase the bioavailability, reduce systemic absorption, and reduce the necessity for frequent administration resulting in improved patient compliance. Nanoemulsions may be utilized to beat a number of these issues. Dilutable nanoemulsions are potent drug delivery vehicles for ophthalmic use due to their varied benefits as sustained effect and high ability showing drug penetration to the deeper layers of ocular structure and also the aqueous humor. Ammar et al. developed the antiglaucoma drug dorzolamide hydrochloride as ocular nanoemulsion of high therapeutic efficaciousness and prolonged impact. These nanoemulsions showed acceptable physicochemical properties and exhibited slow drug release. Draize rabbit eye irritation test and histologic examination were dispensed for those preparations exhibiting superior properties and revealed that they were nonirritant. Biological analysis of dorzolamide hydrochloride nanoemulsions on normotensive albino rabbits indicated that these products had higher therapeutic efficaciousness, faster onset of action, and prolonged effect relative to either drug solution or the market product. It absolutely was concluded from the study that formulation of dorzolamide hydrochloride in a very nanoemulsion form offered a more intensive treatment of glaucoma, a decrease within the number of applications per day, and a better patient compliance compared to conventional eye drops (Benita and Tamilvanan, 2004).

### *13.7.9 AS VEHICLE FOR TRANSDERMAL DELIVERY*

Drug delivery through the skin to the circulation is convenient for variety of clinical conditions as a result of that there has been a substantial interest in this area (Muller-Goymann, 2004). It offers the advantage of steady-state controlled drug delivery over extended amount of time, with self-administration additionally being attainable, which cannot be the case with duct route. The drug input is terminated at any time by the patient simply by removing the transdermal patch. Their clear nature and thinness of nanoemulsions confer a nice skin feel. An additional advantage is that the total absence of gastrointestinal aspect effects such as irritation and gut ulcers that are invariably related to oral delivery. Stratum drug product are developed for variety of diseases and disorders together with cardiovascular conditions, Parkinsons' and Alzheimer's diseases, anxiety, depression, etc. However, the basic disadvantage that limits the utilization of this mode of administration is that the barrier imposed by the skin for effective penetration of the bioactives. The three routes by that medicine will primarily penetrate the skin are through the hair follicles, sweat ducts, or directly across corneum that restricts their absorption to a large extent and limits their bioavailability. For improved drug pharmacokinetics and targeting, the primary skin barriers have to be compelled to be overcome. Additionally, the locally applied drug distribution through cutaneous blood and channel system must be controlled. Nanosized emulsions are able to simply penetrate the pores of the skin and reach the circulation therefore getting channelized for effective delivery (Dave et al., 2009). Alkaloid has been used for treatment of various types of cancer by oral delivery. Water-in-oil nanoemulsion formulations of caffeine are developed for stratum drug delivery. Comparison of in vitro skin permeation profile between these and liquid caffeine solutions showed important increase in permeability parameters for the nanoemulsion-loaded medicine (Shakee et al., 2010). Use of nanoemulsions in stratum drug delivery represents a very important area of analysis in drug delivery, which boosts the therapeutic efficacy and conjointly the bioavailability of the medicine with none adverse effects. These systems are being employed currently to supply dermal and surface effects and also for deeper skin penetration. Several studies have shown that nanoemulsion formulations possess improved transdermal and dermal delivery properties in vitro, in addition as in vivo (Kemken et al., 1992). Nanoemulsions have improved transdermal permeation of the many drugs over the conventional topical formulations such as emulsions and gels. The nanoemulsions were prepared by the spontaneous emulsification methodology for transdermal delivery of indomathacin. A significant increase within the permeability parameters such as steady-state flux, permeability coefficient, and enhancement ratio was determined in nanoemulsion formulations compared with the conventional NSAID gel (Kreilgaard, 2001).

### *13.7.10 NANOEMULSIONS ROLE IN INTRANASAL DRUG DELIVERY*

Intranasal drug delivery system has currently been recognized as a reliable route for the administration of medication next to parenteral and oral routes. Nasal membrane has emerged as a therapeutically viable channel for the administration of systemic medication and additionally seems to be a favorable way to overcome the obstacles for the direct entry of medication to the target site (Alves et al., 2009). This route is additionally painless, noninvasive, and well tolerated. The nasal cavity is one of the foremost efficient sites owing to its reduced enzymatic activity, high availability of immunoactive sites, and its moderately pervious epithelial tissue (Agu et al., 2005). There are many issues related to targeting medication to brain, particularly the hydrophilic ones and those of high relative molecular mass. This is often owing to the impervious nature of the endothelium that divides the circulation and barrier between the blood and brain (Pardridge, 1999). The olfactory region of the nasal membrane provides an immediate association between the nose and brain, and by the employment of nanoemulsions loaded with drugs, conditions such as Alzheimer's disease, migraine, depression, psychosis, Parkinson's diseases, meningitis, etc. may be treated (Babbar et al., 2008). Preparation of nanoemulsions containing risperidone for its delivery to the brain via nose has been reported. It is inferred that this emulsion is simpler through the nasal route instead of intravenous route. Another application of intranasal drug delivery system in medicine is their use in development of vaccines. Immunity is achieved by the administration of tissue layer matter. Currently, the primary intranasal vaccine has been marketed (Illum et al., 2009). Among the potential delivery systems, the employment of nanobased carriers mostly hold an excellent promise to guard the biomolecules, promote nanocarrier interaction with mucosae, and direct matter to the humor tissues. So the employment of nanoemulsions in intranasal drug delivery system is about to originate important leads to target medication to the brain in treatment of diseases associated with the central nervous system. The intranasal nanoemulsion and gel formulations were developed for rizatriptan salt for prolonged action. Numerous mucoadhesive agents were tried out to form thermotriggered mucoadhesive nanoemulsions. Mucoadhesive gel formulations of rizatriptan were prepared using different ratios of hydroxy propyl methyl cellulose (HPMC) and Carbopol 980. Comparative analysis of intranasal nanoemulsions and intranasal mucoadhesive gels indicated that larger brain targeting may well be achieved with nanoemulsions. Other medications that have been formulated for nasal delivery are insulin and testosterone (Alonsaet al., 2009).

### *13.7.11 ROLE OF NANOEMULSIONS IN PARENTERAL DRUG DELIVERY*

This is one of the common and effective routes of drug administration sometimes adopted for actives with low bioavailability and narrow therapeutic index. Their capability to dissolve greater quantities of hydrophobics, along with their mutual compatibility and ability to shield the medication from reaction and catalyst degradation create nanoemulsions as ideal vehicles to aim parenteral transport. Further, the frequency and dose of injections will be reduced throughout the drug therapy period as these emulsions guarantee the release of medicine during a sustained and controlled mode over long periods of time. In addition, the lack of flocculation, sedimentation, and creaming, combined with a large area and free energy, provide obvious benefits over emulsions of larger particle size, for this route of administration. Their terribly large interfacial area positively influences the drug transport and their delivery, in conjunction with targeting them to specific sites. Major clinical and preclinical trials have thus been applied with parenteral nanoemulsion-based carriers. Nanoemulsions loaded with thalidomide are synthesized wherever a dose as low as 25 mg results in plasma concentrations which might be therapeutic. However, a major decrease within the drug content of the nanoemulsion was discovered at 0.01% drug formulation once in 2 months storage that can be overcome by the addition of polysorbate 80. Chlorambucil, a lipophilic anticancer agent, has been used against breast and ovarian cancer. Its pharmacokinetics and anticancer activity has been studied by loading it in parenteral emulsions prepared by high-energy ultrasonication methodology. Treatment of colon adenocarcinoma in the mouse with this nanoemulsion results in higher growth suppression rate compared to plain drug solution treatment, concluding that the drug-loaded emulsion can be a good carrier for its delivery in cancer treatment. Carbamazepine, a widely used medicament, had no parenteral treatment offered for patients owing to its poor water solubility. Kelmann et al. have developed a nanoemulsion for its intravenous delivery that showed favorable in vitro release mechanics. Parenteral nanoemulsion formulations of the subsequent drugs (diazepam, propofol, dexamethasone, etomidate, flurbiprofen, and prostaglandin E1) are documented as well (Bailey et al., 2002).

The high lipophilicity of benzodiazepine (an anxiolytic and sedative) makes the utilization of solvents (such as propylene glycol, phenyl carbinol and ethanol) for the dissolution of the drug in typical aqueous preparations (Valium® and Stesolid®) necessary, resulting in pain and thrombophlebitis on the patient throughout the injection. the development of a nanoemulsion, commercially obtainable under the name of Diazemuls® (Kabi-Pharmacia) permits for the reduction of those adverse effects, keeping stages of distribution and elimination such as Valium. However, higher doses of Diazemuls are necessary to get a similar result as Valium since this ends up in higher free fraction of plasma diazepam (Doenicke et al., 2001). The solution for IV administration of etomidate (hypnotic short) is attributable to stability issues as its composition contains 35 the concerns propylene glycol (Hypnomidate®). Further, being attributable to the presence of high osmolarity of the solvent, the administration is related to varied adverse effects such as hemolysis, thrombosis, thrombophlebitis, and pain at the area of application. A nanoemulsion containing 2 mg/mL Lipofundin® etomidate in medium-chain lipid named Lipuro-etomidate® (B. Braun) was developed. The emulsion allowed the reduction of the hemolytic and blood vessel sequelae, besides the pain at the time of application (Aarts et al., 1999).

The pharmacokinetics and pharmacodynamics of propofol (anesthetic) are complicated. It shows an initial speedy distribution of about 2–3 min, with high variability between patients, and reduced concentrations to subtherapeutic levels within minutes. However, because of its high lipophilicity, it has a high volume of distribution and its complete elimination from the body can take days. Because of the incidence of anaphylactic effects related to Cremophor EL, present within the original formulation of propofol nanoemulsion as vehicle for this drug-containing composition in soybean oil, glycerol, egg yolk lecithin, and disodium edentate, this vehicle helped to scale back the volume of distribution of the drug, speeding their processes of clearance by the responsible agencies. This formulation conjointly allowed the employment of smallest effective dose have to be compelled to produce the required therapeutic effect, permitting a rapid onset and recovery from anesthesia, in comparison to a nonlipid (ethanol) solution, thereby generating higher security administration, attributable to the lower continuous accumulation of the drug, and eliminating the requirement for constant adjustment of the dose. This product was approved in 1989 within the United States, under the name of Diprivan® 1% or 2% (AstraZeneca/APP Pharmaceuticals). In Brazil, the product is offered as Lipuro 1% (B. Braun) and Diprivan 1% and 2% (AstraZeneca), besides the generic 1% (Eurofarma Labs.). The various generic formulations presently available are established by a further issue of variability in response between people within the induction of anesthesia, apart from the pharmacokinetic characteristics of the drug itself and also the differences in lipoprotein profile of every patient, due to the high binding of propofol to low-density lipoprotein and albumin (Cox et al., 1998). Because of related pain at the injection site and enhanced triglyceride levels when administration is done for long periods, some changes within the formulation of Diprivan adverse effects are projected, as well as some already being marketed as Propofol® Lipuro (B. Braun) as oil core that contains a combination of oils. The addition of more oil to the formulation caused reduction of pain on injection because of enhanced incorporation of the drug within the oily core and the lower quantity of free propofol part in the external binary compound emulsion (Aarts et al., 2004). Various formulations are developed, as an example, the incorporation of upper concentrations of propofol (6%) within the nanoemulsion, or the event of a propofol prodrug in solution (Aquavan®). Furthermore, despite the excellent anti-inflammatory activity of corticoid, the clinical use of corticosteroids is restricted by various side effects. To avoid these drawbacks, lipophilic prodrugs in the body that are gradually hydrolyzed to the active metabolite can be used (thus presenting prolonged anti-inflammatory effect). The advantage is that the use of lower doses than those utilized in conventional water-soluble type (dexamethasone phosphate), reducing the risks of adverse effects. Considering that nanoemulsions are picked up by inflammatory cells of the mononuclear phagocytic system, nanoemulsions were used as a vehicle for lipophilic prodrug of Decadron (palmitate), that is commercially available as Limethason® (Green Cross Co./Mitsubishi Tanabe Pharma Co.). Limethason showed glorious results in the treatment of arthritis, West syndrome, inflammatory diseases, and alternative reaction diseases. Although the solution of dexamethasone phosphate is speedily distributed in water-rich tissues, such as muscles, the nanoemulsion is accumulated primarily in tissues inflamed organs such as liver and spleen. The biodistribution profile is totally different even though the elimination pattern is comparable between the two. Limethason removes over 80% of the phagocytic activity of macrophages at an amount of 0.03 mg/mL. Flurbiprofen (nonsteroidal anti-inflammatory drug, NSAID, for oral use), a lipophilic drug, is employed to treat arthritis and alternative inflammatory diseases associated or not with cancer. The nonavailability of oral and/or numerous gastrointestinal side effects caused by this drug usually requires the utilization of parenteral route. Considering the severe native irritation caused by the metal salt of flurbiprofen, it was developed as a prodrug of flurbiprofen (cefuroxime) and because of the lipophilicity of the latter particularly in soybean oil, it was incorporated in nanoemulsions for parenteral use and is commercially obtainable within the Japanese market since 1992. Administration of Ropion® resulted in a rise in area below the concentration–time curve and reduced clearance in comparison to the solution. The incorporation of the drug into nanoemulsions containing unesterified alkyl oleate, emulsifier, and modified egg yolk led to a lower drug accumulation in organs such as the liver and spleen attributable to the lower uptake by the mononuclear phagocyte system. Prostaglandin E1, that is synthesized in many places of the body, is responsible for numerous physiological effects such as vasodilatation, lowering of blood pressure, angiogenesis, and inhibition of platelet aggregation. When administered for the treatment of various diseases, it has a short half-life; high doses are required, resulting in various adverse effects such as hypotension, diarrhea, local irritation, and pain. During this context, nanoemulsions were created commercially obtainable in 1975, PGE1 complexed to cyclodextrins and, in 1985, prostaglandin E1 incorporated in lipid nanoemulsions (Liple®, Mitsubishi Tanabe Pharma Corporation, Palux®, Taisho Pharmaceutical). Lipid formulations are used to treat cardiovascular diseases as they are accumulated within the walls of injured vessels, transporting the drug to the site of vascular injury, and to protect it from rapid inactivation by the lungs (Labhasetwar and Panyam, 2004; Yokoyama et al., 1996).

### *13.7.12 NANOEMULSIONS AND PULMONARY DRUG DELIVERY*

The respiratory organ is an attractive target for drug delivery as a result of noninvasive administration via inhalation aerosols, rejection of first-pass metabolism, direct delivery to the site of action for the treatment of metastasis diseases, and therefore the convenience of an enormous expanse for local drug action and systemic absorption of drug. Colloidal carriers (i.e., nanocarrier systems) in pulmonary drug delivery offer many benefits such as the potential to achieve relatively uniform distribution of drug dose among the alveoli, accomplishment of improved solubility of the drug from the aqueous solubility, a sustained drug release which consequently reduces dosing frequency, improves patient compliance, decreases incidence of facet effects, and therefore the potential of drug internalization by cells (Mansour et al., 2009). Until now, the submicron emulsion system has not yet been fully exploited for pulmonary drug delivery and extremely very little has been published in this area. Bivas-Benita et al. reported that cationic submicron emulsions are promising carriers for DNA vaccines to the lung since they are ready to transfect pulmonary epithelial cells, that possibly induce cross priming of antigen-presenting cells and directly activate dendritic cells, resulting in stimulation of antigen-specific T cells (Bivas-Benita et al., 2004); thus the nebulization of submicron emulsions are a new and upcoming research area. However, extensive studies are needed for the successful formulation of inhalable submicron emulsions as a result of possible adverse effects of surfactants and oils on lung alveoli function (adverse interactions with lung surfactant). A novel pressurized aerosol system has been designed for the pulmonary delivery of salbutamol using lecithin-stabilized microemulsions formulated in trichlorotrifluoroethane (Lawrence et al., 2000).

### *13.7.13 NANOEMULSION APPLICATION IN GENE DELIVERY*

Emulsion systems are introduced as various gene transfer vectors to liposomes (Liu et al., 1996). Other emulsion studies for gene delivery have described that binding of the emulsion and DNA complex was stronger than liposomal carriers (Yi et al., 2000). This stable emulsion system delivered genes more with efficiency than liposomes (Liu et al., 2010). Silva et al. evaluated factors that influence deoxyribonucleic acid compaction in cationic lipid nanoemulsions [cationic nanoemulsions containing stearylamine (a cationic lipid that presents a primary alkane cluster once in resolution) is in a position to compact genetic material by electricity interactions, and in dispersed systems such as nanoemulsions this lipid anchors on the oil/water interface conferring a positive charge to them] (Andre et al., 2012). The influence of the stearylamine incorporation phase (water or oil), time of complexation, and totally different incubation temperatures were studied. The complexation rate was assessed by electrophoresis migration on agarose gel 0.7%, and nanoemulsion and lipoplex characterization have been done by DLS. The results show that the best DNA compaction process occurs when 120 min of complexation, at temperature (4°C ± 1°C), and after incorporation of the cationic lipid into the liquid phase. Although the zeta potential of lipoplexes was less than the results found for basic nanoemulsions, the granulometry failed to change. Moreover, it absolutely was demonstrated that lipoplexes are appropriate vehicles for gene delivery (Andre et al., 2012).

### *13.7.14 NANOEMULSION IN BIOTERRORISM ATTACK*

Because of their antimicrobial activity, analysis has begun on use of nanoemulsions as a prophylactic-medicated dosage form, a human protective treatment, to prevent the individuals exposed to bioattack such as Anthrax and Ebola (Bharadia et al., 2011). The broad-spectrum nanoemulsions were checked on surfaces by the US Army (RestOps) in December 1999 for decontamination of Anthrax spore. It absolutely was checked once more by RestOps in March 2001 as a chemical decontamination agent (Attama and Charles, 2011). This technology has been tested on gangrene and clostridium botulism spores, and might even be used on contaminated wounds to salvage limbs. The nanoemulsions are often developed into a cream, foam, liquid, and spray to decontaminate a large variety of materials, that is marketed as NanoStat™ (Nanobio Corp.) (Subhashis et al., 2011).

### *13.7.15 NANOEMULSIONS FOR OLIGONUCLEOTIDE DELIVERY*

Antisense oligonucleotides are tested wide within the past few years for the treatment of cancer (Isomura, 2006; Lebedeva et al., 2000). However, poor stability primarily in biological fluids (James, 2007) and low intracellular penetration of those oligonucleotides (ODN) have restricted their therapeutic use. Chemical modifications of the oligonucleotide phosphodiester backbones into phosphorothioate partly enhance the chemical stability to accelerator degradation (Benita et al., 1999; Lysik et al., 2003); however, they do not improve the intracellular penetration. Moreover, drug delivery systems supported liposomes but still suffer from inherent limitations that limit their potential because of stability problems. These limitations include instability and short period, low drug-loading capability, sensitivity to sterilization, and expensive large-scale producing method. It often hypothesized that AN association of ODN molecules with ion oil nanodroplets can reduce considerably the polyanionic character of the ODN molecules. Novel formulations of ion nanoemulsions supported three completely different lipids that are developed to strengthen the attraction of the polyanionic ODN macromolecules to the ion moieties on the oil nanodroplets. The *N*-[1-(2,3-dioleoyloxy)propyl]-*N*,*N*,*N*-trimethylammonium salts (DOTAP) cationic lipid nanoemulsion is capable of retaining ODN despite the high dilution. But 100% of the ODN is changed in distinction to 40–50% with the opposite ion nanoemulsions. The in vitro release kinetic behavior of ODN exchange with physiological anions present within the subtypes steroid sulfatase appears to be advanced and difficult to characterize using mathematical fitting model equations. Any pharmacokinetic studies are required to verify our kinetic assumptions and ensure the in vitro ODN release profile from DOTAP cationic nanoemulsions. Recently, findings between structure of lipid–DNA complexes and their biological activity is gaining a lot of interest (Bochot et al., 2000; Couvreur et al., 2004, 2006). One major downside related to in vivo supermolecule-mediated gene delivery is relatively low transfection potency due to poor stability of the complex upon contact with body fluid (Behar-Cohen et al., 2008). Strong electrostatic interactions between charged lipid–DNA complexes and charged proteins within the blood are responsible for the speedy aggregation of lipid–DNA complexes upon contact with body fluid. One way to overcome the problem of serum instability and to prolong their circulation time within the blood is to guard their surface by adding polyethylene-block-polyethylene glycol (PE-PEG). When 9% or 100% (w/w) PE-PEG is added to the lipidic formulation, the plasma clearance and the liver uptake is significantly reduced. The addition of PE-PEG helps to prolong the circulation of the emulsion within the blood by shielding the positive surface charge and by providing a more hydrophilic surface (Lasic et al., 1997; Monck et al., 2000; Baker and Hong, 1997). However, these enhancements are still not enough to get a DNA/emulsion delivery system that continues to be circulating for an extended period of time (Li et al., 1999; Lundberg et al., 1996).

### *13.7.16 CYTOTOXICITY OF NANOEMULSIONS*

Current analysis is additionally targeted on understanding and taking advantage of the options of tumor microenvironment such as pH and temperature changes. General chemotherapy has been the foremost successful mode of cancer therapy for an extended time. However, infusing therapeutic doses of cytotoxic medicine into the blood stream and achieving the specified concentration within the tumor without manufacturing toxic effects within the healthy body tissues has been the most important challenge in cancer therapy. Similarly, sequence delivery systems encountered the matter of insufficient uptake, cytotoxicity, and undesirable immunogenic aspect effects because of the shortage of safe tissue- or cell-specific vectors. This drawback is being mostly addressed by the appearance of numerous surface-modified nanosized drug delivery systems, which will escape the reticuloendothelial system and reach the target tissue with the help of various target-specific ligands upon general administration. Developing nanocarriers that employ various beneficial properties require the assembly of variety of chemical moieties on one nanosystem. One of the challenges within the formulation of such nanosystems is toxicity. Fang et al. (2009) developed acoustically active nanoemulsions for camptothecin to check the potency of its toxicity toward cancer cells. The nanoemulsions were prepared using liquid perfluorocarbons and oil as oil cores of the inner part. Camptothecin in nanoemulsions with a lower oil concentration exhibited toxicity against melanomas and ovarian cancer cells. Confocal laser scanning research confirmed nanoemulsion uptake into cells. Lysis studies to assess the interaction between erythrocytes and also the nanoemulsions showed less lysis. Employing a 1 mHz ultrasound, an increased release of camptothecin from the system with lower oil concentration may be established, illustrating a drug-targeting effect. Also, native application of toxic doses of perfluorochemical nanoemulsions resulted within the necrosis of cancer cells. Thus, this can be a classical example which is evident that campothecin in nanoemulsion is proved to be extremely cytotoxic in vitro and in vivo. When multifunctional nanoemulsions are developed, these drug delivery systems may enable controlled and targeted release of drugs or therapeutic molecules at the site of action. Such multifaceted, versatile nanocarriers, and drug delivery systems promise a substantial increase within the efficacy of therapeutic applications in pharmaceutical sciences and so the adverse effects will further be reduced (Chesnoy et al., 2001).

## 13.8 MAJOR CHALLENGES OF NANOEMULSION DRUG DELIVERY SYSTEMS

Production of nanoemulsions needs important energy input and though low-energy strategies exist, they are not for industrial-scale production; low-energy ways sometimes need high concentrations of surfactants and customarily do not yield stable nanoemulsions. Nanoemulsions are created on industrial scale via the high-energy methodology that utilizes mechanical devices such as high-pressure homogenizers that are very expensive, extremely energy intensive, and tough to service. This challenge clearly accounts for the low translation of proprietary nanoemulsion formulations into industrial products. There is conjointly the shortage of understanding of the mechanism of production of submicron droplets and therefore the role of surfactants and cosurfactants further as an absence of understanding of the surface chemistry that is involved in production of nanoemulsions (Esquena et al., 2004). For example, few formulation chemists are conscious of the phase inversion temperature (PIT) concept and how this can be usefully applied for the assembly of small emulsion droplets. Finally, there is the fear of introduction of recent systems without full analysis of the value and benefits (Aboofazeli, 2010).

## 13.9 CONCLUSION AND FUTURE PERSPECTIVES

Nanoemulsions are planned for varied applications in pharmacy as drug delivery systems due to their capability to solubilize nonpolar active compounds. Future views of nanoemulsion are very promising in numerous fields of therapeutics or application in development of cosmetics for hair or skin. One of the versatile applications of nanoemulsions is within the area of drug delivery wherever they act as efficient carriers for bioactives, facilitating administration by numerous routes (Azemar et al., 2005). The benefits and applications of nanoemulsions for oral drug delivery are various, where the droplet size is related to their absorption within the GI tract. Because of the renewed interest in flavoring drug formulation, nanoemulsion is also the best delivery platform for these difficult-to-formulate phytopharmaceuticals. The prospects of nanoemulsions lie in the ingenuity of formulation consultants to utilize the benefits of nanoemulsion carriers in overcoming peculiar issues of drug delivery such as absorption, permeation, and stability of each orthodox and herbal drugs (Bibette et al., 1990).

## KEYWORDS

- **nanoscale dispersion**
- **high-pressure homogenization**
- **pulmonary drug delivery**
- **oligonucleotide delivery**
- **nontoxic disinfectant cleaner**
- **mucosal vaccine**

## REFERENCES


Aarts, L. P.; Danhof, M.; Kuks, P. F.; Lange, R.; Langemeijer, H. J.; Knibbe, C. A.; Voortman, H. J. Pharmacokinetics, Induction of Anaesthesia and Safety Characteristics of Propofol 6% SAZN vs Propofol 1% SAZN and Diprivan-10 After Bolus Injection. *Br. J. Clin. Pharm.* **1999,** *47* (6), 653–60.

Aarts, L. P.; Danhof, M.; Naber, H.; Knibbe, C. A.; Kuks, P. F. Long-term Sedation with Propofol 60 mg $ml^{-1}$ vs Propofol 10 mg $ml^{-1}$ in Critically Ill, Mechanically Ventilated Patients. *Acta Anaesthesiol. Scand.* **2004,** *48* (3), 302–307.

Aboofazeli, R. Nanometric Scaled Emulsions (Nanoemulsions). *Iran J. Pharm. Res.* **2010,** *9* (4), 325–326.

Adil, H.; Dilip, A.; Gajendra, S.; Khinchi, M. P.; Gupta, M. K.; Natasha, S. Self-Emulsifying Drug Delivery Systems (SEEDS): An Approach for Delivery of Poorly Water Soluble Drug. *Int. J. Pharm. Life Sci.* **2012,** *3* (9), 991–1996.

Agnieszka, H.; Adam, J.; Kazimiera, A.; Katarzyna, Z. Biocompatible Nanoemulsions of Dicephalic Aldonamide-type Surfactants: Formulation, Structure and Temperature Influence. *J. Colloid Interface Sci.* **2009,** *334*, 87–95.

Agu, R. U.; Kinget, R.; Ugwoke, M. I.; Verbeke, N. Nasal Mucoadhesive Drug Delivery: Background, Applications, Trends, and Future Perspectives. *Adv. Drug Deliv. Rev.* **2005,** *57* (11), 1640–1665.

Ahmad, A.; Koltover, I.; Lin, A. J.; Slack, N. L.; George, C. X. Structure and Structure–Function Studies of Lipid/Plasmid DNA Complexes. *J. Drug Target.* **2000,** *8*, 13–27.

Ahmad, F. J.; Ali, M.; Faiyaz, S.; Khar, R. K.; Shafiq, S.; Sushma, T. Development and Bioavailability Assessment of Ramipril Nanoemulsion Formulation. *Eur. J. Pharm. Biopharm.* **2007,** *66* (2), 227–243.

Alonsa, M. J.; Csaba, N.; Garcia-Fuentes, M. Nanoparticles for Nasal Vaccination. *Adv. Drug Deliv. Rev.* **2009,** *61* (2), 140–157.

Alvarez-Figueroa, M. J.; Blanco-Méndez, J. Transdermal Delivery of Methotrexate: Iontophoretic Delivery from Hydrogels and Passive Delivery from Microemulsions. *Int. J. Pharm.* **2001,** *215*, 57–65.

Alves, G.; Falcao, A.; Fortuna, A.; Pires, A. Intranasal Drug Delivery: How, Why and What for. *J. Pharm. Pharm. Sci.* **2009,** *12*, 288–311.

Al-Zaagi,; Abounassif M. A.; Belal, F.; Gadkariem, M. A. A Stability-indicating LC Method for the Simultaneous Determination of Ramipril and Hydrochloride in Dosage Forms. *J. Pharm. Biomed. Anal.* **2001,** *24*, 335–342.

Andre, L. S.; Anselmo, G. O.; Eryvaldo, S. T. E.; Francisco, A. J.; Lourena, M. V.; Lucymara, F. A. L.; Lucila, C. M. E. Physical Factors Affecting Plasmid DNA Compaction in Stearylamine-containing Nanoemulsions Intended for Gene Delivery. *Pharmaceuticals* **2012,** *5*, 643–654.

Arbor, A. Nanoemulsion Vaccines Show Increasing Promise Oil-based Nasal Vaccine Technique Produces Immunity Against Smallpox, HIV. *AIDS Res. Hum. Retroviruses* **2008,** *24*, 1–9.

Armstrong, R. C.; Bird, R. B.; Hassager, O. *Dynamics of Polymeric Liquids.* Wiley: New York, 1977; Vol. 1, pp 78–82.

Attama, A. A.; Charles, L. Current State of Nanoemulsions in Drug Delivery. *J. Biomater. Nanobiotechnol.* **2011,** *2*, 626–639.

Attwood, D.; Mallon, C.; Taylor, C. J. Phase Studies of Oil-in Water Phospholipid Microemulsions. *Int. J. Pharm.* **1992,** *84*, R5–R8.

Azemar, N.; Garcia-Celma, M. J.; Izquierdo, P.; Nolla, J.; Solans, C. Nano-emulsions *Curr. Opin. Colloid Interface Sci.* **2005,** *10*, 102.

Babbar, A. K.; Kumar, M.; Misra, A.; Mishra, A. K. Intranasal Nanoemulsion Based Brain Targeting Drug Delivery System of Risperidone. *Int. J. Pharm.* **2008,** *358* (1–2), 285–291.

Bailey, P. L.; Guivarch, P. H.; Litman, R. S.; Norton, J. R.; Ward, D. S. Pharmacodynamics and Pharmacokinetics of Propofol in a Medium-chain Triglyceride Emulsion. *Anesthesiology* **2002,** *97* (6), 1401–1408.

Baker, A.; Hong, K.; Papahadjopoulos, D.; Zheng, W. Stabilization of Cationic Liposome-plasmid DNA Complexes by Polyamines and Poly(Ethylene Glycol)-Phospholipid Conjugates for Efficient In Vivo Gene Delivery. *FEBS Lett.* **1997,** *400*, 233–237.

Ball, R. C.; Klein, R.; Lin, M. Y.; Lindsay, H. M.; Meakin, P.; Weitz, D. A.; Universal Diffusion-limited Colloid Aggregation. *J. Phys. Condens. Matter* **1990,** *2*, 3093.

Becher, P. *Emulsions: Theory and Practice*. Reinhold Publishing Corporation: New York, 1965; Vol. 2, pp 122–135.

Behar-Cohen, F.; Benita, S.; Hagigit, T.; Lambert, G.; Nassar, T. The Influence of Cationic Lipid Type on In-vitro Release Kinetic Profiles of Antisense Oligonucleotide from Cationic Nanoemulsions. *Eur. J. Pharm. Biopharm.* **2008,** *70*, 248–259.

Benita, S.; Couvreur, P.; Dubernet, C.; Puisieux, F.; Teixeira, H. Submicron Cationic Emulsions as a New Delivery System for Oligonucleotides. *Pharm. Res.* **1999,** *16*, 30–36.

Benita, S.; Tamilvanan, S. The Potential of Lipid Emulsion for Ocular Delivery of Lipophilic Drugs. *Eur. J. Pharm. Biopharm.* **2004,** *58*, 357–368.

Bharadia, P. D.; Gunjan, J. P.; Modi, D. A.; Pandya, V. M.; Rutvij, J. P. Nanoemulsion: An Advanced Concept of Dosage Form. *Int. J. Pharm. Cosmetol.* **2011,** *1* (5), 122–133.

Bibette, J.; Nallet, F.; Roux, D. Depletion Interactions and Fluid–Solid Equilibrium in Emulsions *Phys. Rev. Lett.* **1990,** *65*, 2470.

Bibette, J.; Leal-Calderon, F.; Poulin, P. Emulsions: Basic Principles. *Rep. Prog. Phys.* **1999,** *62*, 969.

Bibette, J.; Couffin, A. C.; Cates M. E.; Delmas, T.; Piraux, H.; Poulin, P.; Texier, I.; Vinet, F. Nanoemulsion Review. *Langmuir* **2011,** *27*, 1683–1692.

Bielinska, A. U.; Janczak, K. W.; Landers, J. J. Nasal Immunization with a Recombinant HIV gp120 and Nanoemulsion Adjuvant Produces Th1 Polarized Responses and Neutralizing Antibodies to Primary HIV Type 1 Isolates. *AIDS Res Hum Retroviruses* **2008**, *24*, 27181.

Bielinska, A. U.; Janczak, K. W.; Makidon, P. E.; Nigavekar, S. S. Pre-clinical Evaluation of a Novel Nanoemulsion-based Hepatitis B Mucosal Vaccine. *PLOS One* **2008b,** *3*, e2954.

Bivas-Benita, M.; Oudshoorn, M.; Romeijn, S. Cationic Submicron Emulsions for Pulmonary DNA Immunization. *J. Control. Rel.* **2004,** *100* (1), 145–155.

Bochot, A.; Couvreur, P.; Fattal, E. Intravitreal Administration of Antisense Oligonucleotides: Potential of Liposomal Delivery. *Prog. Retin. Eye Res.* **2000,** *19*, 131–147.

Bose, A.; Doyle, P. S.; Gao, Y.; Godfrin, M.; Helgeson, M. E.; Lee, J.; Moran, S. E.; Tripathi, A. *Soft Matter* **2014,** *10*, 3122–3133.

Burggraaf, J.; Cohen, A. F.; Kreilgaard, M.; Kemme, M. J. B.; Schoemaker, R. C. Influence of a Microemulsion Vehicle on Cutaneous Bioequivalence of a Lipophilic Model Drug Assessed by Microdialysis and Pharmacodynamics. *Pharm. Res.* **2001,** *18* (5), 593–599.

Cates, M. E.; Webster, A. J. Osmotic Stabilization of Concentrated Emulsions and Foams. *Langmuir* **2001,** *17*, 595.

Cavello, J. L.; Lyons, G. B.; Rosano, H. L. Mechanism of Formation of Six Microemulsion Systems. In *Microemulsion Systems*. Marcel Dekker: New York, 1987; pp 259–257.

Chakravarthi, V.; Haritha, A.; Koteswara, R. P.; Syed, P. B. A Brief Introduction to Methods of Preparation, Applications and Characterization of Nanoemulsion Drug Delivery Systems. *Indian J. Res. Pharm. Biotechnol.* **2013,** *1* (1), 25–28.

Chang, C. B.; Graves, S. M.; Mason, T. G.; Meleson, K.; Wilking, J. N. Nanoemulsions: Formation, Structure and Physical Properties. *J. Phys. Condens. Matter* **2006,** *18*, 635–666.

Chesnoy, S.; Durand, D.; Doucet, J.; Huang, L.; Stolz, D. B. Improved DNA/Emulsion Complex Stabilized by Poly(Ethylene Glycol) Conjugated Phospholipid. *Pharm. Res.* **2001,** *18*, 1480–1484.

Constantinides, P. P. Lipid Microemulsions for Improving Drug Dissolution and Oral Absorption and Biopharmaceutical Aspects. *Pharm. Res.* **1995,** *12* (11), 1561–1572.

Couvreur, P.; Dubernet, C.; Fattal, E. Smart Delivery of Antisens Oligonucleotides by Anionic pH-sensitive Liposomes. *Adv. Drug Deliv. Rev.* **2004,** *56*, 931–946.

Couvreur, P.; Fattal, E.; Malvy, C.; Toub, N. Innovative Nanotechnologies for the Delivery of Oligonucleotides and siRNA. *Biomed Pharmacother*. **2006,** *60*, 607–620.

Cox, E. H.; Danhof, M.; Knibbe, C. A.; Koster, V. S.; Kuks, P. F.; Langemeijer, H. J.; Lie, A. H. L.; Langemeijer, M. W.; Lange, R.; Tukker, E. E. Influence of Different fat Emulsion-based Intravenous Formulations on the Pharmacokinetics and Pharmacodynamics of Propofol. *Pharm. Res.* **1998,** *15* (3), 442–448.

Dave, K.; Gaur, P. K.; Mishra, S.; Purohit, S. Trans-dermal Drug Delivery System: A Review. *Asian J. Pharm. Clin. Res.* **2009,** *2* (1), 14–20.
Debnath, S.; Satayanarayana; Vijay Kumar G. Nanoemulsion—A Method to Improve the Solubility of Lipophilic Drugs. *Pharmanest—Int. J. Adv. Pharm. Sci.* **2011,** *2* (2–3), 72–83.
Devarajan, V.; Ravichandran, V. Nanoemulsions: As Modified Drug Delivery Tool. *Pharm. Glob. Int. J. Compr. Pharm.* **2011,** *4* (2), 1–6.
Doenicke, A. W.; O'Connor, M. F.; Rau, J.; Roizen, M. F.; Strohschneider, U. Propofol in an Emulsion of Long- and Medium-chain Triglycerides: The Effect on Pain. *Anesth. Analg.* **2001,** *93*, 382–384.
El-Aasser, M. S.; Sudol, E. D. *Emulsion Polymerization and Emulsion Polymers*. Wiley: Chichester, 1997; pp 540–551.
Esquena, J.; Izquierdo, P.; Solans, C.; Tadros, T. Formation and Stability of Nanoemulsions. *Adv. Colloid Interface Sci.* **2004,** *108–109*, 303–318.
Fang, J. Y.; Hung, C. F.; Hua, S. C.; Hwang, T. L. Acoustically Active Perfluorocarbon Nanoemulsions as Drug Delivery Carriers for Camptothecin: Drug Release and Cytotoxicity Against Cancer Cells. *Pharm. Res.* **1998,** *10*, 105–111.
Gabriel, D. A.; Johnson, C. S. *Laser Light Scattering*. Dover Publications Inc.: New York, 1981; pp 42–44.
Gao, Y.; Kim J.; Helgeson, M. E. *Soft Matter* **2015,** *11*, 6360–6370.
Graves, S.; Meleson, K.; Mason, T. G. Formation of Concentrated Nanoemulsions by Extreme Shear. *Soft Mater.* **2004,** *2*, 109.
Grzybowski, B.; Whitesides, G. M. Self-assembly at all Scales. *Science* **2002,** *295*, 2418.
Hansen, F. K.; Ugelstad, J. Particle Nucleation in Emulsion Polymerization: Nucleation in Monomer Droplets. *J. Polym. Sci.* **1979,** *17*, 3069.
Hsiao, L. C.; Doyle, P. S. Celebrating Soft Matter's 10th Anniversary: Sequential Phase Transitions in Thermoresponsive Nanoemulsions. *Soft Matter* **2015,** *11*, 8426–8431.
http://www.echoedvoices.org (accessed Dec 15, 2016).
Hung, C. F.; Hua, S. C.; Hwang, T. L.; Fang, J. Y. Acoustically Active Perfluorocarbon Nanoemulsions as Drug Delivery Carriers for Camptothecin: Drug Release and Cytotoxicity Against Cancer Cells. *Ultrasonics* **2009**, *49*, 39–46.
Illum, L.; Mistry, A.; Stolnik, S. Nanoparticles for Direct Nose-to-brain Delivery of Drugs. *Int. J. Pharm.* **2009,** *379* (1), 146–157.
Isomura, I.; Morita, A. Regulation of NF-kappa B Signaling by Decoy Oligodeoxynucleotides. *Microbiol. Immunol.* **2006,** *50*, 559–563.
James, W. Aptamers in the Virologists' Toolkit. *J. Gen. Virol.* **2007,** *88*, 351–364.
Kemken, J. A.; Muller, B. W.; Ziegler, A.; Influence of Supersaturation on the Pharmacodynamics Effect of Bupranolol After Dermal Administration Using Microemulsions as Vehicle. *Pharm. Res.* **1992,** *9* (4), 554–558.
Khandavilli, S.; Panchagnula, R.; Nanoemulsions as Versatile Formulations for Paclitaxel Delivery: Peroral and Dermal Delivery Studies in Rats. *J. Invest. Dermatol.* **2007,** *127*, 154–162.
Khoshnevis, P.; Lawrence, M. J.; Mortazavi, S. A.; Aboofazeli, R. In-vitro Release of Sodium Salicylate from Water-in-oil Phospholipid Microemulsions. *J. Pharm. Pharmacol.* **1997,** *49* (S4), 47.
Kreilgaard, M. Dermal Pharmacokinetics of Microemulsion Formulations Determined by In-vitro Microdialysis. *Pharm. Res.* **2001,** *18* (3), 367–373.

Kuo, F.; Kotyla, T.; Nicolosi, J. R. Nanoemulsion of an Anti-oxidant Synergy Formulation Containing Gamma Tocopherol Have Enhanced Bioavailability and Anti-inflammatory Properties. *Int. J. Pharm.* **2008,** *363*, 206–213

Labhasetwar, V.; Panyam, J. Sustained Cytoplasmic Delivery of Drugs with Intracellular Receptors Using Biodegradable Nanoparticles. *Mol. Pharmacol.* **2004,** *1* (1), 77–84.

Lasic, D. D.; Frederik; P. M.; Roberts; D. D., Strey, H. H.; Templeton, N. S. Improved DNA: Liposome Complexes for Increased Systemic Delivery and Gene Expression. *Nat. Biotechnol.* **1997,** *15*, 647–652.

Lawrence, M. J.; Rees G. D. Microemulsion-based Media as Novel Drug Delivery Systems. *Adv. Drug. Deliv. Rev.* **2000,** *45*, 89–121.

Lebedeva, I. V.; Stein, C. A. Antisense Oligonucleotides in Cancer: Recent Advances. *BioDrugs* **2000,** *13*, 195–216.

Li, S.; Stolz, D. B.; Tseng, W. C.; Wu, S. P.; Watkins, S. C. Dynamic Changes in the Characteristics of Cationic Lipidic Vectors After Exposure to Mouse Serum: Implications for Intravenous Lipofection. *Gene Ther.* **1999,** *6*, 585–594.

Liu, F.; Yang, J.; Huang, L. et al. Effect of Non-ionic Surfactants on the Formation of DNA/Emulsion Complexes and Emulsion-mediated Gene Transfer. *Pharm. Res.* **1996,** *13* (11), 1642–1646.

Liu, C. H.; Yu, S. Y.; Cationic Nanoemulsions as Non-viral Vectors for Plasmid DNA Delivery. *Colloids Surf. B Biointer.* **2010,** *79* (2), 509–515.

Lundberg, B. B.; Mortimer, B. C.; Redgrave, T. G. Submicron Lipid Emulsions Containing Amphipathic Polyethylene Glycol for Use as Drug-carriers with Prolonged Circulation Time. *Int. J. Pharm.* **1996,** *13*, 119–127.

Lysik, M. A.; Wu-Pong, S. Innovations in Oligonucleotide Drug Delivery. *J. Pharm. Sci.* **2003,** *92*, 1559–1573.

Mansour, H. M.; Rhee, Y.-S.; Wu, X. Nanomedicine in Pulmonary Delivery. *Int. J. Nanomed.* **2009,** *4*, 299–319.

Mason, T. G.; Wilking, J. N. Nonlinear. *Soft Matter Phys. Phys. Rev. E Stat.* **2007,** *75*, 041407.

Mansour, H. M.; Rhee, Y.; Wu, X. Nanomedicine in Pulmonary Delivery. *Nanomed.* **2009,** *4*, 299–319.

Miller, C. A. Spontaneous Emulsification: Recent Developments with Emphasis on Self-emulsification. In: *Emulsions and Emulsion Stability.* Sjoblom, J. (Ed.); CRC Press, Taylor & Francis: Boca Raton, 2006; p 107.

Monck, M.; Lee, D.; Ludkovski, O.; Leng, E. C.; Tam, P. Stabilized Plasmid-lipid Particles for Systemic Gene Therapy. *Gene Ther.* **2000,** *7*, 1867–1874.

Muller-Goymann, C. C. Physicochemical Characterization of Colloidal Drug Delivery Systems such as Reverse Micelles, Vesicles, Liquid Crystals and Nanoparticles for Topical Administration. *Eur. J. Pharm. Biopharm.* **2004,** *58* (2), 343–356.

Pardridge, W. M. Non-invasive Drug Delivery to Human Brain Using Endogenous Blood Brain Barrier Transport System. *Pharm. Sci. Tech. Today* **1999,** *2* (2), 49–59.

Russel, W. B.; Saville, D. A.; Schowalter, W. R. *Colloidal Dispersions*. Cambridge University Press: Cambridge, 1989; p 23.

Shakee, F.; Ramadan, W. Transdermal Delivery of Anticancer Drug Caffeine from Water-in-oil Nanoemulsions. *Colloids Surf. B Biointer.* **2010,** *75* (1), 356–362.

Shakeel, F.; Baboota, S.; Ahuja, A.; Ali, J.; Shafiq, S. Aceclofenac Using Novel Nanoemulsion Formulation. *Pharmazie* **2008,** *63* (8), 580–584.

Shen, H. R.; Zhong, M. K. Preparation and Evaluation of Self Microemulsifying Drug Delivery System (SMEDDS) Containing Atorvastatin. *J. Pharm. Pharmacol.* **2006,** *58*, 1183–1191.

Sintov, A. C.; Shapiro, L. New Nanoemulsion Vehicle Facilitates Percutaneous Penetration In Vitro and Cutaneous Drug Bioavailability In Vivo. *J. Control. Rel.* **2004,** *95*, 173–183.

Subhashis, D.; Satayanarayana, J.; Gampa, V. K. Nanoemulsion—A Method to Improve the Solubility of Lipophilic Drugs. *Pharmanest—Int. J. Adv. Pharm. Sci.* **2011,** *2* (2–3), 72–83.

Tamilvanan, S. Submicron Emulsions as a Carrier for Topical (Ocular and Percutaneous) and Nasal Drug Delivery. *Ind. J. Pharm. Educ.* **2004,** *38* (2), 73–78.

Thomas, G. Process and System for Reducing Sizes of Emulsion Droplets and Emulsions Having Reduced Droplet Sizes. Patent WO2009155353A1, 2009.

Trotta, M. Influence of Phase Transformation on Indomethacin Release from Microemulsions. *J. Control. Release* **1999,** *60*, 399–405.

Walstra, P. *Encyclopedia of Emulsion Technology.* Dekker: New York, 1996; Vol. 4, p 1.

Yi, S. W.; Yune, T. Y.; Kim, T. W., et al. A Cationic Lipid Emulsion/DNA Complex as a Physically Stable and Serum-resistant Gene Delivery System. *Pharm. Res.* **2000,** *17* (3), 314–320

Yokoyama, K.; Watanabe, M. Limethason as a Lipid Microsphere Preparation. *Adv. Drug. Deliv. Rev.* **1996,** *20*, 195–201.

## CHAPTER 14

# NANOCONJUGATE NANOCARRIERS FOR DRUG DELIVERY IN TROPICAL MEDICINE

S. YASRI[1*] and V. WIWANITKIT[2]

*[1]KMT Primary Care Center, Bangkok, Thailand*

*[2]Hainan Medical University, Hainan, China*

*[*]Corresponding author. E-mail: sorayasri@outlook.co.th*

## CONTENTS

## ABSTRACT

The diagnosis and treatment becomes the two important processes in management of diseases. For treatment, there are several tools for therapy. Drug is the thing use aiming at therapeutic effect. To achieve new drug with increased efficacy, the new biotechnology technique can be applied. Nanomedicine technique can be used in this case and the nanoconjugate nanocarriers for drug delivery are the new techniques that should be focused. The application of nanoconjugate nanocarriers for drug delivery for management of medical disorder is possible. In this specific chapter, the authors summarize and discuss on the nanoconjugate nanocarriers for drug delivery for management of tropical diseases. In brief, there are many reports on using nanoconjugate nanocarriers for drug delivery against several tropical diseases including tropical cancers and infections. At present, there are many research groups working on finding new drugs against tropical disease using nanoconjugate nanocarriers for drug delivery technique. The results from published reports are usually favorable; however, the exact use of the nanoconjugate nanocarriers for drug delivery is limited in current clinical practice. It is no doubt that the new technique can be useful in tropical medicine. Nevertheless, the concern on the cost effectiveness and safety of the technique should be raised.

## 14.1 INTRODUCTION

When there is a disease, problem of illness usually causes suffers to human beings or animals. The morbidity and mortality are unwanted outcomes. The diagnosis and treatment becomes the two important processes in management of diseases. For treatment, there are several tools for therapy. However, the most widely use and important tool is "drug." Drug is the thing aiming at therapeutic effect.

To have a successful treatment, there must be these components: (1) good drug, (2) correct drug, (3) appropriate administration, and (4) preferable in in vivo transportation and distribution to the target pathological site. The delivery of drug is an important concern in current pharmacotherapy concept. The delivery of drug becomes the new issues in modern medicine. How to effectively deliver the drug to the wanted site is the big question in management of any disease. With the advent of new biotechnology, the new modalities in drug delivery are continuously developing and the use of nanotechnology is the interesting example.

Basically, nanotechnology is the technology of the "small." The nano-object is the thing in nanoscale, extremely small. When an object is in a nanoscale, there are many new characteristics (e.g., new electrostatic property and new biochemical property). The use of the new characteristics of nano-object is the main concept in application of nano-object in any works. In medicine, the specific new branch of medicine has been launched for a few years. The nanotechnology in drug delivery is very interesting (Fig. 14.1). The early example of the applications of new nanotechnology for drug delivery is in medical endocrinology. The problem of insulin treatment for the patient with diabetes mellitus is common and there have been a long time for researches for searching appropriate technique for insulin delivery. The applied nanoconjugate nanocarriers for insulin delivery can be seen in current medicine and it is the new hope for effective diabetes mellitus management (Khafagy et al., 2007).

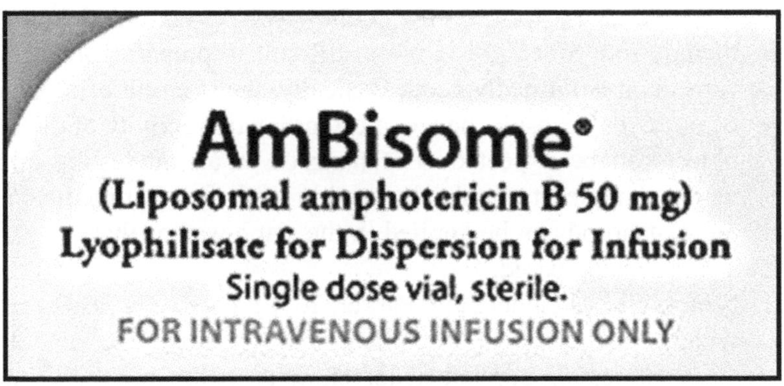

AmBisome®
(Liposomal amphotericin B 50 mg)
Lyophilisate for Dispersion for Infusion
Single dose vial, sterile.
FOR INTRAVENOUS INFUSION ONLY


**FIGURE 14.1** Basic design in applying nanosubstance for carrying drug.

Focusing on the use of nanotechnology for drug delivery, the conjugation of nano-object is the basic application (Jain et al., 2013). Jain et al. (2013) noted that "pharmaceutical and biotechnological research sorts protein drug delivery systems by importance based on their various therapeutic applications." At present, there are many examples of such conjugation application including to

a) Nanocapsules (Kreuter, 1978): This is a basic structure. Basically, the capsule is the specific container form for protective delivery of inside object. The capsule is usually used for drug delivery. Apart from drug tablet, drug capsule is common, and presently widely used in medicine. The nanocapsule is the new nanobased innovation in pharmacology and can be useful in medical therapy.
b) Nanogels nanobiomaterials (Soni and Yadav, 2016): This is also another interesting nano-applied technique. Basically, gel form is the basic form of drug for local application. This is widely use in some braches of medicine such as dermatology. The nanogel is the applied nanobased technique. The nanogel is aimed at the same purpose as basic gel. The local application is the focus. After the application, the penetrating into the target site is the further expected and wanted action of the nanogel.
c) Nanorods (Kumar et al., 2017): Rod is not the widely used technique in clinical pharmacology. Only some specific drugs are prepared in rod form such as contraceptive rod and chemoradiotherapy rod. Since rod is more difficult to prepare comparing to capsule, it is limitedly used. Basically, the strength of rod should be more than capsule, hence, more protective activity of the inside object can be expected. The nanorod is an interesting applied nanotechnique. It can also be used in medicine. For medical treatment, nanorod can be applied in the same way as that the general drug rod is prepared.
d) Nanofabrics (Feinberg and Parker, 2010): Fabribrics is also not the widely used technique in clinical pharmacology. The nanofabrics is an interesting applied nanotechnique. It has presently limited use in medicine. For medical treatment, nanorod can be applied in some very specific cases (such as cancer therapy).
e) Nanotubes (Kumar et al., 2016): Tube is also not the widely used technique in clinical pharmacology. The nanotube is an interesting applied nanotechnique. It has presently limited use in medicine. Similar to nanofabrics, for medical treatment, nanorod can be applied in some very specific cases (such as cancer therapy).

f) Liposomes (Weissig, 2017): Liposome is an interesting useful nanotecnique. The liposome is the preparation of specific lipid-based nano-object that can be used in carrying purpose. The liposome is widely used currently in drug delivery. The consistency of liposome usually controls lipid part which means the permeability in many living cellular compartment; this is the exact usefulness of liposome application as a modern technique for drug delivery.
g) Dendrimers (Rodríguez Villanueva et al., 2016): Dendrimer is a specific complex designed nanoobject. The production of dendrimer is more difficult than that of liposome. The use of dendrimer in nanomedicine can be seen and the pioneer use is usually on cancer diagnosis and cancer therapy.
h) Carbon nanotubes (Kumar et al., 2017): Carbon nanotube is a specific kind of nanotube. The main composition of carbon nanotube is carbon. Since most part of the carbon nanotube is carbon in nature, the application is usually liked to the concept of organic and inorganic substance.
i) Quantum dots (Kagan et al., 2016): Similar to dendrimer, quantum dot is a specific complex designed nano-object. The production of dendrimer is more difficult than that of liposome and dendrimer. The use of quantum dots in nanomedicine can be seen and the pioneer use is usually on cancer diagnosis and cancer therapy. However, most of the applications are in diagnostic purpose.
j) Graphene (Li et al., 2016): Graphene is a specific complex designed nano-object. The use of graphene in nanomedicine can be seen limited. Similar to the case of dendrimer and quantum dot, the pioneer use is usually on cancer diagnosis and cancer therapy.
k) Aquasomes (Jain et al., 2013): Similar to liposome, aquasome is an interesting useful nanotechnique. The aquasome is the preparation of soluble-based nano-object that can be used as carrier. The aquasome is widely used currently in drug delivery. The exact usefulness of aquasome application as a modern technique for drug delivery is very interesting.

As already mentioned, there are many new available nano-objects that can be used in drug delivery. The use of nanoconjugate nanocarriers can be the hope for management of several diseases in the present day. At the first time, the application of nanoconjugate nanocarriers is usually difficult to manage disease such as cancer (Charron et al., 2015; Jin et al., 2014). Hence, there are many reports on using nanoconjugate nanocarriers in oncology. Nevertheless, the application of nanoconjugate nanocarriers can be seen in other group of disease. In this short chapter, the authors will specifically summarize and discuss on the application of nanoconjugate nanocarriers in tropical medicine focusing on tropical diseases that are still the present global health concern.

## 14.2 APPLICATION OF NANOCONJUGATE NANOCARRIERS NANOTECHNOLOGY IN MEDICINE

As already mentioned, the use of nanoconjugate nanocarriers can be seen in medicine (Jain et al., 2014; Jin et al., 2014). In fact, the application of nanotechnology is acceptable for its usefulness. The application can be either in diagnosis or treatment.

a) Diagnosis: The nanoconjugate nanocarriers can be applied as a part of diagnostic reagent or used as a structural compartment of the diagnostic tool in laboratory medicine. Some specific kinds of nanoconjugate nanocarriers can be mainly used in diagnosis. The good example is the quantum rod.
b) Treatment: The nanoconjugate nanocarriers can be applied as technique for drug delivery. Some specific kinds of nanoconjugate nanocarriers can be mainly used in drug formulation. The good examples are the liposome and nanogel.
c) Prevention: The nanoconjugate nanocarriers can be applied in vaccine production technology. Some specific kinds of nanoconjugate nanocarriers can be mainly used in vaccine formulation. The good example is the liposome. Many new vaccines are produced by liposome-based technology.

The use of the nanoconjugate nanocarriers as a part of diagnostic reagent or therapeutic agent can be useful in many ways (Table 14.1). In brief, important advantages of using nanotechnology-based delivery systems are increased efficacy of drug to reach the series target and better basic pharmacological properties, especially for drug stability and solubility. In addition, the nanobased technology can be integrated well with many classical standard biomedical techniques and this can increase the advantage of using nanoconjugate nanocarriers in therapeutic purpose. The mentioned classical biomedical techniques that can be concomitantly used with nanoconjugate nanocarriers are

**TABLE 14.1** The Usefulness of the Nanoconjugate Nanocarriers in Medical Application.

| Purpose | Example |
|---|---|
| Specificity | Quantum dots |
| Permeability | Liposome and nanogel |
| Distribution | Liposome and aquasome |

a) Immunological technique: The immunological technique is usually based on antibody and antigen reaction. The tagging is commonly used in immunological technique and applied in immunodiagnosis and immunotherapy.
b) Radiological technique: The radiological technique is usually based on radioreactive application. Similar to immunological technique, the tagging is commonly used in radiological technique and applied in radiodiagnosis and radiotherapy. Of interest, tagging technique can be applied for manipulation of nanoconjugate nanocarriers and this can result in modern nanoradiodiagnosis and nanoradiotherapy.
c) Molecular biology technique: The molecular biology technique is usually based on DNA and RNA application. Similar to immunological and radiological technique, the tagging is applicable in molecular-based DNA/RNA diagnosis and molecular-based DNA/RNA therapy. Of interest, tagging technique can be applied for manipulation of nanoconjugate nanocarriers and this can result in modern nanomolecular-based DNA/RNA diagnosis and nanomolecular-based DNA/RNA therapy.

As described, it is no doubt that the nanoconjugate nanocarriers technology will be useful in medicine. It is presently applied in several branches of medicine. The most pioneer trails are seen in oncology for management, diagnosis, and treatment of several cancers. After that it can also be applied in several branches of medicine such as dermatology and hematology. In this specific short chapter, the authors will further briefly discuss on the application in tropical medicine, which is the specific branch of medicine concerning tropical disease and public health. Since most of the people in the world are living in the tropical regions, the tropical medicine is the main subject on the big portion of world population. Application of nanotechnology in tropical medicine can be useful for most of our world population.

## 14.3 USEFULNESS OF NANOCONJUGATE NANOCARRIERS FOR DRUG DELIVERY IN SOME IMPORTANT TROPICAL DISEASES

As already mentioned, tropical medicine is specific medicine that is required by more than half of world population. The application of nanotechnology in tropical medicine can be the same ways as already described for general medicine. The application of nanoconjugate nanocarriers technology can be for diagnostic, therapeutic, or preventive purposes. Here, the authors will further discuss in each specific group of important tropical diseases.

### *14.3.1 TROPICAL CANCER*

There are many types of cancers reported in by medical fraternity. Some are common or endemic in the tropical region and they are called tropical cancers. As already noted, the role of nanomedicine in the early phase in mainly applied in oncology, hence, it is no doubt that there are many interesting reports on using nanoconjugate nanocarriers for drug delivery in clinical oncology. For the specific tropical cancer, there are also some interesting reports as will be further described.

a) Cholangiocarcinoma: Cholangiocarcinoma is a specific cancer of the hepatobiliary system. This cancer is reported for the strong association with tropical liver fluke infestation, opisthorchiasis. The high incidence of the cholagiocarcinoma (about 97 cases per 100,000 local population; data from Thai Ministry of Public Health) can be seen in tropical Southeast Asia where the very high incidence of liver fluke (about 3% of local population; data from Thai Ministry of Public Health) can also be seen as well. The liver fluke infestation in this area is relating to the local people eating behavior, intake of uncooked raw fish. The cholangiocarcinoma is usually diagnosed late and the treatment is usually difficult. The application of nanoconjugate nanocarriers for drug delivery for management of cholangiocarcinoma is very interesting.

Basically, the treatment for cholagiocarcinoma is based on surgical resection procedure. The role of anticancer drug is still not confirmed. Another important therapeutic modality for treatment of cholangiocarcinoma is the photodynamic therapy. There are some recent reports on the use of nanoconjugate nanocarriers for drug delivery in photodynamic therapy and other classical therapy for cholangiocarcinoma. The important reports are shown in Table 14.2.

**TABLE 14.2** Some Important Reports on Using Nanoconjugate Nanocarriers for Drug Delivery in Treatment of Cholangiocarcinoma.

| Authors | Details |
|---|---|
| Broekgaarden et al. (2014a) | Broekgaarden et al. (2014a) reported on the success development and in vitro proof-of-concept of interstitially targeted zinc-phthalocyanine liposomes for photodynamic therapy that can be useful in cholangiocarcinoma treatment. |
| Choi et al. (2014) | Choi et al. (2014) reported on effect of 5-aminolevulinic acid-encapsulate liposomes on photodynamic therapy in human cholangiocarcinoma cells. Choi et al. (2014) found that ALA-containing liposomes (Lipo-ALA) increased the uptake efficiency into tumor cells compared to ALA itself, which increased the phototoxic effect and reported for "a positive relationship was evident between small particle size, protoporphyrin IX accumulation, and cell death after Lipo-ALA based photodynamic therapy." |
| Tanaka et al. (2000) | Tanaka et al. (2000) reported on "targeted killing of carcinoembryonic antigen (CEA)-producing cholangiocarcinoma cells by polyamidoamine dendrimer-mediated transfer of an Epstein–Barr virus (EBV)-based plasmid vector carrying the CEA promoter." |
| Towata et al. (2010) | Towata et al. (2010) reported that "hybrid liposomes inhibit the growth of cholangiocarcinoma by induction of cell cycle arrest in G1 phase." |

Viral hepatitis-related hepatocellular carcinoma: The viral hepatitis is a common viral infection. The high prevalence of viral hepatitis can be seen in some tropical regions especially for tropical Asia. The important viral hepatitis, hepatitis B, and hepatitis C are confirmed as blood-born transmitted disease and these viral hepatitis diseases can progress to a more severe complication, hepatocellular carcinoma. The use of nanoconjugate nanocarriers for drug delivery in for management of viral hepatitis and viral hepatitis-related hepatocellular carcinoma can be seen. Basically, the treatment for viral hepatitis-related hepatocellular carcinoma is based on surgical resection procedure. The role of anticancer drug is also accepted as an adjuvant additional therapy. There are some recent reports on the use of nanoconjugate nanocarriers for drug delivery for therapy for viral hepatitis-related hepatocellular carcinoma. The important reports are shown in Table 14.3.

**TABLE 14.3** Some Important Reports on Using Nanoconjugate Nanocarriers for Drug Delivery in Treatment of Viral Hepatitis-related Hepatocellular Carcinoma.

| Authors | Details |
|---|---|
| Cai et al. (2005) | Cai et al. (2005) reported on "preparation of nosiheptide liposomes and its inhibitory effect on hepatitis B virus in vitro." Cai et al. (2005) concluded that "liposomes of nosheptide can be prepared by sodium deoxycholate dialysis and sonication, which ability to inhibit hepatitis B virus HBsAg and HBeAg secreted is better than nosheptide" |
| Govender et al. (2015) | Govender et al. (2015) reported on "effect of poly(ethylene glycol) spacer on peptide-decorated hepatocellular carcinoma-targeted lipoplexes in vitro." Govender et al. (2015) found that the system "systems bind DNA and transfect this cell line with equal efficiency, while transgene expression levels in human embryo kidney cells HEK293 were low and comparable to those achieved in competition assays in HepG2 cells and by lipoplexes decorated with scrambled peptides" |
| Ishihara et al. (1991) | Ishihara et al. (1991) reported on "specific uptake of asialofetuin-tacked liposomes encapsulating interferon-gamma by human hepatoma cells and its inhibitory effect on hepatitis B virus replication" |

### *14.3.2 TROPICAL INFECTION*

There are many infectious infections in medicine. Some are common or endemic in the tropical region and it is called tropical infection. The tropical infections cause a considerable lost each year and it is the public concern for effective management. Similar to cancer management, there are many interesting reports on using nanoconjugate nanocarriers for drug delivery in clinical infectious medicine. For the specific tropical infection, there are also some interesting reports as will be further described.

a) Malaria: Malaria is one of the most important mosquito-borne infections that affects more than 3.2 billion world population annually (data from WHO at http://www.who.int/gho/malaria/en/). The disease is caused by parasite in the group of Plasmodium spp. it is an important blood infection that causes the public health problem in several countries at present. The disease is still prevalent in many tropical countries. The antimalarial drugs are available but the emerging problem of drug resistant malaria can be observed and becomes the big problem in tropical medicine. There are many ongoing researches on searing new antimalarial drug. The use of nanoconjugate nanocarriers for drug delivery in treatment of malaria is very interesting (Aditya et al., 2013). Aditya et al. (2013) noted that "research has been done in nanotechnology and nanomedicine, for the development of new biocompatible systems capable of incorporating drugs, lowering the resistance progress, contributing for diagnosis, control and treatment of malaria by target delivery" to correspond the malarial infection. There are some recent reports on the use of nanoconjugate nanocarriers for drug delivery for malarial therapy. The important reports are shown in Table 14.4.

**TABLE 14.4** Some Important Reports on Using Nanoconjugate Nanocarriers for Drug Delivery in Treatment of Malaria.

| Authors | Details |
|---|---|
| Marques et al. (2016) | Marques et al. (2016) discussed on adaptation of targeted nanocarriers to changing requirements in antimalarial drug delivery. Marques et al. (2016) proposed interesting model as "(1) immunoliposome-mediated release of new lipid-based antimalarials; (2) liposomes targeted to parasitic red blood cell (RBC)s with covalently linked heparin to reduce anticoagulation risks; (3) adaptation of heparin to pRBC targeting of chitosan nanoparticles; (4) use of heparin for the targeting of *Plasmodium* stages in the mosquito vector; and (5) use of the non-anticoagulant glycosaminoglycan chondroitin 4-sulfate as a heparin surrogate for pRBC targeting." Marques et al. (2016) noted that "the tuning of existing nanovessels to new malaria-related targets is a valid low-cost alternative to the de novo development of targeted nanosystems." |
| Portnoy et al. (2016) | Portnoy, et al. (2016) reported on indocyanine green liposomes for diagnosis and therapeutic monitoring of cerebral malaria. Portnoy, et al. (2016) concluded that "liposomal ICG offers a valuable diagnostic tool and a biomarker for effectiveness of CM treatment, as well as other diseases that involve inflammation and blood vessel occlusion." |

**TABLE 14.4** *(Continued)*

| Authors | Details |
|---|---|
| Tyagi et al. (2015) | Tyagi et al. (2015) reported on elastic liposome-mediated transdermal immunization enhanced the immunogenicity of *Plasmodium falciparum* surface antigen, MSP-119. |
| Waknine-Grinberg et al. (2013) | Waknine-Grinberg et al. (2013) discussed on glucocorticosteroids in nanosterically stabilized liposomes. Waknine-Grinberg et al. (2013) reported that the particles "are efficacious for elimination of the acute symptoms of experimental cerebral malaria." |

b) Dengue: Dengue is one of the most important mosquito-borne infections. It is an arbovirus infection that can cause acute febrile illness and can induce hemorrhagic complication due to thrombocytopenia. The disease is still prevalent in many tropical countries (about 390 million new cases annually; data from WHO at http://www.who.int/mediacentre/factsheets/fs117/en/) and causes a number of deaths annually. The antiviral drugs against dengue are not available. The emerging and remerging of dengue around the world becomes the big problem in public health. There are many ongoing researches on searing new antiviral drug against dengue. The use of nanoconjugate nanocarriers for drug delivery in treatment of dengue is very interesting.
There are some recent reports on the use of nanoconjugate nanocarriers for drug delivery for dengue therapy. The important reports are shown in Table 14.5.

c) Leishmaniasis: Leishmaniasis is one of the most important fly-borne infections. It is a blood infection caused by parasite in *Leishmania spp*. that can cause chronic illness. The disease is still prevalent in many tropical countries (between 0.9 and 1.3 million new cases annually; data from WHO at www.who.int/mediacentre/factsheets/fs375/en/). The drugs against leishmaniasis are available. The common drug is paromomycin (Wiwanitkit, 2012). The emerging and remerging of leishmaniasis around the world becomes the big problem in public health. Kayser and Kiderlen (2003) noted that drug that can improve body distribution and target intracellularly persisting pathogens is needed. There are many ongoing researches on searing new antileishmanial drug. The use of nanoconjugate nanocarriers for drug delivery in treatment of leishmaniasis is very interesting. There are some recent reports on the use of nanoconjugate nanocarriers for drug delivery for leishmaniasis therapy. The important reports are shown in Table 14.6. Of interest, the liposomal amphotericin B (AmB) is the first and the only one available new drug against tropical disease that is produced based on nanoconjugate nanocarriers for drug delivery technique (Fig. 14.2).

**TABLE 14.5** Some Important Reports on Using Nanoconjugate Nanocarriers for Drug Delivery in Treatment of Dengue.

| Authors | Details |
|---|---|
| Croci et al. (2016) | Croci et al. (2016) engineered "different compositions of liposomes as ivermectin carriers characterizing and testing them on several cell lines for cytotoxicity." Croci et al. (2016) reported that "the engineered liposomes were less cytotoxic than ivermectin alone and they showed a significant increase of the antiviral activity in all the Dengue stains tested (1, 2, and S221)." |
| Miller et al. (2012) | Miller et al. (2012) reported on liposome-mediated delivery of iminosugars enhances efficacy against dengue virus in vivo. Miller et al. (2012) performed an animal experiment and found that "Liposome-mediated delivery of derivatives-*N*-butyl deoxynojirimycin, in comparison with free NB-DNJ, resulted in a 3-log(10) reduction in the dose of drug sufficient to enhance animal survival." |

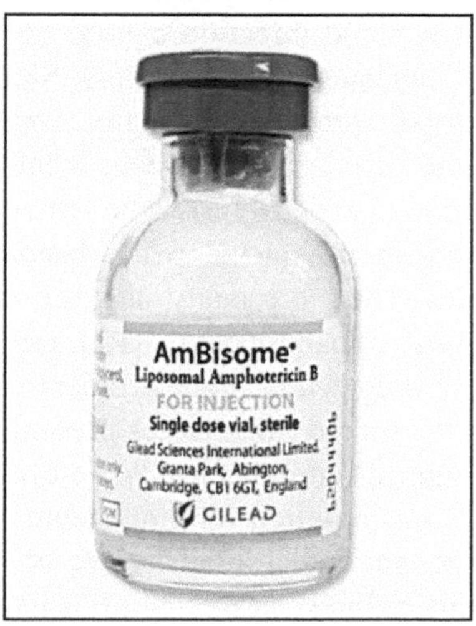


**FIGURE 14.2** Liposomal AmB.

**TABLE 14.6** Some Important Reports on Using Nanoconjugate Nanocarriers for Drug Delivery in Treatment of Leishmaniasis.

| Authors | Details |
|---|---|
| Carvalheiro et al. (2015) | Carvalheiro et al. (2015) reported on the use of hemisynthetic trifluralin analogues incorporated in liposomes for the treatment of leishmanial infections. Carvalheiro et al. (2015) concluded that "free and liposomal TFL-A were active in vitro against leishmanial parasites, and they also exhibited reduced cytotoxicity and hemolytic activity." |
| Daftarian et al. (2013) | Daftarian et al. (2013) reported on targeted and adjuvanted nanocarriers that could lower the effective dose of liposomal amphotericin B and enhances adaptive immunity in murine cutaneous leishmaniasis. |
| Prajapati et al. (2011) | Prajapati et al. (2011) "demonstrated the efficacy and stability of functionalized carbon nanotubes as a delivery mechanism for AmB." |
| Prajapati et al. (2011) | Prajapati et al. (2012) reported on an oral formulation of AmB attached to functionalized carbon nanotubes which is an effective treatment for experimental visceral leishmaniasis. Prajapati et al. (2012) concluded that "this novel formulation of AmB can be administered orally, resulting in 99% inhibition of parasite growth following a 5-day course at 15 mg/kg body weight." |

d) Trypanosomiasis: Trypanosomiasis is another important fly-borne infection. It is a blood infection caused by parasite in *Trypanosoma spp.* that can cause chronic illness. Similar to leishmaniasis, this disease is still prevalent in many tropical countries (about 0.3 million new cases annually; data from WHO at www.who.int/trypanosomiasis_african/en/). The emerging and remerging of trypanosomiasis around the world becomes the big problem in public health. There are many ongoing researches on searing new drug against trypanosomiasis. Romero, and Morilla (2010) noted that new drug with "more efficient pharmacotherapy that (1) eradicates the scarce amastigotes present at the indeterminate/chronic form and (2) employs less toxic drugs than benznidazole or nifurtimox" was required for management of trypanosomiasis. To support that mentioned aim, the use of nanoconjugate nanocarriers for drug delivery in treatment of trypanosomiasis is very interesting. There are some recent reports on the use of nanoconjugate nanocarriers for drug delivery for trypanosomiasis therapy. The important reports are shown in Table 14.7.

**TABLE 14.7** Some Important Reports on Using Nanoconjugate Nanocarriers for Drug Delivery in Treatment of Trypanosomiasis.

| Authors | Details |
|---|---|
| Kuboki et al. (2006) | Kuboki et al. (2006) reported the study on efficacy of dipalmitoylphosphatidylcholine liposome against African trypanosomes. Kuboki et al. (2006) performed an animal experimental study and found that "administration of the DPPC liposome showed a slight but significant reduction in the early development of parasitemia in *T. congolense*-infected mice" and suggested "that parasites were killed by specific binding of the DPPC liposome to the trypanosomes." |
| Oliveira et al. (2014) | Oliveira et al. (2014) performed a study to assess the therapeutic effect of conventional diminazene (C-DMZ) and liposomal encapsulated diminazene (L-DMZ) formulations against trypanosomiasis. Oliveira et al. (2014) found that "the treatment with L-DMZ and C-DMZ led to variable biochemical changes, which defined the functions of the liver and kidneys of treated animals, since the main histopathology alterations were observed in animals treated with liposomes, at their higher dosages." |

e) Filariasis: Filariasis is another important fly-borne infection. It is a blood infection caused by blood parasite. The disease is transmitted by mosquito and this chronic infection is common in many tropical countries (such as in Africa and Southeast Asia) (about 120 million cases in the tropical and subtropical regions; data from WHO at http://www.who.int/lymphatic_filariasis/epidemiology/en/). Long-term infection can result in disability. The emerging and remerging of filariasis around the world becomes the big problem in public health. There are many ongoing researches on searing new drug against trypanosomiasis. The use of nanoconjugate nanocarriers for drug delivery in treatment of filariasis is very interesting. There are some recent reports on the use of nanoconjugate nanocarriers for drug delivery for antifialirial drug purpose. The important reports are shown in Table 14.8.

**TABLE 14.8** Some Important Reports on Using Nanoconjugate Nanocarriers for Drug Delivery in Treatment of Filariasis.

| Authors | Details |
|---|---|
| Ali et al. (2014a) | Ali et al. (2014a) reported on nanocurcumina [nano-CUR (F3)] as a novel antifilarial agent with DNA topoisomerase II inhibitory activity. According to their experiment, Ali et al. (2014a) detected "the antimicrofilarial, antimacrofilarial, anti-wolbachial activity of nano-CUR (F3) over free forms and additionally its strong inhibitory action against the major target filarial parasite enzyme DNA topoisomerase II in vitro." |
| Ali et al.(2014b) | Ali et al. (2014b) reported on the therapeutic efficacy of poly (lactic-*co*-glycolic acid) nanoparticles encapsulated ivermectin (nano-ivermectin) against brugian filariasis in experimental rodent model. According to their study, Ali et al. (2014b) found that the new material "exhibited enhanced microfilaricidal and marginally better macrofilaricidal efficacy than any of the single formulation or drugs combination." |

### *14.3.3 TROPICAL ANEMIA*

There are many anemic diseases in medicine. Some are common or endemic in the tropical region and it is called tropical anemia. The examples of tropical anemia are iron deficiency anemia, which can be due to parasitic infestation or undernutrition and thalassemia, which is a common genetic disorder in many tropical regions (such as Southeast Asia and South Asia).

In hematology, there are many interesting reports on using nanoconjugate nanocarriers for drug delivery against anemic disorder. For the specific tropical anemia, there are also some interesting reports as will be further described.

a) Iron deficiency anemia: Iron deficiency anemia is the common type of anemia and can be seen in any age groups of population around the world. This is an important hematological condition. The problem is due to low body iron. The etiology might be due to poor intake of iron or tropical parasitic intestinal worm infestation (especially for hookworm). Ion therapy is the standard treatment for iron deficiency anemia. The efficacy of the present iron therapy is still the big concern in treatment. As a common disease, there are some interesting reports on the use of nanoconjugate nanocarriers for drug delivery for thalassemia management. The important reports are shown in Table 14.9.

**TABLE 14.9** Some Important Reports on Using Nanoconjugate Nanocarriers for Drug Delivery in Management of Thalassemia.

| Authors | Details |
|---|---|
| Pereira et al. (2014) | Pereira et al. (2014) reported on nanoparticulate iron(III) oxo-hydroxide. Pereira et al. (2014) noted that this particle could deliver safe iron that is well absorbed and utilized in humans. |
| Powell et al.(2014) | Powell et al. (2014) reported on the facile synthesis of tartrate-modified, nanodisperse ferrihydrite of small primary particle size, but with enlarged or strained lattice structure (~2.7 Å for the main Bragg peak versus 2.6 Å for synthetic ferrihydrite). Powell et al. (2014) mentioned that this nanodisperse ferritin-core mimetic efficiently and corrects anemia without luminal iron redox activity. |
| Span et al. (2016) | Span et al. (2016) reported on the success development of a novel oral iron-complex formulation based on hemin-loaded polymeric micelles composed of the biodegradable and thermosensitive polymer methoxy-poly(ethylene glycol)-b-poly[*N*-(2-hydroxypropyl)methacrylamide-dilactate], abbreviated as mPEG-b-p(HPMAm-$Lac_2$). Span et al. (2016) concluded that "The hemin-loaded micelles were stable at pH 2 for at least 3 h which covers the residence time of the formulation in the stomach after oral administration and up to 17 h at pH 7.4 which is sufficient time for uptake of the micelles by the enterocytes. Importantly, incubation of Caco-2 cells with hemin-micelles for 24 h at 37°C resulted in ferritin levels of 2500 ng/mg protein which is about 10-fold higher than levels observed in cells incubated with iron sulfate under the same conditions." |

b) Thalassemia: Thalassemia is the abnormality due to congenital hemoglobin disorder. This genetic disease is very common in many tropical countries (such as Indochina countries). The genetic disorder can result in clinical syndrome including anemia and body structure abnormality. The use of nanotechnology for management of genetic disease is very interesting. As a common disease, there are some interesting reports on the use of nanoconjugate nanocarriers for drug delivery for thalassemia management. The important report is by Capretto et al. (2012). Capretto et al. (2012) noted for mithramycin encapsulated in polymeric micelles (PM-MTH) by microfluidic technology for using as novel therapeutic protocol for beta-thalassemia. Capretto et al. (2012) found that "PM-MTH were able to upregulate preferentially gamma-globin messenger ribonucleic acid production and to increase fetal hemoglobin (HbF) accumulation, the percentage of HbF-containing cells, and their HbF content without stimulating alpha-globin gene expression, which is responsible for the clinical symptoms of beta-thalassemia." Capretto et al. (2012) also proposed that this therapeutic protocol might be effective for the sickle cell anemia.

## 14.4 CONCLUSION

The use of nanotechnology for drug delivery is the new interesting nanomedicine technique.

The nanoconjugate nanocarriers is a nanotechnology that can be useful in new drug development. The main advantages offered by nanotechnology-based delivery systems include increasing efficacy in targeting the desired pathogens or pathological tissue and increasing basic drug properties (such as stability, solubility, and absorption). As already shown and discussed, the application of nanoconjugate nanocarriers for drug delivery for management of tropical diseases is possible. The good examples on application are tropical cancers, tropical infections, and tropical anemia. As noted by Jain et al. (2013), "the effective and potent action of the proteins/peptides makes them the drugs of choice for the treatment of numerous diseases. Major research issues in protein delivery include the stabilization of proteins in delivery devices and the design of appropriate target-specific protein carriers." At present, there are many reports on this issue. The results of the reports are usually favorable. However, it should be noted that there are only many reports in the present day. There is still only one available approved drug for real clinical usage (liposomal AmB). Further researches are still needed. Here are the important concerns for future development of the nanoconjugate nanocarriers for drug delivery for management of tropical diseases.

First, the production of new system requires the advanced nanotechnology, which is still not available in the poor tropical developing countries where the tropical diseases are prevalent. In addition, the drug company usually invests only a little amount of investment for finding new drug in this group.

Second, the safety of the new nanoconjugate nanocarriers for drug delivery for management of tropical diseases should be considered. There is still no proof for the long-term safety of the nanomaterial. The toxicity of the nanomaterial is the big issue for discussion in nanomedicine at present.

Third, the cost-effectiveness of the new coming system should be carefully assessed. If the system is cost effective, the promotion should be implemented. In case that it is not cost effective, how to increase the cost-effectiveness to collaborate with the efficacy of the system is to be considered in the future.

## KEYWORDS

- **nanoconjugate nanocarriers**
- **nanocapsules**
- **liposomes**
- **aquasomes**
- **drug delivery**

## REFERENCES

Aditya, N. P.; Vathsala, P. G.; Vieiram, V.; Murthy, R. S.; Souto, E. B. Advances in Nanomedicines for Malaria Treatment. *Adv. Colloid Interface Sci.* **2013,** 201–202, 1–17.

Ali, M.; Afzal, M.; Abdul Nasim, S.; Ahmad, I. Nanocurcumin: A Novel Antifilarial Agent with DNA Topoisomerase II Inhibitory Activity. *J. Drug Target.* **2014a,** *22* (5), 395–407.

Ali, M.; Afzal, M.; Verma, M.; Bhattacharya, S. M.; Ahmad, F. J.; Samim, M.; Abidin, M. Z.; Dinda, A. K. Therapeutic Efficacy of Poly(lactic-*co*-glycolic acid) Nanoparticles Encapsulated Ivermectin (Nano-ivermectin) Against Brugian Filariasis in Experimental Rodent Model. *Parasitol Res.* **2014b,** *113* (2), 681–691.

Broekgaarden, M.; de Kroon, A. I.; Gulik, T. M.; Heger, M. Development and In Vitro Proof-of-concept of Interstitially Targeted Zinc-phthalocyanine Liposomes for Photodynamic Therapy. *Curr. Med. Chem.* **2014a,** *21* (3), 377–391.

Cal, Q. S.; Huang, H.; Feng, M. Q.; Zhou, P. Preparation of Nosiheptide Liposomes and Its Inhibitory Effect on Hepatitis B Virus In Vitro. *Yao Xue Xue Bao* **2005,** *40* (5), 462–465.

Capretto, L.; Mazzitelli, S.; Brognara, E.; Lampronti, I.; Carugo, D.; Hill, M.; Zhang, X.; Gambari, R.; Nastruzzi, C. Mithramycin Encapsulated in Polymeric Micelles by Microfluidic Technology as Novel Therapeutic Protocol for Beta-thalassemia. *Int J Nanomedicine* **2012,** *7*, 307–324.

Carvalheiro, M.; Esteves, M. A.; Santos-Mateus, D.; Lopes, R. M.; Rodrigues, M. A.; Eleutério, C. V.; Scoulica, E.; Santos-Gomes, G.; Cruz, M. E. Hemisynthetic Trifluralin Analogues Incorporated in Liposomes for the Treatment of Leishmanial Infections. *Eur. J. Pharm. Biopharm.* **2015,** *93*, 346–352.

Charron, D. M.; Chen, J.; Zheng, G. Theranostic Lipid Nanoparticles for Cancer Medicine. *Cancer Treat Res.* **2015,** *166*, 103–127.

Choi, K. H.; Chung, C. W.; Kim, C. H.; Kim, D. H.; Jeong, Y. I.; Kang, D. H. Effect of 5-aminolevulinic Acid-encapsulate Liposomes on Photodynamic Therapy in Human Cholangiocarcinoma Cells. *J. Nanosci. Nanotechnol.* **2014,** *14* (8), 5628–2632.

Croci, R.; Bottaro, E.; Chan, K. W.; Watanabe, S.; Pezzullo, M.; Mastrangelo, E.; Nastruzzi, C. Liposomal Systems as Nanocarriers for the Antiviral Agent Ivermectin. *Int. J. Biomater.* **2016,** *2016*, 8043983.

Daftarian, P. M.' Stone. G. W.; Kovalski, L.; Kumar, M.; Vosoughi, A.; Urbieta, M.; Blackwelder, P.; Dikici, E.; Serafini, P.; Duffort, S.; Boodoo, R.; Rodríguez-Cortés, A.; Lemmon, V.; Deo, S.; Alberola, J.; Perez, V. L.; Daunert, S.; Ager, A. L. A Targeted and Adjuvanted Nanocarrier Lowers the Effective Dose of Liposomal Amphotericin B and Enhances Adaptive Immunity in Murine Cutaneous Leishmaniasis. *J. Infect. Dis.* **2013,** *208* (11), 1914–1922.

Feinberg, A. W.; Parker, K. K. Surface-initiated Assembly of Protein Nanofabrics. *Nano Lett.* **2010,** *10* (6), 2184–2191.

Govender, J.; Singh, M.; Ariatti, M. Effect of Poly(Ethylene Glycol) Spacer on Peptide-decorated Hepatocellular Carcinoma-targeted Lipoplexes In Vitro. *J. Nanosci. Nanotechnol.* **2015,** *15* (6), 4734–4742.

Ishihara, H,; Hayashi, Y.; Hara, T.; Aramaki, Y.; Tsuchiya, S.; Koike, K. Specific Uptake of Asialofetuin-tacked Liposomes Encapsulating Interferon-gamma by Human Hepatoma Cells and Its Inhibitory Effect on Hepatitis B Virus Replication. *Biochem. Biophys. Res. Commun.* **1991,** *174* (2), 839–845.

Jain, A.; Jain, A.; Gulbake, A.; Shilpi, S.; Hurkat, P.; Jain, S. K. Peptide and Protein Delivery Using New Drug Delivery Systems. *Crit. Rev. Ther. Drug Carrier Syst.* **2013,** *30* (4), 293–329.

Jain, K.; Mehra, N. K.; Jain, N. K. Potentials and Emerging Trends in Nanopharmacology. *Curr. Opin. Pharmacol.* **2014,** *15*, 97–106.

Jin, S. E.; Jin, H. E.; Hong, S. S. Targeted Delivery System of Nanobiomaterials in Anticancer Therapy: From Cells To Clinics. *Biomed. Res. Int.* **2014,** *2014*, 814208.

Kagan, C. R.; Lifshitz, E.; Sargent, E. H.; Talapin, D. V. Building Devices from Colloidal Quantum Dots. *Science* **2016,** *353* (6302), aac5523.

Kayser, O,; Kiderlen, A. F. Delivery Strategies for Antiparasitics. *Expert Opin. Investig. Drugs* **2003,** *12* (2), 197–207.

Khafagy, E. S.; Morishita, M.; Onuki, Y.; Takayama, K. Current Challenges in Noninvasive Insulin Delivery Systems: A Comparative Review. *Adv. Drug Deliv. Rev.* **2007,** *59* (15), 1521–1546.

Kreuter, J. Nanoparticles and Nanocapsules—New Dosage Forms in the Nanometer Size Range. *Pharm. Acta Helv.* **1978,** *53* (2), 33–39.

Kuboki, N.; Yokoyama, N.; Kojima, N.; Sakurai, T.; Inoue, N.; Sugimoto, C. Efficacy of Dipalmitoylphosphatidylcholine Liposome Against African Trypanosomes. *J. Parasitol.* **2006,** *92* (2), 389–393.

Kumar, S.; Rani, R.; Dilbaghi, N.; Tankeshwar, K.; Kim, K. H. Carbon Nanotubes: A Novel Material for Multifaceted Applications in Human Healthcare. *Chem. Soc. Rev.* **2017**, *46* (1), 158–196.

Li, D.; Zhang, W.; Yu, X.; Wang, Z.; Su, Z.; Wei, G. When Biomolecules Meet Graphene: From Molecular Level Interactions to Material Design and Applications. *Nanoscale* **2016,** *8*, 19491–19509.

Marques, J.; Valle-Delgado, J. J.; Urbán, P.; Baró, E.; Prohens, R.; Mayor, A.; Cisteró, P.; Delves, M.; Sinden, R. E.; Grandfils, C.; de Paz, J. L.; García-Salcedo, J. A.; Fernàndez-Busquets, X. Adaptation of Targeted Nanocarriers to Changing Requirements in Antimalarial Drug Delivery. *Nanomedicine* **2016,** *S1549–9634* (16), 30162–30169.

Miller, J. L.; Lachica, R.; Sayce, A. C.; Williams, J. P.; Bapat, M.; Dwek, R.; Beatty, P. R.; Harris, E.; Zitzmann, N. Liposome-mediated Delivery of Iminosugars Enhances Efficacy Against Dengue Virus In Vivo. *Antimicrob. Agents Chemother.* **2012,** *56* (12), 6379–6386.

Oliveira, C. B.; Rigo, L. A.; Rosa, L. D.; Gressler, L. T.; Zimmermann, C. E.; Ourique, A. F.; DA Silva, A. S.; Miletti, L. C.; Beck, R. C.; Monteiro, S. G. Liposomes Produced by Reverse Phase Evaporation: In Vitro and In Vivo Efficacy of Diminazene Aceturate Against *Trypanosoma Evansi*. *Parasitology* **2014,** *141* (6), 761–769.

Pereira, D. I.; Bruggraber, S. F.; Faria, N.; Poots, L. K.; Tagmount, M. A.; Aslam, M. F.; Frazer D. M.; Vulpe, C. D.; Anderson, G. J.; Powell, J. J. Nanoparticulate Iron(III) Oxo-hydroxide Delivers Safe Iron That Is Well Absorbed and Utilised in Humans. *Nanomedicine* **2014,** *10* (8), 1877–1886.

Portnoy, E,; Vakruk, N.; Bishara, A.; Shmuel, M.; Magdassi, S.; Golenser, J.; Eyal, S. Indocyanine Green Liposomes for Diagnosis and Therapeutic Monitoring of Cerebral Malaria. *Theranostics* **2016,** *6* (2), 167–76.

Powell, J. J.; Bruggraber, S. F.; Faria, N.; Poots, L. K.; Hondow, N.; Pennycook, T. J.; Latunde-Dada GO, Simpson, R. J.; Brown, A. P.; Pereira, D. I. A Nano-disperse Ferritin-core Mimetic That Efficiently Corrects Anemia Without Luminal Iron Redox Activity. *Nanomedicine* **2014,** *10* (7), 1529–1538.

Prajapati, V. K.; Awasthi, K.; Gautam, S.; Yadav, T. P.; Rai, M.; Srivastava, O. N.; Sundar, S. Targeted Killing of *Leishmania donovani* In Vivo and In Vitro with amphotericin B Attached to Functionalized Carbon Nanotubes. *J. Antimicrob. Chemother.* **2011,** *66* (4), 874–879.

Prajapati, V. K.; Awasthi, K.; Yadav, T. P.; Rai, M.; Srivastava, O. N.; Sundar, S. An oral Formulation of Amphotericin B Attached to Functionalized Carbon Nanotubes Is an Effective Treatment for Experimental Visceral Leishmaniasis. *J. Infect. Dis.* **2012,** *205* (2), 333–336.

Rodríguez Villanueva, J.; Navarro, M. G.; Rodríguez Villanueva, L. Dendrimers as a Promising Tool in Ocular Therapeutics: Latest Advances and Perspectives. *Int. J. Pharm.* **2016,** *511* (1), 359–366.

Romero, E. L.; Morilla, M. J. Nanotechnological Approaches Against Chagas Disease. *Adv. Drug Deliv. Rev.* **2010,** *62* (4–5), 576–588.

Soni G, Yadav KS. Nanogels as Potential Nanomedicine Carrier for Treatment of Cancer: A Mini Review of the State of the Art. *Saudi Pharm. J.* **2016,** *24*, 133–139.

Span, K.; Verhoef, J. J.; Hunt, H.; van Nostrum, C. F.; Brinks, V.; Schellekens, H.; Hennink, W. E. A Novel Oral Iron-complex Formulation: Encapsulation of Hemin in Polymeric Micelles and Its In Vitro Absorption. *Eur. J. Pharm. Biopharm.* **2016,** *108*, 226–234.

Tanaka, S.; Iwai, M.; Harada, Y.; Morikawa, T.; Muramatsu, A.; Mori, T.; Okanoue, T.; Kashima, K.; Maruyama-Tabata, H.; Hirai, H.; Satoh, E.; Imanishi, J.; Mazda, O. Targeted Killing of Carcinoembryonic Antigen (CEA)-producing Cholangiocarcinoma Cells by Polyamidoamine Dendrimer-mediated Transfer of an Epstein–Barr Virus (EBV)-based Plasmid Vector Carrying the CEA Promoter. *Cancer Gene Ther.* **2000,** *7* (9), 1241–1250.

Towata, T.; Komizu, Y.; Kariya, R.; Suzu, S; Matsumoto, Y.; Kobayashi, N.; Wongkham, C.; Wongkham, S.; Ueoka, R.; Okada, S. Hybrid Liposomes Inhibit the Growth of Cholangiocarcinoma by Induction of Cell Cycle Arrest in G1 Phase. *Bioorg. Med. Chem. Lett.* **2010,** *20* (12), 3680–3682.

Tyagi, R. K.; Garg, N. K.; Jadon, R.; Sahu, T.; Katare, O. P.; Dalai, S. K.; Awasthi, A.; Marepally, S. K. Elastic Liposome-mediated Transdermal Immunization Enhanced the Immunogenicity of *P. falciparum* Surface Antigen, MSP-119. *Vaccine* **2015,** *33* (36), 4630–4638.

Waknine-Grinberg, J. H.; Even-Chen, S.; Avichzer, J.; Turjeman, K.; Bentura-Marciano, A.; Haynes, R. K.; Weiss, L.; Allon, N.; Ovadia, H.; Golenser, J.; Barenholz, Y. Glucocorticosteroids in Nano-sterically Stabilized Liposomes Are Efficacious for Elimination of the Acute Symptoms of Experimental Cerebral Malaria. *PLoS One* **2013,** *8* (8), e72722.

Weissig, V. Liposomes Came First: The Early History of Liposomology. *Methods Mol. Biol.* **2017,** *1522*, 1–15.

Wiwanitkit, V. Interest in Paromomycin for the Treatment of Visceral Leishmaniasis (Kala-Azar). *Ther. Clin. Risk Manage.* **2012,** *8*, 323–328.

# CHAPTER 15

# NANOCARRIER-ASSISTED DRUG DELIVERY FOR NEGLECTED TROPICAL DISEASES

BHASKAR DAS[1,2#], MANASHJIT GOGOI[3#], SATAKSHI HAZRA[2], and SANJUKTA PATRA[2*]

[1]*Centre for the Environment, Indian Institute of Technology Guwahati, North Guwahati 781039, Assam, India*

[2]*Department of Biosciences and Bioengineering, Indian Institute of Technology Guwahati, North Guwahati 781039, Assam, India*

[3]*Department of Biomedical Engineering, North-Eastern Hill University, Shillong 793022, Meghalaya, India*

[4]*Department of Biosciences and Bioengineering, IIT Guwahati, India*

[*]*Corresponding author. E-mail: sanjukta@iitg.ernet.in*

## CONTENTS



[#]Equal contribution of authors.

## ABSTRACT

Neglected tropical diseases (NTDs) such as tuberculosis, leishmaniasis, African trypanosomiasis (sleeping sickness), malaria, lymphatic filariasis, dengue, chagas disease, and schistosomiasis affect more than billion people residing in the poorest countries of the world. In spite of this, new research pertaining to tropical diseases is very limited owing to lack of global investment. Conventional pharmacological treatment is compromised by high drug toxicity, increased drug resistance, short half-life of drugs, and inability of drugs to reach intracellular sites infected by parasites. Nanocarrier system is a promising answer to problems associated with conventional drug treatment due to its ability to target specific cells and tissues, high capacity to overcome biological barriers, sustained drug release, improvement in therapeutic index, and pharmacokinetic properties of drugs. The present chapter endorses the current status of application of nano-based carriers in combating the challenges associated with control of some selected neglected tropical diseases. The first part of the chapter discusses about various nanocarriers and their general applications. This is followed by application of these particles in selected neglected tropical diseases as tuberculosis, leishmaniasis, malaria, etc.

## 15.1 INTRODUCTION

Nanomedicine comprises of the multidisciplinary approaches to engineer nanomaterials to meet the demands of medical challenges like sensing, diagnostics, and drug delivery. Nanomaterials are presently being ventured for many applications, like cell and tissue screening, biosensing, and in vivo tumor targeting. Nanoparticles modified with antibodies, peptides, or oligonucleotides (known as aptamers) scaffolded into biologically relevant nanostructures as well as nanoparticle–drug conjugates have been explored as novel drug delivery machineries (Muller and Keck, 2004). Nanoparticles in the size range of 1–100 nm possess unique physical, chemical, optical, electrical, magnetic properties which can be used as potential drug delivery devices. The major highlights of preparing nanoparticles of biomedical importance is the optimization of size, surface properties, and controlled release profile of the drug in targeted sites (Mohanraj and Chen, 2006).

Nanoparticles are widely used in biomedical imaging, diagnostics, and therapeutic applications (Choi and Frangioni, 2010). Gold nanoparticles have targeting, photothermal as well as gene and drug delivery applications (Ghosh et. al., 2008). Gold nanoparticles are being constantly studied to evaluate its role in diagnostics, therapy, and immunology; covering various fields such as analytical methodologies, plasmonic biosensors, drug and gene delivery, photodynamic therapies, and monitoring biodistribution and toxicity of a chemical compound of biomedical importance (Dykman and Khlebtsov, 2012). Similarly, magnetic nanoparticles of 10–50 nm size having magnetite, iron, nickel, and cobalt cores and surrounded by polymeric shells have excellent in vivo and in vitro applications (Akbarzadeh et al., 2012). Modification of nanoparticles with varied surface coatings allow them to use in biomedical applications based on its stability, solubility, and the targeting capabilities. Such novel assemblies tagged to biologically relevant macromolecules lead to the generation of target-specific nanoconjugates. Engineered macromolecular assemblies with covalently attached active chemical/biological entities and/or functional head groups such as drugs, prodrugs, oligonucleotides, antibodies, etc. are designed to target-specific cells/tissues. The merits of using nanoparticles as potent drug carriers include the following:

1. Physical attributes, namely, shape, surface, size, charge, permeability, and chemical properties such as biocompatibility, biodegradability, antigenicity, and toxicity of nanoparticles can be easily manipulated to achieve targeted drug delivery.
2. Size of the nanoparticles is much smaller than animal or parasite cells; so movement of these nanoparticles through cellular system is faster/easier in comparison to microparticles. This increases the therapeutic efficacy of the nanoparticles. Moreover, due to large surface area attachment of drug or any targeting ligand/moiety is easier. By modulating the properties of drug carrier, sustained release of drug at the targeted site can be achieved. Drug activity can be preserved to a longer extent as nanocarriers prevent drugs from prematurely interacting with the biological environment in in vivo settings.
3. Therapeutic efficacy of existing drugs can be enhanced when delivered via nanocarriers due to improved adsorption, distribution, metabolism and excretion (ADME) profiles, reduced drug toxicity, and targeted delivery to the target organism or organ.

4. Different routes of administration such as oral, nasal, parenteral, intraocular, etc. are possible for nanoparticle-based drug systems.

Nanocarriers such as liposome, polymeric nanoparticles, nanosuspensions, solid lipid nanoparticles, nanoemulsions, dendrimers, niosomes, cyclodextrins show immense ability to overcome the problems associated with delivery of conventional drugs which has been diagrammatically shown in Figure 15.1 and described in the following sections.

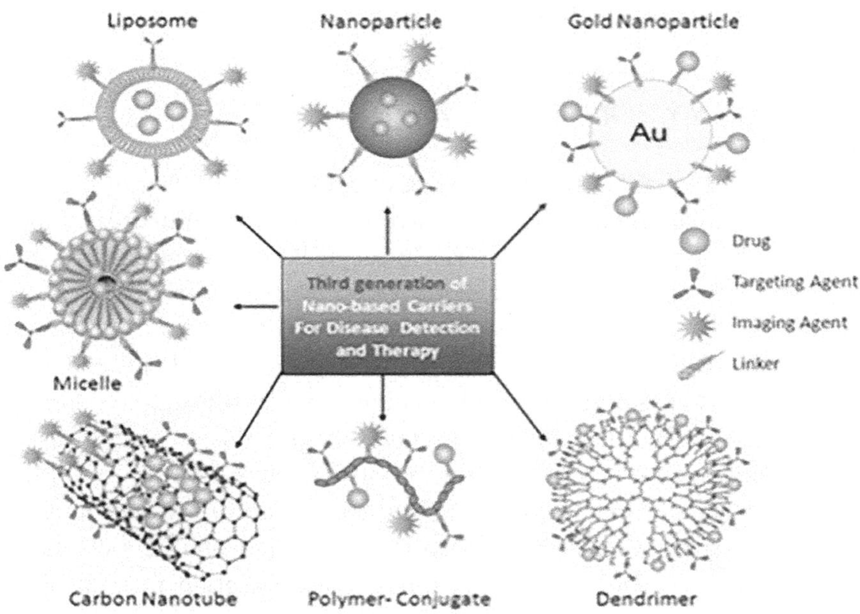


**FIGURE 15.1** **(See color insert.)** Few nanocarriers used in drug delivery. (Adapted from http://slideplayer.com/slide/8751147/.)

### *15.1.1 LIPOSOMES*

Liposome is the most widely used drug delivery system. They are nano- to microsized spherical vesicles made up of phospholipid bilayer(s) (Fig. 15.1). Hydrophilic drugs can be encapsulated in the aqueous core while the membrane encapsulates a hydrophobic drug (Gulati and Wallace, 2012). The liposome surface can be functionalized with stealth material for prolonging blood circulation time. Polyethylene glycol (PEG) is commonly conjugated to the surface of liposomes which forms a hydrating layer and extends its lifetime in blood stream. On the basis of bilayers and size, liposomes can be classified into three types:

a) Small unilamellar vesicles: single lipid layer, size differs from 25–50 nm in diameter.
b) Large unilamellar vesicles: surrounded by single bilayer.
c) Multilamellar vesicles: consist of several lipid bilayers each separated by an aqueous layer.

### 15.1.2 POLYMERIC NANOPARTICLES

Polymeric nanoparticles (PNP) possessing advantages of drug solubilization, stabilization, and targeting have generated wide interest. Based on the technology of production, two types of PNP are reported as nanocapsules and nanospheres. In nanocapsules, drug is distributed homogenously in aqueous or oily solvents, which is surrounded by a polymeric membrane. The main reasons behind popularity of PNP encapsulation of drugs are its high stability, high loading capacity for both hydrophilic and hydrophobic drugs, and administration possibility using various routes. However, defense mechanism could remove PNP by opsonization and phagocytosis. In order to prevent recognition by host immune system, the surface of PNP are modified using hydrophilic side chains as polyethylene glycol leading to prolonged circulation time in blood stream.

### 15.1.3 SOLID LIPID NANOPARTICLES

Solid lipid nanoparticles (SLNs) are nanocrystalline suspensions in the range of 50–1000 nm prepared using solid lipids as fatty acids, triglycerides, steroids, partial glycerides, and waxes. SLNs owing to development from physiological lipids show better tolerability, ability to deliver both hydrophobic and hydrophilic drugs and increased stability of the encapsulated drugs (Bummer 2004; Jenning et al., 2002; Muller and Keck, 2004; Saravanan et al., 2015).

### 15.1.4 NANOEMULSIONS

Nanoemulsions in size of 10–100 nm are thermodynamically stable oil-in-water dispersion. They include advantages as spontaneous generation without high homogenization energy, ability to filter sterilize, and enhanced uptake by lipoprotein receptors in liver after oral administration.

### 15.1.5 DENDRIMERS

Dendrimers are three dimensional macromolecules (shown in Fig. 15.1c) that encapsulate drugs by its dendrimeric core, complexation, and conjugation on the surface for its delivery by a variety of administration routes (Shegokar et al., 2011).

### 15.1.6 NIOSOMES

Cholesterol hydration of charged phospholipids (stearyl amine and dicetylphosphate) and nonionic surfactants (monoalkyl or dialkylpolyoxyethylene ether) results in liposome-like vesicles called niosomes (Smola et al., 2008). Niosomes are similar to liposomes except that the bilayer in case of niosomes is made of nonionic surface active agents, rather than phospholipids in liposomes. Hydrophilic drugs and lipophilic drugs are encapsulated inside the niosome core and hydrophobic provinces, respectively (Kaur and Singh, 2014; Saravanan et al., 2015). The distinctive advantages of niosomes over liposomes include greater stability, higher drug loading capacity, low production costs, no need of special conditions for handling or storage (Karim et al., 2010). Biodegradability, biocompatibility, nonimmunogenicity, and flexibility are their other distinctive features.

### 15.1.7 CYCLODEXTRINS

Cyclodextrins are cyclic oligosaccharides composed of 6–12 D-(+)-glucopyranose units linked by α-(1–4) bonds which are complexed with drugs resulting in controlled site-specific drug delivery, drug safety and stability (Brewster and Loftsson, 2007; Loftsson and Dychene, 2007). A range of cyclodextrins are reported as natural α-, β-, and γ-cyclodextrin molecules and a variety of synthetic derivatives (e.g., hydroxypropyl-γ-cyclodextrin). They are composed of hydrophobic inner cavity for inclusion of hydrophobic drugs and a hydrophilic outer surface. The complexed drugs show significant increase in water solubility compared to the uncomplexed drugs (Sosnik et al., 2010).

### *15.1.8 CARBON NANOTUBES*

The carbon nanotubes (CNT) are nanosized hollow cylindrical structures of carbon. Owing to high toxicity of pristine CNT, the water solubility of functional CNT (f-CNT) was verified in CNT in physiological media. Water soluble CNT carrying ammonium groups finds application in the delivery of therapeutic molecules (Bianco et al., 2005). Carbon nanotubes due to their high cellular uptake, enhanced drug loading, and ability to distinguish healthy cells from abnormal ones allow them to be highly useful as targeted drug delivery apparatus for cancer (Kushwaha et al., 2013). CNT allows introduction of a range of function as targeting molecules, contrast agents, drugs, or reporter molecules on the same tube (Bianco et al., 2005).

## 15.2 OVERVIEW OF NEGLECTED DISEASES AND CURRENT RESEARCH STATUS

Neglected diseases are defined by The World Health Organization (WHO) as "Diseases that are underfunded and have low name recognition, but are major burdens in less developed countries". Chronic infections such as tuberculosis, leishmaniasis, African trypanosomiasis (sleeping sickness), malaria, lymphatic filariasis, dengue, onchocerciasis, chagas disease, and schistosomiasis cause severe health burdens to more than 1 billion people (Trouiller et al., 2002). In spite of large number of clinical cases reported for neglected tropical diseases, they are less researched upon and less funded. The poor health policies significantly reduce pharmaceutical research for development of novel drugs against neglected tropical diseases. The arsenal of conventional drugs against neglected tropical diseases faces a major setback due to development of multidrug-resistance. Further, the conventional pharmacological treatment is compromised by high drug toxicity, increased drug resistance, short half-life of drugs, and inability of drugs to reach intracellular sites infected by parasites. This scenario calls for quest of new drugs which is hampered by lengthy and expensive drug development processes. In this context, the application of nanocarrier systems for the conventional drugs is seen as a promising answer due to its ability to target-specific cells and tissues, high capacity to overcome biological barriers, sustained drug release, improvement in therapeutic index, and pharmacokinetic properties of drugs.

## 15.3 TUBERCULOSIS AND CHALLENGES ASSOCIATED WITH CONVENTIONAL THERAPY

The 2013 WHO global tuberculosis report estimated an unacceptable 1.3 million deaths per year from tuberculosis. The latest 2014 WHO global tuberculosis report revised its estimates and reported half a million more tuberculosis cases than that reported in 2013 report with rise in tuberculosis deaths by 1.5 million. The 2014 WHO report also brought to light the worsening state of the problem of drug resistant tuberculosis, with estimates of 480,000 new cases of multidrug-resistant (MDR) tuberculosis adding to urgent need for efficient diagnosis and control strategies (Zumla et al., 2015). As per WHO, there is shortfall of $1.6 billion per year required for treatment of TB (www.slate.com). Although TB no longer considered a neglected disease, the recent estimates brings to light the necessity to review the present global tuberculosis control strategy and design crucial treatment strategies required to achieve the ultimate goal of complete TB elimination in future. The causative agent of TB is mycobacteria (Gram-positive actinomycetes), mainly *Mycobacterium tuberculosis* which is transmitted by inhalation. The first and second line anti-TB drugs had been reported to inhibit active bacilli growth. Recently, the mismanagement of first-line drugs resulted in the emergence of multidrug-resistant TB (MDR-TB) which compromises the effectiveness conventional anti-TB therapy. This requires the use of second-line drugs which are expensive, less potent, more toxic, poorly permeable, and less stable. Even WHO endorsed DOTS Plus program have failed to effectively manage MDR-TB in high burden areas due to high cost associated with diagnosis and second-line drugs. Owing to the fact that first generation of anti-TB drugs are still effective, any carrier system which can help improve the permeability of these drugs will result in improved effectiveness at lower dose with minimal side effects.

### *15.3.1 NANOCARRIERS TO COMBAT CHALLENGES IN TREATMENT OF MULTIDRUG-RESISTANT TB*

The delivery of antitubercular drugs (ATD) is problematic owing to rich mycolic acid layer in cell wall of *M. tuberculosis* (Saravanan et al., 2015). There is necessity of a carrier system which can improve permeability of ATDs leading to increased drug effectiveness at lower dose with minimal side effects. The call of the hour is to develop novel nanomaterial-based ATD delivery methods with increased bioavailability and site-specific targeting to avoid development of drug resistance. ATDs entrapped in nanocarriers which will act as ATD reservoir continuously supplying the required drug concentration in plasma, thus reducing the toxicity owing to low drug concentration in free plasma at a given time. Nanoparticulate drug delivery system improves the therapeutic index of the drug by delivering the ATDs to the target sites. The potential to develop highly effective TB therapy with existing ATDs using nano-based carriers is of interest owing to dearth of production and approval of new ATDs since last four decades. Klemens et al. (1990) incorporated gentamicin into the liposomes and the encapsulated drug significantly increased anti mycobacterial efficacy coupled to reduction of myocardial infection compared to its free drug counterpart. Targeted delivery of anti-TB drugs as isoniazid (INH) and rifampicin (RIF) were carried out using lung-specific stealth liposomes composed of phosphatidylcholine, dicetyl-phosphate, *O*-steroyl amylopectin, cholesterol, and monosialogangliosides–distearylphophatidylethanolamine–poly(ethylene glycol) 2000. The decreased toxicity of ATDs delivered using liposome-based carrier was due to controlled drug release compared to that in free ATDs. Sharma et al. (2004) showed that PLGA nanoparticles grafted with lectins had extended lifetime in plasma of 6–14 days and reported complete treatment of mycobacterial infection in lungs, liver, and spleen as against uncoated particles. Pandey and Khuller (2004) showed higher sustained release of RIF, INH, and Pyrazinamide (PZA) from poly(lactide-*co*-glycolide) (PLG) nanoparticles resulted in reduction in number of doses to five oral doses (after every 10th day) as opposed to 46 doses of free drug for complete bacterial clearance. The treatment of *Mycobacterium avium* infection by Clofazimine was compromised due to poor solubility. Administration of Clofazimine as a nanosuspension (particle size 385 nm) led to significant control of *M. avium* infection in liver, spleen, and lungs of mice (Peters et al., 2000). SLN loaded with ATDs as RIF, INH, and PZA led to undetectable bacilli in the organs of TB-infected guinea pigs as opposed to 46 doses required using conventional drugs. Nanoemulsion of ramipril administered orally reported sustained drug release for 24 h as compared to conventional capsule and drug suspension (Shafiq et al., 2007). RIF-loaded mannosylated 5th generation (5G) polypropylenimine (PPI) dendrimeric nanocarrier leads to its selective uptake by recognition of lectin receptors on phagocytic cell surface. RIF-dendrimer showed significant increase in phagocytic uptake and decrease in hemolytic effect and toxicity to kidney epithelial cell line as compared to free RIF (Kumar et al., 2006). Mullaicharam and Murthy (2004) reported significant 145-fold increase in accumulation of RIF-loaded niosomes in lungs compared to the corresponding free drug. Rao et al. (2006) reported enhanced antibacterial activity of cyclodextrin complexes against *M. tuberculosis* indicated by reduction in MIC of rifampicin to half. Water insoluble drug nitroimidazole P-824 when complexed with HP-$\gamma$-CD showed activity against multidrug-resistant bacilli.

Further studies show enormous potential for development of improved nanomedicine-based strategies for cost effective and highly efficient combat MDR-TB. It is expected that nanocarriers will revolutionize treatment of MDR-TB by efficient drug delivery systems and will evolve further in future.

## 15.4 MALARIA AND CONVENTIONAL THERAPY

Malaria is a major public health problem in tropical and subtropical countries. Malaria parasites are transmitted from host to host by the bite of an infected *Anopheles* mosquito. Malaria is caused by four *Plasmodium* species, that is, *Plasmodium falciparum*, *Plasmodium vivax*, *Plasmodium malariae* and *Plasmodium ovale* cause malaria in humans. *P. falciparum* and *P. vivax* are the two major species responsible for malaria globally (Duong et al., 2004). Reduction in number of malaria deaths by one-third owing to substantial progress in control strategies, malaria were not considered a neglected disease. However, estimates of WHO malaria report 2016 showed 99% of malaria deaths resulted from *P. falciparum* malaria, whereas *P. Vivax* is estimated to have been responsible for 3100 deaths in 2015 with mostly 86% occurring outside Africa, so there is still much more to be done.

Currently, artemisinin (ART) derivatives, quinine (QN) related drugs, antifolates derivatives, and new class of combination drugs are extensively used to treat malaria. These drugs are used alone or in combination for effective treatment. Recently, high failure rates of artemisinin-based combination therapy (ACT) had been attributed to the development of possible artemisinin resistance (Noedl et al., 2008; Vijaykadga et al., 2006). Moreover, artemisinin derivatives have been reported to show dose, time and route dependent neurotoxicity in laboratory animals (Petras et al., 2000). Quinine-related drugs are highly potent in treating malaria, but they are associated with severe side effects like, cinchonism (auditory symptoms, gastrointestinal disturbances, vasodilation, headache, nausea, and blurred vision) (Taylor and White, 2004), cardiotoxicity, hypoglycaemia (Alkadi, 2007), anaemia, and GI disturbance (Shekalaghe et al., 2010). Antifolate derivatives inhibit the *falciparum* parasites by interfering with folate metabolism, a pathway essential for survival of the parasite via inhibition of enzymes dihydrofolate reductase (DDFR) (Bzik et al., 1987) and dihydropteroate synthase (DHPS) (Lu et al., 2010). Among the new class of WHO approved combination drugs, Artemether/lumefantrine combination is effective against blood stages of *P. vivax*, but is not active against hypnozoites, whereas dihydroartemisinin (DHA) and piperaquine (PQP) combination has the potential to be an effective antimalarial drug against multidrug-resistant falciparum malaria (Ashley et al., 2004; Ashley et al., 2005; Giao et al., 2004; Zwang et al., 2009).

### *15.4.1 NANOTECHNOLOGICAL APPROACHES FOR TARGETING DRUG IN MALARIA THERAPY*

The most important property of nanocarriers developed for delivering antimalarial drugs is the prolong blood circulation time so that they can interact with infected red blood cells (RBCs) and parasite membranes (Mosqueira et al., 2004). In addition to this, the other requisite properties are protection of unstable drugs, cell-adhesion properties, and the ability to be surface-modified by conjugation of specific ligands (Kayser and Kiderlen, 2003).

Malaria infected erythrocytes and the hepatocytes are targeted using nanocarriers either by passive or active targeting strategy (Santos-Magalhães and Mosqueira, 2010). Passive targeting is achieved using conventional nanocarriers (e.g., liposomes, hydrophobic polymeric nanoparticles), or surface-modified long circulating nanocarriers (e.g., PEGylated). In passive targeting, the drug-loaded carriers accumulate at a particular body site due to physicochemical or pharmacological factors (Barratt, 2003; Garnett, 2001); due to the pathophysiological and anatomical features nanocarriers reach the target site(s) and are retained there (Garnett, 2001). Nanocarriers are internalized by mononuclear phagocyte system (MPS) through phagocytosis. It is difficult to target infected host RBCs as they are phagocytically and endocytically inactive. It is noteworthy to mention that conventional nanocarriers administered through the parenteral route are rapidly taken up by MPS cells, delivering the drug inside macrophages. Excess exposure nanocarriers to the phagocytes lead to saturation of phagocytic activity and hence reduction in the effect of antimalarial drug entrapped in the nanocarriers (Scherphof et al., 1997). Interestingly, this type of conventional nanocarriers can easily target *P. vivax* infection in which hypnozoites are the dormant forms of the parasite in the hepatocytes, located side by side with Kupffer cells as shown in Figure 15.2.

Curcuminoid-loaded lipid nanoparticles for parenteral administration were prepared for treating malaria. In order to prepare nanoparticles, trimyristin, tristerin, and glyceryl monostearate were selected as solid lipids and medium chain triglyceride (MCT) as liquid lipid. Size of the nanoparticles was in the range of 120–250 nm and zeta potential data showed that these particles are highly stable. In vivo results showed that pharmacodynamic activity revealed twofold increase in antimalarial activity of curcuminoids entrapped in lipid nanoparticles as compared to free curcuminoids at the tested dosage level (Nayak et al., 2010). Dihydroartemisinin (DHA) loaded solid lipid nanoparticles (DHA-SLNs) with mean particle size of 240.7 nm were prepared for treating malaria. Antimalarial activity of these nanoparticles was enhanced by 24% when compared to free DHA. Results showed that these SLNs were promising for antimalarial therapy (Omwoyo et al., 2016).

The blood circulation time of nanocarriers can be increased by modifying the surface of the nanocarriers with polymers such as poly(ethylene glycol) (PEG). Surface modification delays phagocytosis, and hence changes the biodistribution and pharmacokinetic profile of the drug (Gref, et al., 1995; Lück, et al., 1998;). Passive targeting is useful in malaria therapy to target MPS as long circulating nanocarriers can easily interact with RBCs (Fig. 15.2).

Artemisinin-loaded PEG-grafted polymeric nanospheres or nanoreservoirs were reported to prepare from cyclodextrin derivatives grafted with decanoic alkyl chain (CD-$C_{10}$) for treating malaria. ART-loaded both these formulations were reported to effective in inhibiting growth of P. falciparum and theses formulations were proved to be promising alternatives for injectable ART (Yaméogo et al., 2012).

In active targeting, the surface of the nanocarrier is modified with specific ligands such as carbohydrates, proteins, peptides or antibodies specific to the disease infected cells or tissue, and thereby guide the nanocarriers or drug to accumulate the target site (Torchilin, 2006). In case of malaria, the disease infected erythrocytes in blood and hepatocytes can be targeted by tagging the ligands on the surface of the nanocarriers specific to them (Fig. 15.2).

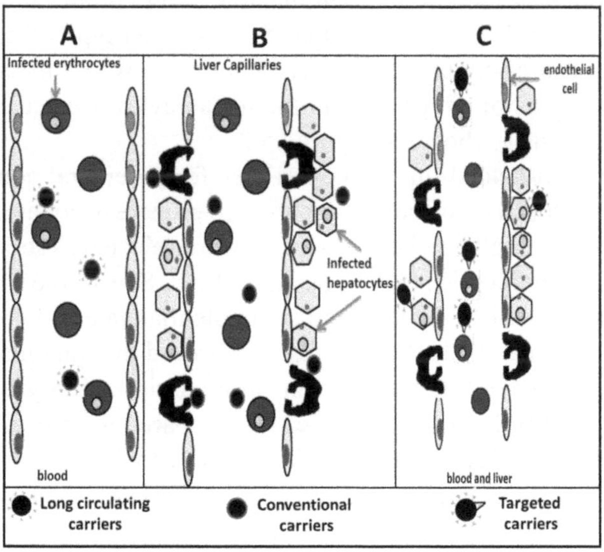


**FIGURE 15.2** Schematic representation of three targeting strategies of antimalarial nanocarriers in *Plasmodium*-infected host: Surface-modified long circulating carriers (A), conventional surface-unmodified carriers (B), and ligand-modified carriers to improve selective targeting to infected erythrocytes and hepatocytes (C). (Adapted from Santos-Magalhães and Mosqueira, 2010. © 2010, Elsevier.)

Nanocarrier-based delivery systems will revolutionize malaria treatment by improving the efficacy of the currently used antimalarial drugs. However, there is a long way to go to design effective drug combinations for better cure rates and reduced side effects.

## 15.5 LEISHMANIASIS AND CONVENTIONAL THERAPY

Leishmaniasis is one of the major NTDs in the tropical and subtropical areas. However, it is becoming more and more common in the developed countries due to increasing immunosuppressed population and the higher frequency of travelers around the world. It is caused by protozoan parasites belonging to the genus *Leishmania*. Visceral leishmaniasis (VL) caused by *Leishmania donovani* is the most threatening. Currently, 0.9–1.3 million people are affected by Leishmaniasis, of which 20,000–30,000 death occur annually (WHO Leishmaniasis report 2016).

### *15.5.1 NANOTECHNOLOGICAL APPROACHES FOR TARGETING DRUG IN LEISHMANIASIS THERAPY*

Oryzalin-loaded liposomes and solid lipid nanoparticles were prepared and their therapeutic efficacy was compared in vitro and in vivo. Results demonstrate the superiority of both oryzalin nanoformulations in reducing the parasitic burden in liver and spleen as compared to the control group (84–91%) and similar to glucantime. Both formulations were showing promising results in treating visceral leishmaniasis (Lopes et al., 2014). Sundar et al. (2010) carried out a clinical trial using liposomal amphotericin B with visceral leishmaniasis infected patients in Bihar, India. A total of 410 patients participated in the trail—304 patients (100%) were treated with liposomal therapy, whereas 106 of 108 patients (98%) were treated with conventional therapy. A single dose of liposomal amphotericin B therapy was found to be as good as conventional therapy and less expensive (Sundar et al., 2010). Amphotericin B loaded PLGA-based nanospheres (AmB-PLGA-NS) were developed for treatment of parasites responsible for visceral Leishmaniasis (Lima et al., 2014). In vivo experiments done on infected BALB/c mice showed that these AmB-PLGA-NS were efficient in treating malaria and they preferentially accumulate in the visceral organs. In vitro results showed that these nanospheres could generate immune response to kill the malaria parasites.

Roychoudhury et al. (2011) prepared sodium stibogluconate (SSG) loaded phosphatidylcholine stearylamine liposomes (PC-SA-SSG) and PC-cholesterol liposomes (PC-Chol-SSG). They compared the therapeutic efficacy of these liposomes formulations with free AmB against SSG-resistant *L. donovani* strains in 8-week infected BALB/c mice. Results showed that a single dose of PC-SA-SSG was effective in curing mice infected with two differentially originated SSG-unresponsive parasite strains at significantly higher levels than AmB, unlike free and PC-Chol-SSG. These liposome formulations were capable of producing immune responses that kill the parasite.

Although nanotechnology based formulations for the leishmaniasis treatment has shown encouraging results, further in vivo and clinical tests are needed to verify challenges arising due to wide difference between the two scenarios. Further research efforts are needed as the area shows enormous promise for development of novel nanodrugs against leishmaniasis.

## 15.6 CHAGAS DISEASE AND CONVENTIONAL THERAPY

Chagas disease, or American trypanosomiasis, is caused by the flagellated parasite *Trypanosoma cruzi* which feeds on blood of humans or animals. Infection is most commonly acquired through contact with the feces of an infected triatomine bug (or "kissing bug"). Chagas disease is endemic throughout much of Mexico, Central America, and South America with an estimated 6–7 million people are infected worldwide as of 2016. Poor housing conditions in rural habitats are conducive to the development and spread of the disease. The disease is primarily characterized by two clinical forms, acute and chronic. Acute phase usually represent symptoms of fever, fatigue, diarrhea, loss of appetite, hepatosplenomegaly, and myocarditis and/or meningoencephalitis in extreme cases. Nearly 30% of those infected patients will develop the chronic form of the disease, which affects nervous system, digestive system and heart. (Sosa-Estani and Segura, 2006).Sometimes an asymptomatic indeterminate stage might comprise a period of several years between the acute and chronic phases. Drugs benznidazole (Garcia et al., 2005) ornifurtimox cannot completely irradicate *T. cruzi* from the body especially in chronically infected patients, and resistance to these drugs has been reported (Buckner et al., 1998).

### *15.6.1 NANOTECHNOLOGICAL APPROACHES FOR TARGETING DRUGS IN CHAGAS DISEASE*

In general, routine drug administration procedures produce peak levels of drug doses at potentially toxic levels. This has led to different side effects and complicated delivery procedures in trypanosomiasis patients co-infected with HIV or other immune system diseases. Hence, the idea of modulating the release of bioactive agents, in the past decades has been in focus. Sufficient light has been thrown on the use of biocompatible as well as biodegradable polymers to design micro and nano-particulate delivery systems as carriers for drugs (Freiberg and Zhu, 2004). Selective delivery of drugs into intracellular sites helps to combat the dissemination of the parasite. Different types of nanodrug delivery vehicles are antibody conjugates (e.g., Mylotarg R®, Tositumomab® or Zevalin®) (Allen, 2002; Damle and Frost, 2003; Milenic et al., 2004), liposomes (e.g., DaunoXome™ or Doxil®/Caelyx®) (Torchilin, 2005) the first anticancer nanoparticle, Abraxane® (O'shaughnessy et al., 2004) and polymer therapeutics (Duncan et al., 2006) have been explored for Chagas disease (Duncan et al., 2005).Liposome made of stearylamine and phosphatidylcholine showed the trypanocidal activity in vitro (Yoshihara et al., 1987). Since encapsulation improves the therapeutic efficiency, the use of polyalkylcyanaoacrylate as a carrier for nifurtimox in the form of nanospheres was shown by Gonzalez-Martin and coworkers in 1998. These nanospheres were active against *T. cruzi* in vitro (Gonzalez-Martin et al., 1998). Allopurinol-loaded nanospheres proposed in 2000 and its trypanocide effect was assessed on chronic Chagas patient but was found unsuitable as sustained drug release system (Gonzalez-Martín et al., 1998; Scherer et al., 1994; Leonard et al., 1996).Poly(ethylene glycol)-*co*-poly(lactic acid) nanoparticles were devised to minimize opsonization and reticulo-endothelial system uptake and passively target the diseased tissues. Itraconazole (IT), ketokonazole (KET) and the fourth generation bis-triazole D0870 (Molina et al., 2001) separately loaded in poly(ethylene glycol)-*co*-poly(lactide) nanospheres, (Gref et al., 1995) of which D0870 showed remarkable trypanocidal activity in vivo (Urbina et al., 2001, 2002).Lipid formulations of amphotericin B cannot be used in vivo as anti-*T. cruzi* agent due to its high toxicity (Horvath et al., 1974), while the same active agent loaded in a lipidic nanocarriers (AmBisome) is less toxic and a good therapeutic alternative (Castro, 1993). To increase the efficacy of drug benznidazole as trypanocidal agent, multilamellar liposomal formulations (conventional and pH-sensitive liposomes) by the modification of its pharmacokinetics have been designed by Morilla et al. (2002).

Nanomedicines against Chagas disease had showed promising results to develop site-specific drug delivery systems targeted to infected tissues resulting in improved results compared to current therapy and thus deserves adequate attention in future.

## 15.7 FILARIASIS OR ELEPHANTIASIS

### *15.7.1 CHALLENGES ASSOCIATED WITH CONVENTIONAL TREATMENT*

Filariasis or Elephantiasis is characterized by thickening of skin, leg tissues, female breasts, and male genitals. Elephantiasis is caused by parasitic worms as *Wuchereria bancrofti*, *Brugia malayi*, and *Brugia timori* which are transmitted to the human body. Currently, there is no reported vaccine to prevent filariasis (elephantiasis). WHO specifies a drug called diethylcarbamazin to kill worms of filariasis in lymph nodes but it leads to side effects such as cell damage and nephrotoxicity. In order to avoid these complications, there arises the need for a carrier which shall ensure site-specific delivery of diethylcarbamazin.

### *15.7.2 NANOTECHNOLOGICAL APPROACHES FOR TARGETING DRUG IN ELEPHANTIASIS*

Recent advances in field of nanotechnology show promise for use of nanorobots for destroying parasitic worms present in lymph nodes as well as filtering larva from blood as follows:

#### *15.7.2.1 NANOROBOTS FOR SPECIFIC AND PERMANENT TREATMENT OF ELEPHANTIASIS*

Nanorobots are theoretical microscopic devices that would work at atomic, molecular and cellular level to fully realize the potential of nanomedicines. Nanorobot is composed of carbon atoms in a diamondoid structure, with use of glucose or natural body sugars and oxygen for propulsion. Manivannan and Rathnakaran (2014) proposed two kinds of nanrobots, one for blood filtration and another to kill and remove parasitic worms from lymph nodes without harmful effects to tissue and metabolism.

#### *15.7.2.1.1 Microbivore Nanorobot for Filtering Larva from Blood*

Microbivores are nanorobots which would patrol bloodstream seeking and digesting unwanted pathogens including bacteria, virus, fungi, or parasitic larva. The significant advantage of a microbivore is that it can treat a parasite even if it has acquired antibiotic resistance. Nanorobot can completely destroy parasites in just 30–45 s compared to long treatment times associated with conventional antibiotics. Nanorobots being 5000 times smaller than blood capillaries could travel along blood, find targeted larva, destroy, and absorb larva of parasitic worms. The architecture of microbivore nanorobots is composed of DNA sensor, CPU and Digester as shown in Figure 15.3. DNA sensors which are coded with the DNA from larva are used for detection of larva in bloodstream.

After finding parasitic larva, the CPU will give the instruction to the Digesting unit. The oligonucleotides in the digesting unit will stop the metabolism of larva. The nanorobot fixed with larva will be excreted from human body. The transvier gives the data about the nanorobots status in blood and larva detected by it.

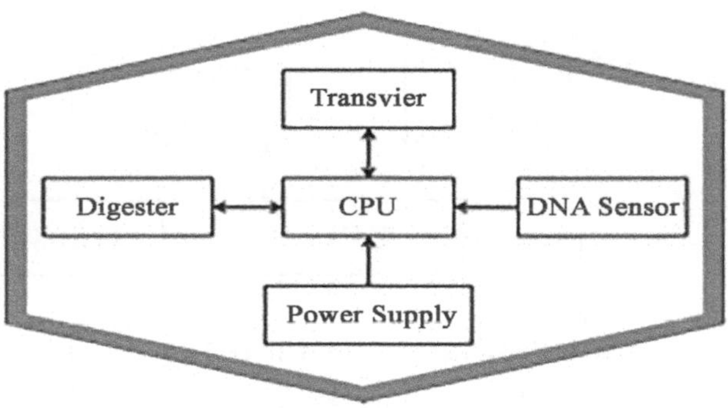


**FIGURE 15.3** The architectural design of a bloodstream microbivore. (Adapted from Manivannan and Rathnakaran, 2014.)

#### *15.7.2.1.2 Nanorobots to Destroy Parasitic Worms in Lymph Nodes*

Filtering larvae from bloodstream may not provide a permanent solution since matured worms may present in lymph nodes. The highly toxic nature of drugs as diethylcarbamazin could be overcome by site-specific targeting of the drug through nanorobots to lymph nodes. To transport these nanorobots to lymph nodes an inlet path and an outlet path for removal of nanorobots with dead worms is of utmost importance. This inlet should be present where nodes start and the outlet should be near to the inlet so that it could be easy way to remove the destroyed worms. The nanorobots will identify the parasites by the preloaded DNA followed by attaching themselves with parasitic worm and injecting the diethylcarbamazin. Finally, these nanorobots controlled by external programmer will enter the outlet and disposed from the body. Figure 15.4 describes the structure of nanorobot in lymph nodes consisting of five important parts as CPU, DNA sensor path tracer, drug injector and holder, and transiver. DNA sensor coded with immobilized DNA detects the L3 larva followed by instruction from CPU to hold the worm by the holder. Then, the diethylcarbamazin is injected into the worm's body and the worm dies within 30 s. The path tracer records the nanorobots motion until it detects the target DNA. The transvier maintains communication with the outlet transmitter receiver to lead the nanorobots carrying the dead worm to the outlet path.

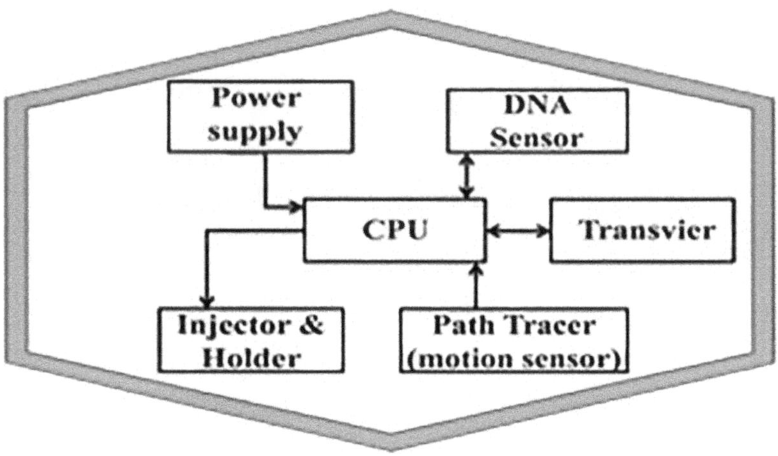


**FIGURE 15.4** Architecture of nanorobots to target filariasis worms in lymph nodes. (Adapted from Manivannan and Rathnakaran, 2014.)

Recent advances in field of nanocarriers holds tremendous promise to exploit nanorobots to overcome the problems associated with conventional drugs in treatment of neglected tropical diseases as in elephantiasis.

## 15.8 SCHISTOSOMIASIS

### *15.8.1 CHALLENGES ASSOCIATED TO CONVENTIONAL TREATMENT*

Schistosomiasis is a neglected parasitic disease and a major health concern in infecting millions of people in 74 countries of Africa, Asia, and South America. It is caused by parasitic flatworms called schistosomes. The current treatment of schistosomiasis involves only two drugs, oxamniquine and praziquantel. Currently, praziquantel remains the drug of choice since the commercial production of oxamniquine had decreased (Khalil et al., 2013). WHO had recommended the oral administration of praziquantel (PZQ) for schistosomiasis. However, the pharmaceutical efficiency of PZQ is compromised by its low water solubility resulting in poor bioavailability. Moreover, the bitter taste of the drug coupled to its high dose requirement makes it unsuitable for pediatric use (Fonseca et al., 2013). Thus, exploitation of nanocarriers as a means to increase efficacy of conventional drugs against schistosomiasis was seen as an alternative.

### *15.8.2 NANOTECHNOLOGICAL APPROACHES TO OVERCOME LIMITATIONS OF CONVENTIONAL TREATMENT*

Polymer nanoparticles are seen as an efficient system for encapsulation of drugs with low water solubility and bitter taste. Polymer nanoparticles could further enhance bioavailability of PZQ. Fonseca et al. (2013) reported encapsulation of PZQ in poly (methyl methacrylate) (PMMA) nanoparticles by miniemulsion polymerization. The nanoencapsulated PZQ showed an extreme high mortality of worms with no cytotoxicity or genotoxicity. Further, the encapsulated drug was efficient at overcoming the problem associated with bitter taste of PZQ. However, following 10 min after dispersion of PZQ in water, bitter taste was felt. Thus, significant efforts to extend the release time to prevent development of bitter taste in the mouth is required in future. The developed nanoencapsulated drug shows great potential to overcome the drawbacks associated conventional schistosomiasis treatment. Modulation of drug release leading to absorption rates as well as masking of bitter taste of conventional schistosomiasis drugs are highly beneficial for pediatric treatment. This technology promises to be an efficient drug delivery system for treatment of schistosomiasis.

Mainardes et al. (2006) developed PLGA nanoparticles as drug carriers for praziquantel. They showed that PLGA nanoparticulate systems increase the pharmaceutical efficiency of praziquantel by localized effect in intestinal membrane for longer time period. Mourão et al. (2005) studied the in vitro effect of praziquantel encapsulated in liposome against schistosomiasis. They reported that praziquantel encapsulated in liposomes caused a more effective decrease in amounts of eggs and parasites in vivo compared to free drug. Frezard et al. (2005) used Meglumine antimoniate in stealth liposomes as an antischistosomial agent. They reported that 55% of parasite elimination by Meglumine antimoniate encapsulated stealth liposomes. Frézard and Melo (1997) encapsulated oxamniquine in liposomes and administered subcutaneously in murine model to test its efficiency in control of *Schistosoma mansoni*. They reported 97% reduction in worms following administration of the liposomal drug which brightens the prospects of using liposomal form of oxamniquine for the treatment of schistosomiasis. Frezza et al. (2007) showed the efficacy of phosphatidylcholine liposomes containing Praziquantel in female mice Swiss-SPF infected with *S. mansoni*. They reported that the liposomal form of the drug is more efficient in treating both mature and immature worms compared to the free drug. Feng et al. (2004) complexed calcium nanoparticles complexed with anti-idiotypic antibody NP30 (CA-NP30 conjugate) were used in treatment of schistosomiasis. They reported remarkable increase in specific antibodies against NP30 in serum levels compared to immunization with NP30 alone. This suggests the ability of nanoparticle CA could act as a vaccine adjuvant of anti-idiotypic antibody NP30 against schistosomiasis. De Araújo et al. (2007) complexed a novel schistosomicidal drug 2-(butylamino)-1-phenyl-1-ethanethiosulfuric acid (BphEA) in a nanoemulsion of cationic or anionic surfactants and reported that BphEA with cationic surfactant had a higher schistosomicidal activity compared to free drug. The increased entrance of the BphEA into the worms establishes it as an efficient delivery system for BphEA.

## 15.9 AFRICAN TRYPANOSOMIASIS

### *15.9.1 OVERVIEW AND CONVENTIONAL THERAPY*

Human African trypanosomiasis or sleeping sickness is caused by protozoa *Trypanosoma brucei gambiense* and *T. brucei rhodesiense*. It is transmitted by glossina flies from human and/or animal reservoirs (Barrett et al, 2003; Brun and Blum, 2012). The antigenic variations compromise the production of conventional vaccines against African trypanosomiasis (Glover et al, 2013). Thus, the only viable option for treatment is HAT chemotherapy involving four drugs, namely, pentamidine, suramin, melarsoprol, and eflornithine/nifurtimox combination therapy (NECT) (Priotto et al., 2009). However, all these drugs suffer from limitations as poor efficacy, acute toxicity, and drug resistance. Recently, utilization of nanocarriers as drug delivery systems is being looked as a promising approach to overcome the problems associated with conventional chemotherapy.

### *15.9.2 NANOCARRIER SYSTEMS TO COMBAT AFRICAN TRYPANOSOMIASIS*

Diminazene aceturate encapsulated in stearylamine (SA)-bearing liposomes showed higher activity, slower absorption, fewer side effects, slower elimination, and longer efficacy as compared to free drugs in *T. brucei* and *T. evansi* infected mouse models. Diminazene aceturate penetrates following fusion between liposomal membrane and the parasite plasma membrane. However, the efficiency of the drugs after intraperitoneal administration may be limited by less stability and inefficient distribution in body not reaching trypanosomes localized in blood, lymph and brain. Activation of complement system and liposome aggregation hampers the pharmacological effect of the drug. To overcome this drawback, modification of liposome surface is carried out by a hydrophilic steric barrier which lowers the interfacial energy. These "stealth" carriers possess high blood residence time in host. Papagiannaros et al. (2005) developed miltefosine liposomal formulation showed enhanced trypanocidal activity and decreased toxicity in vitro compared to free drugs. Nanosuspensions are produced by dispersion of homogenized drug in surfactant which leads to administration of poorly soluble drugs. The surfactants avoid RES recognition prolonging blood circulation time of the nanosuspensions. Melarsoprol nanosuspensions with poloxamer 188 or 407 and mannitol were prepared (Zirar et al., 2008). Tissue distribution in mice showed five- to ninefold high concentrations in liver as that obtained for free drug. However, their size makes it impossible for them to cross the blood–brain barrier and inappropriate for trypanosome-infected animals. For loading hydrophilic drug into hydrophobic systems, the drug was to be transformed to a lipophilic prodrug by grafting a lipophilic group to the drug by a covalent bond or electrostatic bond forming a lipid drug conjugate (LDC). Olbrich et al. (2004) diminazene LDC formulations of particle size between 250 and 450 nm by high-pressure homogenization of stearic acid and fatty acid drug, which forms the matrix, and polysorbate 80 (Tween 80) as surfactant. The emulsifier, Tween 80, was chosen to improve brain targeting properties of the drug by "differential protein adsorption". Tween 80 helps in brain targeting by apolipoprotein (apo) E, apo-I and apo-IV adsorption, which may increase uptake via the LDL receptors at the blood brain barrier. Biodegradable polymers show bright potential for development of sustained release system (SRD) resulting in slow drug release and prolonged prophylactic period. Biodegradable SRDs were prepared by extrusion of poly(DL-lactide) (PLA) and isometamidium chloride or homidium bromide at 25% (drug:polymer wt:wt) under a 3 mm diameter and 3 cm long cylindrical rod. Adult cows exposed to tsetse flies infected with *Trypanosoma congolense* were subcutaneously implanted with SRD of isometamidium or homidium showed significant extension of prophylactic effect (Geerts et al., 1999).

The drawbacks associated with trypanocidal chemotherapies justify the need to design novel nanoparticulate formulations in future for highly efficient and cost effective parasite targeting using conventional trypanocidal drugs.

## 15.10 DENGUE

### *15.10.1 DISEASE OVERVIEW AND CONVENTIONAL CONTROL STRATEGIES*

Dengue virus (DV) infection transmitted to humans by female mosquitoes of *Aedes aegypti* or *Aedes albopictus* affects 2.5 billion people in tropical and subtropical countries. There are no vaccines or antiviral drugs against dengue viruses resulting in controlling of the mosquito vector and prevention of its bite as only effective way to prevent Dengue. Further, extensive use of chemical insecticides leads to serious consequences as mosquito resistance and ecological imbalance. Recently, some studies showed that antivirals as Deoxynojirimycin and its N-alkylated derivatives, flavonoids, fisetin, quercetin and baicalein exhibited activities against dengue virus. Owing to the various existing issues a viable alternative is essential.

### 15.10.2 NANOTECHNOLOGICAL APPROACHES TO CONTROL DENGUE

#### 15.10.2.1 NANOCARRIERS FOR ANTIVIRAL DRUGS

Although application of antivirals to control dengue virus seems to be an interesting approach, these types of compounds commonly show a low bioavailability compromising its therapeutic use. Use of nanocarriers as liposomes, solid lipid nanoparticles, nanoemulsions, and nanocrystals or polymeric nanoparticles is seen as a strategy to overcome the limitations of antiviral drugs. Liposomal formulations include properties of effective antiviral dose, decreased toxicity, and prolonged drug retention. Miller (2012) reported that delivery of iminosugars (all deoxynojirimycin derivatives) via polyunsaturated ER-targeting liposomes (PERLs) showed higher potency against dengue virus at 3-log lower dose as compared to free iminosugars. They prevented accumulation of the dengue virus in organs and serum, inhibited the number of infected cells and viral particles release by primary human monocyte-derived macrophages. Baicalein was incorporated into nanostructured lipid carrier (NLCs) with higher drug level in plasma and longer half-life compared to free drug (Tsai et al., 2012). Silva et al. (2012) showed that administration of Bovine serum albumin nanoparticles (BSA-NPs) with four inactive form of dengue serotypes resulted in induction of anti-DENV IgG antibodies. Similarly, nanoparticles made of chitosan/bacillus Calmette–Guerin (BCG) cell wall components encapsulating the novel dengue nanovaccine produced by UV inactivation of DENV-2 showed ability to generate humoral and cellular immune responses (Hunsawong et al., 2015). Another area of significant interest is use of nanocrystals for improved bioavailability of antidengue drugs. Quercetin, a flavonoid showing antidengue virus activity suffered from disadvantages of low solubility and reduced intestinal absorption. Smith et al. (2011) reported that synthesis of cocrystals led to an improvement in pharmacokinetics of quercetin as compared to that obtained with free drug.

#### 15.10.2.2 NANOINSECTICIDES AS AN ALTERNATIVE FOR CHEMICAL INSECTICIDES

Nanoinsecticides have higher efficiency than conventional insecticides due to higher surface area and increased solubility. Further, lower toxicity of nanoinsectides is an added advantage over conventional insecticides. Duarte et al. (2015) reported significant larvicidal activity of a nanoemulsion containing *Rosmarinus officinalis* essential oil against *A. aegypti* larvae. Nanoparticle biosynthesis had showed significant promise as an alternative to physical and chemical approaches. Banu and Balasubramanian (2015) reported potent larvicidal activity of *Bacillus megaterium*-synthesized silver nanoparticles against *A aegypti*. Similarly, magnetic nanoparticles produced by magnetotactic bacteria *Magnetospirillum gryphiswaldense* show high toxicity to young instars of *A. aegypti* (Murugan et al., 2016). Ghosh and Mukherjee (2013) showed that basil oil nanoemulsion resulted in complete viability loss in larvae following exposure of 90 min. Saxena et al. (2013) reported death of mosquito larvae in water bodies without any ecological toxicity by water soluble carbon nanoparticles (carbon dots) at a dose of 3 μg/mL.

In view of no approved drugs or vaccines against dengue, the use of nanocarrier encapsulated antiviral drugs to combat Dengue Virus (DV) or nanoinsecticides to control mosquito larvae is of paramount importance.

## 15.11 CONCLUSION

Drug-loaded nanocarriers have significant advantages over conventional drug therapy for tropical diseases owing to its virtues of reduced toxicity, enhanced therapeutic efficacy against parasites, better pharmacokinetics of the drug and prolonged drug release in vivo. However, only a few nanocarriers have been used in clinical tests to treat neglected tropical diseases. This is just the beginning of this research area and deserves exploitation of various combinations of drugs in nanocarrier systems to enhance pharmacological effects of conventional drugs to treat neglected tropical diseases. Although nanocarriers offer novel efficient strategies over conventional treatment to control neglected tropical diseases, nanotoxicity is a major area of concern. Biodegradability of nanoparticles, side effects from byproducts and bioaccumulation, change in physicochemical characteristics of material at nanoscale, nanocarriers distribution in the body following systemic administration, development of mathematical and computer models to predict risk and benefits of nanoparticles, safe processes of nanoparticle manufacturing, and disposal and detrimental effects of nanoparticles to environment are few concerns related to the nanomedicine. Scaling up laboratory or pilot technologies of nano drug delivery for commercialization is limited owing to high cost of materials employed and challenges associated to maintaining size and composition of nanomaterials at larger scales. Further, reluctance amongst general public to embrace nanodrugs as an alternative to conventional drug therapy owing to safety concerns is another important challenge that deserves significant attention. The challenges associated with nano drug delivery must be necessarily overcome to develop nanodrugs on the commercial front in near future (Ochekpe et al., 2009; Salouti and Ahangari, 2014). The need of the hour is to create interest in government and pharmaceutical industry to enhance the efficacy of the existing arsenal of conventional drugs against tropical diseases using nanotechnological approaches.

## KEYWORDS


- **nanocarriers**
- **neglected tropical diseases**
- **drug resistance**
- **nanomedicine**
- **nanoparticles**
- **drug delivery**


## REFERENCES


Akbarzadeh, A.; Samiei, M.; Davaran, S. Magnetic Nanoparticles: Preparation, Physical Properties, and Applications in Biomedicine. *Nanoscale Res. Lett.* **2012,** *7*, 144.

Alkadi, H. O. Antimalarial Drug Toxicity: A Review. *Chemotherapy* **2007,** *53*, 385–391.

Allen, T. M.; Sapra, P.; Moase, E.; Moreira, J.; Iden, D. Adventures in Targeting. *J. Liposome. Res.* **2002,** *12*, 5–12.

Ashley, E. A.; Krudsood, S.; Phaiphun, L.; Srivilairit, S.; McGready, R.; Leowattana, W.; Hutagalung, R.; Wilairatana, P.; Brockman, A.; Looareesuwan, S.; Nosten, F. Randomized, Controlled Dose-optimization Studies of Dihydroartemisinin-piperaquine for the Treatment of Uncomplicated Multidrug-resistant Falciparum Malaria in Thailand. *J. Infect. Dis.* **2004,** *190*, 1773–1782.

Ashley, E. A.; McGready, R.; Hutagalung, R.; Phaiphun, L.; Slight, T.; Proux, S.; Thwai, K. L.; Barends, M.; Looareesuwan, S.; White, N. J.; Nosten, F. A Randomized, Controlled Study of a Simple, Once-daily Regimen of Dihydroartemisinin-piperaquine for the

Treatment of Uncomplicated, Multidrug-resistant Falciparum Malaria. *Clin. Infect. Dis.* **2005,** *41*, 425–432.

Banu, A. N.; Balasubramanian, C. Extracellular Synthesis of Silver Nanoparticles Using Bacillus Megaterium Against Malarial and Dengue Vector (Diptera: Culicidae). *Parasitol. Res.* **2015,** *114*, 4069–4079.

Barratt, G. Colloidal Drug Carriers: Achievements and Perspectives. *Cell. Mol. Life Sci.* **2003,** *60*, 21–37.

Barrett, M. P.; Burchmore, R. J.; Stich, A.; Lazzari, J. O.; Frasch, A. C.; Cazzulo, J. J.; Krishna, S. The Trypanosomiases. *Lancet* **2003,** *362*, 1469–1480.

Bianco, A.; Kostarelos, K.; Prato, M. Applications of Carbon Nanotubes in Drug Delivery. *Curr. Opin. Chem. Biol.* **2005,** *9*, 674–679.

Brewster, M. E.; Loftsson, T. Cyclodextrins as Pharmaceutical Solubilizers. *Adv. Drug Deliv. Rev.* **2007,** *59*, 645–666.

Brun, R.; Blum, J. Human African Trypanosomiasis. *Infect Dis Clin North Am.* **2012,** *26*, 261–273.

Buckner, F. S.; Wilson, A. J.; White, T. C.; Van Voorhis, W. C. Induction of Resistance to Azole Drugs in *Trypanosoma cruzi*. *Antimicrob. Agents Chemother.* **1998,** *42*, 3245–3250.

Bummer, P. M. Physical Chemical Considerations of Lipid-based Oral Drug Delivery—Solid Lipid Nanoparticles. *Crit. Rev. Ther. Drug Carrier Syst.* **2004,** *21*, 1–20.

Bzik, D. J.; Li, W. B.; Horii, T.; Inselburg, J. Molecular Cloning and Sequence Analysis of the *Plasmodium falciparum* Dihydrofolate Reductase-thymidylate Synthase Gene. *Proc. Natl. Acad. Sci. U.S.A* **1987,** *84*, 8360–8364.

Choi, H. S.; Frangioni, J. V. Nanoparticles for Biomedical Imaging: Fundamentals of Clinical Translation. *Mol. Imaging* **2010,** *9*, 7290–2010.

Damle, N. K.; Frost, P. Antibody-targeted Chemotherapy with Immunoconjugates of Calicheamicin. *Curr. Opin. Pharmacol.* **2003,** *3*, 386–390.

De Araújo, S. C.; de Mattos, A. C.; Teixeira, H. F.; Coelho, P. M.; Nelson, D. L.; de Oliveira, M. C. Improvement of in vitro Efficacy of a Novel Schistosomicidal Drug by Incorporation into Nanoemulsions. *Int. J. Pharm.* **2007,** *337*, 307–315.

De Castro, S. L. The challenge of Chagas' Disease Chemotherapy: An Update of Drugs Assayed Against *Trypanosoma cruzi*. *Acta. Trop.* **1993,** *53*, 83–98.

Duarte, J. L.; Amado, J. R.; Oliveira, A. E.; Cruz, R. A.; Ferreira, A. M.; Souto, R. N.; Falcão, D. Q.; Carvalho, J. C.; Fernandes, C. P. Evaluation of Larvicidal Activity of a Nanoemulsion of *Rosmarinus officinalis* Essential Oil. *Rev. Bras. Farmacogn.* **2015,** *25*, 189–192.

Duncan, R. Polymer Conjugates as Anticancer Nanomedicines. *Nat. Rev. Cancer* **2006,** *6*, 688–701.

Duncan, R.; Izzo, L. Dendrimer Biocompatibility and Toxicity. *Adv. Drug Deliv. Rev.* **2005,** *57*, 2215–2237.

Duong, S.; Lim, P.; Fandeur, T.; Tsuyuoka, R.; Wongsrichanalai, C. Importance of Protection of Antimalarial Combination Therapies. *Lancet* **2004,** *19*, 1754–1755.

Dykman, L.; Khlebtsov, N. Gold Nanoparticles in Biomedical Applications: Recent Advances and Perspectives. *Chem. Soc. Rev.* **2012,** *41*, 2256–2282.

Feng, Z. Q; Zhong, S. G.; Li, Y. H.; Li, Y. Q.; Qiu, Z. N.; Wang, Z. M.; Li, J.; Dong, L.; Guan, X. H. Nanoparticles as a Vaccine Adjuvant of Anti-idiotypic Antibody Against Schistosomiasis. *Chin. Med. J.* **2004,** *117*, 83–87.

Fonseca, L. B.; Viçosa, A. L.; Mattos, A.; Coelho, P. M. Z.; Araújo, N.; Zamith, H. P. S.; Volpato, N. M.; Nele, M., Pinto, J. C. Development of a Brazilian Nanoencapsulated Drug for Schistosomiasis Treatment. *Vigilância Sanitária em Debate* **2013,** *1*, 85–91.

Freiberg, S.; Zhu, X. X. Polymer Microspheres for Controlled Drug Release. *Int. J. Pharm.* **2004,** *282*, 1–8.

Frezard, F.; Melo, A. L. Evaluation of the Schistosomicidal Efficacy of Liposome-entrapped Oxamniquine. *Rev. Inst. Med. Trop. São. Paulo.* **1997,** *39*, 97–100.

Frezza, T. F.; Madi, R. R.; Banin, T. M.; Pinto, M. C.; Souza, A. L.; Gremião, M. P.; Allegretti SM. Efeito do Praziquantel Incorporado a Lipossomas nos Diferentes Estágios de Desenvolvimento dos ovos de *Schistosoma mansoni*. *Revista de Ciências Farmacêuticas Básica e Aplicada* **2007,** *28*, 209–214.

Garcia, S.; Ramos, C. O.; Senra, J. F.; Vilas-Boas, F.; Rodrigues, M. M., Campos-de-Carvalho, A. C.; Ribeiro-dos-Santos, R.; Soares, M. B. Treatment with Benznidazole During the Chronic Phase of Experimental Chagas' Disease Decreases Cardiac Alterations. *Antimicrob. Agents Chemother.* **2005,** *49*, 1521–1528.

Garnett, M. C. Targeted Drug Conjugates: Principles and Progress. *Adv. Drug Deliv. Rev.* **2001,** *53*, 171–216.

Geerts, S.; Diarra, B.; Eisler, M. C.; Brandt, J.; Lemmouchi, Y.; Kageruka, P.; De Deken, R.; Ndao, M.; Diall, O.; Schacht, E.; Berkvens, D. Extension of the Prophylactic Effect of Isometamidium Against Trypanosome Infections in Cattle Using a Biodegradable Copolymer. *Acta Trop.* **1999,** *73*, 49–58.

Ghosh, P.; Han, G.; De, M.; Kim, C. K.; Rotello, V. M. Gold Nanoparticles in Delivery Applications. *Adv. Drug Deliv. Rev.* **2008,** *60*, 1307–1315.

Ghosh, V.; Mukherjee, A.; Chandrasekaran, N. Formulation and Characterization of Plant Essential Oil Based Nanoemulsion: Evaluation of Its Larvicidal Activity Against *Aedes aegypti*. *Asian J. Chem.* **2013,** *25*, S321–S323.Giao, P. T.; Vries, P. J.; Hung, L. Q.; Binh, T. Q.; Nam, N. V.; Kager, P. A. CV8, a New Combination of Dihydroartemisinin, Piperaquine, Trimethoprim and Primaquine, Compared with Atovaquone–Proguanil Against Falciparum Malaria in Vietnam. *Trop. Med. Int. Health* **2004,** *9*, 209–216.

Glover, L.; Hutchinson, S.; Alsford, S.; McCulloch, R.; Field, M. C.; Horn, D. Antigenic variation in African trypanosomes: The Importance of Chromosomal and Nuclear Context in VSG Expression Control. *Cell Microbiol.* **2013,** *15*, 1984–1993.

Gonzalez-Martin, G. U.; Merino, I.; Rodriguez-Cabezas, M. N.; Torres, M.; Nuñez, R.; Osuna, A. Pharmaceutics: Characterization and Trypanocidal Activity of Nifurtimox-containing and Empty Nanoparticles of Polyethylcyanoacrylates. *J. Pharm. Pharmacol.* **1998,** *50*, 29–35.

Gref, R.; Domb, A.; Quellec, P.; Blunk, T.; Müller, R. H.; Verbavatz, J. M.; Langer, R. The Controlled Intravenous Delivery of Drugs Using Peg-coated Sterically Stabilized Nanospheres. *Adv. Drug Deliv. Rev.* **1995,** *16*, 215–233.

Gulati, V.; Wallace, R. Rafts, Nanoparticles and Neural Disease. *Nanomaterials* **2012,** *2*, 217–250.

Gould, J. E., The Most Neglected Disease. [Online] **2015,** http://www.slate.com (accessed Feb 1, 2016).

Horvath, A. E.; Zierdt, C. H. The Effect of Amphotericin B on *Trypanosoma cruzi* In Vitro and In Vivo. *J. Trop. Med. Hyg.* **1974,** *77*, 144–149.

Hunsawong, T.; Sunintaboon, P.; Warit, S.; Thaisomboonsuk, B.; Jarman, R. G.; Yoon, I. K.; Ubol, S.; Fernandez, S. A Novel Dengue Virus Serotype-2 Nanovaccine Induces Robust Humoral and Cell-mediated Immunity in Mice. *Vaccine* **2015,** *33*, 1702–1710.

Jenning, V.; Lippacher, A.; Gohla, S. H. Medium Scale Production of Solid Lipid Nanoparticles (SLN) by High Pressure Homogenization. *J. Microencapsul.* **2002,** *19*, 1–10.

Karim, K. M.; Mandal, A. S.; Biswas, N.; Guha, A.; Chatterjee, S.; Behera, M.; Kuotsu, K. Niosome: A Future of Targeted Drug Delivery Systems. *J. Adv. Pharm. Technol. Res.* **2010,** *1*, 374.

Kaur, I. P.; Singh, H. Nanostructured Drug Delivery for Better Management of Tuberculosis. *J. Control. Release* **2014,** *184*, 36–50.

Kayser, O.; Kiderlen, A. F. Delivery Strategies for Antiparasitics. *Expert Opin. Investig. Drugs* **2003,** *12*, 197–207.

Khalil, N. M.; de Mattos A. C.; Moraes Moreira Carraro, T. C.; Ludwig, D. B.; Mainardes, R. M. Nanotechnological Strategies for the Treatment of Neglected Diseases. *Curr. Pharm. Des.* **2013,** *19*, 7316–7329.

Klemens, S. P.; Cynamon, M. H.; Swenson, C. E.; Ginsberg, R. S. Liposome-encapsulated-gentamicin Therapy of *Mycobacterium avium* Complex Infection in Beige Mice. *Antimicrob. Agents Chemother.* **1990,** *34*, 967–970.

Kumar, P. V.; Asthana, A.; Dutta, T.; Jain, N. K. Intracellular Macrophage Uptake of Rifampicin Loaded Mannosylated Dendrimers. *J. Drug Target.* **2006,** *145*, 46–56.

Kushwaha, S. K.; Ghoshal, S.; Rai, A. K.; Singh, S. Carbon Nanotubes as a Novel Drug Delivery System for Anticancer Therapy: A Review. *Braz. J. Pharm. Sci.* **2013,** *49*, 629–643.

Leonard, F.; Kulkarni, R. K.; Brandes, G.; Nelson, J.; Cameron, J. J. Synthesis and Degradation of Poly (alkyl α-cyanoacrylates). *J. Appl. Polym. Sci.* **1996,** *10*, 259–272.

Lima, S. A.; Silvestre, R.; Barros, D.; Cunha, J.; Baltazar, M. T.; Dinis-Oliveira, R. J.; Cordeiro-da-Silva, A.; Crucial CD8+ T-lymphocyte Cytotoxic Role in Amphotericin B Nanospheres Efficacy Against Experimental Visceral Leishmaniasis. *Nanomedicine* **2014,** *10*, e1021–1030.

Loftsson, T.; Duchêne, D. Cyclodextrins and Their Pharmaceutical Applications. *Int. J. Pharm.* **2007,** *1*, 329, 1–11.

Lopes, R. M.; Gaspar, M. M.; Pereira, J.; Eleutério, C. V.; Carvalheiro, M., Almeida, A. J.; Cruz, M. E. Liposomes versus lipid nanoparticles: Comparative Study of Lipid-based Systems as Oryzalin Carriers for the Treatment of Leishmaniasis. *J. Biomed. Nanotechnol.* **2014,** *10*, 3647–3657.

Lu, F.; Lim, C. S.; Nam, D. H.; Kim, K.; Lin, K.; Kim, T. S.; Lee, H. W.; Chen, J. H.; Wang, Y.; Sattabongkot, J.; Han, E. T. Mutations in the Antifolate-resistance-associated Genes Dihydrofolate Reductase and Dihydropteroate Synthase in *Plasmodium vivax* Isolates from Malaria-endemic Countries. *Am. J. Trop. Med. Hyg.* **2010,** *83*, 474–479.

Lück, M.; Paulke, B. R.; Schröder, W.; Blunk, T.; Müller, R. H. Analysis of Plasma Protein Adsorption on Polymeric Nanoparticles with Different Surface Characteristics. *J. Biomed. Mater. Res. A.* **1998,** *39*, 478–485.

Mainardes, R. M.; Chaud, M. V.; Gremião, M. P.; Evangelista, R. C. Development of Praziquantel-loaded PLGA Nanoparticles and Evaluation of Intestinal Permeation by the Everted Gut Sac Model. *J. Nanosci. Nanotechnol.* **2006,** *6*, 3057–3061.

Manivannan, R.; Rathnakaran, S. Nanorobots to Find a Permanent Solution for Lymphatic Filariasis or Elephantiasis. *Future Sci.* [Online] 2014. htps://futuresciencemanivannan.files.wordpress.com /2014/08/filaria.pdf (accessed Jan 20, 2017).

Milenic, D. E.; Brady, E. D.; Brechbiel, M. W. Antibody-targeted Radiation Cancer Therapy. *Nat. Rev. Drug Discov.* **2004,** *3*, 488–499.

Miller, J. L.; Lachica, R.; Sayce, A. C.; Williams, J. P.; Bapat, M.; Dwek, R.; Beatty, P. R.; Harris, E.; Zitzmann, N. Liposome-mediated Delivery of Iminosugars Enhances Efficacy Against Dengue Virus In Vivo. *Antimicrob. Agents Chemother.* **2012,** *56*, 6379–6386.

Mohanraj, V. J.; Chen, Y. Nanoparticles: A review. *Trop. J. Pharm. Res.* **2006,** *5*, 561–573.

Molina, J.; Urbina, J.; Gref, R.; Brener, Z.; Júnior, J. M. Cure of Experimental Chagas' Disease by the Bis-triazole DO870 Incorporated into 'Stealth' polyethyleneglycol–polylactide Nanospheres. *J. Antimicrob. Chemother.* **2001,** *47*,101–104.

Morilla, M. J.; Benavidez, P.; Lopez, M. O.; Bakas, L.; Romero, E. L. Development and in vitro Characterisation of a Benznidazole Liposomal Formulation. *Int. J. Pharm.* **2002,** *249*, 89–99.

Mosqueira, V. C.; Loiseau, P. M.; Bories, C.; Legrand, P.; Devissaguet, J. P.; Barratt, G. Efficacy and Pharmacokinetics of Intravenous Nanocapsule Formulations of Halofantrine in *Plasmodium berghei*-infected Mice. *Antimicrob. Agents Chemother.* **2004,** *48*, 1222–1228.

Mourão, S. C.; Costa, P. I.; Salgado, H. R.; Gremião, M. P. Improvement of Antischistosomal Activity of Praziquantel by Incorporation into Phosphatidylcholine-containing Liposomes. *Int. J. Pharm.* **2005,** *295*, 157–162.

Mullaicharam, A. R.; Murthy, R. S. Lung Accumulation of Niosome-entrapped Rifampicin Following Intravenous and Intratracheal Administration in the Rat. *J. Drug Deliv. Sci. Technol.* **2004,** *14*, 99–104.

Muller, R. H.; Keck, C. M. Challenges and Solutions for the Delivery of Biotech Drugs: A Review of Drug Nanocrystal Technology and Lipid Nanoparticles. *J. Biotechnol.* **2004,** *113*, 151–170.

Murugan, K.; Wei, J.; Alsalhi, M. S.; Nicoletti, M.; Paulpandi, M.; Samidoss, C. M.; Dinesh, D.; Chandramohan, B.; Paneerselvam, C.; Subramaniam, J.; Vadivalagan, C. Magnetic Nanoparticles are Highly Toxic to Chloroquine-resistant *Plasmodium falciparum*, Dengue Virus (DEN-2), and Their Mosquito Vectors. *Parasitol. Res.* **2016,** 1–8.

Nayak, A. P.; Tiyaboonchai, W.; Patankar, S.; Madhusudhan, B.; Souto, E. B. Curcuminoids-loaded Lipid Nanoparticles: Novel Approach Towards Malaria Treatment. *Colloids Surf. B* **2010,** *81*, 263–273.

Noedl, H.; Se, Y.; Schaecher, K.; Smith, BL.; Socheat, D.; Fukuda, M. M. Evidence of Artemisinin-resistant Malaria in Western Cambodia. *N. Engl. J. Med.* **2008,** *359*, 2619–2620.

Ochekpe, N. A.; Olorunfemi, P. O.; Ngwuluka, N. C. Nanotechnology and Drug Delivery Part 2: Nanostructures for Drug Delivery. *Trop. J. Pharm. Res.* **2009,** *8*, 275–287.

Olbrich, C.; Gessner, A.; Schröder, W.; Kayser, O.; Müller, R. H. Lipid–drug Conjugate Nanoparticles of the Hydrophilic Drug Diminazene—Cytotoxicity Testing and Mouse Serum Adsorption. *J. Control. Release* **2004,** *96*, 425–435.

Omwoyo, W. N.; Melariri, P.; Gathirwa, J. W.; Oloo, F.; Mahanga, G. M.; Kalombo, L.; Ogutu, B.; Swai, H. Development, Characterization and Antimalarial Efficacy of Dihydroartemisinin Loaded Solid Lipid Nanoparticles. *Nanomedicine* **2016,** *12*, 801–809.

O'shaughnessy, J. A.; Blum, J. L.; Sandbach, J. F.; Savin, M.; Fenske, E.; Hawkins, M. J.; Baylor-Charles, A. Weekly Nanoparticle Albumin Paclitaxel (Abraxane) Results In Long-term Disease Control in Patients with Taxane-refractory Metastatic Breast Cancer. *Breast Cancer Res. Treat.* **2004,** *88*, S65.

Pandey, R.; Khuller, G. K. Subcutaneous Nanoparticle-based Antitubercular Chemotherapy in an Experimental Model. *J. Antimicrob. Chemo.* **2004,** *54*, 266–268.

Papagiannaros, A.; Bories, C.; Demetzos, C.; Loiseau, P. M. Antileishmanial and Trypanocidal Activities of New Miltefosine Liposomal Formulations. *Biomed. Pharmacother.* **2005,** *59*, 545–550.

Peters, K.; Leitzke, S.; Diederichs, J. E.; Borner, K.; Hahn, H.; Müller, R. H.; Ehlers, S. Preparation of a Clofazimine Nanosuspension for Intravenous Use and Evaluation of its Therapeutic Efficacy in Murine *Mycobacterium avium* Infection. *J. Antimicrob. Chemo.* **2000,** *45*, 77–83.

Petras, J. M.; Young, G. D.; Bauman, R. A.; Kyle, D. E; Gettayacamin, M.; Webster, H. K.; Corcoran, K. D.; Peggins, J. O.; Vane, M. A.; Brewer, T. G. Arteether-induced Brain Injury in *Macaca mulatta*. I. The Precerebellar Nuclei: The Lateral Reticular Nuclei, Paramedian Reticular Nuclei, and Perihypoglossal Nuclei. *Anat. Embryol.* **2000,** *201*, 383–397.

Priotto, G.; Kasparian, S.; Mutombo, W.; Ngouama, D.; Ghorashian, S.; Arnold, U.; Ghabri, S.; Baudin, E.; Buard,V.; Kazadi-Kyanza, S.; Ilunga, M. Nifurtimox-eflornithine Combination Therapy for Second-Stage African *Trypanosoma brucei* Gambiense Trypanosomiasis: A Multicentre, Randomised, Phase III, Non-inferiority Trial. *Lancet* **2009,** *374*, 56–64.

Rao, V. M.; Nerurkar, M.; Pinnamaneni, S.; Rinaldi, F.; Raghavan, K. Co-solubilization of Poorly Soluble Drugs by Micellization and Complexation. *Int. J. Pharm.* **2006,** *319*, 98–106.

Roychoudhury, J.; Sinha, R.; Ali, N. Therapy with Sodium Stibogluconate in Stearylamine-bearing Liposomes Confers Cure Against SSG-resistant *Leishmania donovani* in BALB/c Mice. *PLOS One* **2011,** *6*, e17376.

Salouti, M.; Ahangari, A. Nanoparticle based Drug Delivery Systems for Treatment of Infectious Diseases. In *Application of Nanotechnology in Drug Delivery*, 1st ed.; Sezer, A. D., Ed.; InTech Publishing House: Vienna, 2014; pp 155–192.

Santos-Magalhães, N. S.; Mosqueira, V. C. Nanotechnology Applied to the Treatment of Malaria. *Adv. Drug Deliv. Rev.* **2010,** *62*, 560–575.

Saravanan, M.; Duche, K.; Asmelash, T.; Gebreyesus, A.; Negash, L.; Tesfay, A.; Hailekiros, H.; Niguse, S.; Gopinath, V.; K Barik S. "Nano-Biomaterials"—A New Approach Concerning Multi-Drug Resistant Tuberculosis (MDR-TB). *Pharm. Nanotechnol.* **2015,** *3*, 5–18.

Saxena, M.; Sonkar, S. K.; Sarkar, S. Water Soluble Nanocarbons Arrest the Growth of Mosquitoes. *RSC Adv.* **2013,** *3*, 22504–22508.

Scherer, D.; Robinson, J. R.; Kreuter, J. Influence of Enzymes on the Stability of Polybutylcyanoacrylate Nanoparticles. *Int. J. Pharm.* **1994,** *101*, 165–168.

Scherphof, G. L.; Velinova, M.; Kamps, J.; Donga, J.; van der Want, H.; Kuipers, F.; Havekes, L.; Daemen, T. Modulation of Pharmacokinetic Behavior of Liposomes. *Adv. Drug Deliv. Rev.* **1997,** *24*, 179–191.

Shafiq, S.; Shakeel, F.; Talegaonkar, S.; Ahmad, F. J.; Khar, R. K.; Ali, M. Development and Bioavailability Assessment of Ramipril Nanoemulsion Formulation. *Eur. J. Pharm. Biopharm.* **2007,** *66*, 227–243.

Sharma, A.; Sharma, S.; Khuller, G. K. Lectin-functionalized Poly (Lactide-co-glycolide) Nanoparticles as Oral/Aerosolized Antitubercular Drug Carriers for Treatment of Tuberculosis. *J. Antimicrob. Chemo.* **2004,** *54*, 761–766.

Shegokar, R.; Al Shaal, L.; Mitri, K. Present Status of Nanoparticle Research for Treatment of Tuberculosis. *J. Pharm. Pharm. Sci.* **2011,** *14*, 100–116.

Shekalaghe, S. A.; ter Braak, R.; Daou, M.; Kavishe, R.; van den Bijllaardt, W.; van den Bosch, S.; Koenderink, J. B.; Luty, A. J.; Whitty, C. J.; Drakeley, C.; Sauerwein, R. W. In Tanzania, Hemolysis After a Single dose of Primaquine Coadministered with an Artemisinin is Not Restricted to Glucose-6-phosphate Dehydrogenase-deficient (G6PD A−) Individuals. *Antimicrob. Agents Chemother.* **2010,** *54*, 1762–1768.

Silva, E. F.; Orsi, M.; Andrade, Â. L.; Domingues, R. Z.; Silva, B. M.; de Araújo, H. R.; Pimenta, P. F.; Diamond, M. S.; Rocha, E. S.; Kroon, E. G.; Malaquias, L. C. A Tetravalent Dengue Nanoparticle Stimulates Antibody Production in Mice. *J. Nanobiotechnol.* **2012,** *10*, 13.

Smith, A. J.; Kavuru, P.; Wojtas, L.; Zaworotko, M. J.; Shytle, R. D. Cocrystals of Quercetin With Improved Solubility and Oral Bioavailability. *Mol. Pharm.* **2011,** *8*, 1867–1876.

Smola, M.; Vandamme, T.; Sokolowski, A. Nanocarriers as Pulmonary Drug Delivery Systems to Treat and to Diagnose Respiratory and Nonrespiratory Diseases. *Int. J. Nanomedicine* **2008,** *3*, 1–19.

Sosnik, A.; Carcaboso, Á. M.; Glisoni, R. J.; Moretton, M. A.; Chiappetta, D. A. New Old Challenges in Tuberculosis: Potentially Effective Nanotechnologies in Drug Delivery. *Adv. Drug Deliv. Rev.* **2010,** *62*, 547–559.

Sosa-Estani S., Segura, E. L. Integrated Control of Chagas Disease for Its Elimination as Public Health Problem: A Review. *Mem. Inst. Oswaldo Cruz.* **2015,** *110*, 289–98.

Sundar, S.; Chakravarty, J.; Agarwal, D.; Rai, M.; Murray, H. W. Single-dose Liposomal Amphotericin B for Visceral Leishmaniasis in India. *N. Engl. J. Med.* **2010,** *362*, 504–512.

Taylor, W. R.; White, N. J. Antimalarial Drug Toxicity. *Drug Saf.* **2004,** *27*, 25–61.

Torchilin, V. P. Recent Advances with Liposomes as Pharmaceutical Carriers. *Nat. Rev. Drug Discov.* **2005,** *4*,145–160.

Trouiller, P.; Olliaro, P.; Torreele, E.; Orbinski, J.; Laing, R.; Ford, N. Drug Development for Neglected Diseases: A Deficient Market and a Public-health Policy Failure. *Lancet* **2002,** *359*, 2188–2194.

Tsai, M. J.; Wu, P. C.; Huang, Y. B.; Chang, J. S.; Lin, C. L.; Tsai, Y. H.; Fang, J. Y. Baicalein Loaded in Tocol Nanostructured Lipid Carriers (Tocol NLCs) for Enhanced Stability and Brain Targeting. *Int. J. Pharm.* **2012,** *423*, 461–470.

Urbina, J. A. Chemotherapy of Chagas Disease. *Curr. Pharm. Des.* **2002,** *8*, 287–295.

Urbina, J. A. Specific Treatment of Chagas Disease: Current Status and New Developments. *Curr. Opin. Infect. Dis.* **2001,** *14*, 733–741.

Vijaykadga, S.; Rojanawatsirivej, C.; Cholpol, S.; Phoungmanee, D.; Nakavej, A.; Wongsrichanalai, C. In vivo Sensitivity Monitoring of Mefloquine Monotherapy and Artesunate–Mefloquine Combinations for the Treatment of Uncomplicated Falciparum Malaria in Thailand in 2003. *J. Trop. Med.* **2006,** *11*, 211–219.

WHO Malaria Report 2016. http://www.who.int/malaria/en (accessed Feb 7, 2017).

WHO Leishmaniasis Report 2016. http://www.who.int/mediacentre/factsheets/fs375/en (accessed Feb 9, 2017).

Yaméogo, J. B.; Gèze, A.; Choisnard, L.; Putaux, J. L.; Gansané, A.; Sirima, S. B.; Semdé, R.; Wouessidjewe, D. Self-assembled Biotransesterified Cyclodextrins as Artemisinin Nanocarriers–I: Formulation, Lyoavailability and In Vitro Antimalarial Activity Assessment. *Eur. J. Pharm. Biopharm.* **2012,** *80*, 508–517.

Yoshihara, E.; Tachibana, H.; Nakae, T. Trypanocidal Activity of the Stearylamine-bearing Liposome In Vitro. *Life Sci.* **1987,** *40*, 2153–2159.

Zirar, S. B.; Astier, A.; Muchow, M.; Gibaud, S. Comparison of Nanosuspensions and Hydroxypropyl-β-Cyclodextrin Complex of Melarsoprol: Pharmacokinetics and Tissue Distribution in Mice. *Eur. J. Pharm. Biopharm.* **2008,** *70*, 649–656.

Zwang, J.; Ashley, E. A.; Karema, C.; D'Alessandro, U.; Smithuis, F.; Dorsey, G.; Janssens, B.; Mayxay, M.; Newton, P.; Singhasivanon, P.; Stepniewska, K. Safety and Efficacy of Dihydroartemisinin-piperaquine in Falciparum Malaria: A Prospective Multi-centre Individual Patient Data Analysis. *PLOS One* **2009,** *4*, e6358.

Zumla, A.; George, A.; Sharma, V.; Herbert, R. H.; Ilton, B. M.; Oxley, A; Oliver, M. The WHO 2014 Global Tuberculosis Report—Further to Go. *Lancet Glob. Health* **2015,** *3*, e10–e12.

# CHAPTER 16

# SELF-ASSEMBLY OF SUCROSE AND TREHALOSE ALKYL ETHERS INTO NANOPARTICLES AND NANORODS UNDER AQUEOUS CONDITIONS

JUN-ICHI KADOKAWA*

*Department of Chemistry, Biotechnology, and Chemical Engineering, Graduate School of Science and Engineering, Kagoshima University, 1-21-40 Korimoto, Kagoshima 890-0065, Japan*

*E-mail: kadokawa@eng.kagoshima-u.ac.jp*

## CONTENTS

## ABSTRACT

In this chapter, the hierarchically organized self-assembly of sucrose and trehalose alkyl ether amphiphiles to construct nanoparticles and nanorods under aqueous conditions is described. Because amphiphiles contain antagonistic hydrophilic and hydrophobic moieties in the same molecule, their regularly organized self-assembly properties in water help in the formation of controlled aggregate morphologies. Carbohydrate-based amphiphiles composed of hydrophilic carbohydrate moieties and hydrophobic carbon chains have been extensively studied, and a large variety of self-assembly morphologies have been observed. As described in this chapter, the self-assembly properties of disaccharide alkyl ethers such as sucrose and trehalose derivatives have been investigated under aqueous conditions. These amphiphiles formed regularly organized nanoparticles and nanorods under aqueous conditions depending on their chemical structures. Their hierarchical self-assembly processes have been investigated using several analytical methods.

## 16.1 INTRODUCTION

Amphiphilic molecules, known as amphiphiles contain antagonistic, hydrophilic, and hydrophobic moieties in the same molecule. In aqueous media, such molecules typically self-assemble into regularly organized aggregates such as spherical micelles, cylindrical micelles, spherical vesicles, and planar bilayers depending on the molecular shape and solution conditions (Fig. 16.1) (Fuhrhop and Wang, 2004; Ge and Liu, 2009; Kwon, 2003; Shimizu, 2002; Shimizu et al., 2005). Glycolipids, which are natural amphiphiles, composed of carbohydrates as the hydrophilic moieties exhibit important in vivo functions in biological systems (Corti et al., 2007; Kitamoto et al., 2009). Hence, synthetic carbohydrate-based amphiphiles have been extensively studied, and a large variety of self-assembly morphologies have been observed (Fuhrhop and Wang, 2004; Shimizu, 2002; Shimizu et al., 2005). For example, the self-assembly properties and applications of disaccharide fatty acid esters as the surfactants have been investigated, where the hydrophilic carbohydrate moieties are typically connected to the hydrophobic long-chain alkyl fatty acids through ester linkages (Kralova and Sjoblom, 2009).

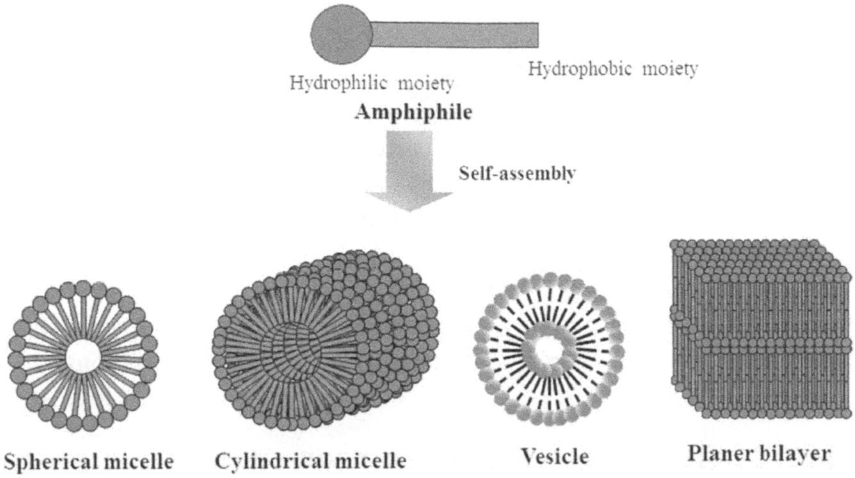


**FIGURE 16.1** Representative morphologies of regularly organized aggregates from amphiphiles.

Moreover, sucrose fatty acid esters are one of the most well-known synthetic carbohydrate-based amphiphiles; they have practically been used as the additives in various food industries in the form of emulsifiers, because sucrose is the representative naturally occurring small carbohydrate, that is, disaccharide (Awad and Sato, 2001, 2002; Hodate et al., 1997; Katsuragi et al., 2001; Queneau et al., 2008). The self-assembly properties of sucrose fatty acid esters depend on the length and type of fatty acid chains (Goodby et al., 2007; Molinier et al., 2006).

Trehalose is a nonreducing disaccharide similar to sucrose in which the two glucose residues are linked through an α,α-1,1′-glucosidic linkage (Ohtake and Wang, 2011). Because trehalose is industrially produced from starch via an enzymatic approach and available at lower costs, it is regarded as a new renewable resource comparable to sucrose (Feofilova et al., 2014). Naturally occurring glycolipids containing a trehalose residue are found in nature, for example, trehalose 6,6′-dimycolate, the most widely studied trehalolipid (Franzetti et al., 2010).

Compared to the case of disaccharide fatty acid ester amphiphiles, there are fewer studies on the self-assembly properties of amphiphilic disaccharide ethers, in which the carbohydrate residues such as sucrose and trehalose contain alkyl chains connected by ether linkages (Gagnaire et al., 2000; Queneau et al., 2001). Because an ether linkage is more stable than an ester linkage under aqueous conditions, new ether-linked carbohydrate amphiphiles have potential applications as emulsifiers and additives in various food industries. Furthermore, the author believes that ether-linked carbohydrate amphiphiles exhibit self-assembly properties different from those of ester-linked carbohydrate amphiphiles, because of the simpler structure of the ether linkage and the exclusion of the possibility of hydrogen bonding by the carbonyl group.

With this viewpoint and background, the self-assembly properties of sucrose and trehalose alkyl ethers are described in this chapter. These amphiphiles formed regularly organized nanostructures such as nanoparticles and nanorods under aqueous conditions by simple operations.

## 16.2 SYNTHESIS OF SUCROSE AND TREHALOSE ALKYL ETHERS

Two types of sucrose and trehalose alkyl ether amphiphiles were synthesized from sucrose or trehalose, respectively, that is, sucrose or trehalose 6-*O*-(6′-*O*-)monoalkyl ethers and trehalose 6,6′-di-*O*-alkyl ethers. In the synthesis of sucrose monoalkyl ethers with different chain lengths, the products were obtained as the mixtures of 6-*O*- and 6′-*O*-alkyl ethers because of the following order of reactivity of primary hydroxy groups in sucrose: 6>6′>>1′ where the reactivity between 6 and 6′ is not significantly different; however, the reactivity of 1′ is much lower than that of the others.

Sucrose 6-*O*- and 6′-*O*-alkyl ether mixtures with different chain lengths were synthesized as follows (Fig. 16.2) (Kanemaru et al., 2010; Tanaka et al., 2013). The common intermediate **1** with a hydroxy group at the 6- or 6′-position was first prepared by successive tritylation at the 6- or 6′-position of sucrose, benzylation of the other hydroxy groups, and detritylation (Mach et al., 2001; Ohtake, 1970; Sofian and Lee, 2001, 2003). Then, etherification of the intermediate was performed with octyl, decyl, dodecyl, tetradecyl, and hexadecyl bromides in the presence of NaH in DMF at high temperatures. The products were debenzylated by catalytic hydrogenation, affording sucrose 6-*O*- and 6′-*O*-octyl, decyl, dodecyl, tetradecyl, and hexadecyl ether mixtures (**C8-**, **C10-**, **C12-**, **C14-**, and **C16-mixtures**). The molar ratios of 6-ether to 6′-ether in the products were almost 1:1 or higher because of the higher reactivity of the 6-position than that of the 6′-position.

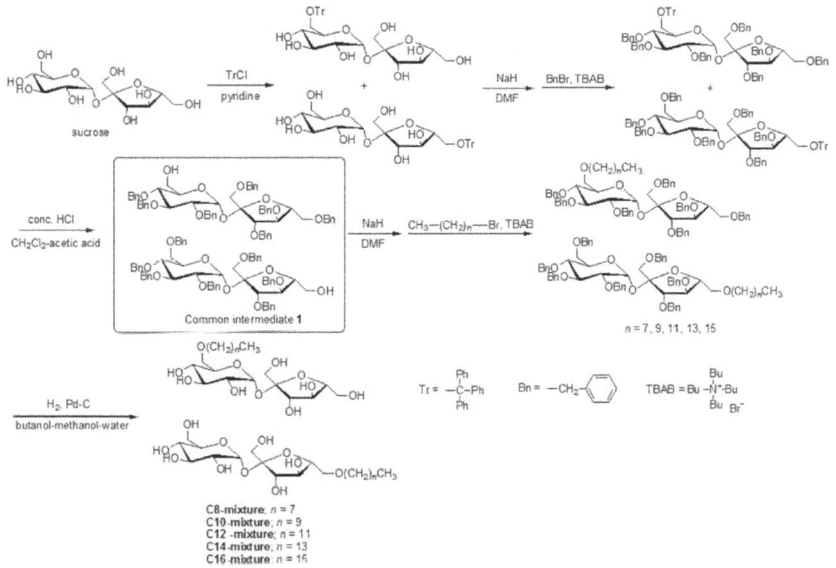


**FIGURE 16.2** Synthesis of sucrose 6-*O*- and 6′-*O*-alkyl ether mixtures from sucrose.

In the following study, the author found that monoetherification of the common intermediate **2** bearing two hydroxy groups at the 6- and 6′-positions, prepared by the successive ditritylation of sucrose, benzylation of the other hydroxy groups, and detritylation, with octyl or hexadecyl bromide predominantly occurred at the 6-position under the selected conditions (Fig. 16.3) (Ohkawabata et al., 2012). Furthermore, the major products 6-ether derivatives could be separated from the minor products 6′-ether derivatives by simple silica gel column chromatography, isolating sucrose 6-*O*-octyl and hexadecyl ethers (**C8-** and **C16-mono(S)**s).

Trehalose 6-*O*-alkyl and 6,6′-di-*O*-alkyl ethers were synthesized from the common intermediate **3** bearing two hydroxy groups at the 6- and 6′-positions (Fig. 16.4) (Kanemaru et al., 2012a). The intermediate was prepared by successive ditritylation at the 6- and 6′-positions of trehalose, benzylation of the other hydroxy groups, and detritylation. To isolate the first ditritylated derivative from the crude products containing unreacted trehalose by a simple silica gel column chromatography, the products were once acetylated using a mixture of acetic anhydride/pyridine. Then, 6,6′-di-*O*-trityltrehalose hexaacetate was isolated from the acetylated products by silica gel column chromatography, followed by deacetylation, to obtain the desired ditrityl trehalose. Monoetherification of the resulting intermediate

**3** with octyl, decyl, dodecyl, tetradecyl, and hexadecyl bromides, followed by debenzylation, produced the five trehalose 6-*O*-alkyl ethers with different chain lengths (**C8-**, **C10-**, **C12-**, **C14-**, and **C16-mono(T)**s). On the other hand, the dietherification of the common intermediate **3** with octyl bromide was successfully achieved under different conditions, followed by debenzylation, yielding trehalose 6,6′-di-*O*-octyl ether (**C8-di**) (Fig. 16.4) (Kanemaru et al., 2012b). Trehalose 6,6′-di-*O*-butyl and hexyl ethers were synthesized by the same procedure.

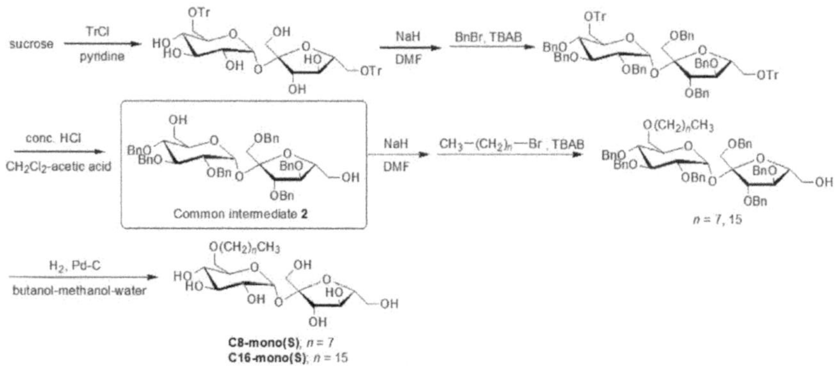


**FIGURE 16.3** Synthesis of sucrose 6-*O*-octyl and 6-*O*-hexadecyl ethers from sucrose.

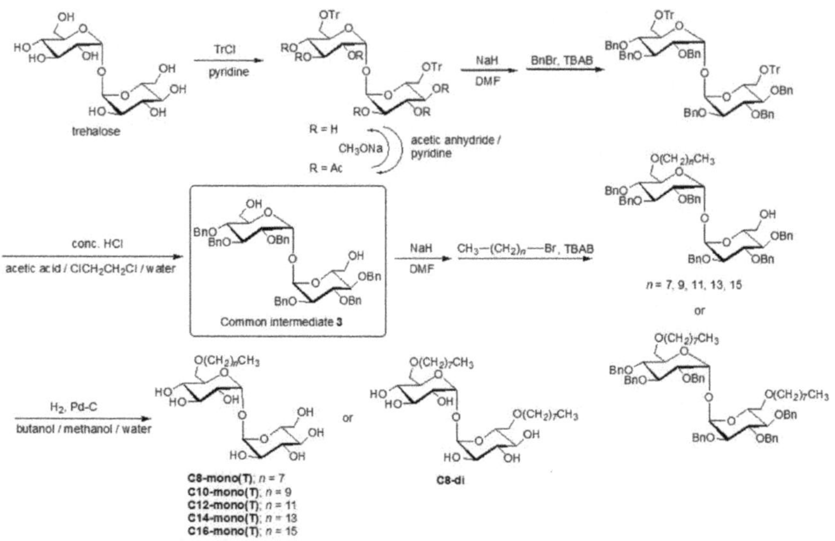


**FIGURE 16.4** Synthesis of trehalose 6-*O*-alkyl and 6,6′-di-*O*-alkyl ethers from trehalose.

## 16.3 SELF-ASSEMBLY PROPERTIES OF SUCROSE ALKYL ETHERS UNDER AQUEOUS CONDITIONS

The self-assembly properties of sucrose 6-*O*- and 6′-*O*-alkyl ether mixtures were investigated under aqueous conditions (Kanemaru et al., 2010; Tanaka et al., 2013). The scanning electron microscopy (SEM) images of their samples on aluminum plates prepared by drying a dispersion of the materials in water ($1 \times 10^{-5}$ mol/L) showed particle-like nanoaggregates (Fig. 16.5). The average diameters of the nanoparticles from **C8-**, **C10-**, **C12-**, and **C14-mixtures** were 169, 104, 173, and 156 nm, respectively, whereas nanoparticles with a significantly different average diameter of 51 nm and larger aggregates were formed by further assembly of the nanoparticles, as detected in the SEM image of **C16-mixture**. Moreover, the standard deviation in the diameters of this sample was smaller than that in the diameters of the other particles (4.34 and 16–45, respectively). The hierarchical structures of the nanoparticles were further confirmed by transmission electron microscopy (TEM) measurements (Fig. 16.6). The TEM image of **C8-mixture** showed vesicle-type particles with diameters of ca. 150 nm, whereas the TEM image of **C16-mixture** showed some particles with diameters smaller than 50 nm, which did not show the vesicle-like morphology. The SEM and TEM results indicate that the self-assembly of **C16-mixture** was completely different from that of the mixtures of other derivatives under aqueous conditions.

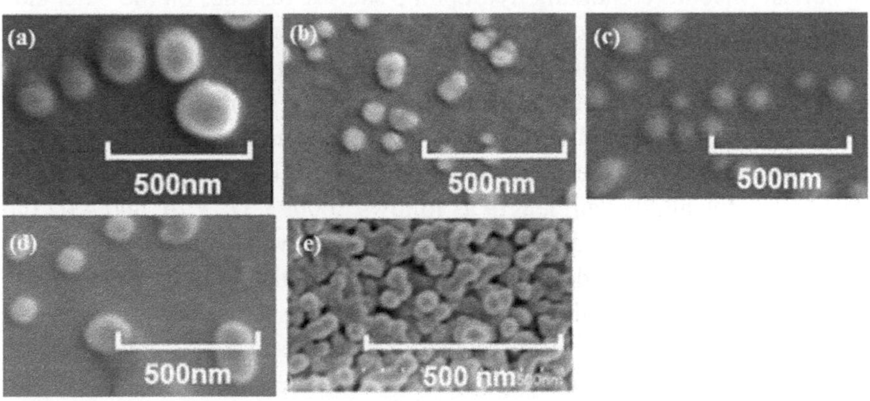


**FIGURE 16.5** SEM images of samples prepared from dispersions of sucrose 6-*O*- and 6′-*O*-alkyl ether mixtures in water ($1.0 \times 10^{-5}$ mol/L): (a) **C8-**,(b) **C10-**, (c) **C12-**, (d) **C14-**, **and** (e) **C16-mixtures**.

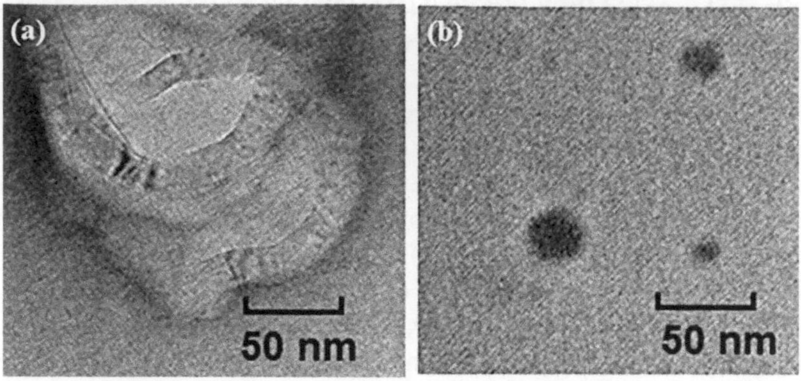


**FIGURE 16.6** TEM images of samples prepared from dispersions of sucrose 6-*O*- and 6′-*O*-alkyl ether mixtures in water ($1.0 \times 10^{-5}$ mol/L): (a) **C8-** and (b) **C16-mixtures**.

The powder X-ray diffraction (XRD) of these derivatives was measured in the drying state from aqueous dispersions to confirm the self-assembly processes in greater detail. The XRD profile of **C16-mixture** exhibited diffraction peaks assignable to the (111), (200), (311/222), (400), and (600) Bragg reflections of a face-centered cubic (FCC) structure (El-Safty and Hanaoka, 2004; Liu et al., 2008; Tang et al., 2007). The diameter of a sphere in the FCC structure was calculated to be 5.1 nm from the XRD pattern. As the molecular sizes of the sucrose and hexadecyl chains of **C16-mixture** were calculated to be ca. 3.0 nm in length, the mixture potentially formed spherical micelles containing hydrophilic sucrose residues on the outer side and hydrophobic hexadecyl chains on the inner side, resulting in a diameter of ~6 nm, consistent with the XRD result. On the other hand, the XRD profiles of the mixtures of other derivatives showed diffraction peaks ascribable to (001) and (002) Bragg reflections of the lamellar patterns. The XRD patterns showed that the width of each lamellar layer increased with an increase in the alkyl chain length as follows: 3.00, 3.37, 3.76, and 4.02 nm for **C8-**, **C10-**, **C12-**, and **C14-mixtures**, respectively. These XRD results of the mixtures of four derivatives indicated that the lamellar planes were first probably formed with alternating hydrophilic sucrose and hydrophobic alkyl chain layers, which further constructed the vesicle-type nanoparticles.

The dynamic light scattering (DLS) of the materials was also measured to evaluate the self-assembly properties in aqueous dispersions. The DLS profile of the aqueous dispersion of **C16-mixture** in $1.0 \times 10^{-3}$ mol/L showed an average diameter of 7.1 nm corresponding to a spherical micelle. On the other hand, the DLS results of the aqueous dispersions of **C8-**, **C10-**, and **C12- mixtures** in the same concentration showed the formation of larger aggregates with average diameters of 106.4, 126.5, and 102.8 nm, respectively, probably corresponding to the vesicle-like particles formed from the lamellar planes. Interestingly, the DLS profile of the aqueous dispersion of **C14-mixture** in the same concentration showed much smaller particles with an average diameter of 6.8 nm similar to **C16-mixture**. This result indicates that **C14-mixture** formed spherical micelles at this concentration in water, not consistent with the hierarchical structure in the solid state as detected in the SEM and XRD results. The average diameter increased to ~82 nm when the DLS of the aqueous dispersion of **C14-mixture** was analyzed at a higher concentration ($1.0 \times 10^{-2}$ mol/L). It was further supported by the DLS measurement; the diameters of the particles in the aqueous dispersions of **C14-mixture** reversibly changed depending on the concentration. Therefore, the XRD profile of this mixture in the solid state prepared by drying the aqueous dispersion showed a lamellar pattern.

Based on the analytical results, the following self-assembly processes are proposed for the formation of nanoparticles under aqueous conditions (Fig. 16.7). **C16-mixture** forms spherical micelles with a diameter of ~5–7 nm in water. The micelles regularly organize according to the FCC structure during the drying to construct the nanoparticles with a diameter of ~50 nm (Sathe et al., 2009). Several nanoparticles further assemble to form larger aggregates. On the other hand, the self-assembly of **C8-**, **C10-**, and **C12-mixtures** under aqueous conditions was completely different; the vesicle-like particles were primarily formed based on the lamellar planes, which are constructed with the self-organized alternating hydrophilic sucrose and hydrophobic alkyl layers. Furthermore, **C14-mixture** showed concentration-induced micelle–lamellar transition behavior in the aqueous dispersion (Gummel et al., 2011; Takahashi et al., 2010).

The two pure sucrose derivatives, that is, **C8-mono(S)** and **C16-mono(S)**, were also subjected to self-assembly under aqueous conditions (Ohkawabata et al., 2012). In the SEM image of a sample of **C16-mono(S)** on an aluminum plate, which was prepared by drying its dispersion in water ($1.0 \times 10^{-5}$ mol/L), nanoparticles were observed with an average diameter of 51 nm (Fig. 16.8a). Although this value is similar to that observed in a sample of **C16-mixture** (Fig. 16.5e), the standard deviations in the diameters of both the samples were significantly different, 12.8 and 4.34 for the pure sample and mixture, respectively.

Moreover, the SEM image of **C16-mono(S)** did not show larger aggregates on further assembly of several particles as observed in **C16-mixture**. The TEM image of **C16-mono(S)** showed vesicle-like particles, which further assembled to form larger aggregates (Fig. 16.9a), whereas such a vesicle morphology of the particles was not observed in the TEM image of **C16-mixture**(Fig. 16.6b). The SEM image of a sample of **C8-mono(S)** showed particle-type nanoaggregates (Fig. 16.8b), but the average diameter (117 nm) was larger than that of **C16-mono(S)**. The TEM image of **C8-mono(S)** showed vesicles and their assemblies, constructed from the lamellar planes of 10 odd alternating black (hydrophilic sucrose) and white (hydrophobic octyl) layers (Fig. 16.9b). It was also observed that the lamellar planes at the interfacial area between the two vesicles were fused.

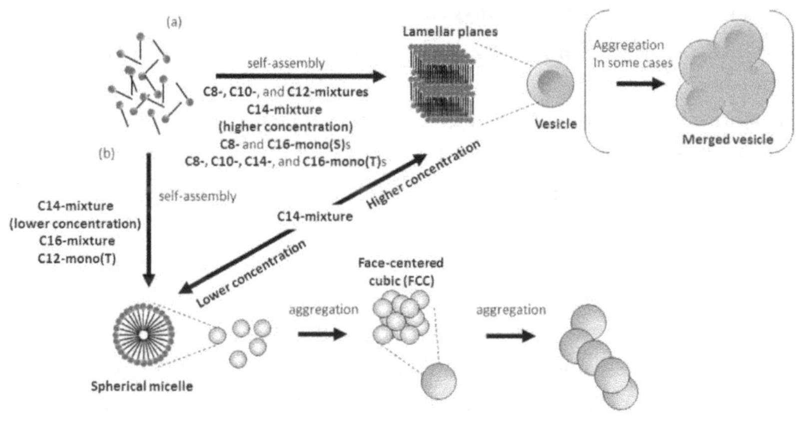


**FIGURE 16.7** Plausible self-assembly processes to form (a) lamellar planes and vesicles and (b) spherical micelles.

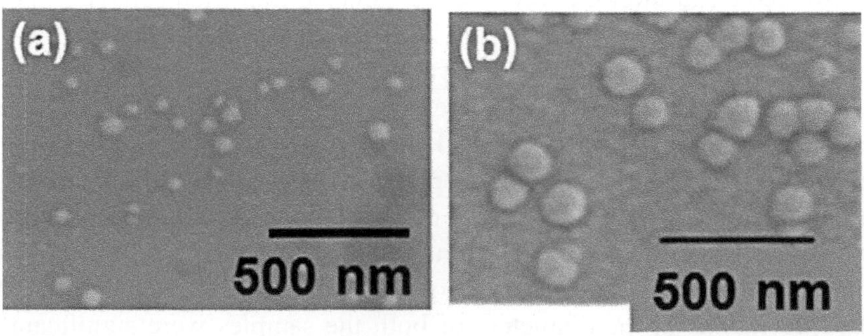


**FIGURE 16.8** SEM images of samples prepared from dispersions of sucrose 6-*O*-alkyl ethers in water ($1.0 \times 10^{-5}$ mol/L): (a) **C16-mono(S)** and (b) **C8-mono(S)**.

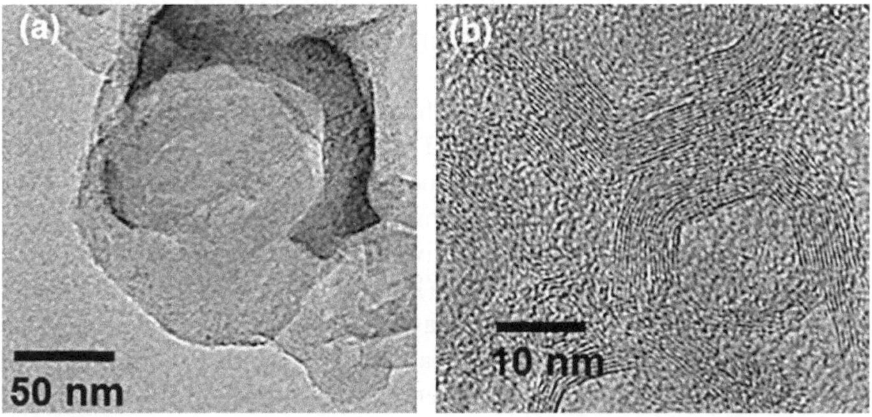


**FIGURE 16.9** TEM images of samples prepared from dispersions of sucrose 6-*O*-alkyl ethers in water ($1.0 \times 10^{-5}$ mol/L): (a) **C16-mono(S)** and (b) **C8-mono(S)**.

The XRD profiles of **C8-** and **C16-mono(S)**s showed lamellar diffraction patterns, and the widths of the lamellar layers of the former and latter derivatives were calculated to be 2.9 and 3.9 nm, respectively. The DLS results of the dispersions of **C8-** and **C16-mono(S)**s ($1.0 \times 10^{-5}$ mol/L) exhibited monomodal profiles with average diameters of ca. 100 nm.

The analytical results indicate that the self-assembly of **C8-** and **C16-mono(S)**s under aqueous conditions occurred equally similar to that of the **C8-**, **C10-**, and **C12-mixtures** (Fig. 16.7). Furthermore, in this case, some of the vesicles were fused further by the fusion of the planes at the interfacial area. The particles observed in the SEM images and DLS profiles probably corresponded to a sole vesicle or fused vesicles. The comprehensive results of the self-assembly processes of sucrose alkyl ethers under aqueous conditions indicate that the mixing of the stereoisomer sucrose 6′-*O*-hexadecyl ether with 6-*O*-ether induced a self-assembly property different from that of the latter derivative alone; however, such a mixing did not affect the self-assembly property of the octyl derivative mixture (**C8-mixture**). This also indicates that the subtle difference in the chemical structures of the derivatives strongly affects the self-assembly properties under aqueous conditions.

## 16.4 SELF-ASSEMBLY PROPERTIES OF TREHALOSE ALKYL ETHERS UNDER AQUEOUS CONDITIONS

The self-assembly properties of trehalose 6-*O*-monoalkyl ethers (**C8-**, **C10-**, **C12-**, **C14-**, and **C16-mono(T)**s) under aqueous conditions were investigated in the same manner as sucrose alkyl ethers (Kanemaru et al., 2012a). The SEM images of the samples on aluminum plates prepared by drying the dispersions of the derivatives in water ($1.0 \times 10^{-5}$ mol/L) showed nanoparticle morphologies (Fig. 16.10). The average diameters of the particles of **C8-**, **C10-**, **C14-**, and **C16-mono(T)**s decreased in the following order: 136.4, 114.1, 97.4, and 78.1 nm (Fig. 16.10a,b,d,e). On the other hand, the average diameter of the particles of **C12-mono(T)** was much smaller, 58.3 nm (Fig. 16.10c). The TEM images of the former four samples showed several vesicle-type particles with diameters of ca. 30–60 nm as shown in Fig. 16.11b for **C16-mono(T)**; these particles were constructed from the lamellar planes formed by the self-organization of the alternating hydrophilic trehalose and hydrophobic dodecyl layers. The image also showed that the planes at the interfacial area between two vesicles were fused, probably stabilizing the larger aggregates in water. Although the TEM image of **C12-mono(T)** showed that some particles with diameters of ca. 15–17 nm formed aggregates with sizes of ca. 60 nm, vesicle morphology was not observed in the particles (Fig. 16.11a). The results indicate that the self-assembly of **C12-mono(T)** was completely different from that of the other derivatives under aqueous conditions.

The XRD profiles of the former four derivatives showed diffraction peaks assignable to (001) and further peaks assignable to (002), (003), and/or (004) Bragg reflections of the lamellar patterns. From the XRD patterns, the widths of the layers for **C8-**, **C10-**, **C14-**, and **C16-mono(T)**s were calculated to be 3.1, 3.3, 3.8, and 3.9 nm, respectively. On the other hand, the XRD profile of **C12-mono(T)** showed diffraction peaks assignable to (111), (200), and (311/222) Bragg reflections of the FCC structure. The diameter of a sphere in the FCC structure was calculated to be 5.1 nm. Because the size of the derivative was calculated to be ca. 3 nm, the diameter of the micellar sphere constructed with the outer hydrophilic trehalose moieties and the inner hydrophobic dodecyl chains was estimated to be ca. 6.0 nm. This value is consistent with that obtained from the XRD analysis. The DLS results of the dispersions of all the derivatives in water

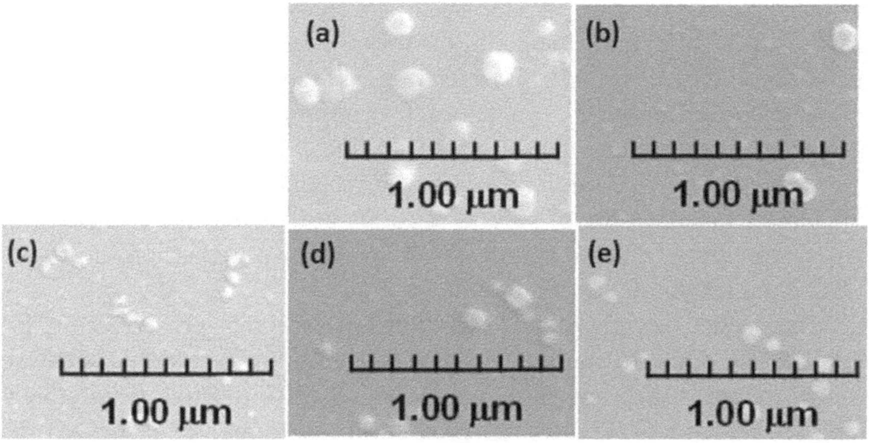


**FIGURE 16.10** SEM images of samples prepared from dispersions of trehalose 6-*O*-alkyl ethers in water ($1.0 \times 10^{-5}$ mol/L): (a) **C8-**,(b) **C10-**, (c) **C12-**, (d) **C14-**, and (e) **C16-mono(T)**.

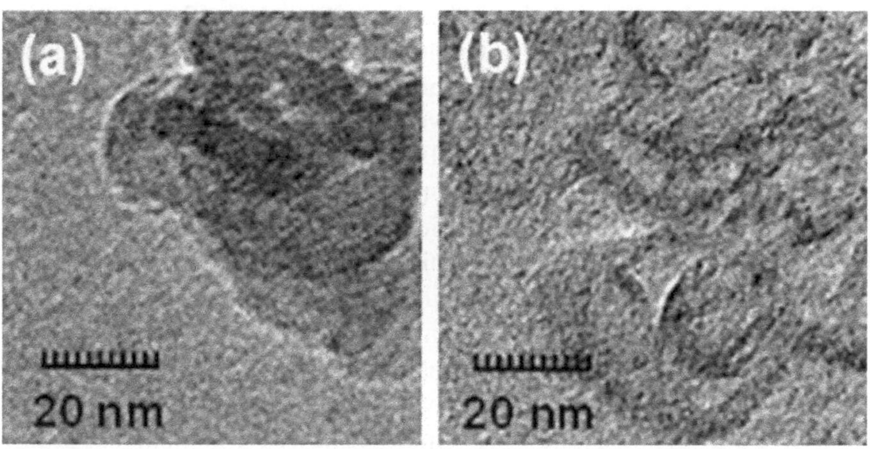


**FIGURE 16.11** TEM images of samples prepared from dispersions of trehalose 6-*O*-alkyl ethers in water ($1.0 \times 10^{-5}$ mol/L): (a) **C12-** and (b) **C16-mono(T)**.

($1.0 \times 10^{-5}$ mol/L) showed monomodal profiles. The average diameters of **C8-**, **C10-**, **C14-**, and **C16-mono(T)**s decreased in this order: 123.6, 123.3, 104.4, and 84.4 nm, respectively. These values are consistent with those calculated from the SEM images. The analytical results indicate that **C8-**, **C10-**, **C14-**, and **C16-mono(T)**s formed the vesicles under aqueous conditions; they fused to form stable nanoaggregates with ca. 100-nm diameters. The average diameter of **C12-mono(T)** in the DLS profile was much smaller than that observed in the SEM image, 12.9 and 58.3 nm, respectively. The former probably corresponded to either a spherical micelle or the assembly of several micelles in the solution state, and the latter represented further aggregates from the micelles according to the FCC organization in the dried state.

The results strongly indicate that the self-assembly processes of **C8-**, **C10-**, **C14-**, and **C16-mono(T)**s and **C12-mono(T)** were similar to those of **C8-**, **C10-**, **C12-**, and **C14-mixtures** and **C16-mixture**, respectively (Fig. 16.7). Therefore, the DLS profiles in the former case showed the presence of stable aggregates obtained by the fusion of several vesicles in water, because they were constructed by the fusion of planes in the interfacial areas. On the other hand, in the latter case, the DLS result indicates that the spherical micelles were present in individual form, or only some of them assembled in the aqueous dispersion.

Similar to the other trehalose ether derivative, the self-assembly property of **C8-di** was also investigated under aqueous conditions (Kanemaru et al., 2012b). The SEM image of a sample on an aluminum plate prepared by drying a dispersion of **C8-di** in water ($1.0 \times 10^{-5}$ mol/L) showed rod-shaped nanoaggregates with ca. 30–100 nm in width and ca. 300–800 nm in length (Fig. 16.12a). The TEM image of the same sample showed a hollow nanorod of ca. 30 nm in width and ca. 150 nm in length (Fig. 16.12b); this was constructed from the lamellar planes by the self-organization of the alternating hydrophilic trehalose and hydrophobic octyl layers. The above microscopic results indicate that the self-assembly of **C8-di** under aqueous conditions was completely different from that of sucrose and trehalose alkyl ether derivatives. The DLS result of the dispersion of **C8-di** in water ($1.0 \times 10^{-5}$ mol/L) showed a monomodal profile with an average size of 99.4 nm. This value probably corresponded to the average size in terms of the widths and lengths of the nanorods. On the other hand, the SEM images of other dialkyl derivatives with butyl and hexyl chains showed uncontrolled morphologies of the nanoaggregates (Fig. 12c,d).

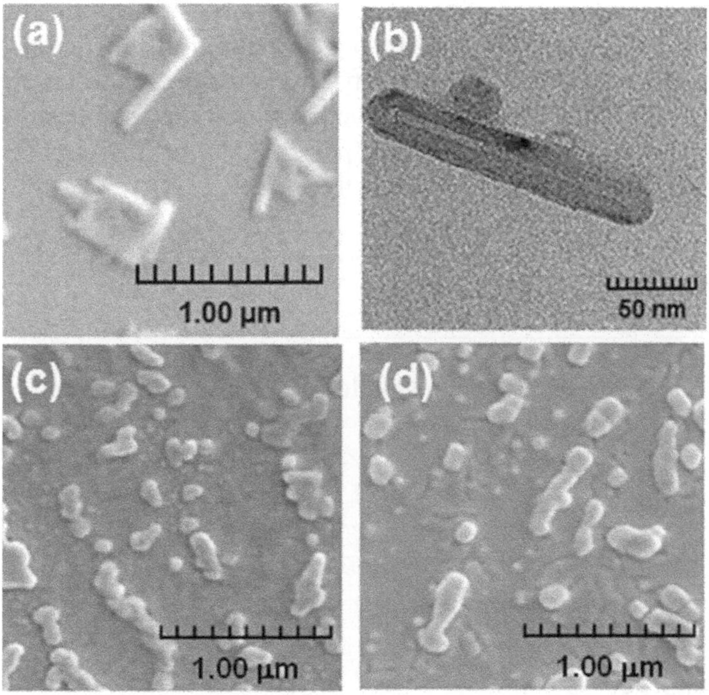


**FIGURE 16.12** SEM and TEM images of samples prepared from dispersions of trehalose 6,6′-di-*O*-octyl ether (**C8-di**) in water ($1.0 \times 10^{-5}$ mol/L) (a and b, respectively) and SEM images of samples prepared from dispersions of trehalose 6,6′-di-*O*-butyl and hexyl ethers in water ($1.0 \times 10^{-5}$ mol/L) (c and d, respectively).

## 16.5 CONCLUSION

This chapter describes an overview of the self-assembly properties of sucrose and trehalose alkyl ether amphiphiles under aqueous conditions to construct hierarchically organized nanoparticles and nanorods. Two types of self-assembly processes occurred from the sucrose and trehalose monoalkyl ethers to construct the nanoparticles: (1) primary formation of spherical micelles that regularly organized further according to the FCC structure during the drying to construct the nanoparticles and (2) construction of vesicle-like particles from the lamellar planes, constructed with the self-organized alternating hydrophilic sucrose or trehalose and hydrophobic alkyl layers. On the other hand, the self-assembly property of

**C8-di** under aqueous conditions was completely different; this molecule formed lamellar planes by the self-organization of the alternating hydrophilic trehalose and hydrophobic octyl layers that were further organized to construct hollow nanorods. This study shows the usefulness of these molecules as amphiphiles with great potential for practical applications in food, biomedical, and other fields in the future.

## KEYWORDS


- **carbohydrate-based amphiphile**
- **disaccharide alkyl ether**
- **self-assembly**
- **sucrose**
- **trehalose**


## REFERENCES


Awad, T.; Sato, K. Effects of Hydrophobic Emulsifier Additives on Crystallization Behavior of Palm Mid Fraction in Oil-in-Water Emulsion. *J. Am. Oil Chem. Soc.* **2001**, *78*, 837–842.

Awad, T.; Sato, K. Acceleration of Crystallisation of Palm Kernel Oil in Oil-in-Water Emulsion by Hydrophobic Emulsifier Additives. *Colloid. Surf. B* **2002**, *25*, 45–53.

Corti, M.; Cantu, L.; Brocca, P.; Del Favero, E. Self-assembly in Glycolipids. *Curr. Opin. Colloid. In. Sci.* **2007**, *12*, 148–154.

El-Safty, S. A.; Hanaoka, T. Microemulsion Liquid Crystal Templates for Highly Ordered Three-dimensional Mesoporous Silica Monoliths with Controllable Mesopore Structures. *Chem. Mater.* **2004**, *16*, 384–400.

Feofilova, E. P.; Usov, A. I.; Mysyakina, I. S.; Kochkina, G. A. Trehalose: Chemical Structure, Biological Functions, and Practical Application. *Microbiology* **2014**, *83*, 184–194.

Franzetti, A.; Gandolfi, I.; Bestetti, G.; Smyth, T. J. P.; Banat, I. M. Production and Applications of Trehalose Lipid Biosurfactants. *Eur. J. Lipid Sci. Technol.* **2010**, *112*, 617–627.

Fuhrhop, A. H.; Wang, T. Y. Bolaamphiphiles. *Chem. Rev.* **2004**, *104*, 2901–2937.

Gagnaire, J.; Cornet, A.; Bouchu, A.; Descotes, C.; Queneau, Y. Study of the Competition Between Homogeneous and Interfacial Reactions During the Synthesis of Surfactant Sucrose Hydroxyalkyl Ethers in Water. *Colloid. Surf. A* **2000**, *172*, 125–138.

Ge, Z. S.; Liu, S. Y. Supramolecular Self-Assembly of Nonlinear Amphiphilic and Double Hydrophilic Block Copolymers in Aqueous Solutions. *Macromol. Rapid Comm.* **2009**, *30*, 1523–1532.

Goodby, J. W.; Gortz, V.; Cowling, S. J.; Mackenzie, G.; Martin, P.; Plusquellec, D.; Benvegnu, T.; Boullanger, P.; Lafont, D.; Queneau, Y.; Chambert, S.; Fitremann, J. Thermotropic Liquid Crystalline Glycolipids. *Chem. Soc. Rev.* **2007**, *36*, 1971–2032.

Gummel, J.; Sztucki, M.; Narayanan, T.; Gradzielski, M. Concentration Dependent Pathways in Spontaneous Self-assembly of Unilamellar Vesicles. *Soft. Matter.* **2011**, *7*, 5731–5738.

Hodate, Y.; Ueno, S.; Yano, J.; Katsuragi, T.; Tezuka, Y.; Tagawa, T.; Yoshimoto, N.; Sato, K. Ultrasonic Velocity Measurement of Crystallization Rates of Palm Oil in Oil-Water Emulsions. *Colloid. Surf. A* **1997**, *128*, 217–224.

Kanemaru, M.; Yamamoto, K.; Kadokawa, J. Self-assembling Properties of 6-*O*-alkyltrehaloses Under Aqueous Conditions. *Carbohydr. Res.* **2012a**, *357*, 32–40.

Kanemaru, M.; Yamamoto, K.; Kadokawa, J. Self-assembling Property of 6,6′-Di-*O*-octyltrehalose Under Aqueous Conditions. *Chem. Lett.* **2012b**, *41*, 954–956.

Kanemaru, M.; Kuwahara, S. Y.; Yamamoto, K.; Kaneko, Y.; Kadokawa, J. Self-assembly of 6-*O*- and 6′-*O*-hexadecylsucroses Mixture Under Aqueous Conditions. *Carbohydr. Res.* **2010**, *345*, 2718–2722.

Katsuragi, T.; Kaneko, N.; Sato, K. Effects of Addition of Hydrophobic Sucrose Fatty Acid Oligoesters on Crystallization Rates of N-hexadecane in Oil-in-Water Emulsions. *Colloid. Surf. B* **2001**, *20*, 229–237.

Kitamoto, D.; Morita, T.; Fukuoka, T.; Konishi, M.; Imura, T. Self-assembling Properties of Glycolipid Biosurfactants and their Potential Applications. *Curr. Opin. Colloid. In* **2009**, *14*, 315–328.

Kralova, I.; Sjoblom, J. Surfactants Used in Food Industry: A Review. *J. Disper. Sci. Technol.* **2009**, *30*, 1363–1383.

Kwon, G. S. Polymeric Micelles for Delivery of Poorly Water-soluble Compounds. *Crit. Rev. Ther. Drug.* **2003**, *20*, 357–403.

Liu, X. H.; Huang, R.; Zhu, J. Functional Faceted Silver Nano-hexapods: Synthesis, Structure Characterizations, and Optical Properties. *Chem. Mater.* **2008**, *20*, 192–197.

Mach, M.; Jarosz, S.; Listkowski, A. Crown Ether Analogs from Sucrose. *J. Carbohydr. Chem.* **2001**, *20*, 485–493.

Molinier, V.; Kouwer, P. H. J.; Fitremann, J.; Bouchu, A.; Mackenzie, G.; Queneau, Y.; Goodby, J. W. Self-organizing Properties of Monosubstituted Sucrose Fatty Acid Esters: The Effects of Chain Length and Unsaturation. *Chem. Eur. J.* **2006**, *12*, 3547–3557.

Ohkawabata, S.; Kanemaru, M.; Kuawahara, S.; Yamamoto, K.; Kadokawa, J. Synthesis of 6-*O*-Hexadecyl- and 6-*O*-Octylsucroses and their Self-Assembling Properties under Aqueous Conditions. *J. Carbohydr. Chem.* **2012**, *31*, 659–672.

Ohtake, S.; Wang, Y. J. Trehalose: Current Use and Future Applications. *J. Pharm. Sci.* **2011**, *100*, 2020–2053.

Ohtake, T. Studies of Tritylated Sucrose. I. Mono-*O*-tritylsucroses. *Bull. Chem. Soc. Jpn.* **1970**, *43*, 3199–3205.

Queneau, Y.; Chambert, S.; Besset, C.; Cheaib, R. Recent Progress in the Synthesis of Carbohydrate-based Amphiphilic Materials: The Examples of Sucrose and Isomaltulose. *Carbohydr. Res.* **2008**, *343*, 1999–2009.

Queneau, Y.; Gagnaire, J.; West, J. J.; Mackenzie, G.; Goodby, J. W. The Effect of Molecular Shape on the Liquid Crystal Properties of the Mono-*O*-(2-hydroxydodecyl) Sucroses. *J. Mater. Chem.* **2001**, *11*, 2839–2844.

Sathe, B. R.; Shinde, D. B.; Pillai, V. K. Preparation and Characterization of Rhodium Nanostructures Through the Evolution of Microgalvanic Cells and Their Enhanced Electrocatalytic Activity for Formaldehyde Oxidation. *J. Phys. Chem. C* **2009**, *113*, 9616–9622.

Shimizu, T. Bottom-up Synthesis and Structural Properties of Self-assembled High-axial-ratio Nanostructures. *Macromol. Rapid. Comm.* **2002**, *23*, 311–331.

Shimizu, T.; Masuda, M.; Minamikawa, H. Supramolecular Nanotube Architectures Based on Amphiphilic Molecules. *Chem. Rev.* **2005**, *105*, 1401–1443.

Sofian, A. S. M.; Lee, C. K. Synthesis and Reactions of Halodeoxy Sucrose. Part 2: Synthesis and Taste Properties of 4,1′,6′-trihalodeoxy Sucrose Analogues. *J. Carbohydr. Chem.* **2001**, *20*, 191–205.

Sofian, A. S. M.; Lee, C. K. Synthesis and Taste Properties of 4,1′,4′,6′-tetrahalodeoxysucrose Analogues. *J. Carbohydr. Chem.* **2003**, *22*, 185–206.

Takahashi, Y.; Kondo, Y.; Schmidt, J.; Talmon, Y. Self-assembly of a Fluorocarbon-hydrocarbon Hybrid Surfactant: Dependence of Morphology on Surfactant Concentration and Time. *J. Phys. Chem. B* **2010**, *114*, 13319–13325.

Tanaka, K.; Ohkawabata, S.; Yamamoto, K.; Kadokawa, J. Self-assembling Properties of 6-*O*- and 6-*O*-Alkylsucrose Mixtures Having Different Chain Lengths Under Aqueous Conditions. *J. Carbohydr. Chem.* **2013**, *32*, 259–271.

Tang, J. W.; Zhou, X. F.; Zhao, D. Y.; Lu, G. Q.; Zou, J.; Yu, C. Z. Hard-sphere Packing and Icosahedral Assembly in the Formation of Mesoporous Materials. *J. Am. Chem. Soc.* **2007**, *129*, 9044–9048.

# INDEX

## I

Milton Keynes UK
Ingram Content Group UK Ltd.
UKHW022052141024
449569UK00031B/1608

9 781774 631638